STRUCTURE AND FUNCTION OF BIOLOGICAL MEMBRANES

MOLECULAR BIOLOGY

An International Series of Monographs and Textbooks

Editors: Bernard Horecker, Nathan O. Kaplan, Julius Marmur, and Harold A. Scheraga

A complete list of titles in this series appears at the end of this volume.

STRUCTURE AND FUNCTION
OF BIOLOGICAL MEMBRANES

Edited by *LAWRENCE I. ROTHFIELD*

UNIVERSITY OF CONNECTICUT
HEALTH CENTER
FARMINGTON, CONNECTICUT

 *1971*

ACADEMIC PRESS *New York and London*

ACADEMIC PRESS, INC.
111 Fifth Avenue, New York, New York 10003

United Kingdom Edition published by
ACADEMIC PRESS, INC. (LONDON) LTD.
Berkeley Square House, London W1X 6BA

LIBRARY OF CONGRESS CATALOG CARD NUMBER: 71-154389

PRINTED IN THE UNITED STATES OF AMERICA

CONTENTS

v

4. The Molecular Organization of Biological Membranes

S. J. Singer

5. The Concept of Periplasmic Enzymes

Leon A. Heppel

STRUCTURE–FUNCTION RELATIONSHIPS
IN BIOLOGICAL MEMBRANES

6. Enzyme Reactions in Biological Membranes

L. Rothfield and D. Romeo

LIST OF CONTRIBUTORS

Numbers in parentheses refer to pages on which the authors' contributions begin.

D. Chapman, Department of Chemistry, Sheffield University, Sheffield, England (13)

G. H. Dodd, Unilever Research Laboratory, Welwyn, Herts., England (13)

J. M. Fessenden-Raden, Section of Biochemistry and Molecular Biology, Division of Biological Sciences, Cornell University, Ithaca, New York (401)

V. Ginsburg, National Institute of Arthritis and Metabolic Diseases, National Institutes of Health, Bethesda, Maryland (439)

O. Hayes Griffith, Institute of Molecular Biology, and Department of Chemistry, University of Oregon, Eugene, Oregon (83)

Leon A. Heppel, Section of Biochemistry and Molecular Biology, Division of Biological Sciences, Cornell University, Ithaca, New York (223)

Patricia Jost, Institute of Molecular Biology, and Department of Chemistry, University of Oregon, Eugene, Oregon (83)

A. Kobata, National Institute of Arthritis and Metabolic Diseases, National Institutes of Health, Bethesda, Maryland (439)

E. C. C. Lin, Department of Microbiology and Molecular Genetics, Harvard Medical School, Boston, Massachusetts (285)

M. J. Osborn, Department of Microbiology, University of Connecticut Health Center, Farmington, Connecticut (343)

E. RACKER, Section of Biochemistry and Molecular Biology, Division of Biological Sciences, Cornell University, Ithaca, New York (401)

D. ROMEO, Istituto di Chimica Biologica, Universita di Trieste, Trieste, Italy (251)

LAWRENCE ROTHFIELD, Department of Microbiology, University of Connecticut Health Center, Farmington, Connecticut (3, 251)

S. J. SINGER, Department of Biology, University of California at San Diego, La Jolla, California (145)

ALAN S. WAGGONER, Department of Molecular Biophysics and Biochemistry, Yale University, New Haven, Connecticut (83)

PREFACE

The decision to publish a volume on membrane structure and function at this time reflects a general feeling that membranes have moved to the center of the biological stage. This involves two main factors. First, there is an increasing awareness that membranes hold a central position in a multitude of biological phenomena of great interest, including neural function, cell and nuclear division, biological transport, energy metabolism, macromolecular synthesis, and normal and abnormal cell–cell interactions related to differentiation and to cancer biology. Second, in recent years there have been highly significant advances in the application of biochemical and biophysical techniques to problems of membrane biology, making it possible to begin to explain membrane phenomena at the molecular level in a meaningful way.

The organization of the volume follows from these facts. The main theme concerns the explanation of membrane phenomena at the molecular level with strong emphasis on information obtained from biochemical and biophysical approaches. There is relatively little discussion of the behavior of intact cells, and the orientation thereby differs markedly from that of classic membrane physiology.

The book is not a superficial survey of membrane biology. The presentations involve discussions of key areas in considerable depth, and the volume presupposes that the reader is conversant with basic biochemistry and cell physiology. It is broadly aimed at biologists and biochemists, and hopefully will be appropriate for graduate and postdoctoral students as well as for active investigators and teachers in the biological sciences.

The volume is divided into three general sections, an Introduction and two general sections emphasizing, respectively, the structure and functions of membranes. In the second section, Newer Approaches to the

Study of Membrane Structure, the emphasis is on the overall molecular organization of membranes. This section deals with a nameless area which can perhaps be called molecular anatomy. It describes the details of the supramolecular organization of membranes as they have emerged from the use of physical and biochemical techniques. This section concludes with critical discussions by S. J. Singer, who addresses the important and controversial question of the nature of the specific interactions of membrane lipids and proteins, and by L. A. Heppel, who discusses the localization of enzymes in a newly defined cellular domain, the periplasmic space. In the third section of the book, Structure–Function Relationships in Biological Membranes, the level of discussion is more biological and biochemical and concerns the relationship of the molecular organization of membranes to specific membrane functions. Beginning with a general discussion of catalytic membrane proteins, this section proceeds to a consideration of the role of membranes in a variety of important cellular functions, and concludes with a critical look at the structural and functional organization of several specific and well-studied membrane systems in eukaryotic cells.

The successful completion of this volume is due largely to the generous cooperation of the authors, all active contributors to the advance of knowledge in this rapidly developing field. It also reflects the valuable assistance of Miss Mary Jane Goff of the University of Connecticut. Finally, the book would not have emerged in its present form without the continued advice and constructive criticism of my colleague and friend, Dr. Mary Jane Osborn.

LAWRENCE I. ROTHFIELD

STRUCTURE AND FUNCTION
OF BIOLOGICAL MEMBRANES

Introduction

1

BIOLOGICAL MEMBRANES: AN OVERVIEW AT THE MOLECULAR LEVEL

LAWRENCE ROTHFIELD

What is a membrane? In a symposium held in 1961, the great physiologist Homer Smith discussed at length the arguments for the existence of the plasma membrane as a structural entity. He concluded, reasonably enough, that the evidence was overwhelmingly in favor of this idea, and he found untenable the older idea that the boundary between the cell and the outside world was merely an interface between two continuous phases (Smith, 1962). Today it seems extraordinary that such an argument ever could have taken place. The existence of the plasma membrane and other biological membranes is a basic foundation of modern cellular biology, and the idea that internal membrane systems constitute the basic cytoskeleton and circulatory system of the cell is firmly established. Membranes have been isolated free of other cellular components, weighed, fractionated, analyzed, photographed, and studied by an infinite and wondrous variety of techniques. Unfortunately, despite this extensive body of knowledge, it is still difficult to give a simple definition of a biological membrane. Most biologists will, however, agree with the following description:

> Biological membranes are continuous structures separating two aqueous phases. They are relatively impermeable to water-soluble compounds, show a characteristic trilaminar appearance when fixed sections are examined by electron microscopy, and contain significant amounts of lipids and proteins.

The fact that membranes contain large amounts of lipids, particularly phospholipids, distinguishes them from most other cellular structures, and it is not surprising that until recent years membrane research focused primarily on the lipid components. This focus on membrane lipids to the relative neglect of the proteins was encouraged by the fact that the most obvious role of membranes is their barrier function. This requires that unregulated passage of water-soluble materials be prevented, and the lipids, with their markedly hydrophobic hydrocarbon chains, seemed logical candidates for this task.

Perhaps an equally important reason for the intensive study of membrane lipids in the 1950's and 1960's was the emergence of powerful and simple tools to investigate the biochemical and physical properties of these molecules. The refinement of column chromatographic techniques utilizing silicic acid and other supports in the mid-1950's permitted the convenient isolation of large amounts of pure lipids, while thin-layer chromatography allowed their rapid and accurate identification. Gas–liquid chromatography of lipids was developed during the same period, and this permitted the accurate identification and quantitation of the fatty acid and other hydrocarbon components of membrane lipids. The availability of these analytic tools rapidly led to a firmly based understanding of lipid chemistry in general, and to the nature of the lipid components of biological membranes in particular. During this period lipid chemists also developed techniques for the organic synthesis of phospholipids (reviewed by van Deenen and de Haas, 1964). The availability of well-defined synthetic and natural lipids, together with the extensive body of knowledge of the biochemistry of lipids that emerged from the revolution in analytical lipid chemistry described above, made it possible to use sophisticated physical techniques to examine the behavior of lipids and to study in detail the ways in which lipid molecules arrange themselves in aqueous environments.

It has long been known that phospholipids and glycolipids (i.e., glycerophosphatides and sphingolipids) form ordered multimolecular structures when placed in aqueous surroundings (see Dervichian, 1964, for review). This reflects the fact that these molecules are classical *amphipaths,* containing in the same molecule both a nonpolar and a polar region. In Fig. 1 these are indicated by the nonpolar hydrocarbon chains (R^1 and R^2) and by the more polar regions (phosphoryl-R^3), respectively. Amphipathic molecules of this type form multimolecular complexes when placed in a polar environment such as is found in the intracellular and extracellular milieu of biological systems. As discussed in greater detail in Chapter 4, this reflects the fact that the association of nonpolar residues of one lipid molecule with the nonpolar portion of its neighbor is energetically much more favorable than is the interaction of the nonpolar portion of any molecule with the polar water molecules of

Erratum

STRUCTURE AND FUNCTION
OF BIOLOGICAL MEMBRANES

Edited by

Lawrence I. Rothfield

Figure 1, page 5 – structure of Sphingolipids should be as follows:

$$
\begin{array}{c}
\text{O} \\
\| \\
\boxed{R^1}-CO-CH_2 \\
\text{O} \\
\| \\
\boxed{R^1}-CO-CH \\
\qquad CH_2-OP-\boxed{R^3} \\
\qquad\qquad \underset{O^-}{|}
\end{array}
$$

$$
\begin{array}{c}
\boxed{R^2} \\
| \\
HC \\
\diagdown CH \\
| \\
O \qquad HOCH \\
\| \qquad | \\
\boxed{R^1}-C-NH-CH \qquad O \\
\qquad\qquad CH_2-OP-\boxed{R^3} \\
\qquad\qquad\qquad \underset{O^-}{|}
\end{array}
$$

Glycerophosphatides   Sphingolipids

Fig. 1. Structure of glycerophosphatides and sphingolipids: R^1 and R^2 represent the hydrocarbon portions of the acyl chains and the sphingosine group, respectively; R^3 represents the polar residues (e.g., —OH, ethanolamine, choline, oligosaccharides, etc.).

the solvent; similarly, the interaction of water molecules with adjacent water molecules in the solvent phase is energetically much more favorable than is the interaction of water molecules with the hydrocarbon regions of the lipids. As a result, the concentration of single lipid molecules in true solution is aqueous solvents is vanishingly small. Instead, various types of multimolecular structures are formed by phospholipids and sphingolipids in aqueous environments. In all of these structures the molecules are arranged to *minimize* interactions between hydrocarbon residues of the lipid and the water molecules of the solvent.

The most common of these structures is the bimolecular leaflet (see Fig. 2), and under most conditions this lamellar structure is the only phase seen when single phospholipids or mixtures of naturally occurring lipids are studied. The stability and prevalence of this lamellar phase has strongly influenced theories of membrane structure by supporting the idea that the basic framework of membranes also may be a bimolecular

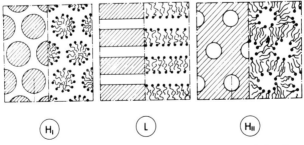

Fig. 2. Structure of liquid crystalline structures from phospholipid–water systems. In the left half of each diagram the noncross-hatched areas indicate the location of water molecules. Abbreviations: H_I, "hexagonal I" phase; L, lamellar phase; H_{II}, "hexagonal II" phase. (From Luzatti *et al.*, 1966.)

leaflet of phospholipid. However, careful investigation by low-angle x-ray diffraction has shown that several other organized structures can also occur when the conditions of study are changed (Luzzati *et al.*, 1966; Gulik-Krzywicki *et al.*, 1967). These include tubular structures arranged with fatty acid chains directed into the interior of the tube ("hexagonal I") as well as the reverse structure in which the interior of the tube is filled with the aqueous solvent while the hydrocarbon chains are directed outward ("hexagonal II," see Fig. 2). The ability of phospholipids to change their structural organization from one type to another is illustrated in the phase diagram in Fig. 3. The type of ordered structure varies with the conditions of study; higher temperatures and high concentrations of lipid, for example, appear to favor the "hexagonal II" structure while lamellar structures are found under most other conditions.

This emphasizes the key role of the environment in determining the manner in which lipid molecules arrange themselves. Since biological membranes contain other molecular species in addition to amphipathic lipids, these other molecules (e.g., proteins) may play a major role in determining the organization of membrane lipids *in situ*. Two possibilities should be considered. (a) The functional groups of the peptide chains may change the local environment and thus induce lipid molecules to organize themselves into a particular structure. For example, one might imagine a cluster of acidic amino acids in a particular portion of the membrane, coming from one or many peptide chains. By virtue of local pH changes this might favor a specific structural arrangement of the phospho-

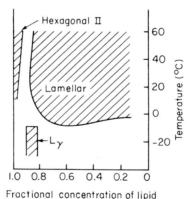

Fig. 3. Phase diagram of low-angle x-ray diffraction study of mixed mitochondrial lipids in water. The cross-hatched areas indicate regions in which only a single phase was seen. Other areas indicate transition regions in which more than one phase was seen. Lγ indicates a complex lamellar phase interpreted as showing a much greater degree of order in the hydrocarbon chains than in the lamellar phase seen at higher temperatures. (Adapted from Gulik-Krzywicki *et al.*, 1967).

lipids (for example, a "hexagonal II" structure). It should be pointed out that reversible transitions in lipid organization could be induced by this type of mechanism. Thus, conformational changes in membrane proteins may be induced by specific physiological effectors (e.g., the binding of a transportable substrate to a protein carrier). These conformational changes might then affect the local environment and induce local phase transitions in membrane lipids as described above, resulting in physiologically pertinent alterations in local membrane structure. (b) Proteins or other molecules may become part of the continuous "lipid phase." Appropriate folding of peptides can expose nonpolar residues which would be expected to associate with the hydrocarbon portions of membrane lipids for the same reasons outlined above for lipid–lipid interactions.[*] The mixture of protein and lipid molecules would then be viewed as a ternary system (protein–lipid–water) and the types of structural organization might be quite different from the binary lipid–water systems discussed above.

For these reasons there has been an increasing awareness that the elegant and important studies during the 1960's of lipid–water systems must be extended to more complex systems whose composition more nearly approaches that of the native membrane. Several advances in this direction, such as x-ray studies of cytochrome c–phospholipid–water systems by Luzzati and his co-workers (Gulik-Krzywicki et al., 1969), are discussed in Chapters 2–4.

The lipid components of membranes are not rigid structural entities. Instead, it is likely that the hydrocarbon chains of lipid molecules are in constant motion in most, if not all, physiological circumstances. This is consistent with the well-established "liquid crystalline" nature of the ordered lipid structures found in lipid–water systems. The mobility of the hydrocarbon chains in lipid–water systems and in membranes has been revealed by a variety of physical techniques as discussed in Chapters 2 and 3. It can be viewed simplistically as movement of a random nature within the backbone of, for example, individual hydrocarbon chains of phospholipid molecules in bimolecular leaflets or membranes. This implies that the nonpolar interior of these structures is fluid in nature. The implications for membrane function are many, including the possibility of relatively unrestricted movement of nonpolar compounds, or of amphipathic molecules whose nonpolar portion is anchored in the nonpolar interior of the membrane, within the plane of the membrane structure. In general, mobility is favored by higher temperatures (see Fig. 3), by greater degrees of unsaturation of the hydrocarbon chains, and by shorter

---

[*] This concept is extensively discussed in Chapter 4.

chain lengths. The change from a less ordered (i.e., more mobile) to a more ordered structure when the temperature is lowered is revealed by changes in several physical parameters (see Chapters 2 and 3). For example, in low-angle x-ray diffraction studies of the lamellar (i.e., bilayer) phase of lipid–water systems, diffuse reflections ascribable to the hydrocarbon chains change to sharper reflections when the temperature is lowered beyond a transition temperature, and this is interpreted to reflect a change from a less ordered to a more ordered structure. The transition temperature varies, depending on the type of hydrocarbon chains in the phospholipid. Similar transitions have been seen in plasma membranes, varying characteristically depending on the nature of the fatty acyl components of the membrane phospholipids (Engelman, 1970). Thus the concept that the interior of organized lipid structures is fluid in nature at physiological temperatures, which developed from studies of pure lipid–water systems, appears also to hold true for biological membranes, implying a degree of similarity in the organization of the lipid molecules in the two situations. In considering functional aspects of membrane biology, the possible relationships between changes in fluidity and alterations in function in localized portions of the membrane must be kept in mind.

Studies of lipid–water systems have profoundly influenced thinking about the molecular structure of biological membranes in at least two ways: first, by emphasizing the almost obligatory formation of ordered multimolecular structures when lipids are placed in aqueous surroundings and, second, by revealing the highly fluid nature of the hydrocarbon chains in these structures. The overall picture is of a combination of long-range order and short-range disorder.

However, the marked emphasis on the behavior of simple lipid–water systems has had both good and bad effects. On the positive side of the ledger is a continuously improving understanding of the physicochemical factors that affect the interactions of lipid molecules with themselves and with their environment. This is clearly a prerequisite to a meaningful understanding of the structural and functional role of these molecules in membranes. On the other hand, detailed consideration of the other major macromolecular constituent, the membrane proteins, has lagged behind. There is general agreement that membrane proteins must play a key role in membrane structure and function. In considering the specificity of a variety of membrane functions, such as transport, the major answers will almost certainly lie in the nature of the protein components. Despite this, a somewhat disproportionate emphasis has been placed on the role of lamellar and other lipid phases in membranes, with a less rigorous consideration of the factors involved in fitting proteins into the membrane structure. There are several reasons for this, including the much greater complexity of the potential interactions of peptides which can, at the same time, interact with portions of the same protein molecule, with other proteins, with lipids, and with the aqueous solvent. Only recently have the major triumphs of modern protein chemistry reached a stage where knowledge of the physicochemical behavior of proteins and of the multiple factors affecting their interactions makes it possible to explore their role in membranes in depth. Another problem has been the difficulty in obtaining membrane proteins in purified form. Until recently this made it impossible to study simple systems of the type that have yielded so much information about membrane lipids. Finally, the use of synthetic model peptides in studies of protein–lipid interactions has not been exploited to any considerable degree. However, although detailed studies of membrane proteins are still at an early stage, the past few years have seen major advances in this area, both in the identification of specific proteins that are involved in specific membrane functions and in the examination of the structural characteristics of membrane proteins *in situ*.

The major unanswered questions of the molecular biology of membranes can be summarized as follows: What is the molecular architecture of biological membranes, or in more specific terms, how are protein and lipid molecules arranged in biological membranes? How does this molecular organization affect and regulate specific membrane functions? By what mechanism are the individual molecules which comprise membranes synthesized and how are they assembled to form complete membranes?

Luckily both the theoretical background and the experimental techniques are now at hand to permit an attack on these basic problems of membrane structure and function. This volume is an attempt to begin to answer these questions at the molecular level.

References

Dervichian, D. G. (1964). *Progr. Biophys. Mol. Biol.* **14**, 265.
Engelman, D. (1970). *J. Mol. Biol.* **47**, 115.
Gulik-Krzywicki, T., Rivas, E., and Luzzati, V. (1967). *J. Mol. Biol.* **27**, 303.
Gulik-Krzywicki, T., Schechter, E., Luzzati, V., and Favre, M. (1969). *Nature (London)* **223**, 1116.
Luzzati, V., Reiss-Husson, F., Rivas, E., and Gulik-Krzywicki, T. (1966). *Ann. N. Y. Acad. Sci.* **137**, Art. 2, 409.
Smith, H. W. (1962). *Circulation* **26**, 987.
van Deenen, L. L. M., and de Haas, G. H. (1964). *Advan. Lipid Res.* **2**, 168.

Newer Approaches to the
Study of Membrane Structure

2

PHYSICOCHEMICAL PROBES OF MEMBRANE STRUCTURE

D. CHAPMAN and G. H. DODD

ABBREVIATIONS

ANS	1-Anilinonaphthalene-8-sulfonate
CD	Circular dichroism
DSC	Differential scanning calorimetry
DTA	Differential thermal analysis
NMR	Nuclear magnetic resonance
ORD	Optical rotatory dispersion
PMR	Proton magnetic resonance

I. INTRODUCTION

The structure of cell membranes is now widely recognized as an outstanding problem of present-day molecular biology. Uncertainty about the structure of membranes naturally implies uncertainty about the many mechanisms associated with membrane function. Explanations for the processes of ion and molecular transport, the details at the molecular level of the nerve impulse, and the way in which energy transport molecules are organized must remain incomplete while membrane structure is being debated.

The introduction of the electron microscope has revealed the existence not only of the plasma membrane but also of many additional membrane systems in cells (e.g., membranes associated with the endoplasmic reticulum, with the mitochondrion, and with the cell nucleus). On the basis of electron microscope data some workers have suggested that all cell membranes have a common structural feature, while others pointing to the varying biochemical data of lipid to protein content in membranes suggest that this is probably not the case.

A. Biochemical Data on Membranes

The compositions of some common membranes in terms of lipid and protein content are given in Table I. It can be seen that the lipid to protein ratio varies considerably from one type of membrane to another; the problem becomes more complex when one considers that in addition to lipids and proteins the complete membrane also contains sugar molecules, metal ions, water, and sometimes cholesterol.

When one looks more closely at the analytical data, it is apparent that the situation has even more complications. Usually animal cell membranes contain not one lipid class but many (e.g., a membrane may contain phosphatidylcholines, phosphatidylserines, and phosphatidylethanolamines). Furthermore, associated with each lipid class there is a distribution of fatty acids varying in chain length and degree of unsaturation. Usually the saturated fatty acids are found to be esterified at the 1-position on the phosphoglyceride and unsaturated fatty acids at the 2-position.

Rouser and co-workers have summarized their considerable analyses of many membrane systems as follows:

1. All animal cell membranes contain phospholipids. The same classes of phospholipid are found in vertebrates and invertebrates. Some membranes (e.g., myelin) contain glycolipids whereas others do not. Only certain membranes contain sterol.

2. Plasma (cell surface) membranes, membranes of the endoplasmic reticulum, nuclear membranes, and mitochondrial membranes from the same species have different compositions. All differ quantitatively and to some extent qualitatively in the classes of lipids present. For example, plasma membranes or elaborations of these (such as myelin) appear to

TABLE I

PROTEIN AND LIPID COMPOSITION OF ANIMAL AND BACTERIAL MEMBRANES

	Molar ratio[a]			Area ratio[b] (protein/lipid)	Reference
	Amino acid	Phospholipid	Cholesterol		
Myelin	264	111	75	0.43	O'Brien and Sampson (1965)
Erythrocyte	500	31	31	2.5	Maddy and Malcolm (1965)
Bacillus licheniformis	610	31	0	4.8	Salton and Freer (1965)
Micrococcus lysodeikticus	524	29	0	4.3	Salton and Freer (1965)
Bacillus megaterium	520	23	0	5.4	Weibull and Bergström (1958)
Streptococcus faecalis	441	31	0	3.4	Shockman and co-workers (1963)
Mycoplasma laidlawii	442	25.2	2.3	4.1	Razin et al. (1965)

[a] Data are calculated from the percentage compositions given in the references indicated, using the appropriate molecular weights.
[b] The approximate area occupied in a monomolecular film, assuming, that an average amino acid occupies 17 Å^2 (Weibull and Bergström, 1958), a phospholipid molecule 70 Å^2 (Bear and co-workers, 1941; Vandenheuvel, 1963), and a cholesterol molecule 38 Å^2 (Vandenheuvel, 1963).

contain most of the glycolipid of the cell. Even where qualitative differences in distribution of lipid classes are not seen, as in endoplasmic reticulum and nuclei, the quantitative distribution of lipid classes is different and characteristic for each membrane.

3. Plasma membranes, as shown by studies of mammalian erythrocytes, exhibit large species variations in composition. The proportions of the different phospholipids vary greatly and the total amount as well as the types of both ceramide polyhexosides and gangliosides is very different in different species. Data from whole organs indicate that plasma membranes from different cell types of the same species may vary in composition.

4. Mitochondria from bovine heart, kidney, and liver contain diphosphatidylglycerol (cardiolipin), phosphatidylcholine, and phosphatidylethanolamine as the major phospholipids in the approximate molar ratio 1/4/4, but the inner and outer membranes of the mitochondria differ in composition. Pure outer membrane preparations contain all the cholesterol of the mitochondrion and appear to be devoid of diphosphatidylglycerol. On the other hand, mitochondria from different organs of one species appear to have the same phospholipid distribution, and species variation in mitochondrial phospholipid composition among vertebrates is at most slight. The total amount of mitochondrial lipid from different organs is variable and appears to correlate with the proportion of cristae in the mitochondria; those containing more cristae also have more phospholipid.

5. The fatty acid composition of each class of lipids from different organelles and organs of one species, as well as from different species, is variable. This is true even when the classes of lipids are the same in the different structures. Individuality is thus expressed most clearly in differences in fatty acid composition.

B. The Structure of Membranes

Some of the major unanswered questions about membrane organization are the following. Why is a cell membrane made up of combinations of molecules, including lipid, protein, sugar, cholesterol, and water? Why is there a variation in class of lipid or in fatty acid composition, and why is cholesterol present in some membranes and not others? How are the constituent molecules organized into the final functioning membrane?

With structures containing so many different components, clearly a variety of structural arrangements is theoretically possible. It is therefore not surprising to find at present a wide range of speculations dealing with these questions. These speculations often depend upon the emphasis given to one of the components present. Thus an emphasis on lipid and its importance in determining structure leads to one possibility (i.e., the bilayer model) while an emphasis on protein leads to the subunit model. In between there is a range of suggested structures dependent upon modifications of lipid by protein and vice versa. The major emphasis of some of these models is on the predominant type of interaction thought to be involved, such as electrostatic interaction, hydrophobic interaction, and hydrogen bonding (see Chapter 4 for a more extensive discussion).

So as to limit the area of speculation, a great deal of experimental information about the organization of components in the membrane is clearly required. The most direct information should be obtainable by examination of cell membranes themselves. However, in view of the many components present in intact membranes, studies of model systems involving interactions of selected components are also required. This information can then be used to interpret the results obtained with the more direct approach.

II. PROBES OF MEMBRANE STRUCTURE

In recent years a number of physical techniques have become available which enable one to probe into the organization of the components of the membrane. Some of these techniques actually involve the introduction of a probe into the system (e.g., spin labels and fluorescent labels); others use the existing nuclei as natural probes (i.e., nuclear magnetic methods). Thermal techniques such as differential thermal analysis use the thermal properties and phase behavior of the membrane constituents to provide information about their organization.

We shall now discuss some of these methods and point to the conclusions which are emerging from their application. In those techniques which require the separation of the cell membrane from the cell contents prior to examination, we assume for the present that this itself does not cause a major change in the organization of the membrane components.

A. x-Ray Methods

It is not our intention to review here the application of x-ray methods to membrane structure. It has been amply reviewed elsewhere (Finean, 1969; Chapman, 1966) and is mentioned briefly in Chapters 1 and 4 of this volume. The technique is of course a very powerful one and continues to make important information available. Recent x-ray studies have been made of lipid–protein complexes (i.e., model membrane structures) by two groups of independent workers (Shipley et al., 1969; Luzzati and Faure, 1969) with similar results. The complexes are formed between

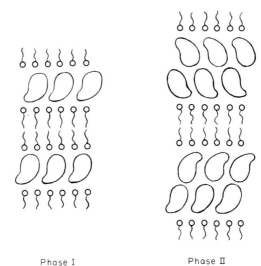

Phase I   Phase II

Fig. 1. A schematic diagram illustrating the way in which certain proteins and lipids, e.g., cytochrome *c* and phosphatidylserine–water systems, are organized (Shipley *et al.*, 1969). A similar arrangement occurs with ferricytochrome *c*–phosphatidylinositol (Luzzati and Faure, 1969). With both systems two lamellar phases are formed. In phase I there is a single layer of protein between the lipid bilayers, while in phase II there are two layers of protein between the bilayers. The main interaction is believed to be electrostatic between the polar groups of the lipid and the amino acids of the protein.

cytochrome *c* and phospholipids or between lysozyme and phospholipids. The cytochrome *c* complexes appear to form two lamellar phases, one containing a single protein layer and the other a double protein layer (Fig. 1). These complexes are presumed to be formed by electrostatic interaction between the basic groups of the protein and the acid groups of the phospholipid.

Certain dimensional similarities exist among the long spacings observed when these model complexes in water are compared with different membrane species, e.g., myelin. Whether this is significant of some underlying structural similarity remains to be seen.

B. Infrared Spectroscopy

1. Introduction

The infrared absorption spectrum of a molecule in the 2–15 μ region is associated with absorption of electromagnetic radiation arising from vibrations of the atoms of the molecule. Certain bands have been associated empirically with functional groups present in the molecule (e.g., a band between 5 and 6 μ occurs when there is a C=O group present and a band at 3 μ occurs when an OH group is present).

The main application of the infrared spectroscopic technique in membrane research has been to obtain information about protein conformation in the membrane. One difficulty, however, is that water gives strong absorption bands which can "blot out" the underlying spectrum of the membrane material. Hence the water content is often reduced prior to examination, or alternatively the H_2O is replaced by D_2O, which has a different absorption spectrum.

2. Studies of Membranes

Infrared spectroscopy has been applied to the study of certain membranes, including myelin (Chapman, 1965; Maddy and Malcolm, 1965, 1966) and the plasma membranes of erythrocytes and Ehrlich ascites carcinoma cells (Wallach and Zahler, 1966). The broad diffuse bands in the spectrum show the fluidity of the lipid chains in these structures. Some studies (Chapman and co-workers, 1968c) have pointed out the importance of a band at 720 cm^{-1} indicative of an all-planar configuration of the lipid chains.

The position of certain bands, referred to as amide I and amide II, in the infrared spectra of peptide chains differs depending upon the conformation of these chains (Maddy and Malcolm, 1965). Thus the amide I band is located at 1652 cm^{-1} and is associated with an α-helical and/or random coil conformation of peptide chains. The amide II band at about 1535 cm^{-1} does not allow distinction between the α- and β-conformation. On the other hand, a band at 1630 cm^{-1} is correlated with a β-conformation.

The infrared spectra of membranes of erythrocytes (Maddy and Malcolm, 1966; Chapman et al., 1968a,b) and Ehrlich ascites carcinoma cells (Wallach and Zahler, 1966) do not show a strong band at 1630 cm^{-1}, suggesting that there is no extensive β-structure in these membranes.

Extraction of lipids from the membrane abolishes the band at 1740 cm^{-1} due to the carbonyl stretching vibration (νC=O) of the fatty acid esters. There is also some reduction of the two amide bands due in part to the extraction of sphingomyelin.

Examination of the infrared spectra of rat liver mitochondria and their inner and outer membranes (Wallach, 1969) showed a distinct shoulder near 1630 cm^{-1} indicative of the presence of β-structure and another shoulder near 1690 cm^{-1} consistent with the β-conformation being in an antiparallel pleated sheet. Spectra of "structural protein" extracted from mitochondria showed a preponderance of β-structure.*

Wallach and co-workers (Wallach, 1969) have emphasized the importance of infrared spectroscopy for the study of protein conformation of cell membranes. They commented that in contrast to the infrared studies, measurements of optical rotatory dispersion (ORD) and circular dichroism (CD) have led some authors to conclude that the proteins of the mitochondrial membrane lack the β-conformation. Recent studies of CD bands and ORD extrema of certain β-structures show lower intensities than the commonly used reference compound poly-L-lysine and can be displaced by over 10 nm to the red of the positions found with this polypeptide model. These spectral shifts can place the CD minimum of the β-structure at, or to the red of, the position of the $(n - \pi)$ band of α-helical polypeptides. They concluded that it is not possible at present to arrive at conclusions concerning helical content in mitochondrial membranes in the absence of reliable infrared spectroscopic data.†

3. Conclusions

Infrared spectroscopy is useful for studying cell membranes and provides valuable information about the structure of the proteins present. Other regions, particularly the long wavelength region where slower vibrational modes occur, may also give useful information about membrane and water organization. This region may also provide many useful transparent regions free from absorptions arising from water vibrations.

C. Nuclear Magnetic Resonance Spectroscopy

1. Background and Theory

a. Theory of NMR Spectroscopy. Nuclear magnetic resonance (NMR) spectroscopy is a powerful method of probing biological systems and has recently been applied to lipids, lipoproteins, and membrane systems. With this technique, particular nuclei can be examined (primarily the ^1H and ^{31}P) while the remaining nuclei are transparent. In essence the different nuclei can act as natural probes of the molecular motion of structural components. The technique basically consists of measuring the energy required to change the alignment of nuclei in an applied magnetic field.

Many atomic nuclei possess a nonzero nuclear spin, I. The following

---

Editor's note: It should be noted that there is considerable doubt about the existence of mitochondrial "structural protein" as an entity (see Chapter 9).

† *Editor's note:* See Chapter 4 for a somewhat different view as well as for a more extended discussion of protein conformation in membranes.

nuclei are of principal interest: ^{1}H ($I = \frac{1}{2}$), ^{2}H ($I = 1$), ^{13}C ($I = \frac{1}{2}$), ^{14}N ($I = 1$), ^{31}P ($I = \frac{1}{2}$), ^{17}O ($I = \frac{5}{2}$). Some nuclei such as ^{12}C and ^{16}C have a zero spin and cannot be studied by NMR spectroscopy. In a constant magnetic field (H_0), several discrete orientations of the magnetic moment are possible.

The energy difference (ΔE) between any two adjacent spin states, or the Zeeman splitting, is

$$\Delta E = \mu H_0 / I$$
$$= 2\,\mu H_0 \qquad\qquad \text{when} \quad I = \frac{1}{2} \quad (1)$$

where μ is the magnetic moment.

Thus absorption of energy by the spins can occur when a radio frequency field H_1, of the proper frequency v_0, is applied perpendicular to H_0. That is, when

$$h v_0 = \Delta E = 2\,\mu H_0 \qquad\qquad (2)$$

where h is Planck's constant. It follows from Eq. (2) that NMR data can be expressed in either field units (H_0) or frequency units (v_0). The results are frequently expressed in terms of resonant frequency (v_0) in MHz (cycles per second $\times 10^6$), at a given applied field strength. For protons in a field of 10^4 G, v_0 is 42.58 MHz. The corresponding frequencies for other nuclei are shown in Table II.

TABLE II
RESONANT FREQUENCIES (v_0) OF SOME IMPORTANT MAGNETIC
NUCLEI IN A MAGNETIC FIELD OF 10^4 G

Nucleus	I	v_0(MHz)	Relative sensitivity at constant field
^{1}H	1/2	42.58	1.00
^{2}H	1	6.54	9.64×10^{-3}
^{13}C	1/2	10.71	1.59×10^{-2}
^{14}N	1	3.08	1.01×10^{-3}
^{17}O	5/2	5.77	2.92×10^{-2}
^{19}F	1/2	40.06	0.834
^{31}P	1/2	17.24	6.64×10^{-2}

b. Spin-Lattice Relaxation. Since transitions from the lower spin state occur more frequently than the reverse transitions, the absorption of energy tends to equalize the populations of each level and, in the limit, radio frequency saturation may occur. This is normally hindered by *spin-lattice relaxation*. By this radiationless process, energy of the spins in the upper state is disposed of to the "lattice" of surrounding nuclei via certain oscillating magnetic fields in a time (T_1) of 10^{-5}–10^4 seconds.

c. Effects of Dipole–Dipole Interactions with Neighboring Nuclei. In an actual assembly of spins the dipole–dipole interactions, which depend on local nuclear structure, lead to a broadening of the energy levels since the absorption of energy is determined both by H_0 and by the local field due to neighboring magnetic nuclei. Thus the resonant frequency in a rigid sample varies from nucleus to nucleus, although the distribution of frequencies usually has complete symmetry about v_0. As an example of the magnitude of the effects of neighboring nuclei, the local field of a pair of protons can amount to 10–20 G (or 42–84 kHz) (Andrew, 1958; Abragam, 1961; Slichter, 1963).

d. Theory of Nuclear Magnetic Relaxation. Relaxation times derive from the process of spin-lattice relaxation and reflect the loss of energy from spins in the upper state to the "lattice" of surrounding nuclei. In an actual sample this occurs by the synchronous fluctuations in time of the local fields produced by thermal motions of the neighboring nuclei. Relaxation times can be measured and can aid in the interpretation of molecular reorientation and structure (Bloembergen *et al.*, 1948; Bloembergen, 1961; Kubo and Tomita, 1954).

e. Changes in Line Width. The finite line widths observed in a real sample of spins are therefore a manifestation of the broadening of energy levels due to two factors: (a) dipole–dipole interactions with neighboring nuclei (see above) and (b) finite relaxation times due to the effects of neighboring nuclei on spin-lattice relaxation.

In a solid, the dipolar interactions are determined by the relative positions of the constituent nuclei. The manner in which such interactions are modified by structural and motional features determines the width and shape of the corresponding absorption lines. The second moment of the absorption line can be computed in the case of a rigid assembly of magnetic nuclei in simple molcecules, and by comparison with observed magnitudes at various temperatures, information on molecular structure and reorientation can be deduced (Andrew, 1958; Abragam, 1961). This is due to the fact that *the observed line widths* (Δh) *and second moment are reduced in the presence of sufficiently rapid molecular reorientation.* For many proton-containing systems, the pertinent frequencies v_c are greater than 60 kHz. If the reorientation is, in addition, sufficiently incoherent, then extreme narrowing results from the efficient reduction of the average local dipolar field. For example, liquid water present in biological materials will give a narrow line superimposed on a broad background signal due to the host lattice protons.

For the wide lines found for solids, the shape of the absorption line is commonly presented in the form of the first derivative of the absorption line. Thus a measurement of the line width between points of maximum

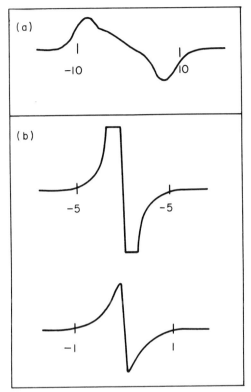

Fig. 2. The PMR spectra of a polycrystalline phospholipid illustrating the reduction of line width by increasing molecular motion for 1,2-dimyristoyl-DL-phosphatidyl-ethanolamine at (a) −196°C, (b) 120°C. Two different field sweeps are shown (Chapman and Salsbury, 1966).

and minimum slope can be made without the difficulties imparted by the sloping and noisy baselines that usually accompany broad absorption lines. Examples of proton magnetic resonance (PMR) derivative absorption spectra for different rates of molecular reorientation are given in Fig. 2.

In a nonviscous liquid state, narrow proton absorption lines ($\Delta v \lesssim 1$ Hz) are observed due largely to an efficient averaging of local fields. In the limit of extreme narrowing, very fine splittings can be observed under high-resolution conditions. If H_0 is homogeneous and constant over the sample volume to better than 1 part in 10^8, the effects of the slightly different effective magnetic fields experienced by nonequivalent nuclei can be observed.

f. Chemical Shifts. A diamagnetic moment due to orbital electronic circulation of the electrons associated with a given nucleus is opposed to the constant field H_0 and, further, is proportional to H_0. This diamagnetic shielding can be reduced by paramagnetic currents and can be further modified by the presence of π electrons in, for example, aromatic ring systems. Thus a variety of factors can affect the local magnetic field, and the actual field experienced by a given nucleus is, in general,

$$H = H_0(1 - \sigma) \tag{3}$$

where σ is a dimensionless shielding constant depending on the electronic environment of the nucleus. This changes the applied field strengths at which the nucleus resonates, resulting in *chemical shifts* of the absorption spectrum. Therefore high-resolution NMR spectra show separate signals associated with the different functional groups in the molecule examined, e.g., methyl alcohol shows three main signals at different τ values corresponding to the CH_3, CH_2, and OH groups (Pople *et al.*, 1959; Emsley *et al.*, 1965, 1966; Chapman and Magnus, 1966). The chemical shifts can then be used to quantify the different field positions of resonance of a given chemical group (H) and that of a reference (H_{ref}). With respect to a chosen reference, the chemical shift is given by

$$\delta = \frac{H - H_{ref}}{H_{ref}} \times 10^6 \text{ ppm} \tag{4}$$

Because of the relatively small magnitude of the absolute shift, the absolute value is multiplied by 10^6, thereby giving the chemical shift in parts per million (ppm), indicating a shift in the positon of resonance of a given number of gauss per million gauss. Similarly, the frequency can be used to calculate the chemical shift

$$\delta = \frac{\text{shift in frequency}}{\text{oscillator frequency}} \times 10^6$$

A common proton reference taken as $\delta = 0$ ppm is the resonance of the equivalent methyl groups of tetramethylsilane. An alternative scale gives $\tau = 10$ ppm for this resonance.

Higher resolution under optimum conditions permits the observation of further fine structure due to *spin-spin splitting*, which gives information on other groups of nuclei, adjacent or close to the one being observed. This splitting is field independent, unlike the chemical shift, and is the result of interaction of nuclear moments via the bonding electrons for nonequivalent nuclei.

High-resolution spectra have been obtained with phospholipids dissolved in organic solvents, showing all the chemically shifted proton groupings (Chapman and Morrison, 1966) (Fig. 3).

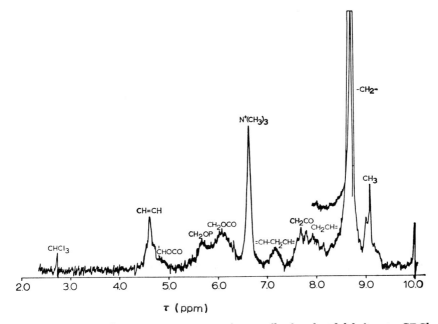

Fig. 3. High-resolution PMR spectrum of egg yolk phosphatidylcholine in CDCl₃ at 60 MHz. Internal reference is TMS. (From Chapman and Morrison, 1966.)

It is important to note that in all resonance experiments the rate of motion from one nonequivalent site to the other must be slower than the resonant frequency in order to observe the individual environments of the two sites. This is of particular importance, for example, in the very rapid molecular exchange of protons in aqueous systems in which a single narrow line is expected to lie midway between the two chemically shifted positions.

2. Methods and Experimental Details

As described above, the observation of proton resonance for a sample immersed in a constant field of 10^4 G demands a stable radio frequency of 42.6 MHz. According to the character of the sample and the nature of the problem, three basic experimental techniques are distinguishable: (a) high-resolution NMR, (b) low-resolution or wide-line NMR, and (c) pulsed NMR.

In all cases a stable electromagnet, permanent magnet, or superconducting magnet provides the steady field H_0, with provision for linear variation over various ranges. The frequency of the applied field depends on the instrument and is usually kept constant (e.g., 40, 60, 100, 120 MHz). Current flow occurs at the position of resonance and is measured directly. For high-resolution NMR, the field in addition is usually kept homogeneous to 1 part in 10^9 over a typical sample volume of 0.3 ml by means of sample spinning and field shaping coils. Precalibrated charts for proton resonance facilitate the routine extraction of chemical shift data when combined with sophisticated field stabilization techniques (Emsley et al., 1965, 1966).

The problem of chemical shift measurement in biological systems (Cohn, 1963) demands the use of a suitable inert reference compound, thereby avoiding the need to measure the absolute field strength. The reference in nonaqueous solvents can normally be tetramethylsilane as mentioned above, while in aqueous systems the sodium salt of 2,2-dimethyl-2-silapentane-5-sulfonic acid (DSS) is commonly used. Both of these compounds give a signal in a region in which very few other compounds absorb. An external reference in a coaxial tube may also be used but, for precise measurements, bulk susceptibility corrections are required (Emsley et al., 1965, 1966). Sample sizes and concentrations are currently reduced to a minimum due to the advent of the time-average computer.

The use of D_2O in place of light water removes the obscuring effect of an intense OH absorption, since the resonance of deuterium is many MHz removed from the proton resonance, but the possible modification of active processes due to the stronger structure of heavy water should not be forgotten (Kavanau, 1964).

To reduce the complicated spectra derived from overlapping chemical shifts and spin-spin mutiplets, the sample solution can be examined at a higher field H_0 (to which only the chemical shift is proportional) or by double irradiation methods, by which interacting spins can be effectively decoupled. In addition, the sensitivity increases at higher fields and the current application of superconducting magnets for the resolution of highly shielded groups in proteins is promising.

The weak proton resonances from low concentrations of biological macromolecules in water (replaced by D_2O) are now regularly enhanced by computing techniques (see below) and by the general progress toward higher radio frequencies. The advent of the Varian 220 MHz NMR spectrometer, which employs a superconducting magnet, now makes possible the study of interactions pertinent to biochemical systems in the range of 1–5 Å with a sensitivity of 120:1. For example, a concentration of 1–2% protein in D_2O gives a useful spectrum from only one scan (Chapman and Kamat, 1968; McDonald and Phillips, 1967).

For many biological samples, the breadth of the NMR lines and the limited availability of biological specimens severely restrict the signal/ noise ratio at practical concentrations. Increased values of the signal/

noise ratio can be attained by a time-averaging computer (CAT) by which the coherent fluctuations in the incoherent noise are enhanced according to $n^{1/2}$, where n is the number of successive sweeps.

Pulsed NMR methods have been particularly applied to the study of relaxation processes, diffusion processes, and to some extent, chemically shifted fine structure (Abragam, 1961). A variety of different trains of radio frequency pulses is possible but an adequate discussion of the methods involved is not feasible here.

3. Studies of Phospholipids and Phospholipid–Water Systems

In order to interpret the proton magnetic resonance (PMR) spectra of membrane systems it is necessary to understand the behavior of simple lipid systems. The following discussion therefore concerns studies of anhydrous lipids both in the anhydrous state and in the presence of water. These studies provide a conceptual framework for the interpretation of membrane spectra.

a. Anhydrous Systems. Wide-line PMR studies of various phosphatidylethanolamines at liquid nitrogen temperature (Chapman and Salsbury, 1966) in the solid state show broad resonance lines about 15 G wide (i.e., $\Delta v \approx 60$ kHz). As the temperature increases, a gradual narrowing of the line width occurs until, at a particular transition temperature, a considerable fall in line width takes place (Fig. 4). The reason for this sudden drop in the line width to about 0.1 G appears to be the

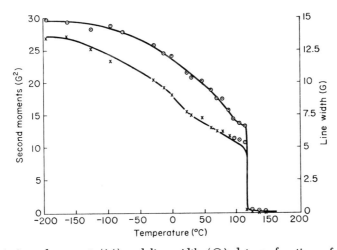

Fig. 4. Second moment ($\times$) and line width ($\odot$) data as functions of temperature of 1,2-dimyristoyl-DL-phosphatidylethanolamine. (From Chapman and Salsbury, 1966.)

onset of a mesomorphic condition. In this state, considerable molecular motion of the hydrocarbon chains of the lipid occurs, and this motion almost averages the magnetic dipole–dipole interactions to zero. This averaging process involves translational modes as well as rotational oscillation of the methylene groups of the hydrocarbon chains. These movements are particularly marked at the ends of the chains and also involve a marked departure from the all-planar character of the hydrocarbon chains as shown by the infrared spectra of the lipids in this mesomorphic phase (Chapman *et al.*, 1966). Spin-lattice relaxation measurements (Chapman and Salsbury, 1966) provide an indication that the motion of the polymethylene chains approaches, but is not as great as, that of a liquid hydrocarbon of a similar chain length.

A similar behavior has also been observed for 1,2-distearoyl-L-phosphatidylcholine (Salsbury and Chapman, 1968). In addition, reorientation of the $N^+(Me)_3$ protons occurs at low temperature and results in a reduction of line width which is not observed for the phosphatidylethanolamines in the same temperature range.

b. Aqueous Systems. Wide-line PMR studies of phosphatidylcholine–water systems demonstrated changes in molecular reorientation caused by increasing hydration (Veksli *et al.*, 1969). In the liquid crystalline phase of 1,2-dipalmitoyl-L-phosphatidylcholine, the presence of a broad component ($\Delta h \approx 1.5$ G) in addition to the previously observed narrow line ($\Delta h \approx 10^{-2}$ G) was observed. The width of the narrow line for any degree of hydration is greater than that of an isotropic liquid.

Other phase changes occur with phospholipids when small amounts of water are present in the liquid crystalline phase. With phosphatidylcholines a phase change from a lamellar to a cubic phase takes place, and the associated changes in line width have been followed by high-resolution PMR at 60 MHz (Penkett *et al.*, 1968).

The PMR spectra of egg yolk phosphatidylcholine containing less than 5% water, at different temperatures, are shown in Fig. 5. Below 25°C the spectrum (Fig. 5a) shows a "narrow" component superimposed on a broad component. The narrow component probably arises from the mobile fraction of the long-chain methylene and methyl protons. Above 25°C the material becomes liquid crystalline, the spectrum intensity suddenly increases and, between 25° and 90°C, a single line is observed (Fig. 5b). A broad component (~ 1 G or 4 kHz) for egg yolk phosphatidylcholine is observed by first-derivative presentation of the absorption line, but this is not clearly demonstrated in the pure absorption mode spectra (Salsbury and Harris, 1968).

At 90°C a high-resolution spectrum is observed (Fig. 5c) in which the

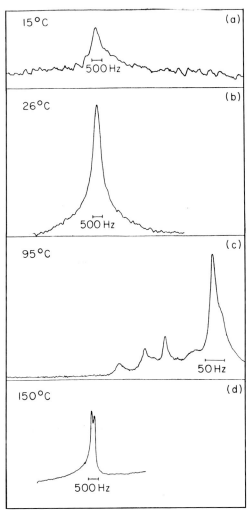

Fig. 5. High resolution PMR spectra of egg yolk phosphatidylcholine at different temperatures. (a) Below liquid crystalline transition temperature, (b) in lamellar phase, (c) in cubic phase, (d) above cubic phase. (From Penkett *et al.*, 1968.)

chemically shifted absorptions are clearly delineated. The drastic narrowing to a high-resolution spectrum is also observed in the first-derivative spectra, in which the broad component is not observed above a certain temperature for dipalmitoylphosphatidylcholine (Veksli *et al.*, 1969). Above 110°C the line width increases (Fig. 5d) and obscures the chemically shifted fine structure at higher temperatures. This change in egg yolk phosphatidylcholine takes a considerable time to occur (15 minutes at 150°C). The transition is reversible on cooling, but if the material is held at 200°C for 15 minutes the changes are no longer reversible.

The abrupt change from a broad line spectrum to a highly resolved multiline NMR spectrum of egg yolk phosphatidylcholine occurs at the temperature at which a phase change from a liquid crystalline lamellar structure to a cubic type of organization takes place (Small, 1967; Reiss-Husson, 1967). A change therefore occurs from an anisotropic to an isotropic phase. The abrupt change to a broadened spectrum which occurs at temperatures above 115°C is also associated with a change of liquid crystalline phase from the cubic phase to a different mesophase (Small, 1967). Similar results have been obtained with dipalmitoylphosphatidylcholine (Chapman et al., 1967).

The observation of a high-resolution NMR spectrum below the capillary melting point of a substance is not unique. It has been observed with plastic crystalline cyclohexanol (Suga and Seki, 1962) and trimethylacetic acid (Suga and Seki, 1963). The explanation for this behavior is associated with the considerable self-diffusion which occurs with these spherical molecules. A similar situation is observed with cyclohexane (Andrew and Eases, 1953). However, the appearance of a high-resolution spectrum followed by its disappearance as the temperature increases is unusual.

As illustrated in Fig. 6 the spin-lattice relaxation time measured from the rate of recovery from saturation of the 1 kHz line width appears to show no abrupt change with increasing temperature. However, Fig. 6 also illustrates that, in contrast, the line width shows considerable change in the same temperature range (Penkett et al., 1968). The relaxation time measurements indicate that no great change in orientation frequencies of the order of 10^7 Hz occurs for the protons responsible for the 1 kHz line width.

The origin of the residual line width in the liquid lamellar phase at first was not clear and it was suggested that in addition to the dipolar effects some anisotropic effect was present, giving rise to the field dependence of the lines. Henson and Lawson (1970) suggested that self-diffusion through internal magnetic field gradients might contribute to the spin–spin relaxation process. Recent pulse studies in our laboratory, however, show that the major effect is indeed dipolar in its origin. It may be that chemical shift anisotropy effects may be related to the field dependence effects. The T_1 values appear to be related to some spin-diffusion process, possibly related to the molecular motion of the $N(CH_3)_3$ protons of the choline group. Hence, in the liquid crystalline phase, the intense narrow line width of 10^{-2} G is assigned to the rapidly reorienting

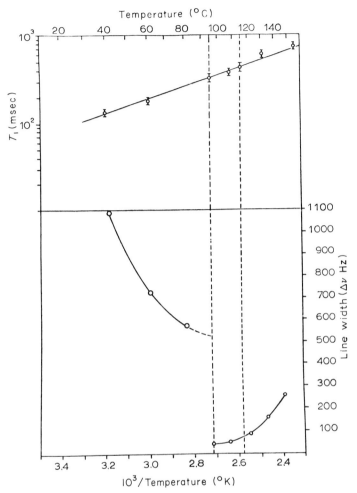

Fig. 6. Variation of narrow absorption line width ($\Delta\nu$) and spin-lattice relaxation time (T_1) with temperature for egg yolk phosphatidylcholine (from Penkett *et al.*, 1968). The dotted lines show the temperature range in which the cubic phase occurs.

protons of the lipid chains. The intermediate line $\Delta h \approx 1.5$ G is assigned to the $N(CH_3)_3$ group. It is observed with the gel phase and narrows just below the transition temperature (gel to liquid crystal). This indicates that the $N^+(CH_3)_3$ group must be undergoing rapid reorientation with a correlation frequency $\nu_c > 10^5$ Hz. Signals corresponding to these two groups of protons are just resolved with egg yolk lecithin using a high resolution NMR spectrometer and can be more clearly observed at 220 MHz. A few studies of lipid–cholesterol–water systems have been made. This work has shown that lines at 2.3, 3.4, and 5.8 G are observed with model lipid–cholesterol–water systems. The aim is to determine which lines originate from the protons of the cholesterol system and which lines originate from the hydrocarbon chains of the lipid as a consequence of chain inhibition and increased dipolar interactions.

c. Sonicated Lipid–Water Dispersions. Sonication of lipids to improve their dispersibility in water has often been used in biochemical studies. It is therefore of interest to study the NMR spectra of such sonicated dispersions in D_2O.

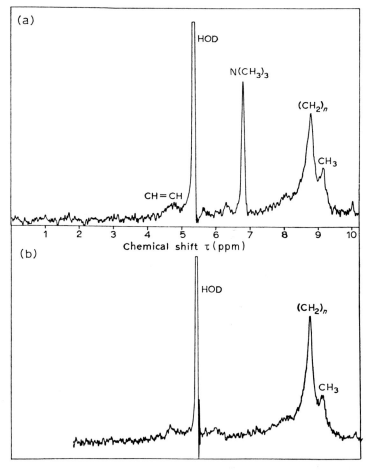

Fig. 7. High-resolution PMR spectra at 60 MHz of sonicated dispersions of (a) egg yolk phosphatidylcholine, (b) phosphatidylserine. (From Penkett *et al.*, 1968.)

Spectra obtained with sonicated dispersions of phosphatidylcholine and phosphatidylserine are shown in Fig. 7. Similar spectra are occasionally obtained from phosphatidylcholine without sonication, but only after lengthy homogenization. Increasing the time of sonication of a coarse dispersion of egg yolk phosphatidylcholine causes the high-resolution signals to grow at a steady rate. The intensity of the alkyl chain signal and the choline signal as a function of time of sonication is shown in Fig. 8. After 20 minutes no further variation in the intensities is observed. The appearance of the dispersion changes from cloudy white to almost optically clear as the signal intensity grows.

The fact that a high-resolution NMR spectrum is obtained with sonicated lipids could imply that the process of sonication destroys *all* the lamellar or bilayer structure and that a completely disordered system is formed, somewhat akin to a liquid. However, this does not appear to be the case. The sonication process appears to break up the large coarse aggregates into small spherical particles of size about 200–800 Å (Chapman *et al.*, 1968a) in which there is a single bilayer or sometimes two bilayers forming the outer shell. These sonicated lipid dispersions have become increasingly popular as model membrane systems (Thompson and Henn, 1970). This means that we have a model membrane system

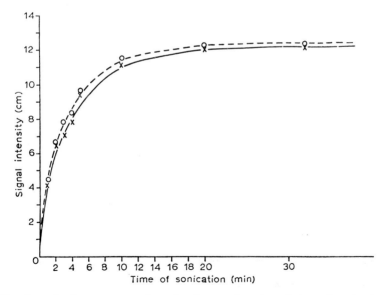

Fig. 8. Peak absorption signal amplitude as a function of time of sonication for egg yolk phosphatidylcholine in D_2O (concentration $= 50\%$ by weight) for the chemically shifted absorptions: —✕—, $(CH_2)n$; - - - O - - -, $N^+(Me)_3$. (From Penkett *et al.*, 1968.)

providing high-resolution signals for all the proton groupings in the lipid molecule. In principle we can study the interaction of drug molecules, polypeptides, and other molecules with particular groups of the lipid by studying the differential line broadening in the spectrum.

Why do these particles give rise to a high-resolution spectrum? This could be due to one or all of the following. (a) Although the sonicated lipid system is still a bilayer system, there has been some modification of molecular motion in the system. It is interesting to note that spin label studies of coarse and sonicated phosphatidylcholine show little difference in hydrocarbon chain reorientation rates (Hubbell and McConnell, 1968; Barratt et al., 1969). (b) The smaller and more symmetrical particles, while maintaining the bilayer system, have sufficient isotropic and rapid particle reorientations to average out any dipolar interactions. We are studying the effects of increasing the viscosity of the solvent in which the particles are spinning to check these points.

Calculations show that the molecular reorientation rates are of the order required for line narrowing to occur and that this may be an important contributory factor in revealing the high-resolution signals.

d. Phospholipid–Cholesterol Interactions. An interesting application of NMR studies of sonicated lipid dispersions is the study of phospholipid–cholesterol interactions (Chapman and Penkett, 1966). This has led to the conclusion that the presence of cholesterol results in some restriction on mobility of the lipid chains. The conclusion has been supported by recent diffusion experiments with liposome systems (De Gier et al., 1968). Wide line studies are also consistent with this idea.

As shown in Fig. 9 the high-resolution NMR spectrum of mixtures of

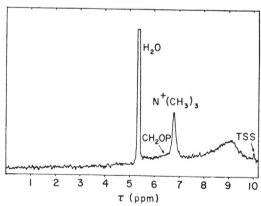

Fig. 9. The PMR spectrum of dispersion in D_2O of egg yolk phosphatidylcholine and cholesterol (1:1). (From Chapman and Penkett, 1966.)

egg yolk phosphatidylcholine and cholesterol dispersed by sonication in water contrasts with that illustrated for egg yolk phosphatidylcholine alone in Fig. 7. The sharp $(CH_2)_n$ and CH_3 signals from the lipid chains are broadened for a 1:1 molar ratio of phosphatidylcholine and cholesterol. The $N^+(Me)_3$ peak, on the other hand, remains sharp in the spectrum. Thus there is a differential line broadening of chemically shifted fine structure for a 1:1 molar ratio of phospholipid and cholesterol (Chapman and Penkett, 1966). This suggests that the presence of cholesterol in the dispersion reduces the amplitudes and/or frequencies of the molecular motion of the hydrocarbon chains of the lipid. There is less effect on the molecular motion of the $N^+(Me)_3$ group. It may be that the cholesterol, because of its flat bulky shape, restricts the molecular motion of the lipid chains in the lipophilic region but, due to its single hydroxyl group, is not able to interact to a similar extent with the polar $N^+(Me)_3$ group in the hydrophilic region. The absence of signals from the cholesterol structure suggests that in the lipid dispersion the cholesterol molecule cannot exhibit isotropic motion.

e. Phospholipid–Polypeptide Interactions. Both valinomycin and alamethicin are examples of cyclic molecules which induce transport of ions across biological (Pressman, 1968) and artificial membranes (Muller and Rudin, 1968; Henderson *et al.*, 1969). Their action differs in two important ways; first, valinomycin shows potassium/sodium ion specificity whereas alamethicin transports both ions with the same ease and, second, alamethicin (but not valinomycin) induces in black lipid films electrical properties analogous to excitability in nerve membranes. High-resolution NMR spectroscopy has been used to investigate the interactions of valinomycin and alamethicin with phospholipids in model membrane systems and in natural membranes. The interaction of the related molecule gramicidin S with phospholipids has also been studied. Gramicidin S is similar to alamethicin and valinomycin in many ways, being a cyclic surface-active polypeptide antibiotic containing 10 amino acid residues (alamethicin contains 19 amino acid residues and valinomycin contains 6 amino acid and 6 hydroxy acid residues). However, gramicidin S does not induce ion transport across mitochondrial membranes.

Nuclear magnetic resonance studies of the interaction of valinomycin, alamethicin, and gramicidin S with aqueous dispersions of phosphatidylserine and egg yolk phosphatidylcholine show two distinct types of behavior, exemplified by valinomycin and alamethicin on the one hand and gramicidin S on the other. Figure 10 shows how the 60 MHz proton spectrum of a sonicated dispersion of 1% phosphatidylserine in D_2O is altered by the addition of a small quantity of alamethicin (molar ratio 100:1);

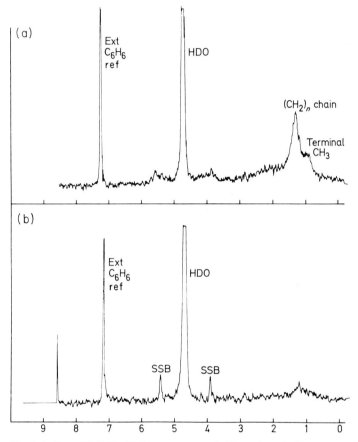

Fig. 10. (a) The 60 MHz PMR spectrum of 1% sonicated dispersion of phosphatidylserine in D_2O; (b) as (a) but with phosphatidylserine + alamethicin, molar ratio 100:1 (Finer *et al.*, 1969).

the hydrocarbon chain signal is drastically broadened, so that it becomes lost in the baseline. This effect is due to a reduction in motion of the lipid chains, leading to increased correlation times and hence reduced relaxation times. A similar effect is observed with phosphatidylcholine. Valinomycin and alamethicin interact with the hydrophobic parts of the lipids, so that each molecule by some cooperative effect appears to inhibit the motion of considerable numbers of lipid chains (for example, judging from the spectral behavior, one alamethicin molecule affects 600 molecules of phosphatidylserine). In contrast to this, the addition of gramicidin S to a sonicated dispersion of phosphatidylcholine does not produce any reduction in hydrocarbon chain signal, although the choline signal is slightly reduced at high polypeptide concentrations. This result is interpreted as showing that gramicidin S, which does not transport ions, interacts primarily with the head groups of the lipid molecules rather than the hydrocarbon chains.

4. Nuclear Magnetic Resonance Studies of Serum Lipoproteins

The serum lipoproteins are an interesting system of lipid and protein whose PMR spectra have been studied (Chapman et al., 1969a,b). The details of these spectra are to be contrasted with the spectra of sonicated membrane systems (see Fig. 11).

5. Membranes

When considering the application of NMR spectroscopy to the study of cell membranes, one must recall the results of the studies of simple lipid–water systems which showed that when lipids are in a bilayer arrangement but are present in large aggregates, a high-resolution NMR spectrum is not obtained.

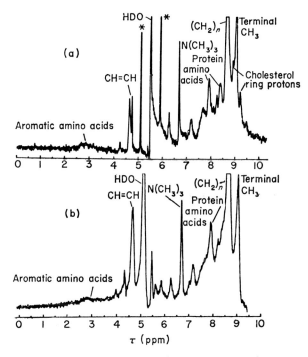

Fig. 11. The 220 MHz PMR spectrum of human serum α-lipoprotein, subfaction HDL_2 in D_2O at temperatures of approximately (a) 50°C, (b) 13°C. Asterisks indicate spinning side bands (Chapman et al., 1969a,b).

a. Myelin. Myelin is considered by many workers using x-ray techniques to be predominantly in a bilayer form and in an aggregate type of structure. Furthermore the myelin membrane contains an unusually high proportion of lipid (see Table I). One might therefore expect that the myelin membrane would give a wide-line spectrum, and this is the case. The PMR spectrum in D_2O is shown in Fig. 12. Interpreting a spectrum of this type is extremely difficult without attempting a detailed interpretation of the line width. One can, however, compare the wide-line spectrum of the *myelin membrane* with that of the *total lipid*, (i.e., phospholipid and cholesterol) extracted from the membrane and dispersed in D_2O (Fig. 12). Such a comparison shows very good agreement.

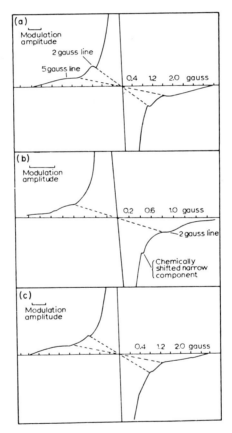

Fig. 12. Wide line 60 MHz PMR spectrum of (a) Wet myelin–95% D_2O, 10 G sweep (b) Wet myelin–95% D_2O, 5 G sweep. (c) Total lipid extract from myelin–95% D_2O, 10 G sweep. Dotted lines are used to indicate different PMR lines (Jenkinson *et al.*, 1969).

Since independent x-ray studies show that the myelin lipid when dispersed in water is certainly in a bilayer structure, this good agreement in PMR spectra is consistent with the lipids also being in a bilayer structure in the membrane. (The lines in the spectra at 2 and 5 G can be analysed to provide information about lipid–cholesterol interaction and inhibition of chain motion.) We have seen that sonication of lipid aggregates breaks these down into small particles which enable high-resolution spectra to be obtained. While there is clearly a possible danger of affecting membrane organization by sonication, it seems worthwhile to see what happens in this situation. As seen in Fig. 13, sonication of myelin membranes results in a high-resolution NMR pattern similar to that of the sonicated lipid itself.

A comparison of the spectrum of the myelin fragments with that of sonicated total lipid and of phospholipid without cholesterol shows that the cholesterol influences the mobility of hydrocarbon chains while the cholesterol molecules themselves are prevented from having complete isotropic motion.

b. Erythrocyte Membranes. Wide-line NMR studies have been carried out with intact erythrocyte membranes and such studies are still in progress. A simple comparison of the wide-line spectrum with the spectrum of the total lipid of the erythrocyte membrane can be made, but there is a large amount of protein present in the membrane and unequivocal interpretations are therefore more difficult.

When the membrane is placed in a powerful denaturing material such as trifluoroacetic acid, high-resolution signals are observed from all the lipid groups e.g., the (CH_2) and (CH_3) groups and the $N^+(CH_3)_3$ groups. Signals associated with the denatured protein are also observed and can be assigned to particular amino acids (Chapman et al., 1968b).

Experiments on the interaction of other molecules with the intact erythrocyte membranes are also possible and a number have been performed. In this case, high-resolution NMR spectra can be observed from either the interacting molecule or the lipid and protein material released from the membrane. Thus the effects of detergents such as sodium dodecylsulfate and lysophosphatidylcholine have been studied. These studies show that sodium dodecylsulfate releases lipid from the membrane and the high-resolution spectra of the lipid can be seen. The effect of lysophosphatidylcholine is to form an additive complex with the membrane, thereby changing the expected NMR spectrum of the lysophosphatide. If the membrane is treated with phospholipase C, high-resolution signals are observed from the protons of the diglyceride chains.

Sonication of the erythrocyte membranes results in the appearance of

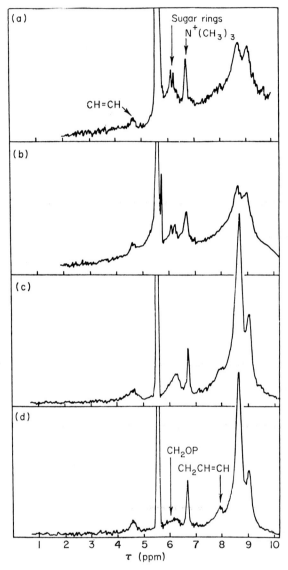

Fig. 13. The PMR spectra at 60 MHz of dispersions in D_2O at 65°C of (a) myelin, (b) total lipid extract from myelin, (c) the cholesterol-free lipid (phospholipid and galactolipid), (d) the phospholipid extract (Jenkinson *et al.*, 1969).

reproducible high-resolution signals. The PMR spectrum of erythrocyte membrane fragments in aqueous dispersion is shown in Fig. 14a (Chapman *et al.*, 1968b) Peaks occur at 6.3 ppm due to CH_2OC and CH_2OP protons of sugar or lipid, at 6.7 ppm due to $N^+(CH_3)_3$ protons

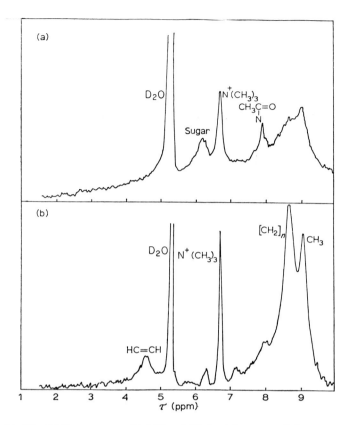

Fig. 14. The PMR spectra at 60 MHz of 5% (w/v) sonicated dispersions in D_2O of (a) erythrocyte membrane fragments, (b) total equivalent lipid (phospholipid and cholesterol). (From Chapman and Kamat, 1968.)

in phosphatidylcholine and sphingomyelin, and at 7.8 ppm due to NCOCH$_3$ protons in N-acetyl groups in sugars or protein. Signals expected from protons in the CH=CH (at 4.7 ppm) or (CH$_2$)$_n$ groups (at 8.7 ppm) of the lipid chains are not present in the spectrum. Neither are there any signals from the aromatic amino acids. Hence sonication does not appear to cause a marked denaturation of the membrane protein. The following conclusions have been reached.

1. Some of the protons in the sugar residues are situated in a local environment which allows sufficient freedom of movement to give rise to the narrow signals at 6.3 and 7.8 ppm. Consistent with this conclusion is the fact that the external surface of the erythrocyte membrane contains a large number of short oligosaccharide chains attached to protein (Watkins, 1966; Eylar et al., 1962).

2. The N$^+$(CH$_3$)$_3$ protons are in an environment similar to that ob-

served with sonicated phosphatidylcholine–D_2O and have similar freedom of molecular motion.

3. The lipid hydrocarbon chains in the membrane fragments are restricted in their molecular freedom. This may arise from lipid–lipid, lipid–cholesterol, or lipid–protein interactions. The fact that the total lipid, including cholesterol, gives a PMR spectrum as shown in Fig. 14b suggests that some restriction on the chain movements in the membrane fragments may be caused by lipid–protein interactions as well as by some interaction with cholesterol. The absence of any cholesterol signals suggests that the cholesterol is in some anisotropic environment.

Can one conclude anything about the structure of the erythrocyte membrane from these spectroscopic studies? The wide-line NMR studies are consistent with the presence of some lipid bilayers. If one assumes that the sonication process does not cause a major reorganization of the membrane components (a number of studies have been made of sonication effects on the erythrocyte membrane using a variety of techniques; the main conclusion is that the membrane breaks into small vesicles having intact membrane structure), then it is apparent that polar groups of sugars of lipids and some acetyl groups can move freely, giving high-resolution signals. On the other hand, the lipid chains and hydrophobic amino acid groups (e.g., the aromatic amino acids) of the membrane proteins are restricted in their mobility, and the cholesterol present is inhibited in its isotropic molecular freedom.

These broad conclusions are consistent with the idea developed many years ago from surface chemistry studies and from more recent studies of protein structure that the basic stable energetic system will be one in which the main hydrophobic groups are in an interior site away from water, while hydrophilic groups are in exterior sites in the water phase.

It is also of great interest that the $N^+(CH_3)_3$ group associated with phosphatidylcholine and sphingomyelin is free to move. (There is no change in the area of this signal when the membrane fragments are treated with detergent.) This is consistent with the idea that regions of the membrane containing phosphatidylcholine or sphingomyelin do not interact electrostatically with membrane protein. If there are any electrostatic interactions with proteins, they may involve negatively charged lipids such as phosphatidylserine. Experiments with phospholipid–polypeptide systems show that chain-broadening effects can be caused by such lipid–protein interactions as well as by phospholipid–cholesterol interactions.

However, sonication may disturb the original membrane organization in some way and, therefore, other nuclear resonance techniques such as the study of ^{13}C nuclei and pulsed methods are also being applied to membrane systems.

6. The Study of Water

The use of NMR to study the structure of water has been stimulated by the extreme current interest in the intermolecular structure of water and the nature of water in biological systems.

Salsbury and Chapman (1971) are currently employing deuteron resonance for the study of D_2O associated with liquid crystalline structures formed by phospholipids. The electric quadrupole moment of the deuteron leads to a splitting of the deuteron NMR line shape due to interactions with the electric field gradient at the sites of the deuterium nuclei. In most deuterated hydrates of salts and in polycrystalline heavy ice, the quadrupole coupling constants are found to be significantly greater than 200 kHz, although in the presence of flipping (180°) motions about the DOD bisector, this may be reduced to nearly 200 kHz. The heavy water associated with phosphatidylcholine in its lamellar liquid crystalline state gives a value of only 1–2 kHz, which is a function of temperature and hydration. Thus a residual motional anisotropy is indicated in which the deuterons are equivalent through rapid rotation and/or exchange but are orientated to a limited extent between the lipid bilayers up to a total of 21 moles of D_2O per mole of phospholipid.

A reduction of the line splitting just prior to the gel to liquid-crystal transition point indicates that there is a rearrangement of water structure involving a two-phase system consistent with some rearrangement of the polar head group. This links up with the observation showing the narrowing of the line associated with the $N(CH_3)_3$ group and the pretransitional thermal peak (see page 70).

7. Conclusions

Nuclear magnetic resonance spectroscopy has been shown to have great potential for the study of lipid molecules. Information about the molecular motion of lipids in the solid state, in the mesomorphic state, and in various solvents including water has been obtained. When the technique is applied to the study of the structure of cell membranes there are some difficulties of unequivocal interpretation, but these should be resolved by further studies. Useful information has been found of relevance to the organization of the membrane components. Further studies of lipid–polypeptide interactions may also aid in interpreting this information in terms of membrane structure and function.

D. Electron Spin Resonance Spectroscopy

Electron spin resonance spectroscopy is another powerful physical technique for the study of the structure and function of membranes and is discussed in detail in the following chapter.

E. Fluorescent Probes

1. Introduction

Fluorescence techniques have been applied in the membrane field to a lesser extent than have the magnetic resonance methods. As yet only a few exploratory fluorescence studies have been made. In this section, therefore, we shall begin by examining the theoretical and experimental basis for fluorescence techniques as used in membranology. It will be convenient at this stage to discuss the use of fluorescent probes for studying conformational changes of proteins. We shall then discuss the fluorescence studies on membrane transitions which have been reported and indicate the possible future use of these techniques for studying the dynamic aspects of membrane behavior.

2. Fluorescence Techniques

The basic light-emission processes of membrane systems are essentially similar to those found in simpler chemical systems; that is, emission can occur from singlet or triplet excited states produced by introducing the energy in the form of light or chemical reactions. Some familiarity with the basic principles of luminescence techniques is necessary before the potential of this technique in membranology can be put in perspective. A short review of these principles is given in the next section. Reviews of luminescence techniques with particular reference to biological systems (Radda and Dodd, 1968; Stryer, 1968; Weber and Teale, 1965), and chemical systems (Hercules, 1966) are available.

a. Fluorescence and Phosphorescence. The essential events involved in the production of excited electronic states are shown in Fig. 15a. When a ground-state molecule (i.e., a molecule as it normally exists in a bottle on the shelf) interacts with light of the appropriate energy a particular electron in the molecule is promoted to a higher electronic state. The time sequence involved in these transitions is shown in Fig. 15c. The particular electronic state to which the electron is excited depends on the energy of the incident radiation.

The relationship between the energy involved in the transition and the wavelength of light required is given by the expression

$$E = h\nu = h\frac{c}{\lambda}$$

where E is the energy of transition in kcal/mole, h is Planck's constant (6.62×10^{-27} erg sec), ν is the frequency of the light, c is the velocity of light, λ is the wavelength of the light.

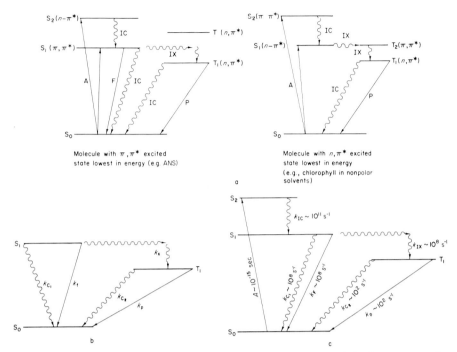

Fig. 15. Energy relationships in luminescence processes. (a) Comparison of luminescence processes for molecules having lowest n,π^* and π,π^* states: A, absorption; F, fluorescence; P, phosphorescence; IC, internal conversion; IX, intersystem crossing; S_0, ground state; S_1, lowest excited singlet state; S_2, higher excited singlet state; T_1, lowest triplet state; T_2, higher triplet state. (b) Summary of rate constants for excited-state processes: k_{c_1}, internal conversion for lowest excited singlet state; k_{c_2}, internal conversion for lowest triplet state; k_p, phosphorescence; k_f, fluorescence; k_x, intersystem crossing. (c) Relative rates of the steps involved in the luminescence processes ($s^{-1} = sec^{-1}$) described in the preceding diagrams.

Table III shows the range of energies used in biological work. It is important to remember that biological systems such as proteins and membranes can be damaged easily by high-intensity ultraviolet light.

If the electron is excited to a high-energy state the electron is demoted to the lowest vibrational level of the lowest electronic state within about 10^{-11} seconds. The energy lost in this electron descent is degraded as heat. From the lowest excited state the electron can return to the ground state by four main pathways.

1. After an interval the electron can return to the ground state with the emission of light. This process is called *fluorescence*. Because the electron is returning from a lower vibrational state (and possibly a lower electronic state) than the one to which it was originally excited, the

TABLE III

THE ENERGY RELATIONSHIPS OF RADIATION USED IN FLUORESCENCE WORK

Wavelength (nm)	Energy (kcal/mole)	Approx. color	Region of typical excitation maxima	Region of typical emission maxima
200	14.3			
250	114.4		Phenylalanine	
300	95.3		Tryptophan, tyrosine	
350	81.7		NADH, ANS, Vit A	Tryptophan, tyrosine
400	71.5	Visible limit		
450	63.5	Violet-blue	Flavins	NADH
500	57.2			Flavins, Vit A
		Green		
550	52	Yellow		
600	47.7			
		Orange		
650	44.0		Chlorophyll	Chlorophyll
700	40.8	Red		

energy of the fluorescent light will be less than the absorbed light and hence the emission will take place on the long-wavelength side of the absorption band.

2. The second possible descent route for the electron is called *internal conversion*. The electron returns to the ground state but the energy of the excited electron is degraded as heat. The details of this process are poorly understood.

3. Another possible route for the electron is called *intersystem crossing* and occurs when the electron is transferred to the lowest triplet state (see below). This requires a reversal of the electron spin. After a comparatively long interval (10^{-2}–10 seconds) the electron returns to the ground state with the emission of light. This process is called *phosphorescence*.

4. The fourth pathway is known as *energy transfer*. The energy of the excited state is transferred to another molecule (see below).

The rates of the possible reactions leading from an excited state can be characterized by the rate constants as shown in Fig. 15b. The relative rates of these competing processes will determine the amount (efficiency) of the fluorescence. The fluorescence quantum yield (Q_f) is defined as

$$Q_f = \frac{\text{number of quanta emitted}}{\text{number of quanta absorbed}}$$

and can be expressed in terms of the rate constants shown in Figs. 15b and c as

$$Q_f = \frac{k_f}{k_f + k_p + k_{c_1} + k_{c_2}}$$

It is important to distinguish between the terms electron state and electron orbital. An *electron orbital* is defined as that volume element of space in which there is a high probability (99.9%) of finding an electron. It is calculated from a one-electron wave function and is assumed to be independent of all other electrons in the molecule. *Electronic states* concern the properties of all the electrons in all of the orbitals. Since interactions between electrons are quite significant, the movement of an electron from one orbital to another will result in a change in the electronic state of the molecule. It is important to consider the states of the molecule involved rather than only the orbitals involved in such an electron promotion.

The electronic states of organic molecules can be grouped into two broad classes: singlet states and triplet states. A singlet state is one in which all of the electrons in the molecule have paired spins. Triplet states are those in which one set of electron spins has become unpaired—that is, all electrons in the molecule except two have paired spins.

The two most important singlet states in organic molecules are π,π^* and n,π^*, respectively. In the former an electron is promoted from a ground-state π orbital to a vacant π orbital of higher energy. The well-known ultraviolet absorption bands of the aromatic amino acids are due to transitions of this type. An n,π^* state results from excitation of an electron in a nonbonding n orbital to an excited π orbital. The nonbonding electrons localized on the oxygen atom of carbonyl groups is a good example of this type of chromophore. The electron distribution of the excited state differs greatly from the ground state; for example, the oxygen atom has a lower electron density in the excited state and is a much stronger acid. Thus ground-state phenol has a pK_a of 10.02 while the pK_a of the excited state is 5.7.

The geometry of the molecule in the excited state may differ radically from its geometry in the ground state. An interesting example of the change in geometry which can occur on excitation is given by acetylene. In the ground state, acetylene is linear, with a C—H bond length of 1.06 Å. However, in the excited state the H—C— bond angle becomes 120° compared with 180° in the ground state. Effectively what happens is that after excitation the carbon atoms undergo a rehybridization from sp to sp^2 and adopt an ethylene configuration with the hydrogens in the *trans* position. Geometrical changes of this type may be important in membrane-bound carotenoids.

An important determinant of the fluorescence quantum yield of a molecule is the nature of the lowest singlet excited state. The fluorescence behavior of a molecule depends on the nature of the lowest excited state and not on the state to which it was originally excited. Some properties of the two most important excited states in organic molecules are shown in Table IV. Molecules with an n,π^* lowest excited state may have a very low fluorescence quantum yield because of the small difference in energy between the lowest singlet state and the lowest triplet state. Intersystem crossing with the consequent production of phosphorescence instead of fluorescence is thereby facilitated (Fig. 15a). The longer lifetime of the n,π^* excited state also assists singlet–triplet crossing.

TABLE IV

COMPARISON OF n,π^* AND π,π^* SINGLET STATE OF FLUORESCENT PROBES

Property	n,π^* State	π,π^* State
ϵ_{max}	$10–10^3$	$10^3–10^5$
Lifetime (sec)	$10^{-7}–10^{-5}$	$10^{-9}–10^{-7}$
Singlet–triplet split (kcal/ mole)	Generally 5–6	Generally 15–30
Rate of intersystem crossing	Greater than for fluorescence	Of the same order as fluorescence
Dipole moment	Maybe a decrease	Generally an increase of up to 18 D

The energy of the excited state may be transferred to a second molecule in two important ways. In *collisional quenching* there is "contact" between the excited molecule and the quencher. The mechanism can be depicted as

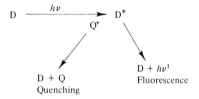

where D is the ground-state molecule, D^* is the excited-state molecule, and Q is the quenching molecule. A less interesting case occurs when the quencher can interact with the ground-state molecule D.

Energy transfer that occurs over distances larger than the contact distance of molecular collision is called *noncollisional energy transfer* and can be depicted as

$$D \overset{h\nu}{\to} D^* \qquad \text{(excitation)}$$
$$D^* + A \to D + A^* \qquad \text{(transfer)}$$
$$A^* \to A + h\nu^1 \qquad \text{(emission or radiationless deactivation)}$$

where D and D^* are as before and A and A^* represent the ground and the excited states of an acceptor, respectively. This type of transfer can occur over distances of 15–100 Å and arises from a vibrational coupling interaction between the excited states of the donor and the excited states of the acceptor.

These processes are true nonradiative transfer processes and should not be confused with the process of emission of radiation by one molecule and reabsorption of emitted radiation by another.

b. The Fluorescence of Proteins. The fluorescence of proteins may derive from the aromatic amino acid residues or from bound small molecules. The terms "intrinsic fluorescence" and "extrinsic fluorescence" have been applied in an analogous way to Cotton effects observed in the study of the optical rotatory dispersion (ORD) of proteins.

Intrinsic Protein Fluorescence. *General Principles.* Of the twenty or so amino acids that are found in proteins, only three aromatic residues have chromophores that absorb and emit light in a convenient spectral range. The spectral properties of tryptophan, tyrosine, and phenylalanine are summarized in Table V. The attachment of the amino acid side chains does not greatly alter the spectral characteristics of the parent compounds, indole, phenol, and benzene.

The quantum yield of tyrosine and tryptophan fluorescence in proteins and polypeptides is considerably lower than those of the free amino acids. The chief cause of the reduced quantum yield is collisional quenching of the excited rings by protons from side chain —COOH and —NH$_3^\oplus$ groups.

The fluorescence of tyrosine is relatively insensitive to the polarity of solvent in contrast to that of tryptophan. One can illustrate solvent effects

TABLE V
ABSORPTION AND EMISSION SPECTRA OF AROMATIC AMINO ACIDS

Amino acid	Absorption		Emission	
	λ_{max} (nm)	ϵ (liter/mole·cm)	λ_{max} (nm)	Quantum yield
Phenylalanine	257	200		
	206	9000	282	0.035
	187	58,000		
Tyrosine	275	1200		
	222	8000	303	0.21
	192	47,000		
Tryptophan	280	5500		
	220	32,000	350	0.20
	196	21,000		

on the latter using acetyltryptophanamide, this being a more realistic model for a tryptophan residue which is incorporated in a polypeptide chain. The fluorescence emission spectrum of this molecule is similar to

Acetyltryptophanamide

that of tryptophan and its intensity of fluorescence is sensitive to relatively small changes in the solvent (Fig. 16). Additives which decrease the dielectric constant of water result in enhancement of fluorescence, while an increase in dielectric constant has the opposite effect. Urea, a common protein-denaturing agent, enhances fluorescence despite the fact that its dielectric increment is positive. Some direct interaction may be involved.

The intensity of the amino acid fluorescence is also significantly tem-

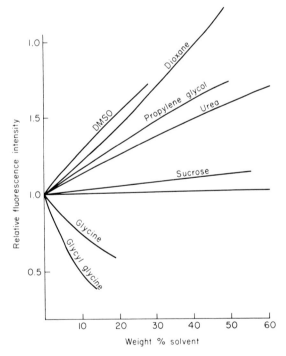

Fig. 16. Solvent effect on the fluorescence intensity of acetyltryptophanamide: DMSO, dimethyl sulfoxide. (After Radda and Dodd, 1968.)

perature dependent. This is not an unusual observation, but it is important for interpreting temperature effects on protein and membrane fluorescence. The temperature dependence can be understood in terms of three independent processes which contribute to the return from the excited state to the ground state. Two of these (fluorescence and one nonradiative process) have negligible activation energies. The other nonradiative process has an activation energy which can be interpreted as the energy difference between the lowest level of the first singlet excited state and the region where the potential energy surfaces of the excited and ground states cross. This activation energy, then, can be calculated and is 8.1 kcal/mole for tryptophan and 7.1 kcal/mole for tyrosine.

The quantum yield of fluorescence of proteins is very low (in the region of 0.8–7%) (Weber and Teale, 1965) and this may be due to quenching processes of the type described for the aromatic amino acid residues. Simple proteins can be divided into two groups on the basis of their fluorescence properties.

Proteins that contain only phenylalanine and tyrosine (of the aromatic amino acids) exhibit fluorescence spectra which are indistinguishable from that of free tyrosine, with $\lambda_{max} = 303$ nm. This reflects the insensitivity of tyrosine fluorescence to the polarity of the environment due to the low polarity of its excited state. Proteins in this class include insulin, tropomyosin, zein, and ribonuclease.

Proteins that contain both tyrosine and tryptophan show a fluorescence spectrum which is characteristic of tryptophan only. (Most globular and membrane proteins belong to this class.) The maxima of emission depend on the protein. For instance λ_{max} for chymotrypsin in 332 nm and for pepsin 346 nm, and it is likely that in pepsin, tryptophan residues are in contact with water while in chymotrypsin they are largely buried in the hydrophobic interior of the protein. The lack of tyrosine fluorescence in these proteins is independent of the protein conformation, as tyrosine emission is not observed in either native or denatured forms. The reasons for this are still unexplained.

The Influence of Conformation and Environment on the Intrinsic Fluorescence of Proteins. A knowledge of the detailed stereochemistry of proteins in solution not only is of intrinsic chemical interest but is a major factor in understanding the mechanism of enzyme catalysis and the possible arrangement of proteins in membranes and other structures.

Two approaches have been used in evaluating the correlation between the fluorescence properties of free or membrane-bound proteins and their structure. In the first, generally referred to as *solvent perturbation,* the action of relatively inert solvents, which are unlikely to alter the conformation of proteins, is observed (Steiner *et al.,* 1963). Alternatively, by bringing about structural transformations in the protein in a predictable direction (e.g., denaturation by urea, heat or pH changes), changes in fluorescence emission are recorded.

The idea behind solvent perturbation is basically simple. The native forms of globular proteins are relatively compact and impermeable to solvent. It is therefore expected that the electronic states of aromatic residues inbedded in the interior of the protein should be insensitive to solvent composition. In contrast, residues on the protein surface will be influenced by the solvent. Clearly there must be intermediate stages where the aromatic residues lie in cavities that are accessible only to solvent molecules of certain dimensions. Most organic solvents that influence the fluorescence of tryptophan (cf. Fig. 16) can be used to explore the environment of this chromophore in native proteins.

The effect of urea on tryptophan-containing proteins is highly variable (Steiner *et al.*, 1963). We have seen that it enhances the fluorescence of acetyltryptophanamide, while with proteins, in some cases a large decrease and in others an increase in the fluorescence emission is observed. In cases where pronounced quenching is observed (e.g., bovine serum albumin, lysozyme) the intrinsic exalting effect of urea is countered by the effects arising from the loss of organized structure during denaturation.

Many of the details of these observations are not yet clearly understood but the examples serve to illustrate the potential usefulness of fluorescence methods in structural studies.

Phosphorescence in Proteins. Protein phosphorescence contains emission from both tyrosine and tryptophan. In most of the proteins studied a significant fraction of the energy in excited tyrosine residues is not transferred to tryptophan. Phosphorescence, therefore, is a promising method for studying tyrosine emission, which is not the case for fluorescence.

Extrinsic Protein Fluorescence. *Covalently Bound Probes: Polarization of Fluorescence.* The polarization of fluorescence of covalently bound probes was originally introduced by Weber as a method for measuring the shape and size of a protein. The polarization is a function of (a) the viscosity and temperature of the solvent, (b) the size, shape, and rigidity of the protein, (c) the excited-state lifetime of the probe, and (d) the freedom of rotation of the fluorescent label. The mean rotational relaxation time of the protein can be determined by studying the polarization of fluorescence as a function of viscosity. This can lead to an estimate of the shape of the protein in suitable instances. These aspects of the technique have been extensively reviewed (Weber, 1963; Steiner and Edelhoch, 1962). Covalently bound probes have not often been used for detecting protein or membrane conformational changes.

Noncovalently Bound Probes. This technique was introduced by Weber and colleagues in the early 1950's (Weber and Laurence, 1954) but its full potential as a method for detecting very small perturbations in protein structure has been realized only in the last three years. 1-Anilino-naphthalene-8-sulfonate (ANS) is the best-known example of this class of probe (Fig. 17). It has a very low (0.004) fluorescence quantum yield in solution (Stryer, 1965). When it binds to a protein or membrane the quantum yield of the dye changes dramatically up to 200-fold and the emission maximum is blue shifted by 20–70 nm (Stryer, 1965) (Fig. 18).

Protein or membrane transitions of state are reflected by a change in the parameters of bound ANS (Dodd and Radda, 1967). The dye has a marked ability to detect small ligand-induced subunit interactions in complex allosteric proteins like glutamate dehydrogenase (Dodd and Radda, 1971). Although this enzyme is not a lipoprotein, the dynamic interactions of the six protomers in the enzyme oligomer mimic some of the cooperative transitions thought to be of importance in membranes. Thus it is not surprising that ANS, which has been used for detecting the $\bar{R}$ for allosteric systems, also detects changes in the state of membranes.

It is important to understand the mechanism of fluorescence changes

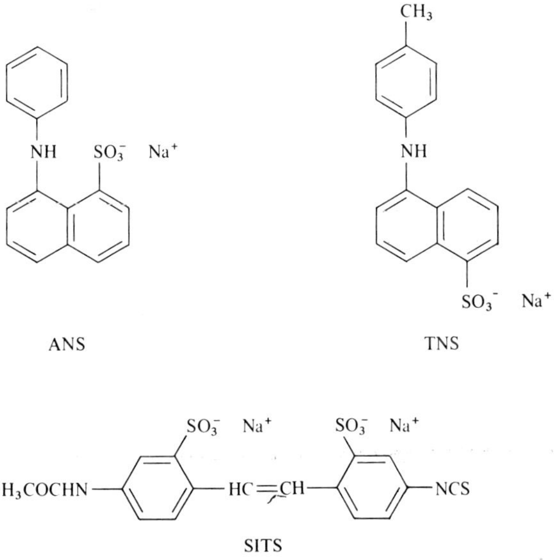

Fig. 17. The structure of common fluorescent probes.

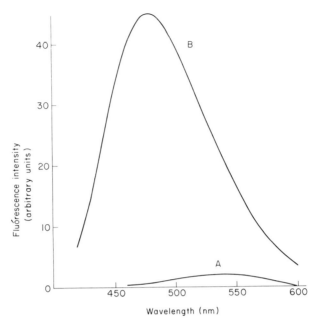

Fig. 18. Fluorescence emission of ANS (20 μM) in buffer and in the presence of erythrocyte stroma. A, ANS in isotonic sodium phosphate buffer, pH 7.4. B, ANS with erythrocyte stroma (0.5 mg/ml) in pH 7.4. (After Freedman and Radda, 1969.)

of ANS when it binds to a pure protein or to a membrane. The nature of the molecular changes in the immediate environment of ANS in membranes is inferred directly from our knowledge of the factors responsible for these fluorescence changes. The two changes that have to be accounted for are the increase in quantum yield and the shift of the emission maximum. The main factors involved are solvent viscosity, temperature, proton transfer to the excited state, the rates of intersystem crossing and internal conversion, possible inversion of the energy levels of π-π^* and n-π^* states, and the polarity of the medium. Despite the extensive use of ANS-type dyes in the last two years the significance of the above effects is not known with certainty in any instance.

Increasing the solvent viscosity at a constant dielectric constant increases the quantum yield of ANS-type dyes (McClure and Edelman, 1966; Oster and Nishijima, 1964). It is the microscopic rather than the bulk viscosity which is the significant parameter (Oster and Nishijima, 1964). The effect is smaller than that produced by decreasing the dielectric constant and is attributable to suppression of vibronic deactivation of the excited state by intramolecular motion between the phenyl and naphthyl rings (Weber and Laurence, 1954; Oster and Nishijima,

1956). This is the "loose-bolt" effect of Lewis and Calvin (1939), and it is important in every instance in which ANS-type dyes bind to macromolecules. Temperature variation affects the "loose-bolt" effect.

A deuterium solvent isotope effect on the fluorescence of ANS and related compounds has been found (Förster and Rokes, 1967; Stryer, 1966). The fluorescence quantum yield of ANS in D_2O is slightly greater than in H_2O. The fluorescence of ANS is lower in carboxylic acids than in the corresponding alcohols (McClure and Edelman, 1966). These observations can be interpreted in terms of a proton transfer to the excited state of the dye. Since such proton transfer results in a lower quantum yield, it is not of importance in the ANS–membrane interactions. The importance of intersystem crossing in reducing the fluorescence quantum yield of ANS-type compounds is not known.

Fluorescence enhancement arising from solvent-induced inversion of $n\text{-}\pi^*$ and $\pi\text{-}\pi^*$ excited states is exhibited by aromatic carbonyl compounds and by nitrogen heterocycles (Hercules, 1966). The solvent dependence of the fluorescence is in the opposite direction of that found with ANS. Hence this mechanism does not play a part in ANS fluorescence changes.

The major effect on ANS fluorescence arises from solvation of the excited state and is related to the dielectric constant of the solvent. There are several theories which account for these effects (Lippert, 1967; Wehry, 1967). The long-wavelength band of ANS is a $\pi\text{-}\pi^*$ transition. Both the high positive polarization of ANS in glycerol and the solvent-induced shifts in the emission maximum support this assignation. In ANS-type compounds the $\pi\text{-}\pi^*$ excited state is more polar than the ground state by 10–12 D (Weber, 1961). Therefore polar solvents interact more strongly with molecules in the excited state than with those in the ground state. This excited-state solvation lowers the energy of the excited state and thus decreases the energy difference between the excited and ground states. Hence as the solvent polarizability is increased, a red shift of the emission maximum would be expected. This has been experimentally confirmed. A strong solvent interaction with the excited state assists the vibronic dissipation of the excited-state energy and results in a vibrationally excited ground state. Thus in a polar solvent ANS would be expected to have a broad emission band with a low quantum yield, as is found. The fluorescence of ANS is therefore an important method for determining the microscopic polarity in complex heterogeneous biological systems.

In the case of TNS (see Fig. 17) hydrated and anhydrous crystals of the dye exhibit fluorescence spectra similar to those obtained for TNS in water and in organic solvents, respectively (Camerman and Jensen, 1969). The molecular structure of anhydrous TNS suggests extended

56 D. CHAPMAN AND G. H. DODD

resonance over the whole molecule. Thus far this dye has not been used as extensively as ANS.

3. Fluorescence Studies of Membranes

a. Intrinsic Fluorescence. The fluorescence of plasma membrane vesicles prepared from Ehrlich ascites carcinoma cells has been reported (Wallach and Zahler, 1966). The salient features of the fluorescence spectra are shown in Table VI.

The fluorescence spectra in these solvents resemble those of tryptophan; emission is blue shifted by 15 nm compared to the fluorescence of free tryptophan (λ_{max} of 350 nm). This shift indicates that the tryptophan residues are buried in a nonpolar environment (Section II,E,2,b, above). Addition of lysolecithin does not alter the emission spectrum. Since lysolecithin tends to disrupt the structure of lipid molecules, the hydrophobic environment of the tryptophan residues probably consists of nonpolar amino acid side chains rather than lipid chains. However, in a 9:1 2-chloroethanol:water mixture, the membrane tryptophan emission shifts to longer wavelengths and a shoulder due to tryptophan fluorescence appears at 310 nm. The maximum fluorescence in this solvent is about ⅔ that found in aqueous suspension. This is similar to what is found in fluorescence studies of solvent perturbation of globular proteins (see Section II,E,2,b, above). The solvent mixture exposes the tryptophan residues to a more polar environment and alters the tryptophan energy transfer. This may result from the 2-chloroethanol/water inducing a change in the structure of the protein.

The membrane proteins of both beef heart mitochondria and human erythrocytes contain all three possible aromatic amino acid residues and exhibit emission from the tryptophans. No extensive studies of the fluorescence properties have been reported.

The sensitivity of fluorescence measurements for detecting weak com-

TABLE VI

FLUORESCENCE PROPERTIES OF ERLICH ASCITES CARCINOMA CELL PLASMA MEMBRANE VESICLES

Solvent	λ_{max}	
	Excitation (nm)	Emission (nm)
$2 \times 10^{-2} M$ Tris–HCl (pH 8.2)	277–278	335
$2 \times 10^{-2} M$ Tris–HCl (pH 8.2) + lysolecithin	277–278	335
$0.1 N$ HCl	277–278	335
2-Chloroethanol:water	272–277	335–342
(9:1, v/v)	277	310 (shoulder)

plex formation is illustrated in studies on membrane "structural proteins" (Edwards and Criddle, 1966). The "structural protein" from beef heart mitochondria was shown to form stoichiometric complexes with cytochromes b, a, and c_1 by ultracentrifugal and electrophoretic methods (Edwards and Criddle, 1966). These methods however failed to detect an interaction between the "structural protein" and cytochrome c. It was found that cytochrome c quenced the "structural protein" fluorescence and the complex formation was followed using fluorescence titrations. The interaction appears to be electrostatic in nature and is markedly dependent on the ionic strength of the solvent. The fluorescence quenching method was also used to demonstrate formation of a 1:1 complex between the "structural protein" and myoglobin (Edwards and Criddle, 1966).* A similar study has been carried out on components of the human erythrocyte membrane (Wasemiller et al., 1968).

b. Extrinsic Membrane Fluorescence. Covalent Labeling. A fluorescent label for the outer components of plasma membranes was developed by Maddy (1964) following a suggestion by Mitchell that the vectorial aspects of membrane permeability could be experimentally investigated by using a reagent that would covalently react with an intact cell without permeating or damaging it and would therefore label only those components accessible from outside the permeability barrier. Subsequent fractionation of the cell would reveal only those reactive components available at the time of the reaction.

Maddy synthesized the fluorescent label 4-acetamido-4'-isothiocyanostilbene-2,2'-disulfonic acid, SITS (Fig. 17). Intact erythrocytes were reacted with the dye and ghosts were prepared from the labeled intact cells. The uptake of SITS was shown to be due to covalent binding rather than adsorption by comparing it with the uptake of the nonreactive analog 4,4'-bis(acetamido)stilbene-2,2'-disulfonic acid. Although this compound has no reactive group, it would be expected to be adsorbed in the same way as the isothiocyanate. However, when the erythrocytes were treated with this analog and the ghosts were prepared in a manner identical to that used for the SITS ghosts, they showed no fluorescence. The fluorescence of the SITS ghosts on the contrary could not be removed by washing. The SITS label was not removed from the cell surface by changing the pH to 11; this suggests that it labeled an amino, histidyl, or guanidyl group.

The total amount of SITS taken up by the cells remained constant at about 5×10^5 molecules SITS per erythrocyte when the concentration

---

* *Editor's note:* See Chapter 9 for a discussion of the nature of "mitochondrial structural protein."

of SITS at equilibrium was varied from about 2-250 μM. This labeling was completed within about 5 minutes. Both the hemoglobin and the globin prepared from reacted cells were free from fluorescence, indicating that the reaction was confined to sites outside the permeability barrier.

The nature of the labeled sites has not yet been identified. Neither the lipids extracted from the ghosts with chloroform/methanol mixtures nor the membrane proteolipids prepared by the method of Folch and Lees (1957) showed any fluorescence. The stilbene appears to be associated with the protein moiety of the cell, perhaps the structural protein. It was shown later that SITS serves as a specific chemical marker for the plasma membrane of the intact cells, and that sulfhydryl groups were also possibly labeled by the dye (Marinetti and Gray, 1967).

It is surprising in view of the extensive work on covalent fluorescent labeling of proteins and enzymes (Chen et al., 1967) that more attention has not been given to the labeling of membranes. However, the complex supramolecular organization of membranes imposes greater restrictions on possible labeling agents. The design of reagents for the fluorescent labeling of plasma membrane has been discussed (Maddy, 1964). A reagent for specific labeling of the membrane exterior must have sufficient hydrophilic groups to prevent its passage through the membrane into the cell interior. The stilbene nucleus of SITS has two major disadvantages. (a) The exciting light induces photochemical *trans-cis* isomerization which results in a fall of fluorescence and (b) the fluorescence emission is in the blue region of the spectrum and could be confused with fluorescence emanating from components of the membrane system.

Noncovalent Labeling. Noncovalently bound fluorescent probes have been used to study membranes of several types. The probes used have been ANS and the related TNS (Fig. 17) and to a lesser extent acridine orange, ethidium bromide, and pyrene derivatives. It would appear that these fluorescent probes are important tools for the further study of membrane function. Very low concentrations of the dyes are needed which should not alter significantly the system under study. Since only bound molecules fluoresce, it is not necessary to remove free dye, and errors due to trapping are reduced. The mechanisms of the change of fluorescence parameters observed when ANS "binds" to membranes are the same as those discussed previously (see p. 53) for ANS binding to proteins.

The dye ANS interacts with erythrocyte membranes and with membrane protein (Freedman and Radda, 1969). The details of the interaction are shown in Table VII. There is a blue shift (of 70 nm) in the

TABLE VII

INTERACTION OF ANS WITH ERYTHROCYTE MEMBRANE PREPARATIONS[a]

Sample	Enhance- ment of fluores- cence	No. of ANS sites (μmole/g protein)	K_{diss} (μM)	Polarization
Intact stroma	170	23	41	0.21
Stroma, pH 3.1	740	44	9	0.23
Sonicated stroma	23	82	10	0.24
Membrane protein	27	89	19	0.20
Membrane protein, pH 3.0	800			
Stroma in 2.7 M NaCl, pH 6.5				0.23

[a] Samples are in isotonic sodium phosphate buffer, pH 7.4, or in isotonic NaCl adjusted to the pH stated. Excitation is at 380 nm, emission at 480 nm. (After Freedman and Radda, 1969.)

emission of ANS and an enhancement of about 100-fold (Fig. 18). These changes indicate that the dye has been transferred to an environment less polar than water.

The polarization of fluorescence for ANS in the membrane is considerably lower than that found for isolated ANS–protein interactions but is similar to that observed for submitochondrial particles. Since polarization of fluorescence is proportional to the rotational correlation time of the emitting molecule, the low polarization value of ANS in a membrane suggests that the dye is localized in a region of high fluidity. Since ANS fluorescence is also enhanced by phospholipid dispersions (Vanderkooi and Martonosi, 1969) and by bile salt micelles (Dodd and Chapman, 1970) the above results are consistent with the dye dissolving in a membrane lipid matrix. Thus the interaction may not be a true binding but may constitute a phase problem.

The dye–membrane interaction is sensitive to changes in ionic strength and is strongly pH dependent, with a pK_a of 3.2–3.5. The pH dependence implicates sialic acid residues which when ionized, interact unfavorably with the negatively charged ANS. The rate of the ANS–membrane interaction is biphasic when the intact stroma is studied, showing a fast phase lasting for about 10 seconds and a slow phase extending beyond 200 seconds. Only the fast phase is observed with sonicated membrane fragments or with membrane protein.

Fluorescent probes like ANS can be used with both intact and sonicated membranes and future studies can be expected to indicate the extent of the disruption of the native membrane structure brought about by sonication. This is also a crucial problem for the interpretation of membrane NMR studies as discussed above (see Section II,C,5 and

Chapman, 1968). With ANS the sonicated membrane shows an eightfold increase in the number of bound dye molecules and a fourfold increase in the K_{diss} of the bound dye (Table VII). The tighter binding of the dye is also reflected in an increase in the polarization of the fluorescence. The ANS fluorescence in the sonicated membranes closely resembles the fluorescence of the dye bound to the lipid-free membrane protein. This suggests that sonication may disrupt the native quaternary structure of the membrane lipoproteins.

The binding of ANS to hemoglobin-free erythrocyte membranes in 20 mM Tris at pH 7.4 is sensitive to cation binding (Rubalcava et al., 1969). Addition of 300 mM NaCl or 3 mM $CaCl_2$ increases the number of bound ANS molecules by approximately a factor of three with a small (20%) decrease in the dissociation constant.

It appears that ANS is an ideal indicator for probing the states of mitochondrial membranes (Azzi et al., 1969). It is sensitive chiefly to the state of the dielectric in which it is bound and is insensitive both to the oxidation-reduction state of the electron transport system and to ionic gradients across the membrane.

The binding of ANS to mitochondrial membrane fragments prepared from beef heart mitochondria following sonications in the presence of EDTA (E-SMP fragments), of Mg^{++} and ATP (M-ASP fragments), or of ammonia and urea (ASU fragments) has been studied (Azzi et al., 1969). ANS dissolves in E-SMP fragments with a 25-fold increase in fluorescence quantum yield and a blue shift in emission maximum. Using the empirical microscopic polarity scale introduced by Stryer (1965), and based on the solvent-induced shifts in ANS fluorescence, it was calculated that the dielectric constant of the interior of the particles was 35. About 200 nmole of ANS were bound per milligram of membrane protein with a K_{diss} of 5.7×10^{-5} M. When 3 μmole of ANS are bound per gram of fragmented membrane the polarization of fluorescence is 0.194, but it falls on increasing the bound ANS concentration. This decrease of polarization is characteristic of energy transfer between ANS molecules, a phenomenon which occurs at a distance of less than 20 Å (Weber and Daniel, 1966).

The ANS fluorescence responds to activity of the electron transport chain. The addition of oxygen to anaerobic membranes induces a rapid oxidation of cytochrome a with a half-life of 500 μsec. The changes in ANS fluorescence parallel the cytochrome a oxidation (and reduction) but are slower by a factor of 4000; the half-time is approximately 2 seconds. On expenditure of oxygen both the cytochrome a and ANS changes are reversed, and again the reduction of cytochrome a (half-time, 1 second) is considerably faster than the decrease of ANS fluorescence (half-time, 7 seconds). Similar changes are found when electron transport is activated by the addition of NADH to aerobic membranes. The possibility that ANS movement, on and off the dye-binding sites, might be responsible for the observed fluorescence changes in such a complex system was excluded by working at dye concentrations in which all of the ANS was bound. Energization of the membranes by the addition of NADH in the absence of oligomycin caused only a small change in ANS fluorescence and in the absorbance of the optical probe bromothymol blue (BTB). After adding oligomycin, which blocks energy-dissipating reactions in the membrane, a further addition of NADH caused large and closely synchronized changes of BTB and ANS. The addition of ATP caused similar changes in cyanide-blocked membrane fragments.

These studies indicate that there is no change in ANS fluorescence which may be directly linked to the oxidation-reduction state of cytochrome a. Thus the energized state of the membrane is not activated simultaneously with the changes in oxidation-reduction states of the electron carriers. Instead it appears that the structural alteration of the membrane may require the turnover of the cytochrome components in order to build up a steady-state concentration of an intermediate of energy conservation. Thus an intermediate step is required between electron transport and alteration in the membrane structure as indicated by the ANS changes. In addition, the relationship between the energy-induced conformational change of the membrane and an energy-utilizing process such as reversed electron transport indicates clearly that the membrane energization precedes the NAD reduction. This correlation may at present be taken to indicate that membrane conformation changes are essential to, rather than secondary indications of, energy conservation.

Gross changes in mitochondrial morphology have been observed in the electron microscope during the change from one respiration state to another (Hackenbrock, 1966; Green et al., 1968). As yet no light-scattering or electron micrographically detectable conformation change has been observed in submitochondrial particles, and thus the change revealed by ANS is at a more elementary level. These fluorescence studies suggest that energy conservation is associated with changes of the conformation of the membrane components of the small-scale type suggested by Chance and colleagues (1967). These small structural changes may be related to the elementary step of energy conservation and may also act as a trigger for the gross structural reorganizations observed in intact mitochondria in light-scattering or electron micrographic studies.

For studies involving the dynamic energy states of membrane binding, local anesthetics such as butacaine have been found to accelerate absorbancy changes of the membrane-bound pH indicator, bromothymol blue, and to accelerate the uptake of Ca^{++}. The possibility that membrane structural changes accompany these alterations of membrane function were explored with ANS (Chance and Lee, 1969). Two ANS binding sites of low polarity were found. The polarity of these sites was further decreased by butacaine; the rates of binding were little affected. A further decrease of polarity was caused by a biphasic reaction with 125 μM Ca^{++} ($t_{1/2} = 5$ msec, 5 seconds). In the presence of 60 μM butacaine, the fast phase is undetectable and the slow phase is accelerated ($t_{1/2} = 2$ seconds). The membrane response time for Ca^{++} in the absence of butacaine may be compared to the earliest times for initiation of Ca^{++} uptake measured by aequorin or murexide. Cytochrome b responds to Ca^{++} at 20 msec and the Ca^{++} uptake rate is maximal shortly thereafter (50 msec). However, the absence of this fast response in the presence of butacaine suggests that fast changes in membrane polarity are associated with the lipid structure itself rather than with the Ca^{++} carrier. The slower phase is associated with the accumulation of cations in the membrane, accompanied by localized membrane alkalinization, as shown by the BTB response. Thus, fast reactions of Ca^{++} with the environmental polarity of the membrane phospholipids suggests fast transitions of membrane structure, as are also required by the response of cytochromes to oxygen pulses.

The information derived from ANS fluorescence changes and nonspecific light-scattering changes in submitochondrial particles has been compared in a series of careful experiments (Chance and Lee, 1969). Significantly, the conformational change reported by light-scattering in submitochondrial particles is dependent upon electron transport but is not related to recognizable energy-coupling events, either as a conformational state related to primary events of energy coupling or as a consequence of subsequent energy-dependent ion movements. The structural or charge reorganizations involved in the energy-coupling reactions, as reported by ANS, do not show nonspecific light-scattering changes and thus may not be associated with the gross structure of the membrane but may be related instead to the molecular properties of the membrane components.

The molecular basis of nerve excitation has been investigated with ANS. Singer and Tasaki (1968) have proposed that the process of excitation is accompanied by conformational changes in macromolecules of the nerve membrane. The intrinsic fluorescence of the nerve shows a transient change during the passage of an impulse. However, the change is too small to elicit details of the membrane rearrangements.

Nerves from lobster, crab, and squid adsorb ANS when it is applied extracellularly. The bound dye has an emission maximum near 450 nm in crab nerve, indicating that the sites at which the dye is bound have a very low dielectric constant. During nerve excitation there was a transient increase in ANS emission. However, the increase was in the range of $2-3.5 \times 10^{-4}$ times the light intensity observed before stimulation and could not be measured without the use of a computer of average transients (CAT). The increase in fluorescence started at about the time of arrival of the nerve impulse at the site of ultraviolet irradiation. The light intensity rapidly rose to its maximum and then gradually declined toward the resting level of fluorescence. Since nerve trunks were used, the precise temporal relationship between the action potential and the transient fluorescent change could not be determined. When ANS is introduced into the interior of a squid axon by the internal perfusion technique, it is localized within the axon interior, indicating that the dye molecules do not penetrate the axon membrane, at least from within. Thus it may be inferred that the extracellularly applied ANS remains outside the axoplasm and consequently that the observed fluorescence changes during excitation do not involve the axoplasm. Although at this stage it is difficult to determine molecular details of the anatomical structure in which the suspected conformational changes take place, the ANS results are consistent with the view that the process of nerve excitation is accompanied by conformational changes in the protein of the nerve membrane (Singer and Tasaki, 1968). Similar experiments using acridine orange instead of ANS have given similar results (Tasaki et al., 1969).

Fluorescence experiments using the fluorochrome pyronine B have been performed with intact frog striated muscles (Carnay and Barry, 1969). Transient changes in fluorescence, the initial phase of which coincided with the action potential, were recorded when a CAT was used. These findings suggest that as in nerve membranes, there is a macromolecular conformational change in muscle membrane during excitation.

The fluorescent probe ANS has also been used for studying isolated excitable membranes (Kasai et al., 1969; Vanderkooi and Martonosi, 1969). It interacts with sarcoplasmic reticulum membranes with the usual enhancement in intensity (Vanderkooi and Martonosi, 1969). The fluorescence of the bound but not the free probe is enhanced when the membranes are titrated with a variety of cations including K^+, Na^+, Cs^+, Ca^{++}, and Mg^{++}. The implied increase in the hydrophobic character of the membranes in the presence of the cations is in agreement with the well-known reduction of the permeability of surface membranes in the presence of high concentrations of Ca^{++}.

Some circular polypeptides including polymixin B sulfate alter the bound ANS fluorescence. These results suggest that the enhancement of ANS fluorescence by the membranes of fragmented sarcoplasmic reticulum may be attributable largely to membrane phospholipids. This conclusion is further supported by the finding that digestion of microsomal membranes with proteolytic enzymes causes only small changes of questionable significance in the fluorescence enhancement. Treatment with phospholipase C, on the other hand, produces considerable reduction in the enhancement of fluorescence by microsomes. These results suggest that well-designed experiments with fluorescence probes of several varieties may be able to distinguish between the various models proposed for membrane structure.

Membrane fragments derived from the innervated excitable surface of cells from the electric organ of *Electrophorus electricus* bind ANS in a hydrophobic environment (Kasai *et al.*, 1969). The temperature dependence of the ANS binding indicates a structural transition occurring around 30°C. This does not correspond to a gross alteration of membrane structure and is interpreted as a rearrangement of components within the membrane.

The polarization of fluorescence of ANS bound to these membranes is insensitive to the solvent viscosity, in contrast to ANS bound to proteins. This may be a characteristic property of ANS bound to a membrane structure. As in the case of other membranes, Ca^{++} increases the affinity of ANS for its membrane sites. *d*-Tubocurarine and Flaxedil, two receptor inhibitors, increase the affinity of ANS for the membrane while two receptor activators, carbamylcholine and decamethonium, have no significant effect on the fluorescence.

4. Fluorescence Studies with Model Membranes

The interaction of ANS with lipid and detergent micelles has provided information which is a useful guide to possible fluorescent probe–cell membrane interactions (Vanderkooi and Martonosi, 1969; Flanagan and Ainsworth, 1968). The polarization of fluorescence of ANS increases on binding to micelles, indicating a lesser degree of rotational freedom for the bound dye.

The results clearly indicate that the binding of the dye to a micelle depends on both nonpolar and electrostatic interactions. Thus ANS binds to micelles of phosphatidylcholines, lysophosphatidylcholines; and cationic detergent, but not to phosphatidylserine or anionic detergents. Dye binding to neutral micelles occurs only if the nonpolar moiety of the probe is above a certain minimum size. For future membrane studies with fluorescent probes it should be possible to build specificity into the probe by a judicious balance of its hydrophobic and electrostatic groups.

In addition to the ability of micelles to bind ANS and other synthetic fluorescent probes, ganglioside micelles bind a number of basic drugs and enhance the drug fluorescence (Albers and Koval, 1962).

Many of the pigments of biological photoreceptor systems are also fluorescent and serve as naturally occuring fluorescent probes. The fluorescence of chlorophyll monolayers spread on water (using decyl alcohol for dispersing chlorophyll) has been studied in great detail (Tweet et al., 1964; Gaines et al., 1965). These monolayers are two-dimensional liquids with no preferred direction in the plane of the films.* From the angular distribution of the fluorescence it was found that the transition moment responsible for the red absorption band makes an angle of $<20°$ with the water surface while that responsible for the blue absorption band describes an angle of $\approx 28°$. Since both of these angles are measured in the porphyrin plane, the chlorophyll molecule must be oriented at the interface with its aromatic plane describing an angle with the water surface and presumably with its hydrophobic phytyl side chain in the organic layer (i.e., the air phase). The excited-state energy from chlorophyll a in these monolayers is transferred to added copper pheophytin a in the monolayer. In mixed films a single-transfer calculation on the Förster model describes the quenching satisfactorily, the range of chlorophyll–quencher interaction being 4 nm. For undiluted monolayers in which chlorophyll interaction is more important, the single-transfer model is not applicable.

Flavins play an important role in many biological photoreceptors. An attempt to elucidate the membrane environment of these pigments has been made by using, as a model system, flavins in a solid matrix (Penzer and Radda, 1967). The flavins were incorporated in thin (2.5×10^{-3} cm) transparent films of methyl cellulose and small shifts in the fluorescence emission were observed. These were compatible with the film providing a less polar environment than an aqueous solution. The fluorescence polarization of flavin mononucleotide and flavin adenine dinucleotide showed that rotational restraint was incomplete. The results suggested that the pigments were localized in channels or cavities with a microscopic viscosity considerably lower than the macroscopic viscosity of the film. Films of this kind provide a simple system in which the photochemical and fluorescence properties of biological pigments can be studied and offer an environment which may be closer than homogeneous aqueous solutions are to the natural ones. Extension of reported studies on pigments in lipid–water systems to include fluorescence studies would appear to be worthwhile for future exploration (Chapman

---

* Monolayer techniques are also discussed in Chapter 6.

and Fast, 1968). The absorption spectra of carotenoids and other pigments incorporated into lipid bilayer membranes have been obtained (Leslie and Chapman, 1967), and the fluorescence of pigments in these membranes is now being studied.

5. Fluorescence Studies on the Membranes of Intact Living Cells

The well-known fluorescent dye fluorescein has been used in an interesting study of the differential permeability of cell membranes (Kanno and Loewenstein, 1964). A small drop (5×10^{-9} ml) of a fluorescein solution was injected into a salivary gland cell of *Drosophila*. No change in any of the properties of the cell was detected as a result of this procedure. The diffusion of the dye throughout the injected cells and into neighboring cells was followed by fluorescence microscopy. The permeability of the junctional surfaces of the cell membranes appears to be high in contrast to the nonjunctional surfaces and intercellular spaces which represent strong diffusion barriers.

The molecular events involved in cellular transport have been investigated using aminonaphthylalanine compounds (Udenfriend *et al.*, 1966). These compounds are actually transported by S37 ascites cells and have fluorescence distinctly different from those of the major constituents of the normal cell. The fluorescence of the labeled amino acids inside the cell showed the same fluorescence parameters as did the free amino acid in aqueous solution. Fluorescent dyes known to bind to intracellular membranes and macromolecules, e.g., ANS and eosin, showed changes in their fluorescence parameters when they were transported into the cells. The intracellular distribution of the amino acids was determined by fluorescence microscopy. Since the fluorescence studies indicated that the amino acids were in true solution within the cell, these experiments gave the first direct evidence for a membrane catalytic pump mechanism for true active transport rather than a mechanism involving binding of the small molecules to sites within the cell.

The differential permeability of intracellular membranes *in vivo* has been studied using tetracyclines. Only the plasma and mitochondrial membranes of monkey kidney tissue-culture cells were permeable to the drugs, and the fluorescent drugs accumulated selectively in the mitochondria (Du Buy and Showacre, 1961). The same method was used to demonstrate that certain protozoa, bacteria, and viruses when phagocytosed by host cells become surrounded by an intracytoplasmic boundary (Du Buy *et al.*, 1964). This boundary is impermeable to the antibiotic and prevents fluorescent labeling of the parasites. When the host cell dies the intracellular parasites become visible, thus indicating that the impermeable boundary is probably of host origin.

The kinetics of fluorescence accumulation in single living mammalian cells have also been studied using quantitative microspectrofluorimetry (Rotman and Papermaster, 1966).

6. Conclusion and Future Prospects

The complex problems of lipid–protein interactions in biological membranes are unlikely to be solved exclusively by the use of a single technique. Rather, each of the many physical methods being applied to the problems will hopefully illuminate a different facet of the structure, and the final picture will be integrated from the evidence emerging from all the possible approaches to the problem. In this light one must ask, what are the distinctive insights which fluorescence methods can offer in this field?

The fluorescence methods in membrane research provide relatively little information about the detailed static structure of the membrane. Information about lipid–tryptophan interactions can be obtained in some cases (Wallach and Zahler, 1966) but NMR is preeminent in the detailed information which it can provide on such interactions (Chapman *et al.*, 1969a,b).

The unique contribution which fluorescence makes lies in the ability of noncovalently bound probes to reflect small conformational perturbations in the membrane. These reporter groups are sensitive to small localized changes in the dielectric state of the environment and can indicate transitions in membrane proteins which are too small to be detected by electron microscopy and light-scattering. Further, the kinetics of even the fastest such changes can be measured with ease, thereby giving a valuable insight into the possible mechanism of these changes. Very low (micromolar) probe concentrations which do not alter either membrane structure or metabolism can be used. A particular advantage of the method is that intact membranes, organelles, and even intact cells can be examined.

The potential of luminescence methods for future membrane research is inextricably linked to instrumental developments. Even more than with magnetic resonance methods, commercially available fluorescence equipment is strictly limited in the ways it can be applied to biological structures such as membranes. Accordingly much of the published work in this field is done on homemade instruments of great sensitivity and versatility. In view of the increasing sophistication of the machines (Freedman and Radda, 1969; Langelaar *et al.*, 1967) being built for protein and membrane research it may soon be necessary for serious workers in these fields to build their own fluorescence equipment.

It can also be expected that a more precise location of the probes in the membranes and a deeper insight into the conformational changes detected will follow when the full power of fluorescence theory is applied to membrane problems.

One can anticipate interesting results from application of the new direct lifetime measuring fluorometers to membranes (Chen *et al.*, 1967). Nonradiative energy-transfer techniques, which have been successfully used in protein chemistry for constructing "spectroscopic rulers" should also be capable of elucidating problems of the orientation of membrane molecules (Stryer and Haughland, 1967; Conrad and Brand, 1968). Perhaps the most exciting developments will come in the field of fluorescence microscopy (Olson, 1960). With these methods one can directly visualize the intracellular distribution of a probe and, by using quantitative techniques, one can hope to study, directly, specific molecular interactions and conformational changes in intracellular organelles of intact cells in tissue culture (MacInnes and Uretz, 1968).

F. Thermal Techniques

1. Introduction

Thermal techniques are now being applied to the study of lipids and membrane systems. A recent review has comprehensively covered a great deal of the present work (Ladbrooke and Chapman, 1969).

The term "thermal analysis" originally referred to the elementary experiment in which the temperature changes of a substance are recorded while it is cooling and plotted against time to yield a curve showing breaks or inflections corresponding to thermal transitions in the sample. The sensitivity of the method was improved by plotting the difference in temperature between the sample and its surroundings, and this led to the development of differential thermal analysis (DTA). In DTA the sample and an inert reference material are heated or cooled at the same rate and the difference in temperature between them is recorded. The differential temperature remains zero or constant until a thermal reaction occurs in the sample. At this point the differential temperature increases until the transition is completed and then decreases again. Thus a peak is obtained on the curve for differential temperature against temperature or time, and the direction of the peak indicates whether the transition is endothermic or exothermic.

During the last ten years the potential value of DTA in organic chemistry has been realized, and sensitive instruments operating in the range $-150°$ to $+500°C$ and requiring only milligram quantities of material have been developed. The reduction in sample size greatly increases the resolution of the technique, permits the use of faster heating rates, and extends the area of application to materials which are only available in small quantities.

Many workers have considered the quantitative aspects of DTA and have attempted to relate the area under the DTA curve to the heat of transition. There are many difficulties, some of which can be eliminated by instrumental design, but a better-founded basis for quantitative studies is provided by the alternative technique of differential scanning calorimetry (DSC) in which the temperature of sample and reference are maintained at an equal level or at a fixed differential throughout the analysis and the variation in heat flow to the sample required to maintain this level during a transition is measured.

2. Presentation of Data

Since the results obtained by DTA and DSC are superficially similar, care is necessary in presenting results in order to maintain a distinction between the two classes of data. The convention is used of presenting DSC data with endothermic changes in an upward direction and marking the temperature scale in degrees Kelvin; DTA curves are drawn with endothermic changes in a downward direction and the temperature scale is marked in degrees centigrade. In this way the DSC results are in conformity with general practice in the presentation of calorimetric data, while the DTA results are consistent with thermometric practice.

3. Phospholipids in Water

Anhydrous phospholipids show an endothermic transition which varies with chain length or unsaturation (Byrne and Chapman, 1964). On addition of water to phosphatidylcholine, the temperature at which the main endothermic transition occurs is lowered and reaches a limiting value when the water content exceeds 20% (Chapman et al., 1967). The limiting transition temperature (T_c) is the minimum temperature at which substantial amounts of water enter the crystal lattice and corresponds to the temperature at which myelin tube formation is observed.

Above the T_c line the lipid–water system exists in a lamellar mesomorphic phase with a maximum uptake of water of 40%. Addition of more water gives rise to a two-phase region with fragments of the lamellar phase dispersed in excess water. This boundary at 40% water is not apparent from the DSC curves, but the curve showing the heat absorbed at the transition indicates that maximum heat absorption is not developed until about 40% water has been added. The transition in water is considerably sharper and well defined compared with the transitions observed with the dry materials, and it is completely reversible. Below the T_c transition a lamellar gel is formed in which the lipid hydrocarbon chains are crystalline. Essentially the same results are obtained with all phospholipids, the T_c lines being disposed along the temperature axis according to the melting point of the fatty acyl residues. (Phase changes in lipid–water systems are discussed in Chapter 1.)

A small "pretransition" peak whose interpretation is not certain is observed with fully saturated phosphatidylcholines in excess water. The temperature interval between this peak and the main endothermic peak increases as the chain length of the phosphatidylcholine decreases.

The DSC curves for the phosphatidylcholine–water system are shown in Fig. 19. An endothermic peak corresponding to the melting of ice is

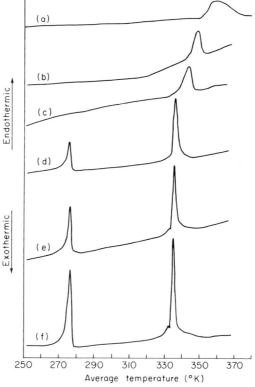

Fig. 19. DSC curves showing the effect of water on the gel to liquid crystalline transition (T_c) for 1,2-distearoyl-DL-phosphatidylcholine. Water content (weight %): (a) 3% (monohydrate), (b) 10%, (c) 20%, (d) 25%, (e) 30%, (f) 40%. A peak at 0°C corresponding to the melting ice is not observed until 25% water is present. When the heat absorbed in the ice-melting transition is plotted against concentration, a straight line is obtained which extrapolates to zero at 20 weight % water (equivalent to 10 molecules of water per molecule of phosphatidylcholine).

not observed until the water content reaches 25%. If the heat absorbed in the ice-melting transition is plotted against concentration, a straight line is obtained which extrapolates to zero at a water content of 20%. This indicates that at this concentration all the water is bound to the phospholipid and does not freeze. This corresponds to a ratio of one molecule of phosphatidylcholine to approximately ten molecules of water. Freezing does not occur even if the mixture is cooled to $-100°C$.

Examination of the cooling curve for a lipid mixture containing excess water shows that freezing of water occurs in two stages. The bulk of the water supercools to $-15°C$ and then freezes, but a second exotherm is obtained at $-50°C$. There is no corresponding endotherm on the heating curve for this transition but examination of the ice-melting peaks at $0°C$ for a mixture which has been cooled to about $-30°C$ as compared with one which has been cooled below $-50°C$ shows that the peak is larger in the latter case. In other words there is a fraction of water present which freezes at $-50°C$ and subsequently melts at $0°C$. This is *in addition to* the unfreezable or bound water and indicates the presence of water of intermediate type.

4. Phospholipid–Cholesterol Interactions

It is difficult to interpret DTA curves for dry mixtures of phosphatidyl-choline and cholesterol. However, if the system is examined in the presence of excess water, then a clearly defined interaction is observed (Ladbrooke *et al.*, 1968). Figure 20 illustrates the effect of cholesterol on the T_c transition of 1,2-dipalmitoyl-L-phosphatidylcholine. Each curve refers to a mixture containing 50 weight % of water. Addition of cholesterol first causes the small pretransition peak to disappear and then causes the T_c transition peak to broaden and decrease in area until a 1:1 molar ratio of phosphatidylcholine to cholesterol is reached and no transition can be observed. This ratio corresponds to the maximum amount of cholesterol which can be introduced before the separation of excess crystalline cholesterol occurs.

Consideration of x-ray diffraction data on this system shows that cholesterol disrupts the ordered array of hydrocarbon chains in the gel phase and this leads to an increased fluidity of the chains. It appears that in the presence of cholesterol the fluidity of the lipid hydrocarbon chains is intermediate between that in the gel and the liquid crystalline phases of the pure phosphatidylcholine. This effect occurs only in the presence of water. On removal of water, phosphatidylcholine and cholesterol crystallize separately.

Equimolar amounts of cholesterol also remove the T_c transition in unsaturated (dioleyl) and mixed-chain (egg yolk) phosphatidylcholines.

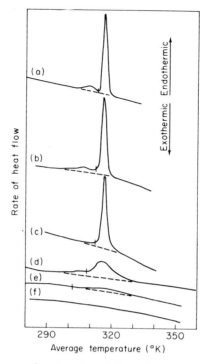

Fig. 20. The DSC curves for 50 weight % dispersions in water of 1,2-dipalmitoyl-L-phosphatidylcholine/cholesterol mixtures containing (a) 0.0, (b) 5.0, (c) 12.5, (d) 20.0, (e) 32.0, (f) 50.0 mole % cholesterol (Ladbrooke *et al.*, 1968).

No specific "complex" needs to be invoked to explain these results since the effect of cholesterol appears to be largely mechanical. With di-behenoylphosphatidylcholine* in which the length of the acyl chains is significantly longer than the cholesterol molecule, the transition is not completely removed in a 1:1 molar mixture with cholesterol. This could indicate that the ends of the hydrocarbon chains "crystallize" in this system. It could also indicate more simply that the *solubility* of cholesterol in the hydrophobic region decreases as the chain length increases, so that the amount of cholesterol actually present in the lipid layers is somewhat less than a 1:1 molar ratio and a small proportion of the lipid molecules crystallizes on cooling.

5. Applications to Biological Membranes

An approach to the study of intact biological membranes by thermal methods can be made. In particular, the following questions have been

* Behenic acid is $CH_3 \cdot [CH_2]_{20} \cdot COOH$.

posed. Do thermal transitions similar to those observed with phospholipids occur in the intact membrane? If so, what is the relationship
between the transition temperature, the membrane environmental temperature, and the phospholipid composition? If they do not occur, is this
because of interactions between phospholipid and cholesterol or between
lipid and protein?

a. Myelin. In a study of beef brain myelin using DTA and DSC
(Ladbrooke *et al.*, 1968) it was found that no thermal transitions associated with the membrane can be detected until the water content is
reduced below a critical value of 20%. At this concentration the ice peak
disappears and an endothermic transition at 35°C first appears (Fig. 21).

On further drying, the heat absorbed in the transition increases and
the transition temperature also increases to a limiting value of 55°C. In
the final states of drying (<5% water) an additional endotherm appears
at 35°C. Comparison of the DTA curves of the fully dried myelin and

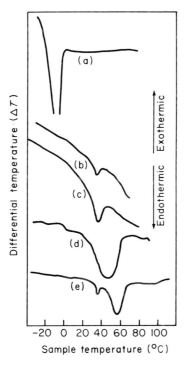

Fig. 21. Dehydration of myelin; DTA heating curves for samples of myelin previously equilibrated at different relative humidities. Approximate final water contents
(weight % water) are (a) 30%, (b) 15%, (c) 10%, (d) 5%, (e) 3%. (Ladbrooke *et al.*,
1968.)

of lipids obtained from myelin by solvent extraction showed that the transition at 35°C is a polymorphic transition of crystalline cholesterol, while that at 55°C is due to a mesomorphic transition of the phospholipid. This behavior is analogous to that of the phosphatidylcholine–cholesterol model system discussed earlier. The 20% bound water is essential to maintain the organization of the lipid layers in myelin, and the presence of cholesterol prevents the lipids from crystallizing, so that no transition is observed in the hydrated myelin. On removal of the bound water, crystallization occurs.

The total lipid extract behaves in a similar way and when dispersed in excess water forms a lamellar liquid crystalline phase with no transition. The cholesterol-free lipid, however, does exhibit a chain-melting transition in excess water in the temperature range 25°–50°C. Thus, in the presence of water, but in the absence of cholesterol, part of the myelin lipid is crystalline at body temperature. The possibility of lipid crystallization occurring in the brain may well prove to be an important factor in the breakdown of myelin in disease.

b. Stratum Corneum Membrane. An interesting study of water binding in stratum corneum membranes of the rat was carried out by Bulgin and Vinson (1967) using DTA. Three endotherms attributable to loss of water were obtained at 103°, 114°, and 135°C. By drying the membrane, the peak at 135°C was shown to correspond to the most strongly bound water. The 103°C peak corresponded to loss of physically sorbed water. Other transitions associated with the lipid (70–80°C) and protein (280°C) were also observed.

c. Erythrocyte Membrane. The erythrocyte membrane is currently being studied in this laboratory. Preliminary results indicate that there are similarities with myelin in that cholesterol inhibits the crystallization of the polar lipids and for this effect there is a requirement for bound water. A difficulty with this membrane is that transitions due to the lipid are significantly smaller than with myelin and are difficult to detect with present instruments. This is partly due to the fact that the lipid content of the membrane is lower than for myelin and also that long-chain saturated lipids, which contribute a large heat effect, are absent. However, the transitions observed are still considerably smaller than would be expected. This may be due to supercooling of the polyunsaturated lipids present in this membrane. We have noticed that when the dry phospholipid fraction is examined on the DTA instrument using a macrotube, two transitions are observed, at −22° and +45°C, but when the same material is examined in a microtube only one transition is observed, at +45°C. When the phospholipid is dispersed in water a very small broad transition is observed in the range 25°–40°C. This seems high and suggests that not all the lipid present is contributing to the peak and the supercooling, particularly of the polyunsaturated lipids, may be occurring. This would explain why the heat absorbed in the transition is apparently so low and is an important point to be considered in interpreting the transitions observed in natural mixed lipid fractions.

d. Other Membranes. Although no thermal studies of mitochondrial membranes have been reported, a detailed x-ray study of the phase behavior in water of a lipid extract from beef heart mitochondria has been made by Gulik-Krzywicki *et al.* (1967). Below 3°C, complex phase regions were found containing domains of crystalline paraffin chains within bilayer leaflets and suggesting that the lipids crystallize gradually as the temperature is lowered. This is precisely the behavior to be expected in a lipid extract containing only a small amount of cholesterol. The molar ratio of phospholipid to cholesterol in beef heart mitochondria is approximately 50/1, compared with a value of 1.5/1 in myelin (Fleischer *et al.*, 1967; O'Brien and Sampson, 1965).

It has been suggested (Chapman, 1967) that one of the functions of the distribution of acyl residues found with phospholipids in biological tissues is to provide the correct fluidity at a particular environmental temperature to match the requirements for diffusion or for the control of metabolic processes in the tissue. Thus in membranes, where rapid metabolic and diffusion processes occur, the transition temperature for the phospholipids will be higher and may be close to body temperature.

In some cases there is a requirement for cholesterol but the prevention of crystallization is probably not the sole role of cholesterol in biological membranes. The fact that cholesterol is retained in the lipid structure when the temperature is below the transition temperature for the phospholipid implies that the mixed lipid–sterol system is particularly stable; this may be important in imparting mechanical strength to membranes such as myelin and that of the red blood cell.

Steim *et al.* (1971) have examined the membranes of *Mycoplasma laidlawii* by differential scanning calorimetry. The membranes of this organism are of particular interest since it has been shown that the composition of the membrane lipids can be changed at will by varying the growth medium. Membranes can be obtained containing no cholesterol and with different degrees of saturation in the fatty acids of the membrane lipids. In all cases a transition is observed in the membrane which corresponds to the transition temperature of the extracted lipids dispersed in water. A second irreversible transition at about 60°C can be attributed to denaturation of the membrane protein.

These results are considered to represent direct verification of the Davson-Danielli model, since comparison of the sizes of the transitions in the membrane and the lipid extract indicates that at least 75% of the lipids in the membranes contribute to the transition and since such a transition would not be expected to occur unless an extensive bilayer arrangement of lipid existed in the membrane. It is assumed that the structure of the lipid extract in water is lamellar and also that all the lipid in the lipid–water system contributes to the observed peak. Even if this is the case it does not support the complete Davson-Danielli model of a bilayer of lipid *sandwiched by protein*. Another explanation is that there are in this membrane extensive regions of lipid bilayer linking lipid–protein regions.

Chapman and Urbina (1970) have also studied the thermal transitions of *M. laidlawii* membranes as well as the extracted lipids. They discuss the various assumptions inherent in such comparisons and point to the difficulties of quantitative assessment of the amount of bilayer present.

6. Conclusions

1. The applications of differential thermal methods in the study of lipids, proteins, and membranes have been examined. While DSC provides a more faithful representation of thermal events in a material and good quantitative data can be obtained, the results are often difficult to interpret; DTA data is useful for providing accurate transition temperatures. Ideally, the two methods should be used in conjunction.

2. Thermal transitions in lipid systems can be readily studied, and consideration of the transition temperatures and enthalpy data provides useful information regarding purity, stability, polymorphism, mesomorphism, hydration, and molecular interactions.

3. Quantitative studies provide reliable data on specific heats and give kinetic parameters which are useful in considering the molecular processes occurring at phase transitions.

4. The techniques can also be applied to proteins and to intact biological membranes and their usefulness in this area has been clearly demonstrated. The sensitivity of current instrumentation is, however, a limiting factor.

III. THE FUTURE

The future will see increased attempts to improve probe systems for the study of membrane structure and function. There are, in some cases, problems regarding the probes which will require attention before unequivocal conclusions can be accepted. To date it can be said that the use of these probe methods has given results which are consistent with the idea that at least some parts of cell membranes contain a bilayer of lipid. Perhaps it can be said that the greater the amount of lipid in a membrane the more likely it will be that some portions contain a bilayer of lipid! It may be possible that the membrane phospholipids connect regions of lipid–protein complexes. These lipid–protein regions may involve some electrostatic interaction between lipid and protein (e.g., phosphatidylserine and protein) and may also allow some hydrophobic interaction with the nonpolar region of the membrane. These lipid–protein complexes may be the most interesting of all for future probe studies. (The possible existence of lipid–protein complexes as discrete structural entities within membranes is discussed further in Chapter 4.)

References

Abragam, A. (1961). "Principles of Nuclear Magnetism." Oxford Univ. Press, London and New York.
Albers, R. W., and Koval, G. J. (1962). *Biochim. Biophys. Acta* **60**, 359.
Andrew, E. R. (1958). "Nuclear Magnetic Resonance." Cambridge Univ. Press, London and New York.
Andrew, E. R., and Eases, R. G. (1953). *Proc. Roy. Soc. Ser. A* **216**, 398.
Aneja, R., and Davies, A. P. (1970). *Chem. Phys. Lipids* **4**, 60.
Azzi, A., Chance, B., Radda, G. K., and Lee, C. P. (1969). *Proc. Nat. Acad. Sci. U. S.* **62**, 612.
Barratt, M. D., Green, D. K., and Chapman, D. (1968). *Biochim. Biophys. Acta* **152**, 20.
Barratt, M. D., Green, D. K., and Chapman, D. (1969). *Chem. Phys. Lipids* **3**, 140.
Bear, S. R., Palmer, K. J., and Schmitt, F. O. (1941). *J. Cell. Comp. Physiol.* **29**, 299.
Bloembergen, N. (1961). "Nuclear Magnetic Relaxation." Benjamin, New York.
Bloembergen, N., Purcell, E. B. M., and Pound, R. V. (1948). *Phys. Rev.* **73**, 679.
Bulgin, J. J., and Vinson, L. J. (1967). *Biochim. Biophys. Acta* **136**, 551.
Byrne, P., and Chapman, D. (1964). *Nature (London)* **202**, 987.
Camerman, A. N., and Jensen, L. H. (1969). *Science* **165**, 493.
Carnay, L. D., and Barry, W. H. (1969). *Science* **165**, 608.
Chance, B., and Lee, C. P. (1969). *FEBS Lett.* **4**, 181.
Chance, B., Lee, C. P., and Mela, L. (1967). *Fed. Proc., Fed. Amer. Soc. Exp. Biol.* **26**, 902.
Chapman, D. (1965). *In* "The Structure of Lipids." Methuen, London.
Chapman, D. (1966). *Ann. N. Y. Acad. Sci.* **137**, 745.
Chapman, D. (1967). *In* "Thermobiology" (A. H. Rose, ed.), p. 123. Academic Press, New York.
Chapman, D., ed. (1968). "Biological Membranes," p. 189. Academic Press, New York.
Chapman, D., and Fast, P. G. (1968). *Science* **160**, 188–189.
Chapman, D., and Kamat, V. B. (1968). "Regulatory Function of Biological Membranes." Elsevier, Amsterdam.

Chapman, D., and Magnus, P. D. (1966). "Introduction to Practical High Resolution NMR Spectroscopy." Academic Press, New York.

Chapman, D., and Morrison, A. (1966). *J. Biol. Chem.* **241**, 5044.

Chapman, D., and Penkett, S. A. (1966). *Nature (London)* **211**, 1304.

Chapman, D., and Salsbury, N. J. (1966). *Trans. Faraday Soc.* **62**, 2607.

Chapman, D., and Salsbury, N. J. (1969). *Progr. Surface Sci.* **3**, 121.

Chapman, D., and Urbina (1970). *FEBS Lett.* **12**, 169.

Chapman, D., Byrne, P., and Shipley, G. G. (1966). *Proc. Roy. Soc., Ser. A* **290**, 115.

Chapman, D., Williams, R. M., and Ladbrooke, B. D. (1967). *Chem. Phys. Lipids* **1**, 445.

Chapman, D., Fluck, D. J., Penkett, S. A., and Shipley, G. G. (1968a). *Biochim. Biophys. Acta* **163**, 255.

Chapman, D., Kamat, V. B., De Gier, J., and Penkett, S. A. (1968b). *J. Mol. Biol.* **31**, 101.

Chapman, D., Kamat, V. B., and Levene, R. J. (1968c). *Science* **160**, 314.

Chapman, D., Leslie, R. B., and Scanu, A. M. (1969a). *Nature (London)* **231**, 260.

Chapman, D., Leslie, R. B., Scanu, A. M., and Hirz, R. (1969b). *Biochim. Biophys. Acta* **176**, 624.

Chen, R., Vurek, G., and Alexander, N. (1967). *Science* **156**, 949.

Cohen, M. H., and Reif, F. (1957). *Solid State Phys.* **5**, 321.

Cohn, M. (1963). *Biochemistry* **2**, 623.

Conrad, R., and Brand, L. (1968). *Biochemistry* **7**, 777.

Das, T. P., and Hahn, E. L. (1958). *Solid State Phys., Suppl.* **1**.

De Gier, J., Demel, R. A., and van Deenen, L. L. M. (1968). "Surface Active Lipids in Foods," p. 39. London.

Dodd, G. H., and Chapman, D. (1970). Unpublished observations.

Dodd, G. H., and Radda, G. K. (1967). *Biochem. Biophys. Res. Commun.* **27**, 500.

Dodd, G. H., and Radda, G. K. (1968). *Biochem. J.* **108**, 5p.

Dodd, G. H., and Radda, G. K. (1971). *Biochem. J.* (in press).

Du Buy, H. G., and Showacre, J. L. (1961). *Science* **133**, 196.

Du Buy, H. G., Riley, F., and Showacre, J. L. (1964). *Science* **145**, 163.

Edwards, D., and Criddle, R. (1966). *Biochemistry* **5**, 583.

Emsley, J. W., Feeny, J., and Sutcliffe, L. H. (1965). "High Resolution Nuclear Magnetic Resonance Spectroscopy," Vol. 1. Pergamon Press, Oxford.

Emsley, J. W., Feeny, J., and Sutcliffe, L. H. (1966). "High Resolution Nuclear Magnetic Resonance Spectroscopy," Vol. 2. Pergamon Press, Oxford.

Eylar, E. H., Madoff, M. A., Brody, O. V., and Omsley, J. L. (1962). *J. Biol. Chem.* **237**, 1192.

Finean, J. B. (1969). *Quart. Rev. Biophys.* **2**, 1.

Finer, E. G., Hauser, H., and Chapman, D. (1969). "Colloquium der Gesellschaft fur Biologische Chemie," Vol. 20, p. 368. Mosbach, Baden.

Flanagan, M. T., and Ainsworth, S. (1968). *Biochim. Biophys. Acta* **168**, 16.

Fleischer, S., Rouser, G., Fleischer, B., Casu, A., and Kritchevsky, G. (1967). *J. Lipid Res.* **8**, 170.

Folch, J., and Lees, M. J. (1957). *J. Biol. Chem.* **191**, 897.

Förster, T., and Rokes, K. (1967). *Chem. Phys. Lett.* **1**, 279.

Freed, J. H., and Frankel, G. K. (1963). *J. Chem. Phys.* **39**, 326.

Freedman, R. B., and Radda, G. K. (1969). *FEBS Lett.* **3**, 150.

Gaines, G., Jr., Tweet, A. G., and Bellamy, W. D. (1965). *J. Chem. Phys.* **42**, 2193.

Green, D. E., Asai, J., Harris, R. A., and Penniston, J. (1968). *Arch. Biochem. Biophys.* **125**, 684.

Griffith, O. H., and Waggoner, A. S. (1969). *Accounts Chem. Res.* **2**, 17.

Gulik-Krzywicki, T., Rivas, E., and Luzzati, V. (1967). *J. Mol. Biol.* **27**, 393.

Gutowsky, II. S., and Tai, J. C. (1963). *J. Chem. Phys.* **39**, 298.

Hackenbrock, C. R. (1966). *J. Cell Biol.* **30**, 269.

Hamilton, C. L., and McConnell, H. M. (1968). In "Structural Chemistry and Molecular Biology" (A. Rich and N. Davis, eds.), p. 115. Freeman, San Fransisco, California.

Henderson, P. J. F., McGivan, J. D., and Chappell, J. B. (1969). *Biochem. J.* **111**, 521.

Henson, J. R., and Lawson, K. D. (1970). *Nature (London)* **225**, 542.

Hercules, D. M., ed. (1966). "Fluorescence and Phosphorescence Analysis." Wiley (Interscience), New York.

Hubbell, W. L., and McConnell, H. M. (1968). *Proc. Nat. Acad. Sci. U. S.* **61**, 12.

Itzkowitz, M. S. (1967). *J. Chem. Phys.* **46**, 3048.

Jackman, L. M. (1959). "Applications of Nuclear Magnetic Resonance Spectroscopy in Organic Chemistry." Pergamon Press, Oxford.

Jenkinson, T. J., Kamat, V. B., and Chapman, D. (1969). *Biochim. Biophys. Acta* **183**, 427.

Jones, G. P., and Douglas, D. C. (1966). *J. Chem. Phys.* **45**, 956.

Kanno, Y., and Loewenstein, W. (1964). *Science* **143**, 959.

Kasai, M., Changeux, J-P., and Monnerie, L. (1969). *Biochem. Biophys. Res. Commun.* **36**, 420.

Kavanau, J. L. (1964). "Water and Solute-Water Interactions." Holden-Day, San Fransisco, California.

Kavanau, J. L. (1966). *Science* **153**, 213.

Kivelson, D. (1957). *J. Chem. Phys.* **27**, 1087.

Korn, E. D. (1966). *Science* **153**, 1491.

Kubo, R., and Tomita, K. (1954). *J. Phys. Soc. Jap.* **9**, 888.

Ladbrooke, B. D., and Chapman, D. (1969). *Chem. Phys. Lipids* **3**, 304.

Ladbrooke, B. D., Jenkinson, T. J., Kamat, V. B., and Chapman, D. (1968). *Biochim. Biophys. Acta* **164**, 101.

Langelaar, J., de Vries, G., and Bebelaar, D. (1967). *J. Sci. Instrum.* **2**, 149.

Leslie, R. B., and Chapman, D. (1967). *Chem. Phys. Lipids* **1**, 143–156.

Lewis, G. N., and Calvin, M. (1939). *Chem. Rev.* **25**, 273.

Lippert, E. (1967). *Z. Elektrochem.* **61**, 962.

Luzzati, V. (1968). In "Biological Membranes" (D. Chapman, ed.), p. 71. Academic Press, New York.

Luzzati, V., and Faure, M. (1969). *Nature (London)* **223**, 1116.

McClure, W. O., and Edelman, G. M. (1966). *Biochemistry* **5**, 1908.

McDonald, C. D., and Phillips, W. D. (1967). *J. Amer. Chem. Soc.* **82**, 6332.

MacInnes, J. W., and Uretz, R. B. (1968). *J. Cell Biol.* **38**, 428.

Maddy, A. H. (1964). *Biochim. Biophys. Acta* **88**, 390.

Maddy, A. H., and Malcolm, B. R. (1965). *Science* **150**, 1616.

Maddy, A. H., and Malcolm, B. R. (1966). See Kavanau (1966).

Marinetti, G., and Gray, G. (1967). *Biochim. Biophys. Acta* **135**, 580.

Muller, P., and Rudin, D. O. (1968). *Nature (London)* **217**, 713.

O'Brien, J. S., and Sampson, E. L. (1965). *J. Lipid Res.* **6**, 537.

Olson, R. A. (1960). *Rev. Sci. Instrum.* **31**, 844.

Oster, G., and Nishijima, Y. (1956). *J. Amer. Chem. Soc.* **78**, 158.

Oster, G., and Nishijima, Y. (1964). *Fortscher. Hoch polym.-Forsch.* **3**, 313.

Penkett, S. A., Flook, A. G., and Chapman, D. (1968). *Chem. Phys. Lipids* **2**, 273.

Penzer, G., and Radda, G. K. (1967). *Nature (London)* **213**, 251.

Pople, J. A., Schneider, W. G., and Bernstein, H. J. (1959). "High Resolution Nuclear Magnetic Resonance." McGraw-Hill, New York.

Pressman, B. C. (1968). *Fed. Proc., Fed. Amer. Soc. Exp. Biol.* **27**, 1283.

Radda, G. K., and Dodd, G. H. (1968). In "Luminescence in Chemistry" (E. J. Bowen, ed.), p. 191. Van Nostrand, Princeton, New Jersey.

Razin, S., Morowitz, H. J., and Terry, T. T. (1965). *Proc. Nat. Acad. Sci. U. S.* **54**, 219.

Reiss-Husson, F. (1967). *J. Mol. Biol.* **25**, 363.

Roberts, J. D. (1959). "Nuclear Magnetic Resonance Applications to Organic Chemistry." McGraw-Hill, New York.

Rotman, B., and Papermaster, B. W. (1966). *Proc. Nat. Acad. Sci. U. S.* **55**, 134.

Rouser, G., Nelson, G. J., Fleischer, S., and Simon, G. (1968). In "Biological Membranes" (D. Chapman, ed.), p. 5. Academic Press, New York.

Rubalcava, B., Martinez de Munoz, D., and Gitler, C. (1969). *Biochemistry* **8**, 2742.

Salsbury, N. J., and Chapman, D. (1971). Unpublished observations.

Salsbury, N. J., and Chapman, D. (1968). *Biochim. Biophys. Acta* **163**, 314.

Salsbury, N. J., and Harris, P. (1968). Unpublished results.

Salton, M. R. J., and Freer, J. H. (1965). *Biochim. Biophys. Acta* **197**, 531.

Sandberg, H. E., and Piette, L. H. (1968). *Agressologie* **9**, 1.

Shipley, G. G., Leslie, R. B., and Chapman, D. (1969). *Nature (London)* **222**, 651.

Shockman, G. D., Kolb, J. M., Bakay, B., Conover, M. J., and Toennies, G. (1963). *J. Bacteriol.* **85**, 168.

Singer, I., and Tasaki, I. (1968). In "Biological Membranes" (D. Chapman, ed.), p. 347. Academic Press, New York.

Slichter, C. E. (1963). "Principles of Magnetic Resonance with Examples from Solid State Physics." Harper, New York.

Small, D. M. (1967). *J. Lipid Res.* **8**, 551.

Steim, J. M., Reinert, J. C., Tourtellotte, M. E., Rader, R. L., and McElhaney, R. D. (1971). *Nature (London)* (in press).

Steiner, R., and Edelhoch, H. (1962). *Chem. Rev.* **62**, 457.

Steiner, R., Lippoldt, R., Edelhoch, H., and Frattali, V. (1963). *Biopolym. Symp.* **1**, 335.

Sternlicht, H. (1967). *Biochemistry* **6**, 2881.

Stone, T. J., Buckman, T., Nordio, P. L., and McConnell, H. M. (1965). *Proc. Nat. Acad. Sci. U. S.* **54**, 1010.

Stryer, L. (1965). *J. Mol. Biol.* **13**, 269.

Stryer, L. (1966). *J. Amer. Chem. Soc.* **88**, 5708.

Stryer, L. (1968). *Science* **162**, 526.

Stryer, L., and Haughland, R. (1967). *Proc. Nat. Acad. Sci. U. S.* **58**, 719.

Suga, H., and Seki, S. (1962). *Bull. Chem. Soc. Jap.* **35**, 1905.

Suga, H., and Seki, S. (1963). *J. Phys. Chem. Solids* **24**, 230.

Tasaki, I., Watanabe, A., Sandlin, R., and Carnay, L. D. (1968). *Proc. Nat. Acad. Sci. U. S.* **61**, 883.

Tasaki, I., Carnay, L., Sandlin, R., and Watanabe, A. (1969). *Science* **163**, 683.

Thompson, T. E., and Henn, F. A. (1970). In "Membranes of Mitochondria and Chloroplasts" (Racker, ed.), p. 1. Van Nostrand-Reinhold, New York.

Tweet, A. G., Gaines, G., Jr., and Bellamy, W. D. (1964). J. Chem. Phys. 40, 2596.

Udenfriend, S., Zactzman-Nirenberg, P., and Guroff, G. G. (1966). Arch. Biochem. Biophys. 116, 261.

Vandenheuvel, F. A. (1963). J. Amer. Oil Chem. Soc. 40, 455.

Vanderkooi, J., and Martonosi, A. (1969). Arch. Biochem. Biophys. 133, 153.

Veksli, Z., Salsbury, N. J., and Chapman, D. (1969). Biochim. Biophys. Acta 183, 434.

Wallach, D. F. H. (1969). Arch. Biochem. Biophys. 131, 322.

Wallach, D. F. H., and Zahler, P. H. (1966). Proc. Nat. Acad. Sci. U. S. 56, 1552.

Wasemiller, G., Abrams, A., and Bakerman, S. (1968). Biochem. Biophys. Res. Commun. 30, 176.

Watkins, W. M. (1966). Science 152, 172.

Weber, G. (1961). In "Light and Life" (W. McElroy and B. Glass, eds.), p. 94. Johns Hopkins Press, Baltimore, Maryland.

Weber, G. (1963). Advan. Protein Chem. 8, 415.

Weber, G., and Daniel, E. (1966). Biochemistry 6, 1900.

Weber, G., and Laurence, D. (1954). Biochem. J. 51, XXXI.

Weber, G., and Teale, F. (1965). In "The Proteins" (H. Neurath, ed.), Vol. 3, p. 445. Academic Press, New York.

Wehry, E. L. (1967). In "Fluorescence" (G. Guilbault, ed.), p. 37. Arnold, London.

Weibull, C., and Bergström, L. (1958). Biochim. Biophys. Acta 30, 340.

3

SPIN LABELING AND MEMBRANE STRUCTURE

PATRICIA JOST, ALAN S. WAGGONER, and O. HAYES GRIFFITH

ABBREVIATIONS AND DEFINITIONS

ESR	Electron spin resonance
G	Gauss, magnetic field strength units
H	Applied magnetic field
$H\|x$, $H\|y$, $H\|z$	The applied magnetic field is parallel to the nitroxide x, y, or z axis
Hz	Hertz, cycles per second (kHz = 10^3 Hz, MHz = 10^6 Hz, GHz = 10^9 Hz)

I. INTRODUCTION

The molecular structure of membranes and model membranes is being investigated with a variety of spectroscopic approaches. One of the interesting and potentially important spectroscopic methods is the use of electron spin resonance in conjunction with paramagnetic reporter molecules ("spin labels"). These paramagnetic reporter groups are analogous to the optically absorbing or fluorescent groups that are so useful in other spectroscopic techniques. The reporter group is either covalently attached to specific functional groups in the system under investigation or diffused into regions of interest (e.g., hydrophobic regions).

The extension of the reporter group technique to electron spin resonance (ESR) investigations was made possible by the synthesis and characterization of a group of organic free radicals, nitroxides, which have the general structure

The N—O group contains the unpaired electron that renders the molecule paramagnetic and hence suitable for detection by ESR. The alkyl side chains (usually methyl groups, indicated here by the short straight lines) are necessary to stabilize the free radical. The remainder of the molecule can contain various chemically reactive groups that are useful for covalent labeling, since the nitroxide moiety is fairly unreactive under many conditions (see Section VII). The spin label introduced into interesting sites of model membranes, membranes, or membrane components reflects the orientation and motion of the spin label and the polarity of its environment. The ESR spectrum is sensitive to all of these variables and to the proximity of other spin labels, so that there is an interesting range of applications to biological problems. Recent review articles on spin labeling (McConnell and McFarland, 1970; Smith, 1971) are available, as well as an article on specific methods (Jost and Griffith, 1971a). A short review article discussing spin labeling of model membranes and biological membranes was written in 1968 (Griffith and Waggoner, 1969). Our purpose in writing this chapter is to summarize the information currently available about membrane structure from spin labeling studies. Discussions of interpretation of ESR spectra, spin labels useful for membrane studies, and technical innovations are included to familiarize the reader with advantages and disadvantages of this relatively new application of electron spin resonance.

II. ELECTRON SPIN RESONANCE SPECTRA OF NITROXIDE FREE RADICALS

A. ESR Spectra in Solution

The appropriate spin Hamiltonian, $\hat{\mathcal{K}}$, for a collection of nitroxide free radicals tumbling rapidly in solution is

$$\hat{\mathcal{K}} = |\beta|g_0 H \hat{S}_z + a_0 \hat{\mathbf{I}} \cdot \hat{\mathbf{S}} \tag{1}$$

where β, g_0, H, $\hat{\mathbf{S}}, S_z$, a_0, and $\hat{\mathbf{I}}$ are the electron Bohr magneton, isotropic g value, laboratory magnetic field, electron spin operator, component of $\hat{\mathbf{S}}$ along H, isotropic electron-nuclear hyperfine coupling constant, and the nitrogen nuclear spin operator, respectively (Carrington and McLachlan, 1967). The first term in $\hat{\mathcal{K}}$, the electron Zeeman term, gives the energy of interaction between the unpaired electron and the magnetic field, H. In the absence of other magnetic interactions, the Zeeman term determines the resonance condition for microwave energy absorption, $h\nu = g_0\beta H$, where ν is the operating frequency of the ESR spectrometer and h is Planck's constant. Since $h = 6.625 \times 10^{-27}$ erg sec, $\beta = 0.9273 \times 10^{-20}$ erg/gauss (G), and $g \sim 2.00$ for most free radicals, it follows that ν (in gigahertz) $= 2.80\, H$ (in kilogauss). Most commercial ESR spectrometers operate at $\nu = 9.5$ GHz (X-band). Hence $H = 9.5/2.8 = 3.4$ kG. Some spectrometers operate at $\nu = 35$ GHz where $H = 12.5$ kG. A second form of the same equation is ν (in megahertz) $= 2.80\, H$ (in gauss). This relation serves as a useful conversion factor (e.g., a coupling constant reported as 42.0 MHz or 42.0 Mc/sec may appear in another article as $42.0/2.80 = 15.0$ G).

The second term in the Hamiltonian takes into account the relatively weak magnetic interaction between the unpaired electron and a nearby nuclear spin (Fermi contact term). This hyperfine term yields the im-

portant rule that the nuclear spin, I, will split the hypothetic single Zeeman line into $2I + 1$ lines. For ^{14}N, $I = 1$ and the result is three lines of
equal intensity separated by the coupling constant a_0 (the nuclear spins of
the ^{16}O and ^{12}C that are also present in the nitroxide group are zero and
therefore do not split these three lines further). Figure 1 illustrates the

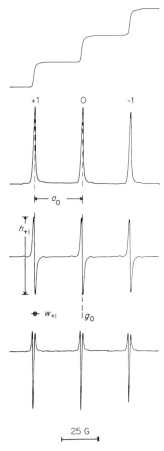

Fig. 1. Methods of presentation of a typical nitroxide ESR spectrum and the
definition of parameters. From top to bottom, the integral of the ESR absorption
spectrum, the ESR absorption spectrum, the first derivative and the second derivative
of the absorption spectrum, respectively. The important a_0 (A_0) and g_0 parameters,
first-derivative line widths (w) and heights (h) are shown. The designation $+1$, 0,
-1 refers to the z component of the nitrogen nuclear spin. To obtain these traces,
the 9.5 GHz first-derivative spectrum of a dilute aqueous solution of the nitroxide
2-doxylpropane (4',4'-dimethyloxazolidine-N-oxyl derivative of acetone) was recorded at room temperature, digitalized, and integrated once or twice or differentiated
using a small digital computer.

various ways the resulting nitroxide ESR spectrum can be displayed. From top to bottom are the integral of the absorption curve, the absorption curve, and the first and second derivatives of the absorption curve. The first derivative is the most common presentation in the literature. Spectra are frequently recorded with the opposite phase from Fig. 1 (producing an upside-down effect). The choice of phase and scan direction results in four possible permutations of the same spectrum. Although all four can be found in the spin labeling literature, the presentation most consistent with current trends in spectroscopy involves positive phase and increasing H from left to right. All spectra presented here follow this convention. The g value determines the center of the spectrum, and a is the distance between two adjacent lines.

Figure 2 is simply the first-derivative spectrum of Fig. 1 recorded at higher spectrometer gain. The satellite lines are due to ^{13}C or ^{15}N in natural abundance (Faber et al., 1967). These weak lines are more a curiosity than an aid in spin labeling studies. It might be desirable in some special line-shape studies to substitute ^{15}N for ^{14}N. For ^{15}N, $I = \frac{1}{2}$ and the result would be two strong lines at the positions indicated in Fig. 2.

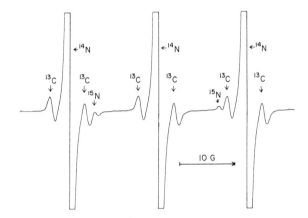

Fig. 2. The same first-derivative spectrum as in Fig. 1 but recorded at higher spectrometer gain. The six ^{13}C lines and the two ^{15}N lines arise from isotopes in natural abundance.

B. ESR Spectra in the Absence of Molecular Motion

1. Anisotropic Effects

The spin Hamiltonian required to describe a collection of *oriented* nitroxides is

$$\hat{\mathcal{K}} = |\beta| \mathbf{H} \cdot \mathbf{g} \cdot \hat{\mathbf{S}} + \hat{\mathbf{S}} \cdot \mathbf{A} \cdot \hat{\mathbf{I}} \qquad (2)$$

where g and $\mathbf{A}$ (also denoted by $\mathbf{T}$) are the g-value tensor and the electron-nuclear hyperfine tensor, respectively. The nuclear Zeeman term has been omitted because its effect is negligible for most 9.5 GHz spin labeling work (Libertini and Griffith, 1970). The second term now includes an electron spin–nuclear spin dipolar interaction in addition to the isotropic interaction of Eq. (1). Both g and A depend on the orientation of the magnetic field and the direction of H is specified in terms of a molecular coordinate system as shown in Fig. 3.

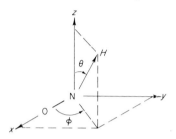

Fig. 3. The molecular coordinate system and angles used to define the direction of the external magnetic field (H). The angle θ is the angle from the direction of the applied magnetic field (H) to the molecular z axis. The angle ϕ is the angle between the x axis and projection of H in the xy plane. The use of the rectangular coordinate system does not imply that nitroxide free radicals are planar.

The spectra obtained along the three molecular axes when the nitroxides are *oriented* in a rigid matrix are of special interest and are shown in Fig. 4. Anisotropy is important in testing membrane models, and the spectra of Fig. 4 define the limits since these are the spectra seen in the absence of molecular motion with oriented samples. *No splitting smaller than that of the top two spectra of Fig. 4 or larger than the third spectrum* ($H \parallel z$) *can be observed* (ignoring solvent effects and excluding dinitroxides). Each three-line spectrum is still characterized by one g value and one coupling constant, but both parameters vary with the orientation of the crystal in the magnetic field. The g values g_{xx}, g_{yy}, g_{zz} measured from all three spectra are different, as can be seen by the distance each center line is shifted from the dashed reference line of Fig. 4. The coupling constants A_{xx} and A_{yy} measured from the top two spectra are obviously very nearly equal but are much smaller than the coupling constant A_{zz} measured from the third spectrum. The actual values are given in Table I along with data on other nitroxides. The agreement is encouraging. The essential features are that $A_{xx} \sim A_{yy}$ and the largest splitting is along the z molecular axis. These parameters are related to the isotropic g_0 and a_0 values through the simple equations $g_0 = (\frac{1}{3})(g_{xx} + g_{yy} + g_{zz})$ and $a_0 = (\frac{1}{3})(A_{xx} + A_{yy} + A_{zz})$.

First-derivative spectra   Absorption spectra

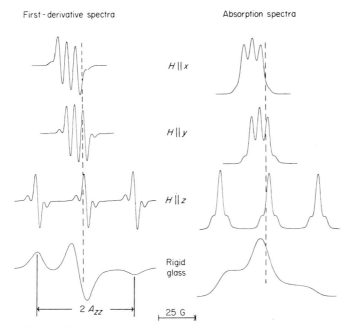

$H \parallel x$

$H \parallel y$

$H \parallel z$

Rigid glass

$\longleftarrow 2 A_{zz} \longrightarrow$ $\longmapsto$ 25 G $\longmapsto$

Fig. 4. The 9.5 GHz ESR spectra of representative spin labels in a rigid matrix. All ESR spectra (except the bottom row) are of the nitroxide 2-doxylpropane (4′,4′-dimethyloxazolidine-N-oxyl derivative of acetone) oriented in the crystal 2,2,4,4-tetramethyl-1,3-cyclobutanedione. The host crystal orients the nitroxide molecules so that their x, y, and z molecular axes are aligned in well-defined directions relative to the crystalline axes. The molecular x axis is parallel to the N—O bond, the z axis is parallel to the nitrogen $2p$ orbital associated with the unpaired electron, and y is perpendicular to the x and z axes. The crystal was rotated until the laboratory magnetic field (H) was parallel to the x ($H \parallel x$), y ($H \parallel y$), and finally the z molecular axis ($H \parallel z$). The small lines symmetrically displaced about the three main lines arise from hydrogen spin-flip transitions. The rigid glass spectrum (bottom row) was obtained using a sample of randomly oriented 12-doxylstearic acid (XXVIII) in egg lecithin at −196°C. In each case the first-derivative spectrum was recorded, digitalized, and then integrated to yield the corresponding absorption spectrum (see Section VI,C). The dashed lines mark the position of a 2,2-diphenyl-1-picrylhydrazyl reference sample ($g = 2.0036$). (Reprinted from Griffith et al., 1971.)

When the magnetic field is not along one of the three molecular axes the corresponding values of g and A are easily calculated from the expressions

$$g = g_{xx} \sin^2 \theta \cos^2 \phi + g_{yy} \sin^2 \theta \sin^2 \phi + g_{zz} \cos^2 \theta \qquad (3)$$

and

$$A = [(A_{xx})^2 \sin^2 \theta + (A_{zz})^2 \cos^2 \theta]^{1/2} \qquad (4)$$

or

$$A = A_{xx} \sin^2 \theta + A_{zz} \cos^2 \theta \qquad (5)$$

where the angles θ and ϕ are defined in Fig. 3 (Libertini and Griffith, 1970). These equations are useful in calculating g and A at an arbitrary orientation of H. In membrane studies, however, they are more often used in computer simulations of ESR spectra or in deriving expressions for anisotropic motion. Equation (3) provides accurate g values although more complicated relations are available (Libertini and Griffith, 1970). In calculating coupling constants, Eq. (4) is more accurate than Eq. (5), but the latter is simpler. Both equations are currently used in spin labeling studies.

TABLE I

PRINCIPAL VALUES OF THE SPLITTING AND g-VALUE TENSORS FOR NITROXIDES ORIENTED IN HOST CRYSTALS[a]

Nitroxide	A_{xx}	A_{yy}	A_{zz}	g_{xx}	g_{yy}	g_{zz}
Di-t-butyl nitroxide (I)[b]	7.6	6.0	31.8	2.0088	2.0062	2.0027
2-Doxylpropane[c]	5.9	5.4	32.9	2.0088	2.0058	2.0022
3-Doxyl-5α-cholestane (XXII)[d]	5.8	5.8	30.8	2.0089	2.0058	2.0021

[a] All splittings are in gauss. All uncertainties are estimated experimental errors for measurements made at room temperature. Values in the complete absence of molecular motion may be outside these ranges. For example, A_{zz} measured at liquid helium temperatures could be 1–2 G larger than numbers reported in this table.

[b] Host: 2,2,4,4-tetramethyl-1,3-cyclobutanedione (TMCB); uncertainties: A, ± 0.1 G; g, ± 0.0001. (Data of Libertini and Griffith, 1970.)

[c] Host: TMCB; uncertainties: A, ± 0.5 G; g, ± 0.0005. (Data of Jost et al., 1971.)

[d] Host: cholesteryl chloride; uncertainties: A, ± 0.5 G; g, ± 0.001. (Data of Hubbell and McConnell, 1969a.)

2. The Rigid Glass Spectrum

The top three spectra of Fig. 4 represent full orientation (i.e., in a host crystal). The other extreme is completely random orientation of nitroxides as illustrated in the bottom spectrum. This general line shape is observed whenever a small concentration of nitroxide is present in a rigid glass, polycrystalline sample, powder, or any randomly oriented spin label (assuming *no* molecular motion). The rigid glass spectrum results from a simple superposition of spectra from randomly oriented spin labels. This is most easily visualized by summing the three individual absorption spectra of Fig. 4 along with spectra of all intermediate orientations to obtain the rigid glass absorption spectrum. The rigid glass spectrum is important because of its ubiquity and because it provides an estimate of A_{zz} as shown in Fig. 4.

C. The Effects of Molecular Motion on ESR Spectra

1. Isotropic Motion

The shape of an ESR spectrum is markedly dependent upon the rotational mobility of the nitroxide. For example, in a nonviscous solvent

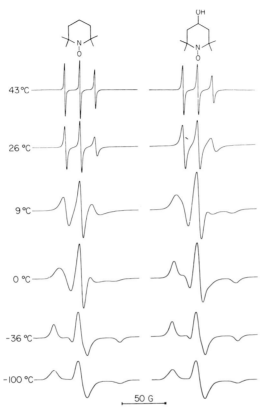

Fig. 5. The effect of viscosity on 9.5 GHz ESR spectra. The structure of the spin labels used are given at the head of the corresponding column of spectra. All spectra were recorded using $5 \times 10^{-4} M$ spin label in reagent-grade glycerol (Mallinkrodt, >95% glycerol by volume, exact concentration not determined). The samples were sealed in quartz tubes to exclude water vapor. (Adapted from Jost and Griffith, 1971a.)

nearly all nitroxides exhibit three sharp lines of nearly equal height (see Fig. 1). The sharp lines are caused by rapid isotropic tumbling motion, which averages away all anisotropic effects discussed above. If the rotational motion is slowed by increasing the solvent viscosity, averaging is incomplete. The result is unequal broadening of the three absorption lines as seen in Fig. 5. Qualitative terms are often used to describe these effects. For example, Fig. 1 illustrates a "freely tumbling" nitroxide. In the left-hand column of Fig. 5, the 26°C spectrum is characteristic of a nitroxide with "slightly hindered' rotation (i.e., a "weakly immobilized" spin label), whereas the 9°C spectrum results from a "moderately immobilized" nitroxide. The bottom spectra of Fig. 5 correspond to "strongly immobilized" nitroxides (rigid glass spectra). The ESR spectrum, however, is much more sensitive to nitroxide mobility than these qualitative terms suggest. The best way to make use of this sensitivity is to describe the motion quantitatively in terms of the shape of the spectrum.

For the range of mobilities where the three ESR lines do not overlap (see top spectra of Fig. 5) theories have been developed that quantitatively relate the line widths to the rotational correlation time τ of the spin label (Kivelson, 1957; Freed and Fraenkel, 1963). A naive definition of τ is simply the average time required for a nitroxide undergoing Brownian motion to rotate through a significant arc (e.g., $40°$). The ESR spectroscopist usually considers τ values of 5×10^{-11} seconds and 10^{-8} seconds as very fast and very slow tumbling limits, respectively. These two values represent approximate outside limits obtainable from ESR spectra. Quantitative correlation time theories based on a rigorous definition of τ, and applicable over the full range of Fig. 5, have been slow in coming. Nevertheless, there is progress in this area and methods for computer simulations of ESR spectra of Fig. 5 have been developed (Itzkowitz, 1967; Alexandrov et al., 1970; Keith and Mehlhorn, 1971; Freed et al., 1971).

Figure 5 points out one additional fact. Nitroxides having similar dimensions may tumble at different rates. As the viscosity of glycerol is increased by cooling, the motion of the alcohol nitroxide 2,2,6,6-tetramethyl-4-piperidinol-1-oxyl is evidently reduced more readily than the motion of 2,2,6,6-tetramethylpiperidine-1-oxyl, which lacks the hydroxyl group. This is particularly evident near $0°C$. Although this behavior is not unexpected, it does provide one more reason why τ values calculated for small molecules from the simple Stokes law relation, $\tau = 4\pi\eta r^3/3kT$ (η is solvent viscosity and r is molecular radius, see Pake, 1962), should not be taken too seriously. In some studies a ratio of peak heights or splittings can be used as a semiempirical measure of progressive changes in a series of ESR spectra. This practice also has pitfalls, but it can be useful if carried out in the proper spirit (see Section IV).

2. Anisotropic Motion

Anisotropic motion in the broadest sense refers to any molecular reorientation that does not occur with equal probability in all directions. Some examples are torsional oscillations, ring inversions, preferential rotation about one axis, flexing-twisting motion, and motion of spin labels rigidly attached to asymmetrically tumbling biomolecules. Anisotropic motion combined with isotropic tumbling at low or intermediate frequencies is complex and beyond the scope of this chapter. However, there are limiting situations that are sufficiently simple to be of general interest. One example is rapid rotation about one axis R, in the *absence of any other motion*. This could be called a spinning-top model. The problem is especially simple when R is along one of the three molecular axes, x, y, or z, of the nitroxide group. Assuming this to be the case, the observed coupling constant and g values are simply averages of the principal values involved. For example, in Fig. 4 if rapid motion occurred about the y axis, the $H \parallel y$ spectrum would remain the same. However, the $H \parallel x$ spectrum and the $H \parallel z$ spectrum would become equivalent, yielding a new three-line spectrum characterized by two parameters $A_\perp$ and $g_\perp$ where $A_\perp = (\frac{1}{2})(A_{xx} + A_{zz}) = (\frac{1}{2})(5.7 + 32.9) = 19.3 \text{ G}$ and $g_\perp = (\frac{1}{2})(g_{xx} + g_{zz}) = (\frac{1}{2})(2.0088 + 2.0022) = 2.0055$. (See data of

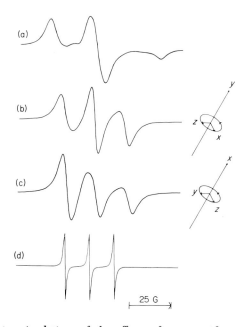

Fig. 6. Computer simulations of the effects of very rapid anisotropic motion on 9.5 GHz ESR spectra. (a) Computed spectrum for a randomly oriented collection of rigid nitroxides; (b) the corresponding spectrum allowing rapid rotation about the nitroxide y axis only; (c) the result for rapid rotation about the x axis only; and (d) the spectrum assuming rapid rotation about all three molecular axes (isotropic spectrum). For convenience, the number of nitroxides included in the calculation was limited to 2000. With a subroutine designed to generate random numbers, all 2000 nitroxides were assigned random orientations. The ESR line positions of each nitroxide were calculated using Eqs. (3) and (4) and were then summed. The individual line widths were 5 G for (a)–(c), and 0.5 G for (d). The line-shape chosen was Lorentzian.

Table I. The terms $A_\perp$ and $g_\perp$ are the A and g parameters observed when the applied magnetic field is perpendicular to R (see above); $A_\parallel$ and $g_\parallel$ are the A and g parameters observed when the applied magnetic field is parallel to R).

The rigid glass type of spectrum will, of course, also be altered because it represents a summation of the three principal spectra plus spectra at intermediate orientations of the magnetic field. Using an appropriate computer program and the six parameters A_{xx}, A_{yy}, A_{zz} and g_{xx}, g_{yy}, g_{zz}, the rigid glass spectrum (a) of Fig. 6 was produced. With the same computer program and the above parameters averaged to reflect rapid rotation only about the y axis, the result is spectrum (b) of Fig. 6. Note that (a) is in good agreement with the experimental rigid glass spectrum of Fig. 4 whereas (b) has a completely different line shape.

If rapid rotation occurred about the x axis instead of the y axis, the $H \parallel x$ spectrum of Fig. 4 would remain the same. However, the $H \parallel y$ and $H \parallel z$ spectra would shift to produce one new spectrum characterized by $A_\perp = (\frac{1}{2})(A_{yy} + A_{zz}) = (\frac{1}{2})(5.7 + 32.9) = 19.3$ G and $g_\perp = (\frac{1}{2}) \times (g_{yy} + g_{zz}) = (\frac{1}{2})(2.0058 + 2.0022) = 2.0040$. The corresponding randomly oriented sample would exhibit spectrum (c) of Fig. 6, which is clearly distinguishable from spectra (a) and (b) (note, for example, the relative height of the low field line).

The other situation of interest is rapid rotation about z. The $H \parallel z$ spectrum would be, of course, the same as in Fig. 4 and the $H \parallel x$ and $H \parallel y$ spectra would shift slightly to the new position defined by $A_\perp = (\frac{1}{2})(5.7 + 5.7) = 5.7$ G and $g_\perp = (\frac{1}{2})(2.0088 + 2.0058) = 2.0073$. The spectrum of the randomly oriented sample differs only slightly from that of Fig. 6a and is not presented. For comparison with spectra (a), (b), and (c), spectrum (d) of Fig. 6 is a computer simulation of rapid isotropic motion based on the parameters $a_0 = (\frac{1}{3})(A_{xx} + A_{yy} + A_{zz}) = (\frac{1}{3})(5.7 + 5.7 + 32.9) = 14.8$ G and $g_0 = (\frac{1}{3})(g_{xx} + g_{yy} + g_{zz}) = (\frac{1}{3})(2.0088 + 2.0058 + 2.0022) = 2.0056$. Rapid rotation about one axis of a nitroxide radical in a single crystal has not yet been observed but a similar effect has been observed in other free radicals (see, for example, Griffith, 1964; Lai et al., 1970).

If the axis of rapid rotation (R) is not parallel to one of the principal axes, the more general relations $A_\parallel = A_{xx} \sin^2 \beta + A_{zz} \cos^2 \beta$ and $A_\perp = (\frac{1}{2}) A_{xx}(1 + \cos^2 \beta) + (\frac{1}{2})A_{zz} \sin^2 \beta$ may be used along with related equations for $g_\parallel$ and $g_\perp$, where β is the angle between R and the z molecular axis. (The first equation follows from Eq. (5) by inspection and the second is obtained directly from the first using the relation $A_{xx} + A_{yy} + A_{zz} = 2A_\perp + A_\parallel$, which follows from properties of the tensor.) For discussions of this and related types of anisotropic motion, see McConnell and McFarland (1970), Seelig (1970), and Keith and Mehlhorn (1971).

Another example of anisotropic motion is a restricted random walk or wobble about some axis, R. For example, one plausible model allows the molecular z axis to execute a rapid walk pointing with equal probability in all directions within a cone about R, specified by an angle γ. This type of motion could result from limited chaotic movement of long-chain lipids in biological membranes. Using Eq. (5), equations relating $A_\parallel$ and $A_\perp$ to the parameter γ are readily obtained (Jost et al., 1971). Corresponding equations for $g_\parallel$ and $g_\perp$ are more troublesome unless large-amplitude motion or rotation is allowed in the molecular xy plane. In any case, the net effect of a small-amplitude wobble about the z axis is to decrease the apparent value of A_{zz} measured from the rigid glass spectrum of Fig. 4 or Fig. 6a. As the angle γ increases, $A_\parallel$ continues to decrease and $A_\perp$ increases until eventually $A_\parallel = A_\perp = a_0$, yielding the solution spectrum (Fig. 6d). The spinning-top model also predicts a decrease in $A_\parallel$ and an increase in $A_\perp$ as β is increased (i.e., as R is tilted away from the z axis). However, the limiting behavior for large values of β is not the same, as is easily seen by examining the above equations. These simple models are useful in understanding some features of spin labeling ESR data. Many other treatments of anisotropic motion exist (see, for example, Falle et al., 1966; Hudson and Luckhurst, 1969).

In liquid crystal studies and related work, the experimental data are frequently related to a 3×3 matrix of order parameters S_{ij} defined by the equation $S_{ij} = (\frac{1}{2})(3 < l_{zi}k_{zi} > - \delta_{ij})$ where z is the magnetic field direction, i and j are molecular axes and l's are direction cosines (Saupe, 1964; Ferruti et al., 1969; Corvaja et al., 1970). In membrane and model membrane studies the most important parameter is the one relating the degree of order of the nitroxide z axis. This parameter, S (i.e., S_3 or S_{33}), may be formally defined as in the above equation or may be simply designated as $S = (A_\parallel - A_\perp)/(A_{zz} - A_{xx})$. For the completely oriented crystalline sample of Fig. 4, $A_\parallel = A_{zz}$ and $A_\perp = A_{xx}$ so that $S = 1$, whereas for a microscopically randomly oriented sample $A_\parallel = A_\perp$ and $S = 0$. Thus, S can be thought of as a normalized anisotropy parameter. The experimental values of $A_\parallel$ and $A_\perp$ are not always easily measured without resort to computer simulations. Experimental examples of anisotropy are mentioned in the model membrane and membrane applications discussed below.

D. Solvent Effects

Both parameters a_0 and g_0 are solvent dependent. The value of a_0 decreases and g_0 increases as the solvent polarity is decreased. For example, di-t-butyl nitroxide (DTBN) in water exhibits $a_0 = 16.7$ G and $g_0 = 2.0056$, whereas DTBN in hexane yields $a_0 = 14.8$ G and $g_0 = 2.0061$ (Kawamura et al., 1967). This effect is apparently general, and the value of a_0 is typically 1 or 2 G smaller in hydrocarbon solvents than in water. The corresponding increase in g_0 is of the order of 0.0005. These shifts are

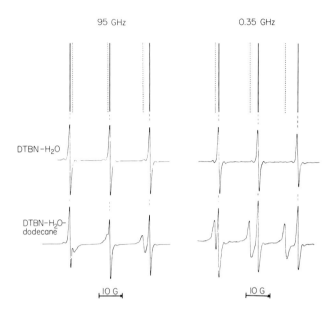

Fig. 7. Solvent effects on 9.5 and 35 GHz ESR spectra of di-*t*-butyl nitroxide (DTBN) at room temperature. Top row: the relative line positions for DTBN in water (solid lines) and hexane or dodecane (dotted lines). Second row: the ESR spectra of a capillary filled with $5 \times 10^{-4} M$ DTBN in water. Third row: the spectra recorded using two samples; capillaries containing $5 \times 10^{-4} M$ DTBN in water and in degassed dodecane (the relative peak heights are not significant since the capillaries were not inserted the same distance into the microwave cavity). (Adapted from Griffith *et al.*, 1971.)

illustrated in Fig. 7. Recalling that g_0 determines the center of the spectrum and a_0 is the distance between adjacent lines, the effect of increasing g_0 is to shift the entire spectrum to lower fields while the small decrease in a_0 contracts the ESR pattern slightly. Thus, when DTBN is present in both an aqueous and hydrocarbon phase, the maximum separation occurs between the two high field lines (see Fig. 7). At 9.5 GHz only partial resolution is obtained, whereas at 35 GHz all six lines of the two superimposed spectra are resolved as shown in Fig. 7. This higher resolution allows the measurement of a_0 and g_0 parameters in an unknown hydrocarbonlike phase.

In studies involving moderately immobilized spin labels, solvent effects can be troublesome because the small changes in the anisotropic parameters A_{xx}, A_{yy}, A_{zz} and g_{xx}, g_{yy}, g_{zz} have not been carefully examined. However, for small rapidly tumbling nitroxides such as DTBN, solvent effects may be exploited much as optical solvent shifts are used to determine the polarity of the immediate environment of optical reporter groups (see Section V).

E. Proximity Effects

The magnetic environment of a nitroxide is perturbed by the presence of nearby nitroxides, molecular oxygen, or other paramagnetic centers (e.g., other free radicals or paramagnetic metal ions). Line broadening by oxygen or high nitroxide concentrations (i.e., $>10^{-3} M$) could yield some information but is more often a source of experimental difficulty (Jost and Griffith, 1971a). A more important situation occurs when two nitroxide free radicals are linked together (e.g., molecules XVIII, XIX, and XX of Fig. 10).[*] The resulting dinitroxide spin labels provide a potentially useful means of sensing relative positions and conformations, even when the membrane sample is not oriented. The Hamiltonian of a dinitroxide spin label is

$$\hat{\mathcal{H}} = \hat{\mathcal{H}}_{\text{Zeeman}} + \hat{\mathcal{H}}_{\text{hyperfine}} + \hat{\mathcal{H}}_J + \hat{\mathcal{H}}_D$$

where the first two terms are similar to the corresponding electron Zeeman and electron-nuclear hyperfine terms of Eqs. (1) and (2), except there are now two unpaired electrons and two nitrogen nuclei. The two new terms are the electron–electron exchange Hamiltonian ($\hat{\mathcal{H}}_J$) and the electron–electron dipole Hamiltonian ($\hat{\mathcal{H}}_D$); $\hat{\mathcal{H}}_J$ arises from electrostatic interactions and the usual form is $\hat{\mathcal{H}}_J = J\hat{\mathbf{S}}_1 \cdot \hat{\mathbf{S}}_2$ where J is the exchange integral and $\hat{\mathbf{S}}_1$ and $\hat{\mathbf{S}}_2$ are the two electron spin operators (Carrington and McLachlan, 1967). This interaction is isotropic and hence is important even when the dinitroxide is rapidly tumbling in solution. In contrast, $\hat{\mathcal{H}}_D$, is a magnetic dipolar interaction and its effects are averaged to zero when the dinitroxide tumbling frequency is large compared to the maximum anisotropy (this is *usually* the case in solutions of low viscosity). The combination of these two electron–electron interactions can lead to very complex line shapes. The analysis is simplified when the dinitroxides tumble rapidly, when the spin labels are partially (or totally) oriented, or when it is possible to freeze the sample to remove complications resulting from intermediate or slow molecular motion.

In solution the spectrum of a dinitroxide depends on the relative magnitudes of J and a_0, as shown diagrammatically in Fig. 8. If the two N—O groups are far apart and $J \ll a_0$, the dinitroxide behaves as two independent monoradicals. Each N—O moiety gives rise to a normal three-line ESR spectrum, and these lines superimpose exactly. When the two nitroxide groups are brought closer together, the condition $J \gg a_0$ is

[*] Roman numerals throughout this chapter refer to structures shown in Fig. 10.

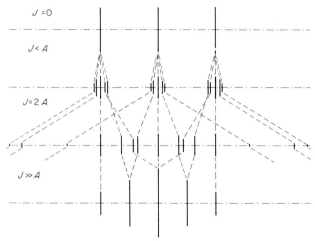

Fig. 8. Calculated ESR line positions of a dinitroxide as a function of the relative magnitudes of the exchange parameter (J) and the isotropic coupling constant (A or a_0). The top spectrum is equivalent to the normal three-line spectrum of a nitroxide radical. (Reprinted from Lemaire *et al.*, 1968.)

eventually reached, in which case there are $2\ nI + 1 = 2(2)1 + 1 = 5$ lines with relative intensities of 1:2:3:2:1, separated by $a_0/2$ G as indicated in Fig. 8. In between the two limits of $J \ll a_0$ and $J \gg a_0$ the spectra are more complex. It is not a simple matter to relate J quantitatively to molecular geometry. However, the observation that the ESR spectrum can be grossly altered by changes in structure or conformation renders J an important parameter in spin labeling (Briere *et al.*, 1965; Lemaire *et al.*, 1968).

The electron–electron dipole term is frequently written in the form $\hat{\mathcal{H}}_D = D(\hat{S}_z^2 - \hat{S}^2/3) + E(\hat{S}_x^2 - \hat{S}_y^2)$ where $\hat{S}_x$, $\hat{S}_y$, $\hat{S}_z$ are components of the total spin operator $\hat{S}$, and D and E are parameters that can be related quantitatively to molecular geometry (Carrington and McLachlan, 1967). A simple example of a dipolar interaction between nitroxides is given in Fig. 9. This spectrum was recorded for a single crystal doped with an unusually high concentration of di-*t*-butyl nitroxide. The strong three-line pattern is the normal spectrum of an isolated nitroxide (see Fig. 4). The two weak five-line patterns marked by arrows arise from occasional neighboring pairs of interacting nitroxides. The five lines in each pattern are a result of $J \gg a$, as discussed above. The dipolar interaction is responsible for the large splitting (distance between arrows). This splitting changes greatly as the crystal is rotated in the magnetic field. From the extrema, the important parameter $|D/g\beta|$ is found to be 126 G (Griffith *et al.*, 1971). Given this number and the point dipole approximation, the

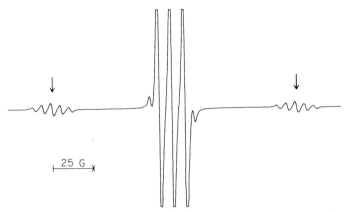

Fig. 9. The 9.5 GHz room-temperature spectrum of di-*t*-butyl nitroxide oriented in a crystal of 2,2,4,4-tetramethyl-1,3-cyclobutanedione. The magnetic field was in the crystalline *ac* plane, approximately 20° and 84° from the *a* and *c* axes, respectively. The concentration of di-*t*-butyl nitroxide was sufficiently high in this crystal that interacting radical pairs can easily be seen (note arrows) symmetrically displaced about the normal three-line ESR spectrum. (Reprinted from Griffith *et al.*, 1971.)

average distance between centers of the two N—O bonds is calculated to be 6.1 Å. Estimates based on x-ray crystallographic data are in the range ~6.6 Å, assuming that the nitroxides occupy certain substitutional sites. The electron–electron dipolar interaction has a fairly long range and the upper practical limit is of the order of 15–20 Å for fixed nitroxides. The spectra are, of course, somewhat more complex when a randomly oriented or partially oriented sample is encountered. However, these problems are tractable and the dipole term will provide quantitative distance and conformational information in future dinitroxide spin labeling experiments.

Nitroxide labeling and proximity effects are not limited to ESR studies. The large magnetic moment of an unpaired electron has a pronounced effect on the relaxation rates of nearby magnetic nuclei. The result is broadening of the nuclear magnetic resonance (NMR) lines of protons or other nuclei in the vicinity of spin labels (or paramagnetic metal ions). The fluorescent properties of aromatic chromophores may also be altered by nearby spin labels. These topics lie outside the scope of this chapter. Fortunately, however, NMR studies involving nitroxides are discussed in several interesting articles (see, for example, Mildvan and Weiner, 1969a,b; Taylor *et al.*, 1969). Techniques include conventional NMR and pulsed NMR (relaxation time measurements). NMR and fluorescence techniques are extensively discussed in Chapter 2. A combination of ESR and NMR can be employed. An interesting example is the study by Mildvan and Weiner (1969b) in which the interaction of a spin-labeled nicotinamide adenine dinucleotide (XVII) with alcohol dehydrogenase was studied. Information regarding the orientation and proximity of the N—O group relative to ethanol in the ternary complex was obtained by measuring line broadening of the methyl and methylene proton NMR peaks. Relaxation effects were also measured and ESR was used to monitor the binding of the spin label. The use of NMR in spin-labeled model membrane systems is not common (see, for example, Waggoner et al., 1967). However, the approach is applicable to model systems and, with courage, to biological membranes.

Nitroxides are capable of reducing the excited-state lifetime of the chromophore, thereby quenching fluorescence (Singer and Davis, 1967; Buchachenko et al., 1967). The interaction leading to fluorescence quenching apparently involves transfer of the excited singlet energy of the chromophore to a nitroxide acceptor (Buchachenko et al., 1967). There are several requirements which have to be satisfied before energy transfer can take place (see Fig. 15 in Chapter 2). One essential requirement for a nitroxide to act as an energy acceptor is that the excited singlet state of the fluorescent chromophore must lie above the lowest excited electronic level of the nitroxide. In other words, the nitroxide must have an absorption band at a longer wavelength than the emission band of the chromophore. Dependence of quenching on the overlap of the two bands has not been studied. The other requirement for quenching is that the nitroxide and the chromophore be sufficiently close. Since the quenching has been found to follow the Stern-Volmer equation (Buchachenko et al., 1967), it is apparent that the interaction is short ranged (perhaps less than 6 Å). Until precise measurements have been made only qualitative statements can be made of nitroxide chromophore proximity. Nevertheless, this interaction offers a possible means of determining the position of spin labels in biological systems containing fluorescent groups such as tryptophan. A tryptophan residue of a protein, for example, may fluoresce less after a nearby site has been tagged with a nitroxide free radical.

III. SPIN LABELS USEFUL IN MEMBRANE STUDIES

The first nitroxide free radicals were prepared around the turn of the century. However, it was not until the early 1960's that useful nitroxides of the general type shown at the top of p. 101 became available.

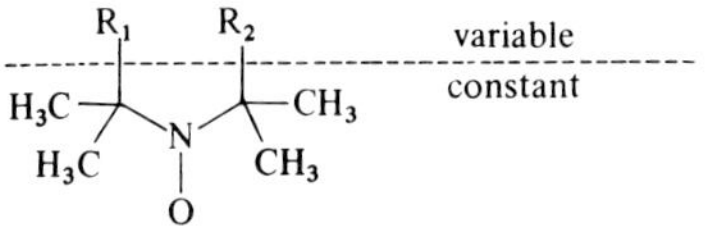

The four methyl groups adjacent to the nitrogen are responsible for the stability of the radical; therefore, the lower part of the molecule usually remains constant. The upper part of the nitroxide shown above is not critical in terms of the ESR signal, but it is the R_1 and R_2 groups that determine the site of binding to a protein or a region of a membrane that the spin label will probe. Through the efforts of E. G. Rozantsev, A. Rassat, M. B. Neiman, A. K. Hoffman, and others the syntheses of a large number of useful nitroxides have been worked out and some are commercially available (e.g., Frinton Laboratories, Syva Associates). Forrester et al. (1968) have written a chapter on nitroxides in general and Rozantsev (1970) has devoted an entire book to the synthesis, reactions, and properties of the above nitroxide radicals.

Another class of nitroxide free radicals was introduced by Keana et al. in 1967. These spin labels are formed by the following reaction

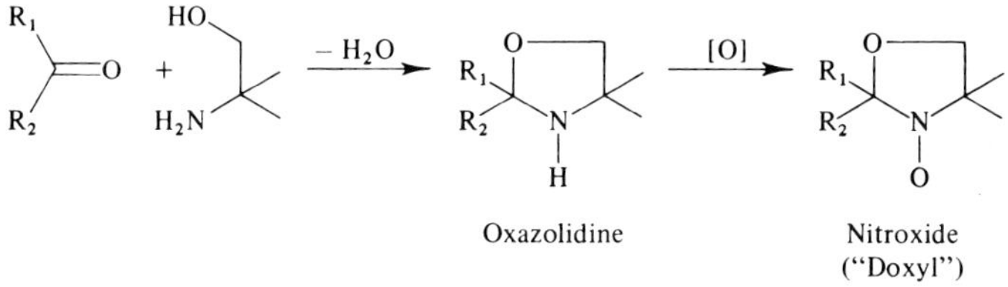

This synthesis is very useful in that it provides a general method of converting ketones into nitroxides. The resulting nitroxides can be named as 4',4'-dimethyloxazolidine-N-oxyl derivatives of the starting ketones, but for simplicity this nitroxide will be referred to as a *doxyl* moiety (e.g., 12-doxylstearic acid).

A comparatively small list of spin labels is presented in Fig. 10 to give the reader an idea of the general types of spin label that have found use in membrane-related experiments. The books of Forrester et al. (1968) and Rozantsev (1970) and the review articles of McConnell and McFarland (1970) and Smith (1971) should be consulted for the preparation, chemistry, and properties of spin labels. The book by Rozantsev is particularly useful because of a very large section presenting detailed syntheses of many of the basic nitroxides.

It is convenient to group the spin labels of Fig. 10 into three arbitrary

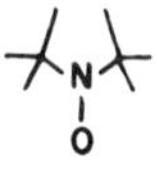

Fig. 10. See page 104 for caption.

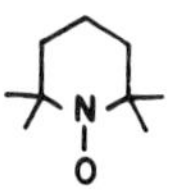

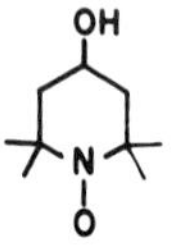

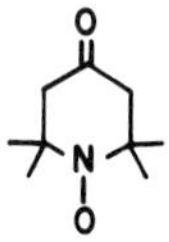

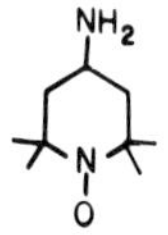

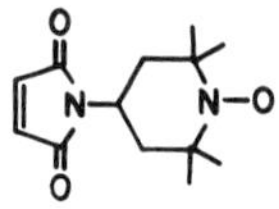

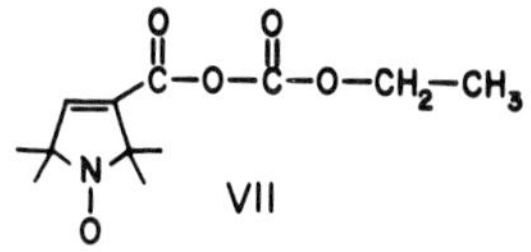

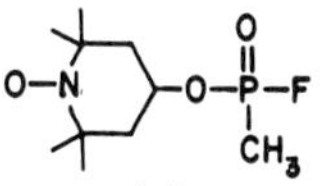

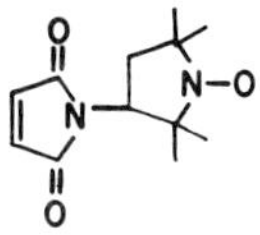

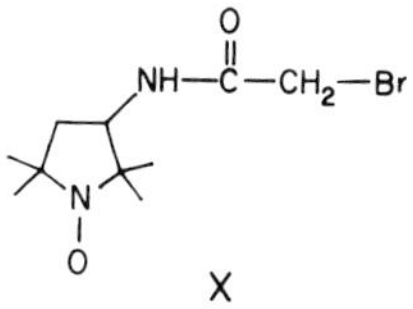

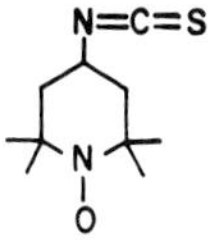

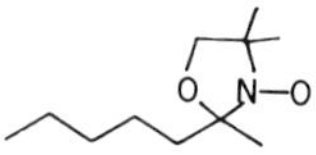

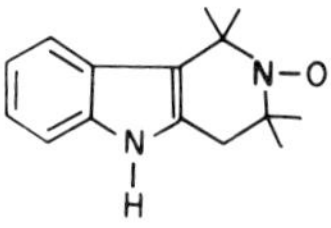

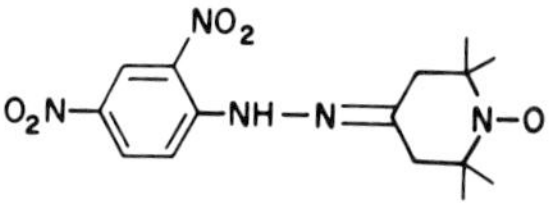

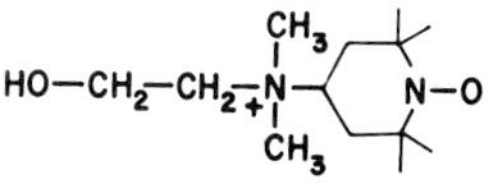

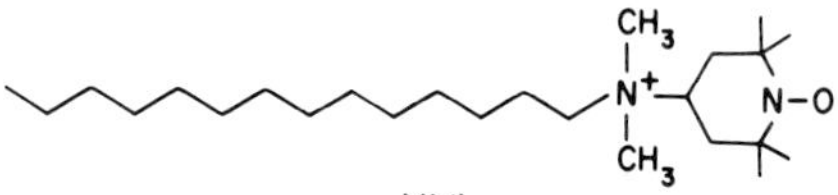

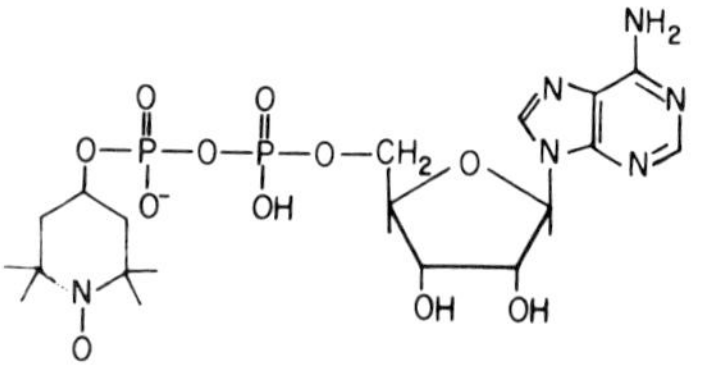

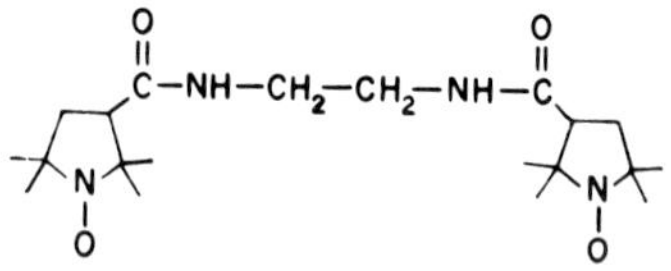

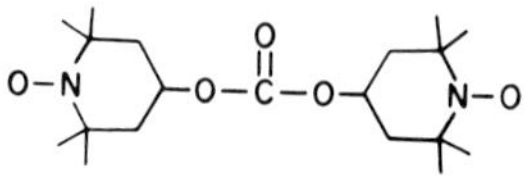

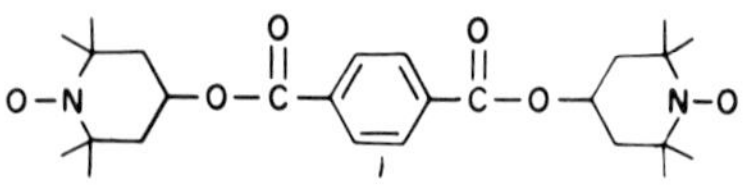

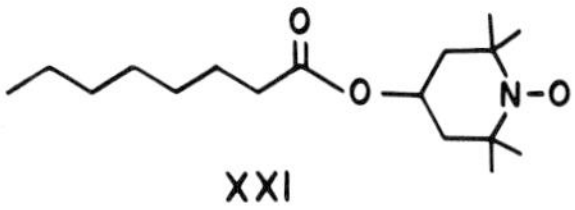

Fig. 10. See page 104 for caption.

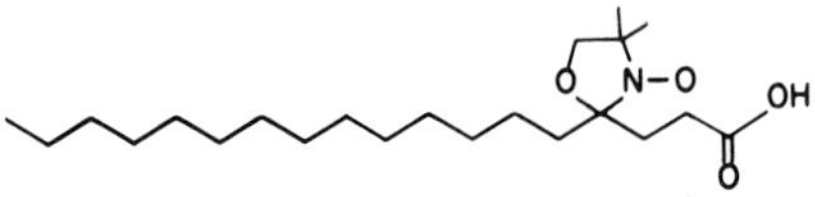

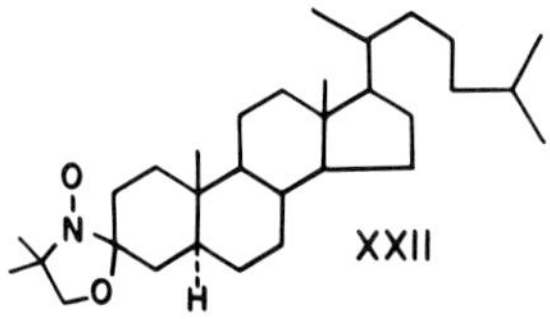

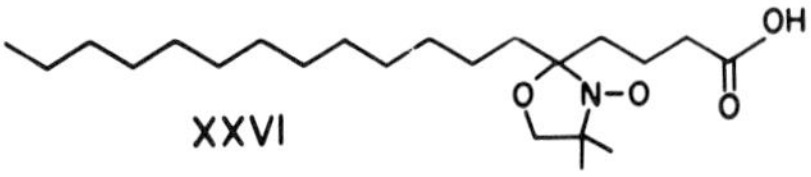

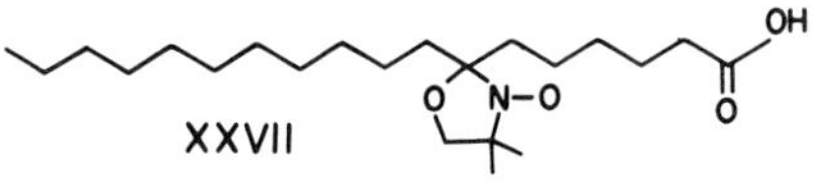

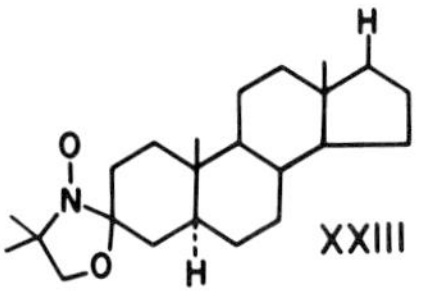

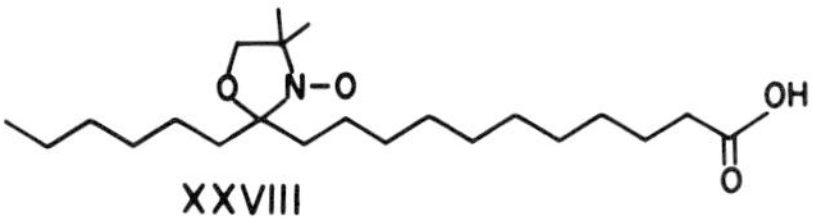

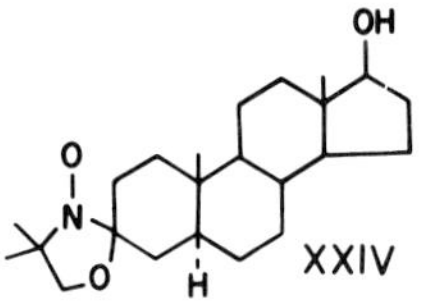

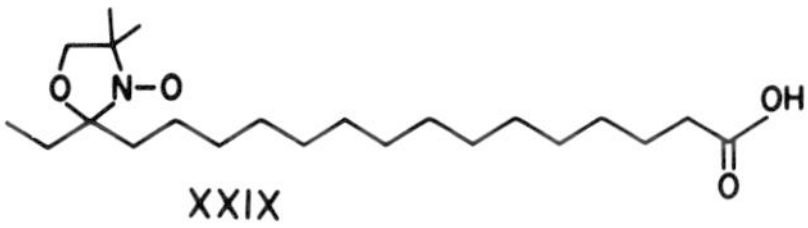

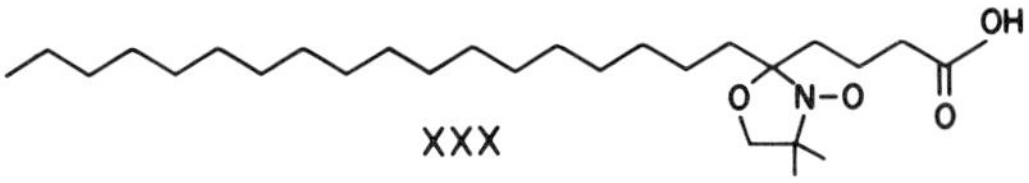

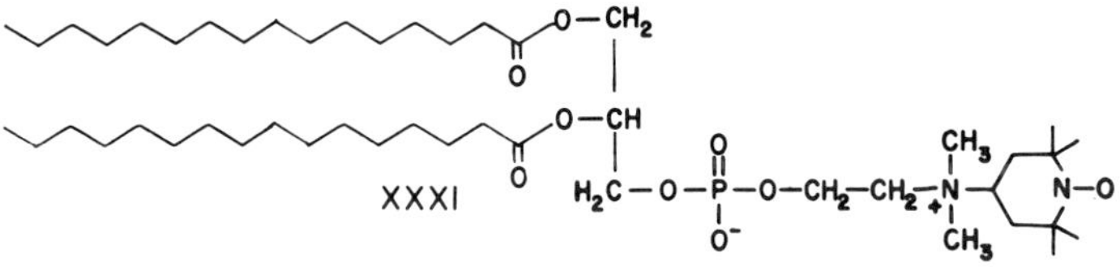

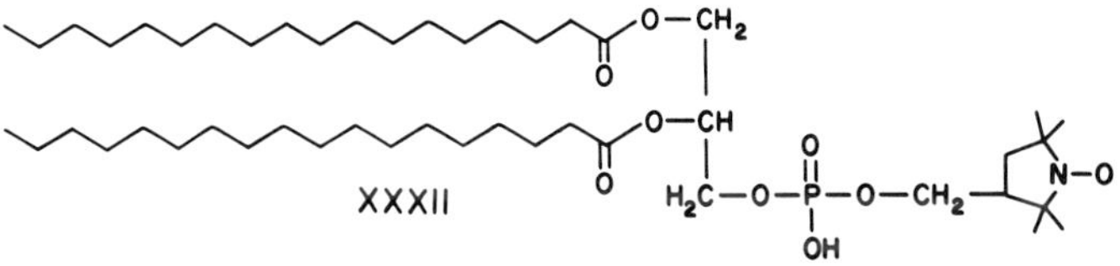

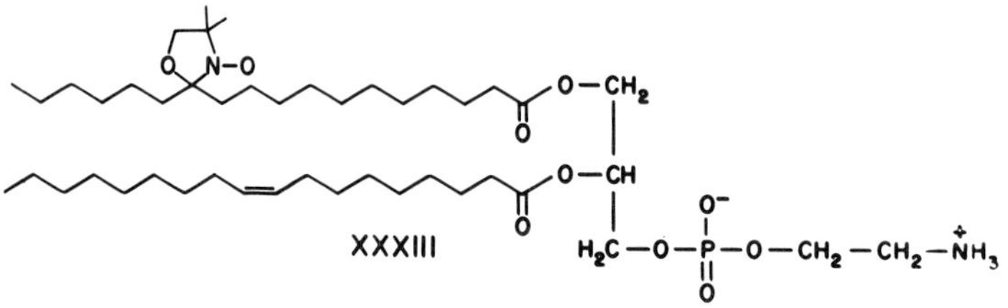

classes: (a) covalently binding spin labels, (b) noncovalently binding nonbiological spin labels, and (c) spin-labeled analogs of biological molecules. Examples of covalently binding spin labels include VI, VII, VIII, IX, X, and XI. For example, the maleimide VI is capable of reacting with sulfhydryl groups

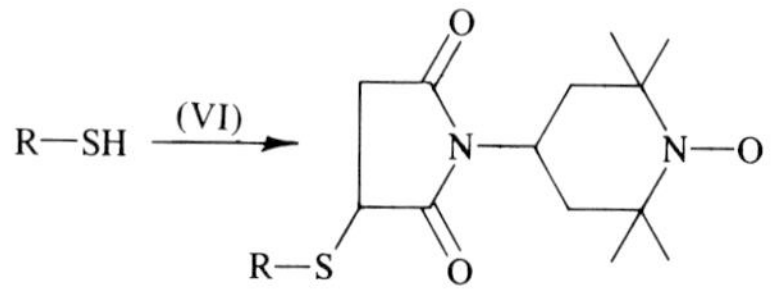

Similarly, the isothiocyanate XI evidently reacts with amines

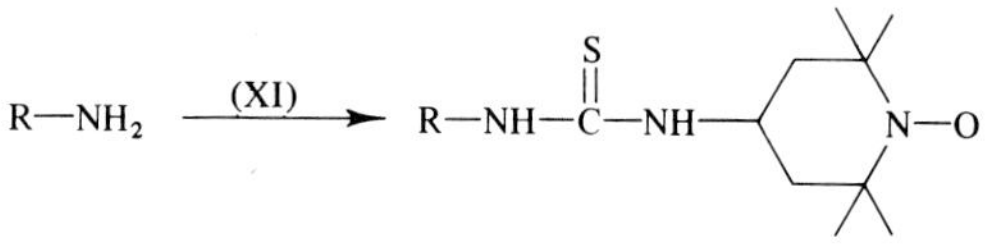

In each case, the R is a protein component. The reactions of maleimides (Gregory, 1955; Griffith and McConnell, 1966; Ohnishi et al., 1966) and isothiocyanates (Lehninger, 1970; Wold, 1967) are by no means totally specific when many functional groups are present. Nevertheless, covalent labeling reagents are useful, especially when the protein of interest can be isolated from the system before labeling.

Fig. 10. Some examples of nitroxide molecules that have been used as spin labels, or as precursors for spin labels, in biological applications. Useful references: I, Hoffman and Henderson (1961); II, Lebedev and Kazarnovskii (1960), Rozantsev and Nieman (1964), Briere et al. (1965); III, Briere et al. (1965); IV, Rozantsev and Nieman (1964), Briere et al. (1965); V, Rozantsev and Kokhanov (1966); VI, Ohnishi et al. (1966); VII, Griffith et al. (1967); VIII, Morrisett et al. (1969); IX, Griffith and McConnell (1966); X, Ogawa and McConnell (1967); XI, Cooke and Morales (1969), Tonomura et al. (1969), McConnell and McFarland (1970); XII, Keana et al. (1967); XIII, Waggoner et al. (1967); XIV, Stryer and Griffith (1965); XV, Kornberg and McConnell (1971); XVI, Hubbell and McConnell (1969a); XVII, Weiner (1969); XVIII, Calvin et al. (1969); XIX, Briere et al. (1965); XX, Briere et al. (1965); XXI, Waggoner et al. (1968); XXII, Keana et al. (1967); XXIII, Hubbell and McConnell (1969a); XXIV, Hubbell and McConnell (1969a); XXV, Hubbell and McConnell (1969b); XXVI, Hubbell and McConnell (1969b), Jost et al. (1971); XXVII, Jost et al. (1971); XXVIII, Waggoner et al. (1969); XXIX, Jost et al. (1971); XXX, Hubbell and McConnell (1969b); XXXI, Kornberg and McConnell (1971); XXXII, Aneja and Davies (1970); XXXIII, Keith et al. (1968). (XXXIII is meant to represent the kinds of phospholipids prepared during biosynthetic incorporation experiments. The specific phospholipids were not separated and characterized.)

Noncovalently binding nonbiological spin labels include I, II, XII and the dinitroxides XVIII, XIX, XX. The small, rapidly tumbling nitroxides I, II, and XII are useful in examining the polarity of accessible regions of membranes and the dinitroxides have been used to probe oriented liquid crystals (see Sections IV and V). Examples of spin-labeled analogs of biological molecules include a spin-labeled nicotinamide adenine dinucleotide, the steroid derivatives XXII, XXIII, and XXIV, the fatty acid derivatives XXV, XXVI, XXVII, XXVIII, XXIX, and XXX, and the phospholipids XXXI, XXXII, and XXXIII. Studies involving these useful analogs account for much of the literature on spin-labeled model membranes and biological membranes (see Sections IV and V).

IV. APPLICATIONS TO MODEL MEMBRANE SYSTEMS

Lipid–lipid and lipid–protein associations in the presence of an aqueous phase occur in membranes. The study of these associations and their interactions at the molecular level has been greatly aided by obtaining information from simple model systems. In such model systems the chemical nature of the participating molecules is well defined and the variables introduced can be more carefully controlled. The immediate aim is to understand molecular association, motion, and organization that is relevant to regions in the much more complex biological membranes. The even more ambitious goal is to apply the techniques developed with model systems directly to biological membranes.

A. Micelles

Although the bimolecular leaflet may well be a prominent feature of lipid structure in membranes, the proposal that globular micelles are also present has received considerable attention (see Lucy, 1968). These proposed small globular micelles are considered to be composed predominantly of phospholipids with the polar end of the molecule at the water–lipid interface. The simplest micellar systems, however, are those in which the lipid consists of a polar head group attached to a single hydrocarbon chain. Such lipids, above a certain critical concentration in water, spontaneously form spherical aggregates with a somewhat hydrophobic interior and can be important in solubilizing added water-insoluble substances. Micelles, or similar structures, could play a role in the transport of hydrophobic molecules.

Spin labeling has been used to study the properties of micelles. Waggoner et al. (1967, 1968) introduced nitroxides such as XIII, XIV, and

XXI into aqueous solutions containing various concentrations of sodium
dodecylsulfate (SDS). Below the critical micelle concentration the probes
tumbled freely, giving spectra similar to those in Fig. 1. As the concentration of SDS was increased, a marked effect on the ESR spectra occurred in the region of the critical micelle concentration (ca. 0.2 weight
% SDS). Above this concentration the spin labels were weakly immobilized and the line shapes were very similar to the 26°C ESR spectrum
in the left column of Fig. 5. No further changes in line shapes occurred
over the concentration range 1–5% SDS. These early observations showed
that nitroxides can be sensitive probes of lipid phase changes. Once
micelle formation occurs, the spin labels are solubilized. Increasing the
SDS concentration above this transition point increases the number of
micelles but does not alter the environment of the spin labels (see Chapter 1 for a brief discussion of phase changes and transition temperatures
in lipid–water systems).

The dinitrophenyl spin label XIV contains a chromophore at the opposite end of the spin label from the N—O group. This permitted independent measurements of solvent effects on both ends of the spin label.
Using solvent shift data and rotational correlation times calculated from
ESR spectra, it was concluded that solubilization is not consistent with
static models (in which the micelle is a *rigid* sphere) but is a dynamic
process (Waggoner *et al.*, 1967, 1968). The solubilized spin labels tumble
rapidly in a relatively polar time-average environment. SDS *micelles* are
apparently much more fluid structures than phospholipid *vesicles*. The
12-doxylstearic acid label XXVIII, for example, tumbles much more
rapidly in SDS micelles than in aqueous dispersions of egg lecithin at
the same temperature (Waggoner *et al.*, 1969).

Ohnishi *et al.* (1970) examined the nature of solubilization with a
dinitroxide similar to XVIII. This dinitroxide also tumbled rapidly in
the micellar solution and exchanged between the water and hydrocarbon
environments. Partition coefficients were calculated from the ESR
spectra. The dynamic character of the micellar solubilization appears to
be general, supporting the idea that SDS micelles have an extremely
fluid structure.

B. Phospholipid Vesicles (Liposomes)

Aqueous dispersions of most naturally occurring phospholipids spontaneously form myelin figures with bimolecular phospholipid lamellae.
The size of the vesicles, hence the number of apparently concentric
lamellae, is dependent on the method of preparation, but the bimolecular
lamellae are characteristically seen. An example of a briefly sonicated
lecithin–water dispersion is shown in Fig. 11. Such preparations were

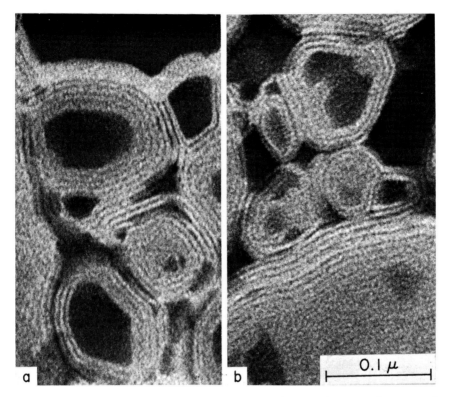

Fig. 11. Electron micrographs of briefly sonicated 5 weight % egg lecithin and water mixtures, negatively stained with 0.5% sodium phosphotungstate (pH 6.5). (a) Without added spin label; (b) parallel preparation containing 150:1 mole ratio of lecithin:12-doxylstearic acid (XXVIII). The structures are representative and were characteristically seen in all fields. Micrographs taken by Mr. John B. Lamb. (Approx. ×280,000.)

first characterized with electron microscopy by Bangham and Horne (1964) and used initially in a study of the diffusion of univalent ions (Bangham et al., 1965). Since then a number of studies have exploited the molecular organization of the liposome to investigate problems relevant to biological membranes [e.g., see Thompson and Henn (1970) for a recent discussion of experimental phospholipid model membranes].

1. Spin-Labeled Phospholipid Vesicles

The introduction of the spin-label steroid XXII (Keana et al., 1967) and the fatty acid spin label XXVIII (Keith et al., 1968; Waggoner et al., 1969) has provided classes of spin labels that can be associated with the lipids in the liposomes. The information in the ESR spectrum complements that from other physical techniques and provides important information about molecular motion and organization in these model membranes.

The usual procedure is to prepare a mixture of lipid spin label and phospholipid (at molar ratios of from 1:150 to 1:500), which is then dispersed in water or aqueous salt solution by agitation or ultrasonication. Physically the presence of the spin label does not seem to alter the gross structural organization of the vesicles. Figure 11 compares negatively stained lecithin–water dispersions in which there is no added spin label (a) to a similar dispersion (b) containing a 150:1 molar ratio of lecithin: 12-doxylstearic acid (XXVIII). Similar results are obtained with equivalent concentrations of a steroid spin label (XXII).

It seems reasonable, both intuitively and from the experimental evidence accumulating in a variety of experiments that will be briefly discussed, that the lipid spin labels are intercalated between the phospholipids and intimately associated with the acyl side chains. Thus the ESR spectrum is influenced by the structure and motion of the ordered phospholipid environment of the lipid spin label.

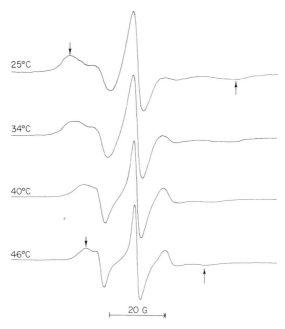

Fig. 12. First-derivative ESR spectra (9.5 GHz) from 5-doxylstearic acid (XXVI) in 20 weight % dipalmitoylphosphatidylcholine, dispersed in water by sonication at 45°–50°C. All spectra are from the same sample, recorded at the temperatures indicated. The arrows indicate the two outside lines, whose positions and amplitude are affected by anisotropic (restricted) molecular motion.

Various positional isomers of the doxylstearic or doxylpalmitic acids have been synthesized (Hubbell and McConnell, 1969b; Seelig, 1970; Jost et al., 1971; Hubbell and McConnell, 1971), as seen in XXV–XXX. The ESR spectrum arises only from the nitroxide moiety, rigidly bonded to the flexible hydrocarbon tail of the fatty acids through a spirane structure. Moving this moiety along the lipid chain allows examination of different regions within the bilayer. The hindered motion of 5-doxylstearic acid (XXVI) in liposomes of the synthetic saturated phospholipid, dipalmitoylphosphatidylcholine, is readily seen in the spectra of Fig. 12. The position and broadness of the outer lines of the top spectrum (indicated by arrows) are interpreted as indicating moderately strong immobilization with some anisotropic motion. Raising the temperature increases the mobility of the probe, as shown in the other spectra of Fig. 12. A semiquantitative treatment of such motion, when the ESR spectrum displays a well-defined high field line, may be found in the recent review of McConnell and McFarland (1970).

2. Effects of Cholesterol

Cholesterol occurs in many cell membranes in appreciable quantities. Its role in membranes is not well understood, although phospholipid–cholesterol interaction has been investigated in a variety of model systems. The presence of cholesterol abolishes the sharp thermal transition (Ladbrooke et al., 1968) in dipalmitoylphosphatidylcholine–water systems. This, together with nuclear magnetic resonance studies (Chapman and Penkett, 1966), has led to a suggestion that the presence of cholesterol keeps the phospholipid chain motion within narrower limits as the temperature varies. Below the transition temperature, cholesterol may increase the fluidity while above the normal transition temperature chain motion would be more restricted (Chapman and Wallach, 1968).

The ESR spectrum is very sensitive to changes in motion, and the effect of adding cholesterol to egg lecithin liposomes containing the 12-doxylstearic acid XXVIII is evident when the two first-derivative spectra in the left column of Fig. 13 are compared. The intensities of the first-derivative spectra have been normalized to reflect the same concentrations of label (i.e., $I_A = I_B$); thus, the amplitude differences are significant as well as the appearance of the lines indicated by arrows. In these liposomes, which are above the transition temperature of the phospholipids, the addition of cholesterol clearly reduces the molecular motion of the spin label as shown by the appearance of the outermost lines in Fig. 13 (Waggoner et al., 1969). The effect of cholesterol below the transition temperature of unsaturated phospholipids can readily be tested, but the

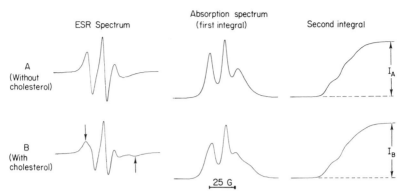

Fig. 13. The effect of cholesterol on the ESR spectrum (9.5 GHz) of spin-labeled liposomes. A, 12-Doxylstearic acid (XXVIII) and egg lecithin (20 weight %) dispersed in water by sonication (lecithin:spin label, 150:1 mole ratio). B, Similar preparation with added cholesterol (lecithin:cholesterol, 1:1 mole ratio). Arrows indicate the outermost lines, which are characteristic of more hindered motion of the spin label. The first-derivative spectra have been replotted, using a small computer, to reflect exactly the same concentration of the spin label, as determined by the values of the double integrals ($I_A = I_B$). See Section VI,C.

data are not yet available. The cholestane spin label XXII also has been used to detect a decrease in motion when cholesterol is present in liposomes (Hsia *et al.*, 1970a).

3. Transition Temperature

When an aqueous phospholipid dispersion is systematically subjected to changes in temperature a transition point is reached, detected by a marked endothermic change. The lipid chains "melt" and exhibit an increase in molecular motion. This phase transition, which is well below the melting point of the lipid, is primarily associated with the hydrocarbon chains which undergo a rapid increase in twisting and flexing (see Chapman and Wallach, 1968, for a brief discussion of the experimental evidence). For dipalmitoylphosphatidylcholine–water systems the transition temperature occurs at 41°C, as determined by differential scanning calorimetry (Ladbrooke *et al.*, 1968; and Chapter 2 of this volume).

The first use of spin labels to detect the physical change occurring at the transition temperature of dipalmitoylphosphatidylcholine (Barratt *et al.*, 1969) involved the use of the small dinitrophenyl nitroxide XIV. As the temperature of heated dipalmitoylphosphatidylcholine–water samples was decreased, a sharp decrease in the motion of the spin labels at the transition temperature was observed.

If doxylstearic acid (XXV–XXIX) do intercalate in the liposomes and faithfully reflect the motion of the surrounding hydrocarbon chains in the bilayers, these spin labels should reflect the change in motion associated with the transition temperature. Extracting a quantitative measure of anisotropic motion from ESR spectra has proven difficult and necessitates simplifying assumptions (see Section II,C,2). A *qualitative* relationship exists between line width and motion. Changes in line width are accompanied by correlated changes in line heights, so that simple internal ratios of measurable aspects of line heights can be used as a crude indication of changes in motion. For a wide variety of line shapes and a reasonable range of line widths, the heights b and c indicated on the spectrum in Fig. 14 are unambiguous. With rapid motion, the ratio c/b has a limiting value of 1.0, and in rigid glasses it approaches 2. It must be stressed that this simple ratio, however, is not a quantitative measure and its variation is undoubtedly influenced by both the degree of anisotropy and the nature of the motion. When the positional isomers 5-, 12-, and 16-doxylstearic acids are present in dipalmitoylphosphatidylcholine dispersions (dispersed by sonication at a temperature above the transition temperature), the ratio c/b varies with temperature. This temperature-dependent variation is shown in Fig. 14. When the nitroxide is near the hydrocarbon end of the spin label (16-doxylstearic acid) there is an abrupt shift in this ratio near the known transition temperature. Although this may be fortuitous, the simplest interpretation is that an abrupt

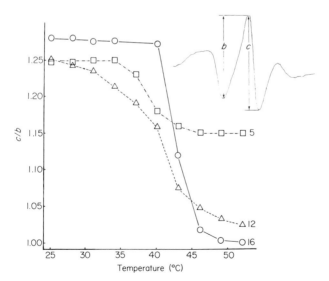

Fig. 14. Variation with temperature of a crude measure of motion (ratio of amplitudes c/b, measured as shown) for the 5-, 12-, and 16-doxylstearic acids (XXVI, XXVIII, XXIX) in 20 weight % dipalmitoylphosphatidylcholine (DPC) (DPC:spin label, 150:1 mole ratio, dispersed in water by sonication above 45°C). Four of the spectra used are shown in Fig. 12.

change has occurred in the motion of the local environment of the spin label. When the nitroxide is nearer the polar end of the spin label, the range is broader and appears to begin at somewhat lower temperatures. Differences in ESR line shapes among the three lipid spin labels in lecithin are much greater than the line shape differences within each temperature series. For this reason absolute magnitudes of c/b should not be used to compare relative motion *among* the three series.

The physical reality underlying these variations cannot yet be interpreted quantitatively with confidence, but the differences in motion seen along the length of the spin-label molecule do tend to support the intuitive inference that the spin labels are intercalated in the lipid bilayer region with the polar end of the acid at the lamellar interface. The line shapes of the ESR spectra indicate that the motion near the polar regions is hindered, while the terminal regions of the hydrocarbon chains are more mobile.

4. Molecular Motion

Molecular motion in dispersions of egg lecithin–cholesterol (2:1 mole ratio) has been examined by Hubbell and McConnell (1971). A series of doxylstearic or doxylpalmitic acids [with the doxyl group at positions 4, 5, 8, and 12 (see XXV, XXVI, XXVIII)] were used to examine the relative motion of small regions along the lipid chains. With cholesterol present, the outer lines (separated by $2A_\parallel$) can be determined at room temperature if the doxyl group is located from C-4 to about C-12 on the long-chain fatty acids. This spectral feature permits an estimation of the $A_\parallel$, and thus the order parameter S (see Section II,C) can be estimated. When S is plotted against the position of the nitroxide moiety there is an exponential decrease in the value of S as the doxyl group is translated along the chain from C-4 to C-12. Although the shape of the plot differs slightly, a similar dependence of the value of S on the doxyl position is seen when the fatty acid spin labels are acylated to lysophosphatidyl-choline to form spin-labeled phosphatidylcholine before incorporating them into the liposomes. These spin-labeled phospholipids were also examined in dispersions of dipalmitoylphosphatidylcholine and the spin label nearest the terminal methylene groups (C-12 in this study) appeared to undergo the largest relative change in motion (as measured by S) as the temperature was raised through the transition point.

Spectra were simulated, based on an assumption of anisotropic molecular motion, and showed good qualitative agreement with the experimental results. The dependence of the order parameter, S, on the position of the doxyl moiety was discussed in terms of a simple molecular model based on probabilities of various conformations of the spin label. Hubbell and McConnell interpreted their results as indicating that the lipid chains of egg lecithin liposomes (in the presence of cholesterol) are relatively "rigid rods" up to about C-8, with greater motion farther along the chains.

5. Dynamic Structure of Liposomes

Transport of molecules or ions across lipid bilayer regions must require some movement of the lipids, either laterally or by flipping from one side of the bilayer to the other. An interesting new use of spin labeling has shown that phospholipid molecules *can* invert from one side of the bilayer to the other in liposomes, and the rate of exchange between the two sides of the bilayer has been determined (Kornberg and McConnell, 1971).

Liposomes of egg lecithin were prepared by prolonged sonication and subsequent centrifugation so that the final dispersion in aqueous buffer contained homogeneous small vesicles bounded by a single bilayer. Exposure to dilute aqueous solutions of ascorbic acid rapidly destroys the paramagnetism of nitroxides accessible to the aqueous phase (Hubbell and McConnell, 1969a). When unlabeled vesicles were formed with the small spin-labeled choline XV trapped inside, and the excess spin label was removed, very little signal was destroyed by ascorbate. This indicated that (under the conditions employed) the vesicles were impermeable to the ascorbate.

Vesicles were then prepared that contained 1 mole % of the phospholipid nitroxide XXXI, which has the nitroxide at the polar end of the molecule. This gave liposomes with the spin label randomly distributed between the two sides of the single bilayer bounding the vesicle. Treatment of these liposomes (spin labeled in the bilayer rather than in the internal aqueous compartment) with ascorbate rapidly destroys part of the ESR signal. This initial rapid signal loss appears to be confined to the spin-labeled lecithin in the external monolayer of the bilayer. The initial rapid signal loss is followed by a slower exponential loss of the signal with a half-time of 6.5 hours at 30°C. This slow signal loss is not accompanied by detectable changes in line shape, although there are line-shape changes associated with the original rapid loss of signal. Control experiments exclude a number of trivial explanations. The combined evidence from the line-shape and intensity changes support the conclusion that vesicles formed with the spin-labeled phospholipid in both sides of the bilayer are rapidly converted, by exposure to ascorbate, to vesicles that contain unreduced spin label only in the internal monolayer. The inner and outer monolayers apparently transversely exchange phospholipid molecules (designated phospholipid "flip-flop" by Kornberg and McConnell) at a steady and measurable rate. This approach offers intriguing possibilities for investigating the possible involvement of phospholipid translocation in the permeation of ions and small molecules across lipid bilayers.

C. Oriented Liquid Crystals

Studies of liquid crystals were among the first applications of nitroxides and dinitroxides. It is well known that molecules such as p-azoxyanisole (PAA) can form phases with properties intermediate between liquids and solids. For example, PAA melts at 118°C forming a liquid crystalline phase which exists in the temperature range 118°–135°C. At 135°C a second phase transition occurs and the solution becomes a normal liquid (i.e., isotropic solution). In a magnetic field the liquid crystalline domains are oriented so that the long axes of PAA are aligned along the magnetic field. Falle et al. (1966) examined liquid crystals of p-azoxyanisole using the dinitroxide XX and a dinitroxide closely related to XIX. The ESR spectra clearly showed that these dinitroxides oriented in the liquid crystalline phase (nematic mesophase). Furthermore, the ESR spectra were sensitive to the nematic–isotropic phase transition. Electron–electron dipolar and exchange interactions provided information regarding the nature of the orientation. Other interesting studies on similar liquid crystals include those of Ferruti et al. (1969) and Corvaja et al. (1970). These liquid crystals do not closely resemble lipid phases. Their importance here lies in the development of methods of obtaining orientation information from the ESR spectra. For example, Ferruti et al. (1969) introduced the use of order parameters in the analysis of ESR data (see Section II,C).

A liquid crystal of more direct relevance to membrane structure has been examined by Seelig (1970). The smectic liquid crystal consisted of a mixture of sodium decanoate, decanol, and water (28:42:30, by weight) and was oriented by shear. This liquid crystal had previously been shown by x-ray crystallography to have a stacked bilayer structure (Mandell et al., 1967). Seelig found that the steroids XXIII and XXIV and a series of long-chain fatty acids with the doxyl nitroxide at positions 4, 5, 8, and 12 were oriented by the bilayers of saturated lipids in the liquid crystals, an observation consistent with intercalation of the spin labels into the bilayer structure. The results indicated rapid anisotropic motion of the oriented spin labels. The quantitative differences between the splittings of the various doxyl fatty acids was marked. For the series of fatty acid spin labels, the degree of ordering was found to decrease with the distance between the carboxyl group and the nitroxide moiety. The ESR spectra were not complicated by the variety of line shapes associated with phospholipid multilayer systems and biological membranes, which simplified determination of values for use in calculating an order parameter.

D. Lipid Multilayers

Natural lecithins dried down out of water, chloroform–methanol, or even chloroform alone tend to orient spontaneously. The resulting lipid films are physically thick, oriented, multilamellar structures. The orientation in these multilayer preparations has been demonstrated by spin labeling (Libertini et al., 1969; Hsia et al., 1970b). The lamellar structure as well as the orientation of the lipid chains has also been studied by x-ray diffraction (Levine and Wilkins, 1971).

1. Orientation

In the first spin labeling study of multilayers, two lipid spin labels, 12-doxylstearic acid and 3-doxyl-5α-cholestane (XXVIII, XXII) were shown to orient in egg lecithin multilayers with the long axis of the lipid molecule perpendicular to the surface supporting the multilayer samples (Libertini et al., 1969). The idealized orientation is shown, using molecular models, in Fig. 15. The z axis of the nitroxide group is parallel to the extended hydrocarbon chain of the fatty acid spin label (A in Fig. 15) and nearly perpendicular to the long axis of the steroid spin label (B in Fig. 15). The splittings showed marked changes depending upon the orientation of the multilayer in the laboratory magnetic field. With the fatty acid spin label present the large splitting was associated with the perpendicular orientation; with the cholestane spin label present, the large splitting was obtained at the parallel orientation. These results indicated that the spin labels tend to orient with the long molecular axes perpendicular to the plane of the multilayer and provide strong evidence for some degree of overall order within the lipids. This ordering is also supported by the results of the x-ray diffraction studies (Levine and Wilkins, 1971).

2. Motion Along Lipid Chains

The multilayer preparation is a model system which is especially favorable for examination of relative motion along the lipid chains. Because of the internal check on orientation, the molecular association of the spin label with the phospholipids is unambiguous; by varying the position of the nitroxide moiety along the length of the stearic acid chain, the relative orientation and motion of successive small regions within the bilayer have been examined at increasing distances from the lamellar plane that is formed by the polar end of the phospholipids (Jost et al., 1971). In this

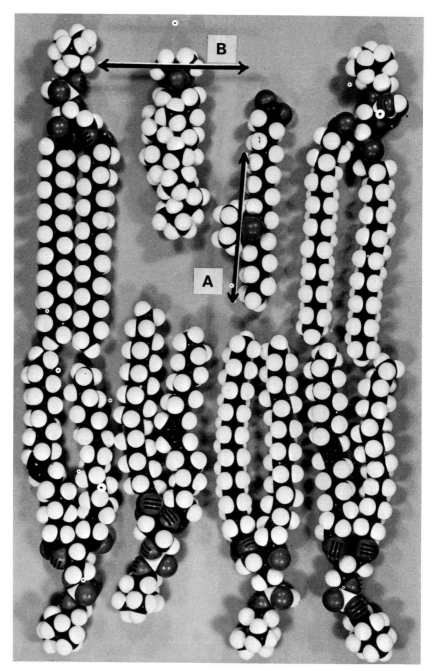

Fig. 15. Space-filling molecular models arranged to approximate the molecular arrangement in a hypothetical bilayer region, showing the probable orientation of two classes of lipid labels: A, Fatty acids; B, steroids. The arrows are along the z axes (direction of maximum splitting) of the nitroxides (12-doxylstearic acid and 3-doxyl-5α-cholestane).

study doxylstearic acids labeled at the 5, 7, 12, and 16 carbons were used. The pairs of spectra shown in Fig. 16 are for multilayers containing these four positional isomers. In each case the spectra from the perpendicular (solid line) and parallel (dotted line) orientations are superimposed. By inspection, it can readily be seen that the degree of anisotropy in the motion changed appreciably along the length of the molecule. The large splitting decreases systematically from the top spectrum to the bottom spectrum. The lines also become progressively narrower in this series. This behavior was maintained over a relatively wide range of temperatures.

Successive degrees of dehydration decreased the motion somewhat, especially near the polar end. Line-shape changes associated with this restriction of motion and a decrease in orientation were simulated for all the doxylstearic acids by assuming a restricted random walk model and a gaussian distribution of orientations about which the motion takes place. The simulations were in good agreement with the experimental spectra.

The very narrow lines of the spectra from the 16-doxylstearic acid, which were relatively uninfluenced by either the state of hydration or the temperature variation from 0°–33°C (this range is above the transition temperature of egg lecithin), are indicative of very large amplitude rapid motion as compared to the other three positional isomers. Relatively speaking, the polar regions up to at least C-12 appear to be anchored and restricted in the range of motion. The terminal region, as represented by C-16, undergoes rapid large-amplitude motion. This is consistent with a highly fluid region involving the terminal segments of the lipid chains near the center of the bilayer.

Inclusion of 50 mole % cholesterol in these multilayers slowed the motion and increased the ordering as expected (Jost and Griffith, 1971b). Preliminary experiments have indicated that the largest relative change is seen at position 12. One interpretation of this observation is that near the polar ends, motion is already so restricted that cholesterol effects on motion are more moderate than farther along the chain; the fluid internal region near the center of the bilayer is affected somewhat, but considerably less than the middle segment of the lipid chain.

Phospholipid multilayers prepared in a slightly different manner (evaporation from chloroform solution onto a quartz surface) have been examined by Hsia et al. (1970b), who incorporated the 3-doxyl-5α-cholestane spin label into egg lecithin containing 10 mole % dicetyl phosphate.

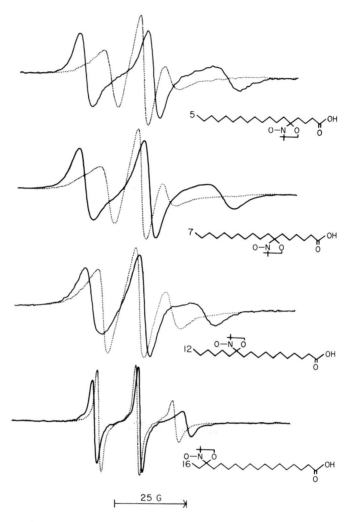

Fig. 16. The ESR spectra (9.5 GHz) of 5-, 7-, 12-, and 16-doxylstearic acids (XXVI–XXIX) in lecithin multilayers formed by drying down aqueous dispersions of 10 weight % lecithin and fatty acid spin label (150:1 mole ratio) on glass slides. The slides were hydrated by equilibration at 81% relative humidity and recorded at room temperature, with the external magnetic field perpendicular (solid line) and parallel (dotted line) to the plane of the glass slide bearing the sample. The spectra have been replotted so that the height of the center peak of all spectra from the parallel orientation are equal. The scale of the spectra from the perpendicular orientation is normalized to the other principal orientation in each pair (see Section VI,C). From Jost et al. (1971).

Dry lecithin films that had previously lacked orientation were found to be oriented after hydration. Orientation without rapid motion about the long axis was seen in unhydrated films containing a 2:1 molar ratio of cholesterol to phospholipid, but fast anisotropic rotation of the spin label occurred on hydration of the films. At this concentration of cholesterol, two phases consisting of phospholipid–cholesterol regions and cholesterol crystals may be present (Bourgès et al., 1967; Lecuyer and Dervichian, 1969), and this complicates comparisons with other multilayer data. However, the presence of water is clearly related to increasing the motion of the probe. This effect of hydration on motion in multilayers has also been reported (Long et al., 1970) using the 12-doxylstearic acid.

Cholesterol increases the spectral anisotropy, but it has been reported that cholesterol derivatives with substituents at the C-3 position do not produce this effect. Added cholesterol increases the order of multilayers of beef brain lipids and cholesterol-free erythrocyte lipids (Butler et al., 1970b), as indicated by the 3-doxyl-5α-cholestane spin label XXII. The maximum ordering effect as measured by the difference in splittings at the two principal orientations was obtained with 20–25 mole % cholesterol. Adding various other (unlabeled) sterols, a qualitative evaluation of the structural requirements was made for this effect. The 3β-hydroxyl group was necessary for the ordering effect and the 3α-configuration was not a substitute. The presence of the hydrocarbon tail on cholesterol was reported to be of less importance in ordering, while the presence or absence of the cholesterol C-5 double bond was stated to make little difference. No quantitative data were reported, but it was concluded that the essential requirements for increasing order in the lipid films is the planar steroid nucleus and the 3β-hydroxyl group.

Lipids from bovine brain white matter were oriented in multilayers (Butler et al., 1970a) by evaporation from chloroform onto quartz. Subsequent hydration was accomplished by brief soaking in various salt solutions. The ratio of the line heights of the first two lines from spectra obtained when the plane of the film was parallel to the magnetic field was used as a measure of the ordering. Using this criterion, the conclusion was that divalent cations were more effective than univalent cations in ordering the lipid films. The ratio of line heights also varied with concentration of the cations present, increasing as the cation concentration was increased, up to a limiting value. No quantitative data were reported except the line height ratios, but the effect appears to vary systematically with concentration for $LaCl_3$, $CaCl_2$, and NaCl.

The phospholipid multilayer system can also be useful in examining changes in lipid structure and motion as a result of chemical treatment, such as those involved in staining for electron microscopy (Jost and Griffith, 1971c). Exposure of spin-labeled lecithin multilayers to osmium tetroxide vapor rapidly blackened the translucent lipid film. When the ESR spectrum was examined, essentially all anisotropy had been abolished and the motion of the lipid spin labels was severely reduced. This effect is illustrated in Fig. 17, in which the spectra in the top row were recorded before and after exposure to osmium tetroxide (A and B, respectively). Spectra C and D are computer simulations based on a simple model of anisotropic molecular motion and a gaussian distribution of orientations (Libertini *et al.*, 1969; Jost *et al.*, 1971). The schematic drawing at the bottom of Fig. 17 suggests a plausible interpretation of the ESR results. Before treatment, the spin labels were oriented and exhibited anisotropic motion that varied along the chain. After treatment, the spin labels were no longer oriented and their motion was severely restricted. This undoubtedly reflects the loss of order and reduction in motion of the surrounding phospholipids.

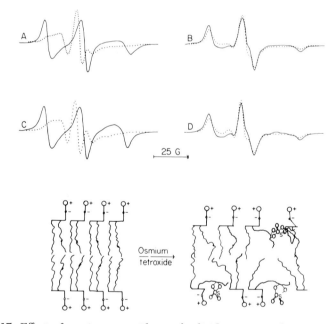

Fig. 17. Effect of osmium tetroxide on the lipid structure of an oriented model membrane system. A, ESR spectra of 5-doxylstearic acid (XXVI) in an oriented egg lecithin multilayer (similar to the preparations in Fig. 16) with the supporting slide perpendicular (solid line) and parallel (dotted line) to the magnetic field; B, the same sample after 4 minute exposure to osmium tetroxide vapors; C and D, computer simulations of A and B, respectively. The spectra are consistent with the lipid structures illustrated schematically at the bottom of the figure.

E. Lipid–Protein Complexes

Lipid–protein complexes have been used as a model system for lipid–protein interactions occurring in membranes. Various proteins, otherwise insoluble in organic solvents, become soluble after the protein is associated with appropriate phospholipids. The first use of such a model system with spin labeling was reported by Barratt et al. (1968). The amino groups in several basic proteins (that lacked sulfhydryl groups) were labeled by reacting the maleimide spin label IX with the protein and then associating the spin-labeled protein with a defined phospholipid. A modest restriction of motion of the highly mobile spin label was observed after formation of the water-insoluble lipid–protein complexes. Solubilization of the complexes in isooctane further restricted the motion. The tentative conclusion was that the changes in motion are due to binding of the phospholipid in the regions of the protein that have reacted with the maleimide spin labels.

Reconstitution of lipoprotein complexes from separated lipid and protein of erythrocyte membranes has been studied by Berger et al. (1970), using both spin-labeled protein (labeled with VI or VII) and lipid spin labels (XXVIII, XXXII) in separate experiments. Composite spectra were obtained from simple mixtures of membrane lipid dispersions and the spin-labeled membrane apoprotein (i.e., treated with neuraminidase to remove sialic acid before labeling). However, when the labeled apoprotein and lipids were mixed under conditions reported to result in lipid–protein recombination (Zwaal and van Deenen, 1971), the broad component of the spectra was much more pronounced. Formation of lipid–protein complexes was interpreted as responsible for lowering the mobility of the spin labels bonded to the protein.

Rottem et al. (1970) reported the results of reaggregation of *Mycoplasma* membrane proteins and membrane lipids in the presence of 5-doxylstearic acid (XXVI) with several different $MgCl_2$ concentrations (a variable that alters the protein/lipid ratio in the aggregate). With increasing $MgCl_2$ concentrations, an increase in the separation of the two outer lines was observed in the moderately immobilized ESR spectrum from the aggregates as the protein/lipid ratio in the aggregates increased. This separation finally attained the value observed when 5-doxylstearic acid is diffused into the native membranes, which have a protein/lipid ratio similar to that seen in the aggregates found at high $MgCl_2$ concentrations. This evidence, combined with the results for native membranes in the same series of experiments (to be discussed in Section V,D), suggests an influence of the protein component on lipid mobility.

Human serum lipoproteins have been spin labeled with maleimide nitroxide (VI) to study the effect of the lipid on the mobility of the protein (Gotto et al., 1970). Spin labeling under the conditions employed did not alter the solubilities, circular dichroism, or immunochemical properties of the preparations. The spin-labeled serum lipoproteins gave composite ESR spectra that included a strongly immobilized component, which was diminished by removal of lipid from the labeled lipoproteins. The broad component of the spectrum was not restored when the delipidized protein was recombined with phospholipid. The conclusion was that the strongly immobilized component of the signal from the spin-labeled lipoprotein may be due to a specific type of lipid–protein interaction that was not restored in reconstituting the delipidized lipoprotein with phospholipid.

V. APPLICATIONS TO BIOLOGICAL MEMBRANES

The applications of spin labeling to biological membranes are becoming more numerous, although the data are sometimes difficult to interpret. In this section, results obtained from spin labeling of membranes are summarized. A review of the growing literature of spin label studies on conformational changes in proteins and nucleic acids, geometry of enzyme active sites and antibody combining sites, and elucidation of enzyme mechanisms has been omitted. Recent reviews (McConnell and McFarland, 1970; Smith, 1971) should be consulted for these studies, many of which involve approaches applicable to intact membranes if sufficient specificity of spin labeling is achieved.

A. Solvent Effects

The dependence of A and g values on the polarity of the chemical environment of the nitroxide (discussed in Section II,D) is potentially useful in studies of membranes. At 9.5 GHz only the high field lines are resolved in the composite spectrum obtained when a small nitroxide is distributed between two sufficiently different environments (see Fig. 7), and this has already found an interesting application in membrane studies. Nitroxides with appropriate solubility properties partition between the aqueous and lipid regions of wet biological samples, as shown in Fig. 18. The ratio of the intensities of the high field lines is proportional to the relative amounts of nitroxide in the two environments. The partitioning of a small nitroxide (II) in wet nerve and in aqueous dispersions of soybean phosphatides (Hubbell and McConnell, 1968) is altered by the addition of several compounds. For example, the relative signal height from the less polar regions (i.e., the height of line A relative to line B in Fig. 18a) of the rabbit vagus nerve was increased by addition of the local anesthetic, tetracaine. Figure 18b is the spectrum obtained when 2-doxylheptane (XII) is added to a wet centrifuge pellet of bovine retinal rod outer segments (Waggoner and Stryer, 1971). The common features of spectra (a) and (b) of Fig. 18 suggest a similar hydrophobic environment in the two membrane preparations.

The phenomenon of changing relative heights of the high field lines with various treatments of the sample can result from several effects. A trivial effect is simply a change in the relative amounts of aqueous phase and membrane present. Other effects include changes in molecular motion, changes in composition that affect solubility or cause small shifts in a_0 and g_0 values, changes in oxygen concentration, or combinations of

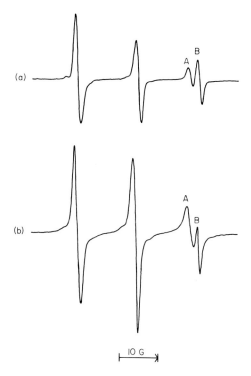

Fig. 18. Partitioning of suitable nitroxides between aqueous and hydrocarbon environments. (a) ESR spectrum of the small nitroxide (II) in a wet rabbit vagus nerve (after Hubbell and McConnell, 1968); (b) 2-doxylheptane (XII) in wet centrifuge pellet of bovine retinal rod outer segments (Waggoner and Stryer, 1971). The two parts of the high field line result from the nitroxides tumbling in hydrophobic regions (A) and aqueous regions (B).

several of these. The development of small-computer applications to ESR (see Section VI) makes it possible to distinguish between some of the effects.

It has been shown that resolution is greatly increased at 35 GHz (Griffith *et al.*, 1971). At 9.5 GHz, for example, the spectrum of di-*t*-butyl nitroxide (I) in a crude beef brain myelin preparation is very similar to those of Fig. 18. However, at 35 GHz all six lines are resolved, allowing the measurement of the important a_0 and g_0 parameters of the nitroxide in the membrane. The measured values are $a_0 = 14.6$ G and $g_0 = 2.0061$. The same nitroxide in hexane yields $a_0 = 14.8$ G and $g_0 = 2.0061$. The numbers are essentially the same, and molecule I diffused into myelin is clearly in a hydrocarbon environment. This approach can be used to characterize the response of the nonpolar regions of membrane preparations to various treatments (e.g., drugs).

B. Biosynthetic Incorporation of Spin Labels

When the methyl 12-doxylstearate was present in the media during growth of *Neurospora crassa*, ESR signal was found in the polar lipid fraction of lipids extracted from the mitochondria (Keith *et al.*, 1968). This, along with control experiments, strongly suggested that this spin-labeled fatty acid can be enzymatically acylated to the glycerol moiety in phospholipids (such as XXXIII). A similar result, i.e., recovery of ESR signal from the polar lipid fraction, was obtained when 12-doxylstearic acid was added to the growth media of *Mycoplasma laidlawii* (Tourtellotte *et al.*, 1970), and the membranes were then isolated and subjected to lipid extraction and separation. *Mycoplasma* readily esterifies exogenous fatty acids into phospholipids and glycolipids. Membranes from cells grown in the presence of stearic acid differ from those grown in the presence of oleic acid, both in lipid composition (McElhaney and Tourtellotte, 1969) and in thermal transitions of the membrane lipids (Steim *et al.*, 1969). When *Mycoplasma* took up the spin-labeled stearic acid in the presence of oleic acid supplements, the ESR spectra (recorded from samples at temperatures between 20° and 60°C) indicated a mobility of the spin label that was somewhat higher than in cells grown in the presence of stearic acid supplements. Spectra obtained from aqueous dispersions of the extracted lipids showed slightly more mobility of the label than spectra from the intact cells, when the two samples were at the same temperature. The lipid mobility appears to be altered slightly by association with protein in the membrane, but these results cannot be easily evaluated since thermally denatured cells gave the same results as the fresh cells. Motion of lipids in *Mycoplasma* membranes labeled by diffusion with doxylstearic acids (Rottem *et al.*, 1970) is discussed in Section V,D.

Mitochondria labeled by biosynthetic incorporation of the methyl ester of 12-doxylstearic acid have been compared with mitochondria labeled with the same methyl ester by diffusion (Keith *et al.*, 1970). The two types of preparation gave very similar spectra but there were qualitative differences in motion after lyophilization of the samples, with a decrease in motion seen only in the spectra from mitochondria labeled by diffusion. There are difficulties in using the biosynthetic approach to spin labeling. Analysis of labeled membrane preparations shows that the spin-label lipid is present in the neutral and free fatty acid fraction as well as the phospholipid fraction (Keith *et al.*, 1968). The ESR spectrum from such membranes is the sum of the various fractions, and this introduces ambiguities. On the other hand, a sizable part of the signal appears to be from spin-labeled phospholipids in the most favorable preparations. The observation that the spin label is taken up by the cell and incorporated into phospholipids provides some evidence that the bulky spin label can serve as a substitute for a normal fatty acid. The resulting ESR spectrum should contain useful information about membrane structure and function. The location of fatty acids diffused into intact membranes may not be the same as the location of the biosynthetically incorporated spin labels.

C. Spin-Labeled Proteins

Membrane proteins can be covalently labeled with some spin labels under physiological or near physiological conditions. In spite of the ease of labeling, it is not easy to obtain useful information from the ESR spectra of such complex preparations.

Drug–membrane interactions have been studied in red cell membranes spin labeled with the maleimide nitroxide VI (Sandberg and Piette, 1968). The maleimide bonded primarily to sulfhydryl groups, as judged by blocking experiments with N-ethylmaleimide. The composite ESR spectrum obtained indicated both weakly and strongly immobilized spin labels. The effect of chlorpromazine was to decrease the weakly immobilized component of the spectrum. The small, but reversible, drug-induced changes in the ESR spectrum may be related to surface or conformational changes in the membrane.

The effects of phenothiazine derivatives on the ESR spectra of spin-labeled bovine red cell ghosts have been extended by Holmes and Piette (1970). The red cell ghosts were labeled with a spin-label analog of iodoacetamide (similar to X). This labeled fewer sites (apparently sulfhydryl groups) than the maleimide label used in the earlier study. The ESR spectrum indicated only moderate slowing of the spin label, and it did not appear to contain a strongly immobilized component, in contrast to the maleimide-labeled membranes. Addition of chlorpromazine to the spin-labeled membranes produced a composite spectrum with a bound component. Various phenothiazine derivatives were compared for their ability to produce this spectral change.

Chapman et al. (1969) also observed a similar composite spectrum when they labeled human red cell ghosts with a spin-label maleimide (IX). The effect of pH on the motion of the spin labels was monitored by using the relative intensity of the narrow component of the low field line. The minimum intensity of this line corresponded to the isoelectric region for the membrane protein. Similar preparations of spin-labeled human red cell ghosts have been used to monitor procedures in isolating red cell protein fractions by Schneider and Smith (1970). Similarities in the ESR spectra before and after solubilization were suggested as a possible criterion for judging the maintenance of structural integrity of solubilized membrane protein fractions.

Fragmented sarcoplasmic reticulum, spin-labeled with either an iodo-acetamide label (similar to X) or isothiocyanate nitroxide (XI), retains the property of ATP-dependent uptake of Ca^{++}, accompanied by ATP hydrolysis (Landgraf and Inesi, 1969). The ESR spectra of membranes labeled with either spin label were similar, and they contained both a prominent broad component and a small narrow component character-istic of composite spectra. Addition of ATP increased the height of the narrow component of the spectrum from the iodoacetamide spin-labeled membranes. The isothiocyanate spin-labeled membrane did not give this ATP-dependent spectral change. With some reservations, this reversible effect was interpreted by the authors as probably due to an ATP-dependent conformational change in the membrane in regions near the site of the iodoacetamide spin label.

These studies, although useful, point out the problem of using spin labels that are capable of reacting with a large number of different pro-teins. In principle, one way to circumvent this difficulty would be to take advantage of the specificity of substrates, coenzymes, or inhibitors for membrane-bound active sites. Spin labels with this kind of specificity have been developed. Spin-labeled haptens have been used in the study of antibody active sites (Stryer and Griffith, 1965; Hsia and Piette, 1969a,b) and substrate analogs have been bound to the active site of chymotrypsin (Berliner and McConnell, 1966; Kosman et al., 1969). The elegant study of the conformation and geometry of the active site of alcohol dehydrogenase (Mildvan and Weiner, 1969a,b) combined ESR and pulsed nuclear magnetic resonance and was based on the binding of a spin-labeled analog of nicotinamide adenine dinucleotide (XVII) to the active site.

This general approach has been applied to membranes by Morrisett *et al.* (1969), who took advantage of the known inactivation of acetylcholinesterase by organophosphonofluoridates. These reagents appear to be highly specific for the active sites of esterases. The spin-label analog VIII was synthesized. This label almost completely abolished the acetylcholinesterase activity of spin-labeled red cell membranes and rat brain nerve ending particles. On the basis of data obtained from control experiments, the authors concluded that the spin label did not appear to attach to other residues or other esterases in the membrane. These two spin-labeled membrane preparations, exhaustively washed to remove unreacted label, had ESR spectra characteristic of a weakly immobilized nitroxide. When the same reaction was carried out with α-chymotrypsin, the ESR spectrum was markedly broadened. The active site of α-chymotrypsin is known to be located in a deep cleft, extending into the interior of the molecule, and the authors suggested that the differences in mobility may be due to differences in the geometry of the active sites. The membrane-bound acetylcholinesterase could possess an active site that is well exposed on the periphery of the molecule. The chemical environment of the spin label could also affect motion, but this is a promising approach to labeling membranes.

D. Lipid Mobility and Orientation

1. Motion of Lipids

The motion of long-chain doxyl fatty acids in *Mycoplasma* membranes has been studied by Rottem *et al.* (1970). The measured parameter used to characterize relative motion was $2A_\parallel$ (designated $2T_m$ in this paper and measured between low and high field peaks in the same manner that $2A_{zz}$ is measured on the rigid glass spectrum of Fig. 4). 5-Doxylstearic acid (XXVI), used in most of the experiments, retains the high field peak through a wide temperature range in many kinds of lipid dispersions (for example, see Fig. 12 in which this spectral feature can be easily measured by recording at higher gain). In the presence of added unsaturated fatty acids (*cis*-Δ^9-octadecenoic) the fatty acid composition of the membrane lipids of *Mycoplasma laidlawii* was dependent on the age of the growing culture and the temperature during growth. Membranes with a higher degree of unsaturation of the lipids were prepared (from cells during early growth or from cells grown at lower temperatures) and compared with membranes containing a lower degree of unsaturated lipids. The spin label was diffused into membranes that had previously been frozen during the preparative procedure, and the excess label was removed by centrifugation. A comparison of the spectra from these spin-labeled membranes showed a difference of several gauss in the measured splitting, especially when the spectra were recorded at temperatures between 5° and 20°C. This indicates that the mobility of the label was measurably higher in membranes with a higher percentage of unsaturated lipids than in comparable membranes containing more saturated lipids.

The *cis* and *trans* isomers of Δ^9-octadecenoic acids differ in their physical properties. The *cis* isomer has a lower melting point, and it forms a more expanded monolayer at an air–water interface (Chapman *et al.*, 1966). In experiments with these two related fatty acids as supplements, the amount of *cis* or *trans* isomers incorporated into the membrane under appropriate growth conditions was about the same. These membranes were spin labeled by diffusion, and the measured splittings differed by as much as several gauss when the two membrane preparations were compared. This indicates that the fatty acid composition of the membrane lipids has a distinct effect on the mobility of the spin label. Membranes containing the *cis* isomer provided an environment in which the spin label was more mobile than when the *trans* isomer was present.

The relative mobilities of the 5-, 8-, and 12-doxylstearic acids diffused into membranes were compared, and there was a definite increase in mobility with the doxyl group farther from the carboxyl end of the spin label. This is similar to the findings in the model systems discussed in Section IV.

The experimental results discussed above support the conclusion that the spin labels reflect the mobility of the unlabeled lipids in the membranes. These observations are also compatible with an assumption that the structure of the hydrophobic regions associated with the spin labels has the properties of a bilayer. The authors emphasized, however, that the distribution of spin labels in the membrane is unknown and that no conclusions can be drawn about the extent or arrangement of the assumed bilayer structure in the membrane preparations.

2. Lipid Orientation

The detection and study of ordered lipid structures is experimentally more difficult in biological membranes than in model systems, both because of the ambiguities introduced by the complexity of the sample and because of the desirability of relatively large, physically oriented samples. However, progress in orienting samples has been made by Hubbell and McConnell (1969b), who spin labeled dog red cells with 3-doxyl-5α-androstan-17β-ol (XXIV) or 12-doxylstearic acid (XXVIII) by diffusing the lipid spin label into the cells. The labeled red cells were oriented by flow, and the ESR spectra showed splittings that were dependent on the orientation of the sample in the laboratory magnetic field. Figure 19 shows the spectra obtained using the 12-doxylstearic acid (the small narrow lines are due to free label in the aqueous solution). The spectral anisotropy can be considered as relatively marked. Not only is the orientation of the cells almost certainly not perfect, but a sizable fraction of the cell surface is not parallel to the faces of the biconcave disc. In spite of these experimental handicaps, the ordering of the lipid spin labels supports the interpretation that the red cells were oriented with respect to the magnetic field and that the hydrophobic regions accessible to the spin labels tend to be ordered. Assuming the most probable orientations of the cells, this ordered region orients the long axis of the spin labels perpendicular to the biconcave surface of the cell. Invertebrate nerve fibers similarly labeled with 5-doxyltricosanoic acid (XXX) or 5-doxylstearic acid (XXVI) also showed some anisotropy, with evidence that the long axis of the lipid label tends to orient perpendicular to the cylindrical surface of the nerve bundle. The evidence from both samples is compatible with the existence of ordered lipids such as in models that assume regions of bilayer organization or other oriented arrangements of lipids.

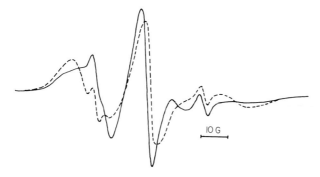

10 G

Fig. 19. The ESR spectra of 12-doxylstearic acid (XXVIII) in flow-oriented red cells with the magnetic field perpendicular (dashed line) and parallel (solid line) to the plane of the hydrodynamic shear. The sharp signals (at the isotropic splitting) are due to the spin label in the aqueous solution. (From Hubbell and McConnell, 1969b.)

VI. TECHNICAL INNOVATIONS: USE OF SMALL COMPUTERS

Methods for the analysis of the ESR spectrum are still being developed. Semiquantitative approaches, several of which have been briefly discussed in earlier sections, are being applied in analyzing data from spin labeling studies. Large central computers are usually used in simulating spectra to test the analytic approach. Further progress in the quantitative analysis of spectra can be anticipated in the next few years.

Another aspect of the problem concerns the availability of spectral information. An important development in this area is the use of small digital computers to collect, store, and treat the data. Some of the more tedious tasks of making measurements from spectra can be made less time-consuming. But more important, it is feasible to obtain otherwise inaccessible information by small-computer treatment of digitalized data.

This brief discussion of the use of the dedicated small computer is based on applications we have found useful in spin labeling studies. The examples selected are intended to illustrate a relatively simple range of applications, most of which are new in ESR. The versatility of this approach will stimulate many additional uses in future spin labeling experiments.

A. Time Averaging

Time averaging is probably the best known use of small computers. It was first used in ESR by Klein and Barton (1963). One example involving nitroxide spin labeling is illustrated in Fig. 20. The top trace represents a normal single-sweep ESR spectrum of rather poor quality (the spectrum was actually digitalized, placed in computer memory, and read out on a digital recorder, but the appearance is the same as if it had been displayed directly). The addition of successive sweeps to the same memory location and division of each point by the number of scans tends to average the random noise voltages, improving the signal to noise ratio. The improvement in signal to noise is proportional to $(n)^{1/2}$ where n is the number of scans. In Fig. 20 the second, third, and fourth traces from the top were recorded after 9, 19, and 40 scans, respectively. The improvement is noticeable even after 9 scans and the small lines stand out much better after 40 scans. The signal to noise ratio for the first scan is roughly 2.7 (see Fig. 20, top trace). After 40 scans the signal to noise ratio should be $2.7(40)^{1/2} = 17$. This is in good agreement with the experimental value of 18 measured from the bottom spectrum of Fig. 20. In principle, if the noise is completely random, the same signal to noise ratio can be achieved by averaging 40 scans with the computer or by scanning only once, but doing it 40 times slower with a long filter time constant. In practice, instrumental drift and occasional nonrandom laboratory noise make the computer approach preferable. Both approaches are, of course, last resorts since considerable time is required to

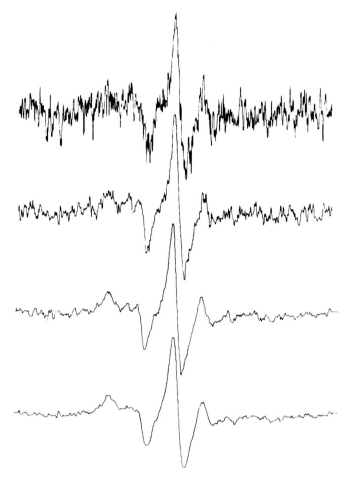

Fig. 20. Increasing the signal-to-noise ratio by computer time averaging. The computer-plotted output, from top to bottom, represents 1, 9, 19, and 40 repetitive scans. The sample was $3.75 \times 10^{-5}\,M$ sodium 5-doxylstearate (XXVI) in an aqueous solution containing 5 weight % egg lecithin at room temperature. The microwave power, modulation amplitude, scan time, scan range, and filter time constant are 5 mW, 1 G, 1.625 minutes, 100 G, and 0 seconds, respectively, with the gain setting constant.

achieve the final signal to noise ratio. (In this case $1.625 \times 40 = 65$ minutes. The system input routines perform boxcar integration on the data received. This technique automatically selects the optimal data filter, and during these scans the instrument filter time constant was turned off.) Wherever possible, the signal to noise ratio should be increased simply by introducing more label or by concentrating the sample.

B. Spectral Titration

Composite spectra are frequently encountered in spin labeling experiments. For example, when a spin-labeled fatty acid is diffused into an aqueous suspension of membranes (see Fig. 19) a complex spectrum is frequently obtained. The observed spectrum is a simple sum of the two spectra arising from equilibrium concentrations of free and bound spin label. The desirability of resolving a complex spectrum into its components is evident in analyzing the solvent effect (Section V,A and Fig. 18). Composite spectra are also encountered in antibody–antigen studies, in probing mechanisms of drug action, and in any system where the label exists in two or more environments. How can the composite spectra be separated into the basic components? This problem warrants special attention because of its universality. The most common source of difficulty is the sharp three-line spectrum of a freely moving nitroxide, which obscures important information contained in the broad spectrum of the bound spin label. This problem is readily solved using a small computer. In the top spectrum of Fig. 21 a typical composite spectrum has been simulated by placing two capillaries in the ESR cavity, one containing a bound nitroxide and the other containing a nitroxide freely tumbling in water. (A biological sample could have been used but since this was apparently the first time a small computer had been used in ESR to attack this problem, it was deemed advisable to use an example in which the individual bound and free spectra were available as controls.) The composite spectrum was first recorded, digitalized, and placed in a computer memory. Then the sharp three-line spectrum of the free nitroxide was recorded and placed in a second computer memory location. In practice, this is usually easy to do since it requires only that the spin label be dissolved in water or other appropriate solvent. The concentration is not important. With the computer it is possible to subtract small increments of the free spectrum from the composite spectrum until the free spectrum is entirely removed. This process is illustrated in Fig. 21, and the fourth spectrum from the top is a pure "bound" spectrum. If the subtraction is carried beyond the "end point" a small free spectrum of opposite phase is observed, as shown in the bottom spectrum of the figure. Thus, the computer can perform a "titration" of the complex spectrum and display both components at the end point.

C. Integration

The principal use of integration is in concentration determinations. For example, if the composite spectrum of Fig. 21 represented an equilibrium between free and bound spin label, it would be useful to determine the equilibrium constant. The first step in concentration determination is to obtain the free and bound spectra from the composite spectra, as discussed above. The next step is to integrate each spectrum twice and compare the results. Double integration is illustrated in Fig. 22. The first spectrum of the bottom row was obtained using two capillaries, one containing the bound nitroxide and the other containing the free nitroxide of the previous sample. The second and third spectra of the bottom row are the free and bound samples run separately. (Actually, a second capillary tube containing only water was run with each sample so that the cavity characteristics would be the same as for the composite spec-

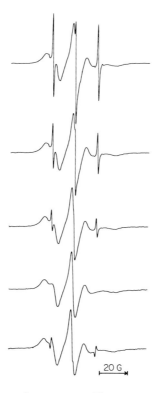

Fig. 21. Spectral subtraction by computer. The top spectrum is a room-temperature composite spectrum. Two capillary tubes were placed in the cavity, one containing an aqueous dispersion of $1.5 \times 10^{-3} M$ 5-doxylstearic acid (XXVI) in 20 weight % lecithin, the second capillary containing $5 \times 10^{-5} M$ ketone nitroxide (IV) in water. The remaining traces are from subtracting increasing amounts of the sharp three-line spectrum from the composite spectrum, with total subtraction occurring in the fourth spectrum. In the bottom spectrum, too much of the sharp three-line spectrum has been subtracted (note the three sharp lines of this spectrum are phase reversed). The microwave power, modulation amplitude, ESR scan time, scan range, and filter time constant are 5 mW, 1 G, 6.5 minutes, 100 G, and 0.3 seconds, respectively.

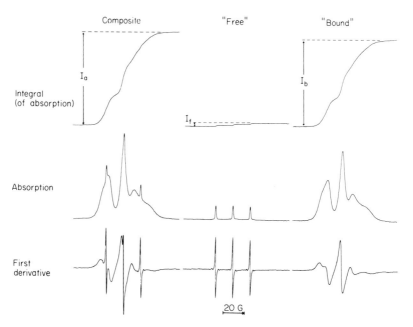

Fig. 22. Integration and double integration by computer. The three first-derivative spectra in the bottom row are, respectively, the composite spectrum, the aqueous sample, and the phospholipid dispersion of Fig. 21. Single integration of these ESR spectra yield the three absorption spectra shown. The corresponding double integrals of the first-derivative spectra are shown in the top row. Instrument settings are the same as in Fig. 21.

trum.) The three spectra were stored in different memory tables (locations) of the computer. Each spectrum was integrated once to generate the ESR absorption curve and again to obtain the area under the absorption curve. Figure 22 gives the results. The quantities I_c, I_f, and I_b are the areas of the composite, free, and bound nitroxide ESR absorption curves, respectively. These quantities should be proportional to the concentrations of nitroxide present. From Fig. 22, $I_c = 236$, $I_f = 7$, and $I_b = 222$, in arbitrary units. The actual concentrations of nitroxide in the two capillaries are $N_f = 5 \times 10^{-5} M$ and $N_b = 1.5 \times 10^{-3} M$. Note that I_c very nearly equals $I_f + I_b$. Furthermore, the ratio $I_b/I_f = 32$ is in good agreement with the known ratio of concentrations $N_b/N_f = 30$. It is also interesting to note that the small amount of free nitroxide present has a profound effect on the overall appearance of the composite spectrum, so that it is tempting to assume that a large proportion of the signal is in the sharp peaks. At equimolar concentrations, however, the bound signal is almost completely obscured by the signal from the freely moving component.

Experimental applications of integration were made in several of the earlier figures in this chapter. The two first-derivative spectra of Fig. 13 have been adjusted to yield the same value for the double integral, so that the amplitudes as well as shapes of the lines are meaningful in comparing the two experimental spectra. In Fig. 16, all spectra have been replotted from digitalized data. The spectra from one principal orientation (dotted line) have arbitrarily been scaled so that all center line heights are the same. In each pair of superimposed spectra the other corresponding principal orientation has been scaled so that the value of the double integral is the same for both orientations of the sample. Rotating an asymmetric sample holder in the cavity often changes the cavity characteristics and invalidates amplitude comparisons, except when such an adjustment has been made. In Fig. 16, appropriate use of quantitative changes in both the amplitudes and splittings are valid within each pair of spectra. Other examples of integration of the ESR spectrum can be seen in Fig. 1 (where differentation of a first-derivative spectrum is included) and Fig. 4.

D. Kinetics

Nitroxide free radicals are remarkably stable compared to most other free radicals. Signal losses, however, occur in the presence of certain oxidizing agents (e.g., potassium permanganate), reducing agents (e.g., ascorbic acid), and enzymes involved in redox reactions. The rate of signal disappearance can provide valuable information regarding the accessibility of the nitroxide. For example, if the nitroxide is buried in a membrane or attached in the hydrophobic cleft of a protein it may not be as readily reduced by water-soluble reducing agents as a nitroxide on the surface of the membrane or protein. To extract this kind of information, one might plot the central peak height of the disappearing nitroxide ESR signal. The implicit assumption is that the width or line shape does not change during the course of the reaction. This assumption is probably reasonable in most homogeneous chemical reactions. In the case of biological or other heterogeneous samples the assumption may not be valid. In any case it has been interesting to compare the kinetic behavior of the ESR peak height with the double integral in a case where there are no obvious line-width changes during the course of the experiment. In Fig. 23 the ESR peak height and double-integral decay curves are plotted for a small water-soluble nitroxide (IV) in the presence of freshly isolated cell membranes (*Mycoplasma laidlawii*) in buffer.

The nitroxide was added at $t = 0$, and the computer collected 300 points over a 10 G span covering the center peak of the sharp three-line spectrum; it then immediately dumped the data onto paper tape, using a fast paper-tape punch. This entire operation of collection and dumping takes less than 90 seconds, so that points can, if necessary, be collected at least every 2 minutes. The data are processed later by feeding the paper tapes back into the computer.

In Fig. 23 the slope of the first-derivative peak amplitude is 35% greater than the slope of the double-integral amplitude. Another interesting feature is that the *apparent* concentration appears to increase in the plot of the first-derivative peak heights, during the same time interval that concentration, as determined by the double integration, is decreasing linearly. One explanation of such an effect is that the lines are originally oxygen broadened and the line shape narrows as the oxygen concentration is reduced, although there may be other effects present. This is the first comparison of this type that has been made involving nitroxides, and it is not possible to say at this point whether the result is general. Cer-

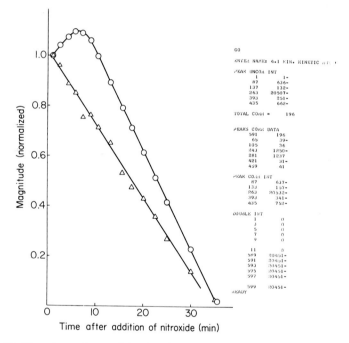

Fig. 23. Time dependence of the first-derivative peak height (○—○) and double integral (△—△) of the center peak. The sample contained, initially, $5 \times 10^{-5} M$ ketone nitroxide (IV) in a buffered aqueous suspension of fresh *Mycoplasma laidlawii* membranes at room temperature. The microwave power, modulation amplitude, scan time, scan range, and filter time constant are 5 mW, 1 G, 6.5 minutes, 100 G, and 0.3 seconds, respectively. The gain setting was constant.

tainly, the marked difference between the slopes of the two curves does inject a note of caution in kinetic experiments.

E. Line Positions, Amplitudes, and Data Storage

The small dedicated computer can easily be programmed to digitalize data points from the first-derivative ESR spectrum. In most of the examples given in this chapter, approximately 1000 points were collected over a 100 G scan. However, collection has been made at much higher resolution. The small computer also has the capability of immediate data treatment and can, for example, perform integrations within less than a minute after the spectrum is collected. In addition, it can scan the data and record the amplitude and field positions of maxima and minima of each curve. The results are printed back on a standard teletype.

This application has recently been found useful in studies of anisotropy in phospholipid multilayers (Jost *et al.*, 1971). In these experiments, the positions and relative magnitudes of the peaks change with systematic changes in the orientations of the sample in the magnetic field. The typical output from the teletype is shown in Fig. 24. The computer first printed out the peak positions in decigauss, measured from the beginning of the scan, and the amplitudes in arbitrary units (the negative sign after a peak height indicates an upward pen deflection). The computer then calculated the average baseline for the first 10 G of the scan and corrected the amplitudes for a zero baseline. From the printout, it is easily seen that the maxima are at 28.6, 45.4, and 59.1 G; the minima are at 33.2, 48.6, and 66.0 G; and the relative amplitudes of the three lines are 727:1440:270 (the term 727 is obtained from $415 + 312 = 727$, for example). The baseline correction is necessary for the next step in which the integration is performed, and the maxima and minima of this calculated absorption spectrum are printed out. The peaks correspond to the points along the baseline where the first-derivative spectrum crosses the baseline; therefore the distance in gauss between these peaks of the absorption spectrum gives the experimental coupling constant, A. The shape of a somewhat similar integral is shown in the right-hand column of Fig. 22. (This program does not collect enough points to give reliable line positions and widths unless the lines are fairly broad. The program used to process the kinetic data in Fig. 23, where the lines were very narrow, collected 300 points over a 10 G span and did not include the feature of converting the x axis from arbitrary units to decigauss, although this could easily have been included in the program.)

Another application has involved monitoring the relative peak heights of spin-labeled lipids in phospholipid vesicles used as model membranes. The spectral peaks and valleys change with temperature due to local melting of the long fatty acid side chains. Computer processing of the experimental data removes the most tedious aspects and provides the measured parameters almost immediately. The data processing time for the program shown in Fig. 24 (multilayer) is 3.8 minutes and during this time samples can be changed, equilibration with a new temperature

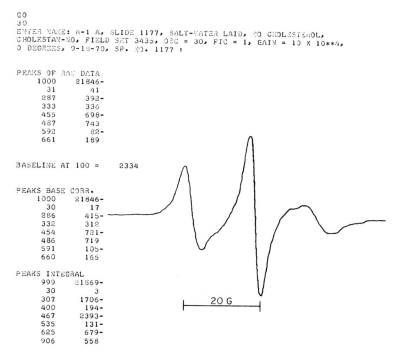

```
GO
30
ENTER NAME: A-1 A, SLIDE 1177, SALT-WATER LAID, NO CHOLESTEROL,
CHOLESTAN-NO, FIELD SET 3435, OSC = 30, FTC = 1, GAIN = 10 X 10**4,
0 DEGREES, 9-18-70, SP. NO. 1177 ↑


PEAKS OF RAW DATA
     1000      21846-
       31         41
      287        392-
      333        336
      455        698-
      487        743
      592         82-
      661        189


BASELINE AT 100 =     2334


PEAKS BASE CORR.
     1000      21846-
       30         17
      286        415-
      332        312
      454        721-
      486        719
      591        105-
      660        165


PEAKS INTEGRAL
      999      21869-
       30          3
      307       1706-
      400        194-
      467       2393-
      535        131-
      625        679-
      906        558
```

Fig. 24. Example of the computer printout obtained from an oriented phospholipid sample containing 3-doxyl-5α-cholestane (XXII). The spectrum is shown on the right (see text for description). The instrument settings are the same as in Fig. 21.

attained, or instrumental parameters adjusted. (This program is started by the symbol ↑, and the computer controls the instrument, collects the spectrum, and automatically proceeds with the calculations.) In the program used to collect the data in Fig. 23, the computer again controls the ESR instrument but is instructed at the beginning of each experiment whether the data are to be processed immediately after collection or stored on tape, depending on how rapidly the data are to be collected.

Storage of reference spectra is another useful aspect of computer collection of data, since the original spectrum fed into the computer from tape can be replotted, or used for spectral subtraction or addition or to compare splittings. Obvious combinations of the procedures already used yield absorption differences spectra analogous to those widely used by optical spectroscopists. The use of an oscilloscope for display makes subtraction of spectra or adjustment of integrations of two ESR absorption spectra to the same value (i.e., equivalent to equal concentration) much more rapid.

The small computer we are using has not as yet been useful in simulations of complex ESR line shapes; these more sophisticated simulations require the use of a large computer. One of several mathematical approaches (Itzkowitz, 1967; Libertini *et al.*, 1969; Alexandrov *et al.*, 1970; Hubbell and McConnell, 1971; Jost *et al.*, 1971) can be used to generate computer simulations of experiment ESR spectra. Comparison of an experimental spectrum with such computer-simulated spectra is greatly simplified if the two spectra, experimental and simulated, are the same in intensity and in the horizontal scale. In theory, the large computer can be asked to match the experimental spectrum in these parameters. In practice, however, an arbitrary intensity is usually included in a series of simulations, and experimental spectra not only vary in intensity but may have been originally collected using different scan ranges.

The problem, then, is to replot the experimental spectra either for comparisons with simulations or for comparison with other closely related experimental spectra adjusting the height and horizontal scale of the spectrum. If the data have been stored on paper or magnetic tape, the operation illustrated in Fig. 25 is simple. The original spectrum (Fig. 25a) is to be compared with a series of simulations (Fig. 25b). For this purpose the spectrum is replotted, adjusting the intensity, horizontal scale, and, in this example, reversing the phase. The instructions given to the small computer are illustrated on the computer printout. Here the

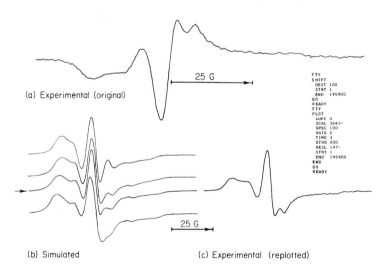

Fig. 25. Replotting experimental data with a small computer. (a) The original experimental spectrum from a dehydrated oriented lecithin preparation containing 5-doxylstearic acid (XXVI); (b) simulations from large computer to be compared; (c) experimental spectrum (a) replotted to match horizontal and vertical scales and phase of spectra (b). The instructions to the small computer are shown on the right. The arrow in (b) indicates the best fit to the experimental spectrum in both splittings and line shape.

conversion was from 100 G/40 cm to 100 G/8 inch. (The spectrum width is multiplied by 100 and divided by 197.) The intensity is scaled by an arbitrary factor calculated from the ratio of the actual magnitude to the desired magnitude. In this case the original spectrum (Fig. 25a) was collected with the phase reversed (necessary because the cavity characteristics were changed by the presence of a thin pyrex chamber), and the minus sign following the scale factor reversed the phase of the replotted spectrum (Fig. 25c). Now the comparison with a series of simulations, shown in Fig. 25b, is made much easier. The arrow points to the simulation which most closely resembles the original spectrum in both the line shape and splittings. Of course, phase reversal alone can be accomplished by simply retracing the spectrum, but adjustment of the other parameters by hand is tedious and lacks accuracy.*

VII. ADVANTAGES AND DISADVANTAGES OF SPIN LABELING

In any reporter group technique, the introduction of the label must perturb its environment as a result of its size and chemical properties. At the molecular level even a small molecule can represent a bulky addition. This is clearly a problem in attempting to assess the results obtained from adding reporter molecules to biological systems. From the large literature accumulating from the use of various reporter group approaches, such as fluorescence, nuclear magnetic resonance (e.g., fluorine probes), optical absorption, and even heavy-atom derivatives in x-ray crystallography, it is clear that no generalizations about the acceptable size or chemical properties of probes can be made. This also applies to spin labeling.

In some cases, the evaluation can be based on maintenance of function in the biological system. For example, the spin-labeled NAD molecule discussed above acts as a competitive inhibitor with respect to NAD (Weiner, 1969) and mimics the behavior of other similar inhibitors. The

---

* The programs for acquiring and processing the data presented above were written in the language CLASS (Conversational Language for Spectroscopic Systems), which is a macro language designed by Mr. Russel Wolfe of Cary Instruments Division of Varian especially for use with chemical instrumentation. It has been implemented for the Varian 620/i computer by the Cary group and was kindly supplied to us by them. The routines for operation of CLASS with spectrophotometers were replaced by routines to drive the ESR spectrometer by Professor Charles Klopfenstein of the University of Oregon, who made available to us the adaptations necessary and developed the programs used by us. All examples of small-computer applications in spin labeling presented here were performed with a Varian 620/i computer, equipped with 8K memory, teletype, and general interface.

biosynthetic incorporation of fatty acid spin labels into mitochondrial phospholipids (Keith *et al.*, 1968) is another example that can be used to illustrate acceptability of a spin label. Although it is difficult to evaluate the distortion introduced by the presence of probes in biological systems, the addition of spin labeling to the battery of the spectroscopic tools available to the biologist promises to add information in appropriately selected systems, which can be combined with information accumulating from other techniques.

Another aspect to be considered is the chemical interaction between the nitroxide moiety and the biological system. For example, it has been shown that, under some extreme experimental conditions, the oxidation of sulfhydryl groups by the spin label can be detected (Morrisett and Drott, 1969).

In addition, certain biological samples cause a rapid loss of signal with some spin labels (for example, see Fig. 23). Although this necessitates a search for conditions that will prevent this process, or at least reduce the rate of signal loss, this is usually not an insurmountable problem.

Electron spin resonance is less sensitive than fluorescence and considerably more sensitive than nuclear magnetic resonance. Although as few as 10^{-13} mole of a very rapidly tumbling nitroxide can be detected, the practical upper limit, without special equipment, is 10^{-10}–10^{-11} mole. If the nitroxide is more immobilized, the concentration requirements are higher, and the usual range of bulk concentrations that are easy to work with are in the range of 10^{-5}–10^{-6} M (using a 10 μl sample volume). The fact that the sample can be optically opaque is often a real advantage.

The ESR spectrum contains information about molecular orientation, motion, and the nature of the chemical environment. Nitroxide–nitroxide interactions can, in principle, be exploited to provide sensitive spectroscopic rulers. Spin labels with greater specificity are becoming available and, combined with suitable biological systems, ESR can provide new information about biomolecular structure.

Acknowledgments

We thank Louis Libertini for helpful comments and for performing the spectral simulations shown in Figs. 6 and 17. We are indebted to Professors Charles Klopfenstein and John Keana for helpful discussions and to Miss Dee Brightman for technical assistance. Financial support by U. S. Public Health Service Grant CA-10337-03 from the National Cancer Institute is gratefully acknowledged. One of us (A. S. W.) is the recipient of an NIH Postdoctoral Fellowship (1-F10-NS02397-01) from the Institute of Neurological Diseases and Stroke.

References

Alexandrov, I. V., Ivanova, A. N., Korst, N. N., Lazarev, A. V., Prikhozhenko, A. I., and Stryukov, V. B. (1970). *Mol. Phys.* **18**, 681.

Aneja, R., and Davies, A. P. (1970). *Chem. Phys. Lipids* **4**, 60.

Bangham, A. D., and Horne, R. W. (1964). *J. Mol. Biol* **8**, 660.

Bangham, A. D., Standish, M. M., and Watkins, J. C. (1965). *J. Mol. Biol.* **13**, 238.

Barratt, M. D., Green, D. K., and Chapman, D. (1968). *Biochim. Biophys. Acta* **152**, 20.

Barratt, M. D., Green, D. K., and Chapman, D. (1969). *Chem. Phys. Lipids* **3**, 140.

Berger, K. U., Barratt, M. D., and Kamat, V. B. (1970). *Biochem. Biophys. Res. Commun.* **40**, 1273.

Berliner, L. J., and McConnell, H. M. (1966). *Proc. Nat. Acad. Sci. U. S.* **55**, 708.

Bourgès, M., Small, D. M., and Dervichian, D. G. (1967). *Biochim. Biophys. Acta* **137**, 157.

Brière, R., Dupeyre, R., Lemaire, H., Morat, C., Rassat, A., and Rey, P. (1965). *Bull. Soc. Chim. Fr.* **1965**, 3290.

Buchachenko, A. L., Khloplyankina, M. S., and Dobryakov, S. N. (1967). *Opt. Spektrosk. (USSR)* **22**, 554.

Butler, K. W., Dugas, H., Smith, I. C. P., and Schneider, H. (1970a). *Biochem. Biophys. Res. Commun.* **40**, 770.

Butler, K. W., Smith, I. C. P., and Schneider, H. (1970b). *Biochim. Biophys. Acta* **219**, 514.

Calvin, M., Wang, H. H., Entine, G., Gill, D., Ferruti, P., Harpold, M. A., and Klein, M. P. (1969). *Proc. Nat. Acad. Sci. U. S.* **63**, 1.

Carrington, A., and McLachlan, A. D. (1967). "Introduction to Magnetic Resonance." Harper, New York.

Chapman, D., and Penkett, S. A. (1966). *Nature (London)* **211**, 1304.

Chapman, D., and Wallach, D. F. H. (1968). *In* "Biological Membranes" (D. Chapman, ed.), pp. 125–202. Academic Press, New York.

Chapman, D., Owens, N. F., and Walker, D. A. (1966). *Biochim. Biophys. Acta* **120**, 148.

Chapman, D., Barratt, M. D., and Kamat, V. B. (1969). *Biochim. Biophys. Acta* **173**, 154.

Cooke, R., and Morales, M. F. (1969). *Biochemistry* **8**, 3188.

Corvaja, C., Giacometti, G., Kopple, K. D., and Ziauddin (1970). *J. Amer. Chem. Soc.* **92**, 3919.

Faber, R. J., Markley, F., and Weil, J. A. (1967). *J. Chem. Phys.* **46**, 1652.

Falle, H. R., Luckhurst, G. R., Lemaire, H., Marechal, Y., Rassat, A., and Rey, P. (1966). *Mol. Phys.* **11**, 49.

Ferruti, P., Gill, D., Harpold, M. A., and Klein, M. P. (1969). *J. Chem. Phys.* **50**, 4545.

Forrester, A. R., Hay, J. M., and Thomson, R. H. (1968). "Organic Chemistry of Stable Free Radicals." Academic Press, New York.

Freed, J. H., and Fraenkel, G. K. (1963). *J. Chem. Phys.* **39**, 326.

Freed, J. H., Bruno, G. V., and Polnoszek, C. (1971). *J. Phys. Chem.* (in press).

Gotto, A. M., Kon, II., and Birnbaumer, M. E. (1970). *Proc. Nat. Acad. Sci. U. S.* **65**, 145.

Gregory, J. D. (1955). *J. Amer. Chem. Soc.* **77**, 3922.

Griffith, O. H. (1964). *J. Chem. Phys.* **41**, 1093.

Griffith, O. H., and McConnell, H. M. (1966). *Proc. Nat. Acad. Sci. U. S.* **55**, 8.

Griffith, O. H., and Waggoner, A. S. (1969). *Accounts Chem. Res.* **2**, 17.
Griffith, O. H., Keana, J. F. W., Noall, D. L., and Ivey, J. L. (1967). *Biochim. Biophys. Acta* **148**, 583.
Griffith, O. H., Libertini, L. J., and Birrell, G. B. (1971). *J. Phys. Chem.* (in press).
Hoffman, A. K., and Henderson, A. T. (1961). *J. Amer. Chem. Soc.* **83**, 4671.
Holmes, D. E., and Piette, L. H. (1970). *J. Pharmacol. Exp. Ther.* **173**, 78.
Hsia, J. C., and Piette, L. H. (1969a). *Arch. Biochem. Biophys.* **129**, 296.
Hsia, J. C., and Piette, L. H. (1969b). *Arch. Biochem. Biophys.* **132**, 466.
Hsia, J. C., Schneider, H., and Smith, I. C. P. (1970a). *Chem. Phys. Lipids* **4**, 238.
Hsia, J. C., Schneider, H., and Smith, I. C. P. (1970b). *Biochim. Biophys. Acta* **202**, 399.
Hubbell, W. L., and McConnell, H. M. (1968). *Proc. Nat. Acad. Sci. U. S.* **61**, 12.
Hubbell, W. L., and McConnell, H. M. (1969a). *Proc. Nat. Acad. Sci. U. S.* **63**, 16.
Hubbell, W. L., and McConnell, H. M. (1969b). *Proc. Nat. Acad. Sci. U. S.* **64**, 20.
Hubbell, W. L., and McConnell, H M. (1971). *J. Amer. Chem. Soc.* (in press).
Hudson, A., and Luckhurst, G. R. (1969). *Chem. Rev.* **69**, 191.
Itzkowitz, M. S. (1967). *J. Chem. Phys.* **46**, 3048.
Jost, P., and Griffith, O. H. (1971a). *In* "Methods in Pharmacology" (C. Chignell, ed.), Vol. 2, Chap. 7. Appleton, New York (in press).
Jost, P., and Griffith, O. H. (1971b). Manuscript in preparation.
Jost, P., and Griffith, O. H. (1971c). Manuscript in preparation.
Jost, P., Libertini, L. J., Hebert, V. C., and Griffith, O. H. (1971). *J. Mol. Biol.* (in press).
Kawamura, T., Matsunami, S., and Yonezawa, T. (1967). *Bull. Chem. Soc. Jap.* **40**, 1111.
Keana, J. F. W., Keana, S. B., and Beethan, D. (1967). *J. Amer. Chem. Soc.* **89**, 3055.
Keith, A. D., and Mehlhorn, R. J. (1971). *In* "The Molecular Biology of Membranes" (F. C. Fox and A. D. Keith, eds.). Sinauer Associates, Stamford, Connecticut (in press).
Keith, A. D., Waggoner, A. S., and Griffith, O. H. (1968). *Proc. Nat. Acad. Sci. U. S.* **61**, 819.
Keith, A. D., Bulfield, G., and Snipes, W. (1970). *Biophys. J.* **10**, 618.
Kivelson, D. (1957). *J. Chem. Phys.* **27**, 1087.
Klein, M. P., and Barton, G. W. (1963). *Rev. Sci. Instrum.* **34**, 754.
Kornberg, R. D., and McConnell, H. M. (1971). *Biochemistry* **10**, 1111.
Kosman, D. J., Hsia, J. C., and Piette, L. H. (1969). *Arch. Biochem. Biophys.* **133**, 29.
Ladbrooke, B. D., Williams, R. M., and Chapman, D. (1968). *Biochim. Biophys. Acta* **150**, 333.
Lai, A. A., Birrell, G. B., and Griffith, O. H. (1970). *J. Chem. Phys.* **53**, 4399.
Landgraf, W. C., and Inesi, G. (1969). *Arch. Biochem. Biophys.* **130**, 111.
Lebedev, O. L., and Kazarnovskii, S. N. (1960). *Zh. Obschch. Khim.* **30**, 1631.
Lecuyer, H., and Dervichian, D. G. (1969). *J. Mol. Biol.* **45**, 39.
Lehninger, A. L. (1970). "Biochemistry." Worth Publ., New York.
Lemaire, H., Rassat, A., and Rey, P. (1968). *Bull. Soc. Chim. Fr.* **1968**, 886.
Levine, Y. K., and Wilkins, M. H. F. (1971). *Nature (London)* (New Biology) **230**, 69.
Libertini, L. J., and Griffith, O. H. (1970). *J. Chem. Phys.* **53**, 1359.
Libertini, L. J., Waggoner, A. S., Jost, P. C., and Griffith, O. H. (1969). *Proc. Nat. Acad. Sci. U. S.* **64**, 13.

Long, R. A., Hruska, F., Gesser, H. D., Hsia, J. C., and Williams, R. (1970). *Biochem. Biophys. Res. Commun.* **41**, 321.

Lucy, J. A. (1968). *In* "Biological Membranes" (D. Chapman, ed.), pp. 233–288. Academic Press, New York.

McConnell, H. M., and McFarland, B. G. (1970). *Quart. Rev. Biophys.* **3**, 91.

McElhaney, R. N., and Tourtellotte, M. E. (1969). *Science* **164**, 433.

Mandell, L., Fontell, K., and Ekwall, P. (1967). *Advan. Chem. Ser.* **63**, 89.

Mildvan, A. S., and Weiner, H. (1969a). *Biochemistry* **8**, 552.

Mildvan, A. S., and Weiner, H. (1969b). *J. Biol. Chem.* **244**, 2465.

Morrisett, J. D., and Drott, H. R. (1969). *J. Biol. Chem.* **244**, 5083.

Morrisett, J. D., Broomfield, C. A., and Hackley, B. E. (1969). *J. Biol. Chem.* **244**. 5758.

Ogawa, S., and McConnell, H. M. (1967). *Proc. Nat. Acad. Sci. U. S.* **58**, 19.

Ohnishi, S., Boeyens, J. C. A., and McConnell, H. M. (1966). *Proc. Nat. Acad. Sci. U. S.* **56**, 809.

Ohnishi, S., Cyr, T. J. R., and Fukushima, H. (1970). *Bull. Chem. Soc. Jap.* **43**, 673.

Pake, G. E. (1962). "Paramagnetic Resonance," Chapter 5. Benjamin, New York.

Rottem, S., Hubbell, W. L., Hayflick, L., and McConnell, H. M. (1970). *Biochim. Biophys. Acta* **219**, 104.

Rozantsev, E. G. (1970). "Free Nitroxyl Radicals." Plenum Press, New York.

Rozantsev, E. G., and Kokhanov, Yu. V. (1966). *Izv. Akad. Nauk SSSR, Ser. Khim.* **8**, 1477.

Rozantsev, E. G., and Nieman, M. B. (1964). *Tetrahedron* **20**, 131.

Sandberg, H. E., and Piette, L. H. (1968). *Agressologie* **9**, 59.

Saupe, A. (1964). *Z. Naturforsch. A* **19**, 161.

Schneider, H., and Smith, I. C. P. (1970). *Biochim. Biophys. Acta* **219**, 73.

Seelig, J. (1970). *J. Amer. Chem. Soc.* **92**, 3881.

Singer, L. A., and Davis, G. A. (1967). *J. Amer. Chem. Soc.* **89**, 158.

Smith, I. C. P. (1971). *In* "Biological Applications of Electron Spin Resonance Spectroscopy" (J. R. Bolton, D. Borg, and H. Swartz, eds.). Wiley (Interscience), New York (in press).

Steim, J. M., Tourtellotte, M. E., Reinert, J. C., McElhaney, R. N., and Rader, R. L. (1969). *Proc. Nat. Acad. Sci. U. S.* **63**, 104.

Stryer, L., and Griffith, O. H. (1965). *Proc. Nat. Acad. Sci. U. S.* **54**, 1785.

Taylor, J. S., Leigh, J. S., Jr., and Cohn, M. (1969). *Proc. Nat. Acad. Sci. U. S.* **64**, 219.

Thompson, T. E., and Henn, F. A. (1970). *In* "Membranes of Mitochondria and Chloroplasts" (E. Racker, ed.), pp. 1–52. Van Nostrand-Reinhold, Princeton, New Jersey.

Tonomura, Y., Watanabe, S., and Morales, M. (1969). *Biochemistry* **8**, 2171.

Tourtellotte, M. E., Branton, D., and Keith, A. (1970). *Proc. Nat. Acad. Sci. U. S.* **66**, 909.

Waggoner, A. S., and Stryer, L. (1971). Unpublished results.

Waggoner, A. S., Griffith, O. H., and Christensen, C. R. (1967). *Proc. Nat. Acad. Sci. U. S.* **57**, 1198.

Waggoner, A. S., Keith, A. D., and Griffith, O. H. (1968). *J. Phys. Chem.* **72**, 4129.

Waggoner, A. S., Kingsett, T. J., Rottschaefer, S., Griffith, O. H., and Keith, A. D. (1969). *Chem. Phys. Lipids* **3**, 245.

Weiner, H. (1969). *Biochemistry* **8**, 526.

Wold, F. (1967). *Methods Enzymol.* **11**, 617.

Zwaal, R. F. A., and van Deenen, L. L. M. (1971). *Chem. Phys. Lipids* (in press).

4

THE MOLECULAR ORGANIZATION
OF BIOLOGICAL MEMBRANES

S. J. SINGER

I. INTRODUCTION

The major constituents of biological membranes are lipids, proteins, and oligosaccharides. Ultimately, it would be desirable to know precisely how the individual molecular species in any specific membrane are arranged in some steady state. Presumably, such knowledge of detailed structures is essential for an adequate understanding of membrane functions, but we are, of course, very far from this knowledge at the present time. The magnitude of the problem is very large, and it becomes more awesome as new information about membranes is acquired. One great complication is the heterogeneity and diversity of the protein and lipid components of many membranes.

The best information available indicates that the proteins of any one type of membrane are grossly heterogeneous. For example, Kiehn and Holland (1968) have shown that mammalian cell membranes from several sources contain a large number of proteins of different molecular weights ranging from less than 15,000 to over 100,000. For any given type of membrane, the distribution of protein was quite reproducible, but it differed for different membranes. No single protein component predominated, although a predominant species might have been expected on the basis of Green and co-workers' hypothesis (Richardson et al., 1963) that a "structural protein" is a principal constituent of membranes. Physical and chemical studies of the proteins of a variety of membranes indicate a broad distribution of heterogeneity (Halder et al., 1966; Green et al., 1968; Rosenberg and Guidotti, 1969; Lenard, 1970).

Among the lipids of membranes, one encounters a truly bewildering degree of complexity. The lipids of any one membrane are of a variety of classes, such as phosphatidylcholine, phosphatidylethanolamine, sphingomyelin, glycolipids, cholesterol, etc. Furthermore, the hydrocarbon chains of any one class can be of a great many chain lengths and several degrees of unsaturation (cf. Van Deenen, 1969) and can exhibit a variety of covalent linkages to the glyceryl phosphate moiety. A striking example of the latter type of heterogeneity is found in erythrocyte membranes: In the human erythrocyte membrane, the phosphatidylethanolamine fraction has 65% of its fatty acid chains in the usual diacyl ester linkage, but 35% in a vinyl ether linkage; in the bovine erythrocyte membrane, the phosphatidylethanolamine has 20% of its fatty acid chains in diacyl ester linkage, 80% as glyceryl ether, and none as vinyl ether; in the porcine erythrocyte membrane, 100% of the fatty acid chains are in diacyl ester linkage, and none in ether linkage (Hanahan, 1969). Eventually, the origins and the structural significance of remarkable differences such as these must be understood. At present, they are a mystery. Such heterogeneity prompts some membranologists to argue that each membrane is unique and that no really useful general pattern of structure is likely to emerge. On the other hand, other investigators, almost as an article of faith, assume that a general pattern of organization of the structural components of membranes does exist and that the heterogeneity and distinctiveness of different membranes can ultimately be understood as variations on a common structural theme. The virtue of the latter viewpoint is that at least it encourages the design of experiments to test its validity, whereas the former is virtually a dead end. It is the working hypothesis of this chapter, therefore, that membranes do exhibit some general structural pattern and that its elucidation would be a useful early step toward a more refined conceptual and experimental analysis of membrane structure, function, and biosynthesis.

Given the present state of obscurity about membrane structure, it is perhaps not surprising that numerous models have been proposed, especially in recent years. Major structural features of membranes are sometimes proposed *ad hoc* or on the basis of irrelevant or unacceptable interpretations of information about simpler systems. One might conclude from all of these efforts that the only constraints on such models lie in the imaginations of their builders. On the other hand, the situation at present is not really so amorphous. A considerable body of information has been acquired in recent years about macromolecular structure and noncovalent interactions, much of it from the explosive development of protein structure studies in the last decade or so. While physicochemical understanding of protein structure is still in a primitive state (cf. Lumry and Biltonen, 1969), certain general principles which qualitatively govern these structures have emerged. Perhaps because many investigators of membrane structure and function are more familiar with lipid than with protein chemistry, the full impact of these principles has not yet been widely felt in membranology. They are relevant not only to proteins, but when suitably generalized, to any macromolecular system in an aqueous environment, and in particular, to membranes. Physico-

chemical information places certain constraints on the visualization of the
ways in which lipids, proteins, and oligosaccharides can be organized in
a membrane and provides a minimal thermodynamic basis for the consideration of models of membrane structure. The first objective of this
chapter, therefore, is to review this information briefly and mostly qualitatively, and then to apply it to the problem of membrane structure.
Second, selected structural studies of membranes are discussed which
contribute further information about membrane structure. Finally, in the
light of these structural considerations some aspects of membrane function and biosynthesis are explored in a preliminary and speculative way.

II. THERMODYNAMICS OF PROTEIN AND OTHER MACROMOLECULAR SYSTEMS

It is of more than historical interest that some important general principles of protein structure were developed on theoretical grounds (Kauzmann, 1959) before the first x-ray crystallographic determinations of
protein structure were made. The x-ray structures have completely confirmed the validity of these principles and have led to the recognition of
others as well. These principles involve equilibrium thermodynamics,
and in order for them to be applicable to proteins, the system consisting
of the protein and its aqueous environment should be in its lowest free-
energy state. In other words, the particular three-dimensional structure
(conformation) adopted by a polypeptide chain in its native state should
represent the thermodynamically most favorable structure, determined
solely by the amino acid sequence and the aqueous environment. That
the structures of some simple proteins are indeed thermodynamically
most favorable has been concluded from the experiments of Anfinsen
and co-workers (Epstein et al., 1963) in which proteins have been completely unfolded and then allowed to refold. Under appropriate conditions, the proteins recover spontaneously a conformation which is essentially indistinguishable from the native one.

A. The Role of the Solvent

The role of the solvent in determining the structure of proteins is
absolutely critical. The full significance of this fact has often not been
adequately appreciated in the past. The principle involved is expressed
chemically by the difference between Figs. 1a and b. In Fig. 1a, the
process of a protein molecule folding into its native conformation is re-

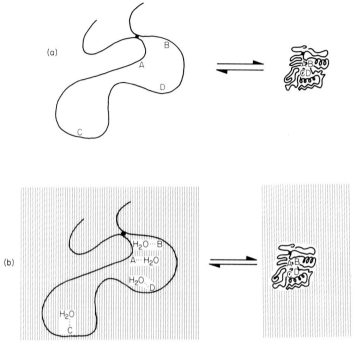

Fig. 1. Schematic distinction between the processes of folding a polypeptide chain in (a) a vacuum and (b) in the solvent water. A, B, C, and D represent amino acid residues on the polypeptide chain (see text).

garded as intrinsic to the molecule itself. Roughly speaking, Fig. 1a suggests that summing up all the interaction energies between groups such as A and B, C and D, etc. can account for the decreased energy of the native folded structure as compared to the unfolded one. In the unfolded state, groups A, B, C, D, etc. are in effect in a vacuum. Figure 1b reminds us, however, that in the unfolded state of the protein, groups A, B, C, D, etc. *are interacting with water molecules;* in the folded state of the protein, groups A and B, C and D, etc. interact with one another, *and furthermore* the water molecules that were formerly interacting with groups A, B, C, D, etc. *must now be interacting with one another* in the bulk aqueous phase. Therefore, the change in free energy, ΔG_{A-B}, resulting from the juxtaposition and interaction of groups A and B in the folded structure is given by Eq. (2) rather than by Eq. (1)

$$\Delta G_{A-B} = G_{A-B} - (G_A + G_B) \tag{1}$$

$$\Delta G_{A-B} = (G_{A-B} + G_{H_2O-H_2O}) - (G_{A-H_2O} + G_{B-H_2O}) \tag{2}$$

These considerations may appear obvious and trivial, but their consequences are very great and have only been adequately appreciated in the last decade.

B. Hydrophobic Interactions

A principal way in which the solvent water influences the structure of proteins is through hydrophobic interactions (Kauzmann, 1959). The result of these interactions is to produce a clustering together of nonpolar residues and structures so as to sequester them from the aqueous environment; for example, they are responsible for the immiscibility of simple hydrocarbons and water. Their precise description, however, is somewhat complicated because a number of factors are involved.

Consider again Figs. 1a and b, and assume that the groups A and B are nonpolar. In terms of Fig. 1a, one would attribute the binding of A to B in the folded state exclusively to van der Waals interactions between A and B. This would be grossly misleading, however. The true physical situation is that in the unfolded state, A and B each exhibits van der Waals interactions with neighboring water molecules, and if the distances between molecules are comparable in the different states, these interactions with water are likely to be of comparable strength to those between A and B. Therefore, the net change in van der Waals interaction energies accompanying the process depicted in Fig. 1b may be small. It is clearly inadequate to say that nonpolar groups cluster together in aqueous solutions because of their van der Waals attractions for one another (Vandenheuvel, 1963). van der Waals interactions play a role, but only as part of a more complex system of interactions implied in Fig. 1b.

Insight into the nature of hydrophobic interactions has come from an analysis of the thermodynamics of solutions of simple hydrocarbons in water (Kauzmann, 1959). These data were available for a considerable time previously (Frank and Evans, 1945), but their relevance to protein structure was not widely appreciated. For example, consider the process

$$CH_4 \text{ (in benzene)} \rightleftarrows CH_4 \text{ (in water)} \tag{3}$$

The change in unitary free energy, ΔG_u,* accompanying this process is unfavorably positive by 2.6 kcal/mole at 25°C, expressing the fact that

* The standard unitary free energy change (Kauzmann, 1959) in a process is the total standard free energy change corrected for any translational entropy terms (cratic entropy) that are not intrinsic to the interaction being studied. If A and B are independent molecules, then the unitary free energy change for the process A + B → AB is obtained as ΔG (total) $- RT \ln \chi$, where χ is the mole fraction of solute in the standard state (generally 1 M concentration).

methane is much less soluble in water than in a nonpolar solvent such as benzene. It is surprising, however, that ΔH for this process is exothermic, -2.8 kcal/mole. From the relation [Eq. (4)] between ΔG_u and ΔH,

$$\Delta G_u = \Delta H - T\Delta S_u \tag{4}$$

it is clear that the enthalpy change for this process makes a favorably negative contribution to ΔG_u. It follows that ΔS_u, the unitary entropy change for the process shown, must be negative; in fact, -18 cal/$^\circ$ mole. In other words, it is the *decrease in entropy* of the system on the right side of Eq. (3) that makes the process in Eq. (3) unfavorable. It has been concluded that this entropy decrease is due to some kind of ordering of the water molecules around the hydrocarbon molecule dissolved in the water (Frank and Evans, 1945; Kauzmann, 1959). This ordering need not result in a fixed cage of water molecules; restrictions on the various degrees of freedom of the water molecules which transiently surround the hydrocarbon molecule may reduce the entropy to a considerable extent.

To a first approximation, these considerations may be extrapolated from simple hydrocarbons to the nonpolar residues of protein molecules (Kauzmann, 1959). In the process of folding the protein molecule in aqueous solution (Fig. 1b), the nonpolar groups are sequestered from contact with water. This is the reverse of the model reaction shown in Eq. (3), and the relevant ΔG_u, ΔH, and ΔS_u values are of opposite sign to those just discussed. Accompanying the folding process, therefore, there is a favorably negative ΔG_u contributed for each buried nonpolar group, presumably due to the positive ΔS_u resulting from the "release" of the ordered water molecules which surround the nonpolar groups in the unfolded state of the protein molecule. The sum of these ΔG_u contributions for all of the buried nonpolar groups in the protein is a large negative number, of the order of magnitude of hundreds of kilocalories per mole, and is very likely the major source of stabilization of the native conformation of protein molecules in aqueous solution (Tanford, 1962).

The x-ray crystallographic studies of protein structure have provided striking confirmation of these thermodynamic predictions. For all proteins so far analyzed, the interior of a folded polypeptide chain shows a high degree of sequestering of the nonpolar amino acid residues (cf. Perutz, 1969). Ionic residues are almost universally excluded from these interior regions. The residues inside are closely packed together with few, if any, holes of atomic dimensions, and the interior is essentially completely free of water molecules. [Water molecules may be present in the crevices *between* the individual polypeptide chains in a protein molecule which contains several chains (Perutz, 1969, and see Section V,A) but generally not in the interior of the individual chains.] The close packing of interior residues, with little empty space between them, is attributable to the fact that such spaces would result in a large diminution in van der Waals interactions (which fall off very rapidly with distance) between the residues in the folded conformation. This would lead to an unfavorable increase in the free energy of the first term on the right side of Eq. (2).

It is in the context of hydrophobic interactions that one aspect of the uniqueness of the solvent water is revealed. For every other simple solvent known, ΔG_u for the process corresponding to that in Eq. (3) is much less positive (Singer, 1962). Even for ethylene glycol, the solvent closest to water in its lack of miscibility with hydrocarbons, ΔG_u is much smaller. As a result, when a protein molecule is dissolved in a solvent other than water, the equivalent of hydrophobic interactions plays a much lesser role, and the protein generally undergoes extensive conformational changes. This is demonstrated below (Section IV,D, Fig. 10). It was pointed out some time ago (Singer, 1962), therefore, that proteins associated with a lipid environment might be expected to adopt conformations which are consonant with that environment. This is discussed further in Sections III,G and IV,E.

C. Hydrogen Bonds

It used to be thought that interpeptide hydrogen bonds ($\diagdown$C$=$O$\cdots$ H$-$N$\diagup$) were a principal source of stabilization of the folded state of protein molecules, i.e., contributed a large negative ΔG to the process shown in Fig. 1. What has been more recently appreciated, however, is that interpeptide hydrogen bonds *formed in aqueous solutions* are at best of only marginal stability (Schellman, 1955; Klotz and Franzen, 1962). This is because the formation of the interpeptide hydrogen bond is better represented by Eq. (6) than by Eq. (5) [analogs of Eq. (2) and (1), respectively]

$$\diagdown C{=}O + H{-}N\diagup \rightleftarrows \diagdown C{=}O\cdots H{-}N\diagup \tag{5}$$

$$\diagdown C{=}O\cdots H_2O + H_2O\cdots H{-}N\diagup \rightleftarrows \diagdown C{=}O\cdots H{-}N\diagup + H_2O\cdots H_2O \tag{6}$$

Experiments with the model compound N-methylacetamide (Klotz and Franzen, 1962) have shown that the formation of an interpeptide hydrogen bond in water solution is accompanied by a ΔG which is about 4 kcal/mole more positive than in a nonpolar solvent such as carbon tetrachloride. This is clearly due to the fact that the hydrogen bonds formed between the $\diagup$C=O group or the $\diagup$N—H group and a water molecule are of strengths similar to the $\diagup$C=O$\cdots$H—N$\diagdown$ and $H_2O \cdots H_2O$ hydrogen bonds, and the overall process in Eq. (6) is not favored.

The situation in Fig. 1b is a little more complicated than the model experiments with N-methylacetamide indicate. In the process of folding, the $\diagup$N—H and $\diagup$C=O groups which were originally exposed to water become sequestered into the less polar interior of the protein molecule. However, further experiments carried out with N-methylacetamide by Klotz and Farnham (1968) have indicated that an interpeptide hydrogen bond has about the same relative free energy in an aqueous or a nonpolar environment. It is concluded therefore that on the average, no significant stabilization of the native conformation results from the formation of interpeptide hydrogen bonds on the inside, rather than on the outside, of the protein molecule.

In view of all the attention that has been given to hydrogen bonds in protein and nucleic acid molecules, the conclusion that hydrogen bonds *per se* cannot account for the stability of the native conformations of these molecules in water solutions appears to be highly paradoxical. It is certainly true that hydrogen bonds play an important role in macromolecular systems; the nature of that role, however, is subtly but significantly different from what had earlier been thought. The paradox is largely resolved in the following way. Let us suppose that the groups A and B in Fig. 1 are a $\diagup$C=O and a $\diagup$N—H group, respectively. In the unfolded state of the protein (left side, Fig. 1b) these groups form strong hydrogen bonds to water molecules. If in the folded state, however, they are sequestered from water molecules (right side, Fig. 1b) but are *not* hydrogen bonded to each other (or to some other suitable hydrogen bond donor or acceptor, respectively), a large energy deficiency results; that is, the hydrogen bonds formed to water molecules on the left side of the equation are incompletely compensated for if hydrogen bonds are not formed on the right side. Therefore, the folded protein will be in a lower free-energy state, other things being equal, *the larger the number of hydrogen bonds formed between donor and acceptor groups in the interior of the molecule.* On the other hand, on the exterior of the protein molecule, it makes little difference whether the $\diagup$N—H and $\diagup$C=O groups are hydrogen bonded to one another or to water molecules.

The distinction made in this section may appear to be more semantic than real. Whether hydrogen bonding is considered to be the primary or a subsidiary factor in providing the free energy of stabilization of the native conformation, the formation of a large fraction of the possible hydrogen bonds in the interior of the protein molecule is predicted in any event. The distinction is important, however. Most likely, hydrophobic interactions are in large part responsible for the fact that many globular proteins are only partially α-helical in their native aqueous conformations. When hydrophobic interactions are relaxed, as in solutions in less polar nonaqueous solvents (see Section IV,D), these proteins become predominantly α-helical, since α-helix is a conformation in which interpeptide hydrogen bonding is maximized. In water solution, a conformation with a lower free-energy state is apparently achieved in which less than the maximum number of interpeptide hydrogen bonds is formed. Other macromolecular phenomena [e.g., the stability of the DNA double helix (Herskovits et al., 1961)] can also be rationalized if hydrophobic interactions, rather than hydrogen bonds, are regarded as one of the primary sources of the conformational stability of macromolecules in water solutions.

D. Hydrophilic Interactions

Among the many important structural features revealed by the x-ray crystallographic studies of proteins, one of the most remarkable is the essentially total exclusion of ionic residues from the interior regions of folded polypeptide chains. With very rare exceptions, the fixed electric charges of the polypeptide chain are all on the exterior surfaces of the molecule in contact with the bulk water phase. In fact, this exclusion of ionic groups from the interior of a protein molecule appears to be a more stringent requirement than the exclusion of nonpolar residues from the exterior surface. If a nonpolar residue occurs in juxtaposition to ionic residues in the amino acid sequence, it may often be constrained to occupy an exterior location on the protein molecule by virtue of the dominant thermodynamic influence of the ionic residues.

The exterior localization of ionic residues, which may seem surprising at first, is to be expected from electrostatic considerations. We may first ask, what are the mechanisms for "burying" a charged group in the low dielectric constant interior of a protein? Burying an isolated electric charge requires a very large expenditure of free energy and need not be considered. More likely mechanisms involve (a) charged groups as ion pairs (i.e., a pair consisting of one negatively and one positively charged ion, one of which may be a counterion), and (b) proton binding or dissociation to remove the charge.

The electrostatic free energy of a fixed ion pair is inversely proportional to the dielectric constant of the medium (cf. Edsall and Wyman, 1958). Physically, this is because an ion pair interacts more strongly with the molecules of polar as compared to nonpolar solvents. A simple model system with which one can estimate the unitary free energy of transfer of an ion pair from water to a nonpolar environment is the process involving glycine in Eq. (7)

$$
\begin{array}{c}
\text{COO}^- \\
/ \\
\text{CH}_2 \quad \text{(in water)} \rightleftharpoons \\
\backslash \\
\overset{+}{\text{NH}_3}
\end{array}
\quad
\begin{array}{c}
\text{COO}^- \\
/ \\
\text{CH}_2 \quad \text{(in nonaqueous solvent)} \\
\backslash \\
\overset{+}{\text{NH}_3}
\end{array}
\tag{7}
$$

For this process ΔG_u may be determined approximately from the solubilities of glycine in water and other solvents as $RT \ln(\chi_w/\chi_s)$ where χ_w and χ_s are the mole fractions of glycine in the saturated solutions in water and in the solvent, respectively, at a particular temperature. (Activity coefficients are assumed to be 1.0 in these calculations.) From the data in Table I, it is clear that ΔG_u is a large positive number for all nonaqueous solvents; the ion pair is at much lower free energy in contact with water than with any other solvent. Therefore, the free energy required to bury an ion pair in the low dielectric constant interior of a protein molecule out of contact with water should be correspondingly large. Summed over as many as fifty ion pairs per molecule, this free-energy term is likely to be as important as any other single factor (Tanford, 1962) in determining protein conformation.

The data in Table I reveal another facet of the unique properties of the solvent water. The interaction of water with ion pairs is not simply attributable to the large dielectric constant of water, since formamide has about the same dielectric constant, yet it requires the free energy of 1.68 kcal/mole to transfer glycine from water to formamide. This suggests that, as with nonpolar solutes, ion pairs dissolved in water induce profound changes in the local structure of liquid water (Franks, 1968).

The second mechanism for burying a charged residue is to discharge it and then internalize it as a polar rather than an ionic group. This process is generally also quite unfavorable thermodynamically. The change in unitary free energy required just to discharge the group is

$$
\Delta G_u = 2.303 \, RT|(\text{pH} - \text{p}K)|
\tag{8}
$$

156

S. J. SINGER

TABLE 1
SOLUBILITY AND FREE ENERGY OF GLYCINE TRANSFER IN VARIOUS
SOLVENTS AT 25°C[a]

Solvent	Solubility (mole/liter)	Log χ[b]	ΔG_u[c] (kcal/mole)
Water	2.886	−1.247	
Formamide	0.0838	−2.476	1.68
Methanol	0.00426	−3.762	3.43
Ethanol	0.00039	−4.638	4.63
Butanol	0.0000959	−5.055	5.19
Acetone	0.0000305	−5.648	6.00

[a] Cohn and Edsall (1943).
[b] The symbol χ is the mole fraction of glycine.
[c] The term ΔG_u is the approximate unitary free energy of transfer of a mole of glycine from water to the solvent in question.

where pK is the negative logarithm of the acid dissociation constant of the group involved. For example, for a carboxyl group of pK 4.5 to be protonated at pH 7.0, $\Delta G_u = +3.3$ kcal/mole at 25°C.

These considerations show that it is thermodynamically necessary for the largest fraction of ionic residues of proteins or other macromolecules to be in direct contact with water rather than sequestered from water.

Similar thermodynamic reasoning indicates that saccharide moieties (ionic or nonionic) of macromolecules are more favorably situated in contact with the bulk aqueous solvent than sequestered from it. The solubilities of simple nonionic sugars are universally much greater in

TABLE II
SOLUBILITY AND FREE ENERGY OF TRANSFER OF SIMPLE SUGARS
IN WATER AND METHANOL[a]

Sugar	Solubility[b]		Temperature (°C)	ΔG_u[c] (kcal/mole)
	Water	Methanol		
Sucrose	0.094	0.0035	15	1.90
Glucose	0.096	0.0023	25	2.20
Lactose	0.0095	0.00045	20	1.80

[a] From Stephen and Stephen (1963).
[b] Expressed as mole fractions in the saturated solutions.
[c] Approximate unitary free energy of transfer from water to methanol, calculated assuming activity coefficients to be unity.

water than in other solvents, even those as polar as methanol (Table II). It therefore requires a substantial amount of free energy to transfer a saccharide from water to methanol. For solvents less polar than methanol, the free energies of transfer would be considerably larger. Therefore, the free energy required to bury a saccharide molecule in the nonpolar interior of a macromolecular structure is very unfavorable.

E. Electrostatic Interactions

Having discussed the strong thermodynamic tendency of the ionic groups of a macromolecule to be in contact with the water phase, we shall next try to determine the role that coulombic attraction between oppositely charged fixed ionic groups (ion-pair bonds) plays in stabilizing its conformation. While satisfactory quantitative estimates of the free energy changes accompanying the formation of an ion-pair bond in water have not yet been made, indirect but cogent evidence suggests that such bonds are of much less significance than hydrophobic interactions in stabilizing protein structures (Kauzmann, 1959). For example, if ion-pair bonds were more important than hydrophobic interactions, the addition of nonaqueous solvents to a solution of a protein in water should, by lowering the effective dielectric constant of the solvent, stabilize the native conformation of the protein molecule. Conversely, if hydrophobic interactions were more important, the addition of a nonaqueous solvent should, by increasing the capacity of the solvent to dissolve nonpolar residues, destabilize the protein conformation. The latter effect is generally observed (cf. Singer, 1962; cf., however, Brandts, 1969).

A macromolecular system in which ion pairs significantly stabilize the structure should be affected by changes in ionic strength. The system should become less stable as the ionic strength is increased, other things being equal, and it should be most stable in the absence of salt. On the contrary, there is little effect of variation in NaCl concentration on the stability of most protein molecules at neutral pH (cf. von Hippel and Wong, 1965), indicating that electrostatic interactions are not dominant. On the other hand, in systems in which ion-pair bonds might well be expected to be important, sufficiently large ionic strengths do disrupt the stable structure, as, for example, the effect of high NaCl concentrations in dissociating the histones and double-stranded DNA from chromosomal nucleoproteins, or in dissociating artificial protein–lipid systems such as ferricytochrome c–phosphatidylinositol (Gulik-Krzywicki et al., 1969).

Although this catalogue of interactions involved in stabilizing the conformations of proteins in aqueous solutions touches upon the major factors that appear to be implicated, it is not complete. There are other possible factors, such as dipole–dipole interactions and protein–water interactions not explicitly taken into account so far (Lumry and Biltonen, 1969) which may also be involved. On the other hand, it is entirely likely that the interactions that have been explicitly considered are sufficient for a qualitative and semiquantitative understanding of protein structure and macromolecular interactions in general.

F. Applications to Soap Micelles and Phospholipid Bilayers

As indicated earlier, these considerations of macromolecular interactions, although most clearly and forcefully evident from studies of proteins, are generally applicable to all macromolecular systems in aqueous environments. In particular, the existence and the thermodynamic properties of soap and detergent micelles and phospholipid bilayers are readily understood only in terms of hydrophobic interactions. The reaction forming a soap micelle may be written as

$$n \; \text{O}\text{(aq}\cdots) \rightleftharpoons \text{OOO} \cdots \text{OOO} \; \text{(aq}\cdots) \tag{9}$$

where the ionic "head" and the nonpolar "tail" of the soap molecule are represented by the symbols 0 and ζ, respectively. The only interactions other than hydrophobic ones that are involved in this reaction are electrostatic. For a single pure soap, with all the ionic heads of the same charge sign, the electrostatic interactions are repulsive and make a large positive contribution to the overall change in unitary free energy. There is also a large decrease in translational entropy (cratic entropy) accompanying the combination of n independent soap molecules into one micelle, which contributes another large positive term to the overall ΔG. Furthermore, ΔH for reaction (9) is often very small (Goddard et al., 1957). Therefore, the overall ΔS_u must be highly positive to overcome all these unfavorable free energy contributions [Eq. (4)]. As pointed out by Kauzmann (1959) such a large positive ΔS_u would be consistent with hydrophobic interactions providing the main driving force for the formation of the soap micelle; the release of water molecules from the nonpolar tails when the latter become sequestered in the interior of the soap micelle is presumably what provides the large positive ΔS_u.

Consistent with these views is the finding that the addition of 20–30 mole % of a nonaqueous solvent such as ethanol to detergent solutions markedly raises the critical micelle concentrations, i.e., drives reaction (9) backward and destabilizes the micelle (Ward, 1940; Ralston and

Hoerr, 1946). The addition of ethanol has the primary effect of reducing the hydrophobic interactions, thereby reducing the stability of the micelle (Herskovits et al., 1961; Singer, 1962). [One might attribute this effect of ethanol, not to a decrease in hydrophobic interactions, but rather to an increase in electrostatic repulsions of the ionic heads of the micelle as the dielectric constant of the solvent is decreased. Conductance measurements show, however, that the net charge on the detergent molecules *decreases* in 30 mole % ethanol, presumably due to the formation of ion pairs with small cations in the solution of lower dielectric constant. If anything, therefore, electrostatic factors themselves would make the micelle more stable, rather than less, in the mixed solvent as compared to water itself (Herskovits et al., 1961).]

With minor adjustments, the same conclusions apply to phospholipids and the bilayers they form in an aqueous environment. In the past, the stability of phospholipid bilayers dispersed in water has often been attributed solely to van der Waals interactions between the nonpolar residues, but this view is untenable for the reasons indicated in Section II,A. Van der Waals interactions are only one factor contributing to the overall hydrophobic interactions. They provide the reason, however, for the closest possible packing of nonpolar residues in the interior of a micelle. Any significant number of holes or gaps of atomic or larger dimensions in the interior of the micelle would make the van der Waals energy contribution to the right side of Eq. (9) smaller than to the left, and would therefore raise the free energy of the components on the right, and destabilize the bilayer.

These considerations emphasize the role of the solvent water in determining the bilayer structures achieved by amphipathic substances such as soaps, detergents, and phospholipids. With a decrease in the water content of such systems to small volume fractions, considerable structural changes may occur, as have been revealed by x-ray diffraction studies (Luzzati et al., 1969). These changes most probably result from the necessity of maintaining the hydrophilic interactions of the polar heads of the lipids with the small volume fraction of water. It has been suggested by Luzzati et al. that these lipid structural changes may be important in the functioning of ordinary membranes. On the other hand, it seems to us unlikely that typical membrane systems in a water-rich environment exhibit such structural transitions, since the bilayer is the structure that permits the maximum exposure of the ionic heads of the phospholipids to the aqueous solvent and the maximum sequestering and van der Waals contact of the nonpolar tails, i.e., maximum hydrophilic and hydrophobic interactions. However, in a highly condensed multilayer lamellar system such as myelin (see Section IV,B), with a very low concentration of water within the interior of the multilayer, the formation of lipid structures other than the bilayer, including perhaps those which have been described for simple detergents (Luzzati *et al.*, 1969), may be favored.

G. Protein Molecules at Interfaces

It has long been known that many water-soluble protein molecules which are compact and globular in water solution are largely unfolded at an air–water or oil–water interface. This may be understood in terms of the thermodynamics discussed in this section. If a protein molecule is constrained to be at such an interface, it cannot maintain its globular conformation and at the same time maximize the hydrophilic interactions of its ionic residues. In the soluble globular conformation, the ionic residues are more or less uniformly distributed over the outer surface of the molecule. If virtually all of these ionic residues must be in contact with the aqueous phase at the interface, however, this can be achieved only if the molecule unfolds (Figs. 2b and c). At the interface, the ionic residues are directed into the aqueous phase, and to maximize hydrophobic interactions, the hydrophobic residues are directed as much as possible into the air or oil phase, subject to steric constraints imposed by the polypeptide chain (Fig. 2c). The protein is therefore spread as a "monolayer," although with a certain amount of buckling out of a completely planar configuration (cf. Cheesman and Davies, 1954).

The fact that many proteins form monolayers at such interfaces has no doubt influenced concepts of how proteins are arranged in membranes, which are loosely thought of as analogous to oil–water interfaces. Two factors must be recognized, however. First, most protein molecules

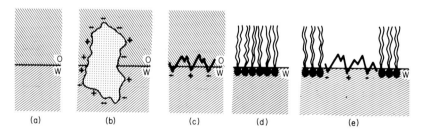

Fig. 2. Schematic representations of (a) an oil (O)–water (W) interface, (b) a globular protein molecule inserted in the oil–water interface, (c) an unfolded protein molecule at the oil–water interface, (d) a phospholipid layer at the air–water interface, and (e) phospholipid and unfolded protein together at the air–water interface (see text for details). The + and − signs represent the ionic residues of the protein.

which spread as "monolayers" at air–water and oil–water interfaces have a stable globular conformation in water solution, and by far the larger fraction of such proteins is in solution as compared to the interface. Thus, such spreading at interfaces does not account for the existence of membrane-specific and membrane-localized proteins. Second, and more important, the surface of a phospholipid micelle is very different from an oil–water interface. This may be appreciated from the following crude analysis (Fig. 2). The ionic heads of the phospholipids must be in contact with the aqueous phase in order to maximize their hydrophilic interactions (Fig. 2d). For the same reason, however, the ionic residues of a protein molecule which is brought to this interface must also be in contact with the aqueous phase. The phospholipids must then *share* the aqueous interface with the polypeptide chain. If the phospholipid molecules are to remain associated with one another, as in a bilayer, the ionic groups of the phospholipids must then move aside to accommodate the contiguous ionic residues of the protein; but if the ionic groups move aside, the fatty acid side chains which are covalently attached to the ionic groups must also move aside. This *removes* the "oil phase" provided by the lipids immediately above the ionic residues of the protein (Fig. 2e). As far as the proteins are concerned, therefore, *there is no oil–water interface parallel* to the surface of a phospholipid bilayer; any interface with the "oil" phase in the bilayer is *perpendicular* (more or less) to the surface of the bilayer. In effect, to satisfy the hydrophobic interactions of its nonpolar residues, the protein must provide its own "oil phase" (i.e., adopt a globular conformation within the phospholipid bilayer).

These qualitative considerations show that as far as protein molecules are concerned, a phospholipid bilayer in water is very different from an oil–water interface. The problem of protein–lipid interactions is discussed in a different way and in more detail in the following section.

III. THERMODYNAMICS OF MEMBRANE SYSTEMS

A. Equilibrium Conditions and Membrane Models

It is known that the lipids and proteins of membranes are held together by noncovalent interactions, and we assume that there is a steady-state structure of a membrane which results from these interactions. In order for the thermodynamic considerations discussed in Section II to be applicable to membrane systems, we need to know whether the steady-state structure of membranes is one of lowest free energy. With simple proteins, it has been demonstrated by the experiments of Anfinsen and co-workers (Epstein *et al.*, 1963) that this condition is satisfied. For membrane systems, the comparable experiments involve the reconstitution of a membrane from its completely dissociated and denatured proteins and lipids. These are very difficult experiments to interpret in an unambiguous manner, mainly because the criteria which are available for judging whether the original membrane structure is completely recovered are inadequate. A number of reconstitution experiments have been carried out (cf. Rothfield *et al.*, 1969), with the recovery of one or more enzymatic activities serving as a criterion for recovery of membrane function. Some enzymatic activities which might be lost upon dissociating a membrane, however, may largely be recovered upon reaggregation of the membrane components without the detailed structure of the membrane having been recovered.

It will be assumed in what follows that at the level of a *domain* within the membrane, the steady-state structure attained is one of lowest free energy. Without being elaborate about it, we shall consider a domain of a membrane to be a limited region containing a small number of protein molecules and associated lipids together with the aqueous environment. This simply evades the issue as to whether the membrane as a whole is instantaneously in its lowest free-energy state, which is a moot point (Katchalski and Oster, 1969). At the level of a domain, therefore, the equilibrium thermodynamics of macromolecular systems discussed in Section II are assumed to apply.

At the molecular level, the following general consequences of these thermodynamic considerations can be recognized:

1. In order to maximize hydrophilic interactions, essentially all the ionic, zwitterionic, and highly polar groups such as sugar residues, which are attached to both the lipids and the proteins in the membrane, should be in contact with the bulk aqueous phase.

2. Models must attempt to maximize hydrophobic interactions of the entire system of lipids and proteins in the membrane. This involves sequestering not only the fatty acid side chains of the lipids from contact with water, but to the maximum extent possible, the nonpolar amino acid residues of the proteins as well. The interior hydrophobic region of the membrane must be highly compact with very few holes or gaps of atomic dimensions or larger (Section II,B).

3. For the reasons given in Section II,C, those potential hydrogen bond donor and acceptor groups of the protein which are sequestered from contact with water should form hydrogen bonds with one another to the maximum extent possible.

These conditions are not highly restrictive and certainly do not of themselves lead to a unique model for the organization of the lipids and proteins of membranes. It is instructive, however, to see how these conditions are satisfied by some of the models that have been proposed.

Before proceeding with this, a few comments about membrane models are in order. In the context of the previous discussion, we shall consider only those models which address themselves to the following questions. How are the lipids and proteins of membranes structured and arranged, and what are the interactions that hold them together? We recognize that there are other ways to analyze membrane models. For example, Stoeckenius and Engelman (1969) have examined membrane models primarily in terms of the following morphological criterion. Is the model a continuum based on a lipid bilayer, or does it invoke an aggregation of subunits? By this morphological test, Stoeckenius and Engelman consider the lipid–protein mosaic model (discussed below) to be only a variation upon the Davson-Danielli-Robertson model, since both share the morphological property that the lipids are arranged in a bilayer, and they consider the Benson model to be an aggregation of subunits. By our structural and thermodynamic criteria, however, the lipid–globular protein mosaic and Davson-Danielli-Robertson models are very different, while the lipid–globular protein mosaic and Benson models have certain features in common, as is demonstrated further on in this section.

All models of the organization of membrane lipids and proteins must of course be highly schematic at our present stage of ignorance about membrane structure. No advocate of a particular model seriously proposes that any membrane is organized precisely according to the schematic representation of the model. The purpose of a model is rather to suggest a *predominant general pattern* of membrane structure, and the question one should ask is whether a membrane is largely organized as a model suggests. On the other hand, some membranologists are, in Abbie Hoffman's apt phrase, engaged in "pouring chicken soup" on this problem, for example, by throwing a few hydrophobic residues of the membrane protein into the interior of the Davson-Danielli-Robertson model, or by inserting a few convenient protein pores in that model to take care of the minor problem of transport through membranes. Such well-intentioned activities do not solve any problems; they simply obscure them. After all, proteins often constitute considerably more than 50% of most membranes, and a rational approach to the structure and location of proteins in the membrane is a first order of business.

Our first object then is to examine whether one or another predominant pattern of membrane structure is most likely, thermodynamically, and second to look for experiments which can unambiguously test the alternatives. After this initial stage, refinements and variations can be introduced that are thermodynamically allowed and that meet the test of experiment.

B. The Davson-Danielli-Robertson Model

The membrane model which is most widely accepted at the present time is the model originally proposed by Gorter and Grendel (1925), elaborated by Davson and Danielli (1952), and refined by Robertson (1964). We shall refer to it as the Davson-Danielli-Robertson model. It is depicted in Fig. 3a. In this model the phospholipids are arranged in a continuous bilayer so that the fatty acid chains occupy the interior core of the membrane. The protein is spread out over both surfaces of the bilayer, largely in monolayer thickness in order to account for the total thickness of the membrane calculated from electron micrographs. [The thickness of a phospholipid bilayer with its fatty acid chains in a non-crystalline state is about 40–45 Å (Stoeckenius, 1962). If the average total thickness of membranes is taken from the density distribution in KMnO$_4$-fixed specimens as about 75 Å (Robertson, 1964), this allows only 15–18 Å for each of the two layers of protein and any associated oligosaccharides in the Davson-Danielli-Robertson model. The thickness of a single monolayer of polypeptide is about 10 Å.] The outer surface of the protein layers is exposed to the bulk water phase. Minor modifications which have been proposed notwithstanding, this model implies the following general structural features. (a) The ionic heads of the phospholipids are largely not in contact with the bulk aqueous phase, but rather with the polar and ionic groups of the protein monolayers. *To account for the fact that most membranes consist of proteins and lipids in a weight ratio considerably greater than unity, the spread protein must essentially com-*

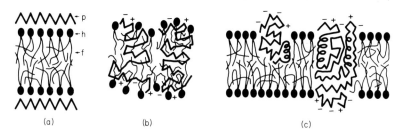

Fig. 3. Schematic representations of three models for the organization of the phospholipids and proteins of membranes. (a) The Davson-Danielli-Robertson model, (b) the Benson lipoprotein subunit model, and (c) the lipid–globular protein mosaic model. Symbols p, h, and f denote the polypeptide chains of the protein, the polar and ionic heads of the phospholipids, and the fatty acid tails of the phospholipids, respectively. The + and − signs represent the ionic residues of the protein.

pletely blanket the ionic heads of the phospholipids. (b) The proteins and lipids interact with one another, and the membrane is stabilized, primarily by electrostatic forces between their ionic groups. (c) The protein monolayers should exhibit a significant amount of β-conformation, or antiparallel pleated sheet, when not in a random conformation. (d) A significantly large fraction of the nonpolar residues of the protein must be exposed to the water. It is relevant to note here that the average amino acid composition of membrane proteins is not clearly distinguishable from that of soluble proteins (see Section III,C). In particular, a substantial fraction of the residues is hydrophobic.

It follows from the discussion in Section II that this is not an arrangement which would lead to the lowest free energy of a membrane in its aqueous environment. The two most serious thermodynamic problems arise because (a) the nonpolar residues of the membrane proteins are largely exposed to, instead of being sequestered from contact with, water molecules. This does not maximize hydrophobic interactions. (b) The ionic groups of the lipids and proteins are largely in contact with one another and out of contact with water; the burying of the ionic groups in a lower polarity environment is very costly in free energy (Section II,D).

C. Electrostatic and Hydrophobic Interactions

The Davson-Danielli-Robertson model is therefore thermodynamically unlikely, but we may ask a more general question which the model raises. Are the interactions that hold the membrane proteins and lipids together *predominantly* electrostatic or hydrophobic? The following experimental facts bear on this question.

1. The average amino acid composition of membrane proteins shows no bias in favor of polar and ionic residues (cf. Woodward and Munkres, 1966; Bakerman and Wasemiller, 1967; Lenaz *et al.*, 1968; Engelman and Morowitz, 1968; Rosenberg and Guidotti, 1969) as compared to other proteins. If anything, the hydrophobic residues are present with somewhat larger than average frequency, particularly in certain fractions of the membrane protein. This is unlike the situation that prevails with nucleoprotein systems, where the proteins have a large content of basic amino acid residues to interact electrostatically with the negatively charged DNA.

2. There is no marked effect of increasing ionic strength in destabilizing or dissociating a large fraction of the membrane proteins from the lipids, as would be expected if electrostatic interactions played a critical role and as is observed with nucleoprotein systems.

3. The largest fraction of the phospholipids of most membranes contains ionic residues that are zwitterionic rather than ionic, e.g., phosphatidylcholine and phosphatidylethanolamine. Such zwitterionic groups would be expected to exert strong electrostatic interactions with proteins only if the latter had precisely juxtaposed pairs of basic and acidic residues to complement the charges on the lipids. Those phospholipids which are not zwitterionic are usually anionic, such as phosphatidylserine, phosphatidylglycerol, and phosphatidylinositol; these might be expected to interact electrostatically mainly with basic proteins. Except in the case of myelin, there do not appear to be substantial amounts of basic proteins associated with membranes. Furthermore, removal of about 60–70% of the polar heads of the membrane phospholipids does not even detectably alter the conformation of membrane proteins (as measured by circular dichroism), let alone result in their dissociation from the membrane (see Section IV,F).

4. Membrane proteins depleted of lipids are generally highly insoluble in aqueous media around neutral pH. If these proteins were highly polar, they would be expected to be quite soluble in water at very low ionic strengths, as are the histones and protamines of nucleoproteins. On the contrary, the solubility and aggregation properties of the bulk of membrane proteins indicate that they are hydrophobic (Richardson *et al.*, 1963); in the absence of lipids, they interact hydrophobically with one another in aqueous solution and aggregate or become insoluble. The function of the lipids appears to be to disperse and "solubilize" membrane proteins by a kind of detergent action, in which the hydrophobic portions of the lipids and proteins interact with one another.

These experimental observations are therefore consistent with the thermodynamic expectations derived from the study of simple proteins (Section II) that hydrophilic and hydrophobic interactions are of predominant importance in the stabilization and organization of membranes. This is not to say that electrostatic interactions play no role whatsoever, but rather that they are relatively much less significant. They are therefore likely to be of more local or specific, rather than general, importance to membrane structure (see Sections III,D and IV,I).

D. Peripheral and Integral Membrane Proteins

Before proceeding further in the discussion of models of membrane structure, it may be useful to think of the proteins associated with membranes as falling into two general categories, *peripheral* and *integral*. In the former category are those proteins which are relatively easily dissociated from membranes by mild treatments and which must, therefore, be attached to the membranes by rather weak bonds. In some cases, these bonds may have a large electrostatic component. An example of a peripheral protein is the bulk of the cytochrome c of mitochondrial membranes, which can be dissociated by high salt concentrations. Similarly, the protein spectrin (Marchesi and Steers, 1968), which may constitute as much as 20% of the protein of hemoglobin-free red blood cell membranes, can be largely removed from the membranes by treatment with chelating agents under mild conditions. Other characteristics of peripheral proteins might be that they contain very little, if any, lipid when they are dissociated from the membrane, and they are soluble and molecularly dispersed in ordinary aqueous buffers. In the intact membrane these peripheral proteins would also be expected to be globular, rather than spread as a monolayer over the membrane surface, in order to maximize their hydrophilic and hydrophobic interactions (Gulik-Krzywicki et al., 1969). One may categorize as integral proteins those which require much more drastic treatments, with reagents such as detergents, bile acids, or protein denaturants, in order to be dissociated from membranes. Usually lipids remain associated with these solubilized proteins, and if the lipids are extracted, the proteins are then highly insoluble or tend to aggregate in aqueous buffers around neutral pH. On the basis of these characteristics, it is clear that although systematic studies of the dissociability of proteins from membranes under conditions of increasing severity have not been reported, the bulk of the proteins of most membranes must be integral.

Some years ago, Green and his colleagues (Richardson et al., 1963) proposed that membranes contain specific *structural* proteins whose function is to provide the matrix for lipoprotein membrane subunits. It was presumed that structural proteins had no other function, that is, they were not enzymes. More recent evidence (Haldar et al., 1966; Lenaz et al., 1968), however, is not consistent with the view that a discrete structural protein characterizes each membrane but rather that membrane proteins are grossly heterogeneous. If most of the membrane protein is integral, it follows that a variety of different membrane proteins can function as integral proteins. Some of the consequences of this hypothesis are considered further in Section IV,J.

E. The Properties of Integral Membrane Proteins

By definition, it is the integral, rather than the peripheral, proteins which are of primary concern to the problem of membrane structure, although any complete description of membranes must adequately account for both categories. What determines that a protein be specifically an integral membrane protein? From the assumption that a local domain of a membrane is in thermodynamic equilibrium in the steady state, the property that distinguishes an integral protein from simple soluble proteins must be its capacity to interact with lipids in water solutions to form lipoprotein structures with lower free energies than the separated lipid and protein. This capacity cannot be fortuitous. Bovine pancreatic ribonuclease and other simple soluble proteins are not membrane proteins and do not normally form lipoprotein complexes with membrane lipids in ordinary aqueous media. The only reasonable explanation is that this is a thermodynamic property of a protein; the amino acid composition and sequence of a protein determines whether it normally exists as a simple soluble protein or whether it associates with lipids as part of a membrane (Lenard and Singer, 1966). Since the average amino acid composition is not highly unusual, it is therefore likely that the amino acid sequences of integral proteins are characteristic and are the primary determinants of their lipophilicity.

The simplest explanation of the difference between soluble proteins and integral membrane proteins in the light of the discussion in Section II is as follows. (a) For soluble proteins, the amino acid sequences allow hydrophobic interactions to be maximized *internally* within the individual polypeptide chain or within the subunits of the soluble protein molecule, while the hydrophilic interactions are simultaneously satisfied *externally;* the presence of sufficient numbers of ionic charges *more or less uniformly distributed over the external surface of the molecule* of a simple protein then allows it to be soluble and monodisperse in the water solution. (b) On the other hand, the amino acid sequences of integral membrane proteins must be characteristically different, allowing especially the hydrophobic interactions of the protein molecule to be maximized by association with the hydrophobic milieu provided by the lipids in the interior of the membrane; when associated with the lipids, the ionic groups of the integral proteins (and of the lipids) must be in contact with the bulk aqueous phase in order to maximize hydrophilic interactions. This suggests that in the conformation adopted by an integral protein in the membrane, *the ionic amino acid residues are asymmetrically distributed over the surface of the protein molecule,* concentrated at those portions of the surface that are exposed to the bulk water and absent from those portions of the surface in contact with the hydrophobic regions of the membrane.

F. The Benson Model

The essential features of these considerations were presented briefly by Singer (1962) and by Benson and Singer (1965). Some of them were then incorporated by Benson (1966) into a specific version of lipid–protein interactions which has come to be known as the Benson model, depicted schematically in Fig. 3b. In this model, the protein is largely globular and in the interior of the membrane to maximize hydrophobic interactions. The lipid molecules, however, are not arranged in a bilayer. Their fatty acid chains are individually intercalated into the folds of the protein chains, with the polar heads of the lipids at the exterior surfaces of the membrane in contact with the water. This structure generates a more or less uniform lipoprotein complex, and it is proposed that such complexes exist as morphological subunits held together by hydrophobic interactions in the plane of the membrane.

Because this structure satisfies the requirements for hydrophobic and hydrophilic interactions, we consider it to be of lower free energy and therefore more likely to exist than the Davson-Danielli-Robertson model. On the other hand, compared to the mosaic model discussed in the following section, it appears to be in too high a free-energy state to constitute the general pattern of organization of most of the lipids and proteins of membranes. Intercalating most of the nonpolar fatty acid chains among the polypeptide chains in the interior of the membrane should, by virtue of separating the polypeptide chains, prevent the formation of the maximum number of interpeptide hydrogen bonds (which the fatty acid chains cannot participate in), and this would be thermodynamically unsatisfactory for reasons given in Section II,C.

G. The Lipid–Globular Protein Mosaic Model

It was essentially on the basis of the thermodynamic considerations of Sections II and III, together with some experimental data on the conformations of the proteins in intact membranes which are discussed in Section IV,E, that Lenard and Singer (1966) proposed a hydrophobic model for the organization of the lipids and the integral proteins (as they are referred to in this chapter) of membranes. Similar ideas were put forward independently by Wallach and Zahler (1966). On the basis of further experiments on the nature of lipid–protein interactions in membranes (Glaser et al., 1970), this model was more sharply delineated and may be designated as a lipid–globular protein mosaic model. It is represented schematically in Fig. 3c. The lipids and the globular integral proteins are arranged in an alternating mosaic pattern throughout the membrane. The hydrophobic portions of the lipids and a large fraction of the nonpolar amino acid residues of the proteins are sequestered from contact with water, mainly in the hydrophobic interior of the membrane, while the ionic groups of the lipids and the charged residues of the proteins are *both* in direct contact with water, predominantly on the

exterior surfaces of the membrane (see, however, Section V,A). Although
not explicitly stated in the original paper (Lenard and Singer, 1966) it
follows from the considerations in Section II,D that saccharide groups,
whether on glycolipids or glycoproteins, must also be essentially completely in direct contact with the bulk aqueous phase. As discussed above,
this arrangement allows for maximum hydrophobic and hydrophilic interactions of the entire system of lipids, proteins, and oligosaccharides.

The phospholipids in this mosaic model are primarily arranged in a
bilayer form similar to that proposed by Davson and Danielli (1952),
with the significant differences that in the mosaic model the ionic groups
of the lipids are exposed to the bulk aqueous phase and the bilayer is not
continuous (Section V,B). The problem arises whether the mosaic model
can account for the fact that membranes are generally considered to be
thicker than a phospholipid bilayer. The following arguments are relevant. First, the nature of the bilayer in the mosaic model need not be
as simple as the representation of Fig. 3c suggests. As mentioned in the
introduction, a great degree of heterogeneity exists among the fatty acid
chains of the lipids, with respect to both chain length and degree of
unsaturation. For many membranes, especially those not containing cholesterol, the fatty acid chains are in a fluid state under physiological
conditions. These considerations suggest that the interior of the membrane in the lipid portions of the mosaic structure may be like a liquid
solution in which other hydrophobic structures such as cholesterol, and
perhaps some of the hydrophobic residues of integral proteins, might be
dissolved. The thickness of this lipid bilayer (the cross-membrane distance between the lipid ionic heads) is therefore not easily predicted. In
support of these suggestions, model black-film lipid bilayer systems have
been found to be thicker than expected for the lipid itself (60–75 Å
instead of 40–45 Å) probably because of the presence between the lipid
monolayers of substantial amounts of the long-chain hydrocarbons used
in the preparation of the films (Bangham, 1968; Henn and Thompson,
1969). Second, the globular integral proteins of the mosaic may protrude
some considerable distance into the aqueous phase beyond the boundary
of the lipid bilayer, and the *average* thickness of the membrane may thus
be greater than that of the bilayer itself.

One of the major differences between the lipid–globular protein
mosaic model and the Davson-Danielli-Robertson model concerns the
proteins of the membrane. In the mosaic model, the proteins are considered to be mostly globular (resembling simple soluble proteins in this
respect), largely intercalated into the membrane interior, and integrally
involved in maintaining the structure of the membrane. The globular
feature was buttressed by the finding that the average conformation of

the proteins in intact membranes is partially α-helical and partially random coil (Section IV,E). The intercalation of globular proteins into the membrane is on a thermodynamic rather than an *ad hoc* basis, and it presumes that hydrophobic rather than electrostatic interactions hold the lipids and proteins together. The degree to which a globular protein molecule may penetrate the interior of the membrane, and in particular whether it completely spans the membrane from one surface to the other, would depend on the size and other structural properties of the protein (see Sections IV,J and V,A). In view of the fact that proteins comprise the largest fraction of the dry weight of most membranes and are most likely responsible for the specific permeability and enzymatic characteristics of membranes, the two models are not only structurally quite different, but lead to quite different functional characteristics as well (Section V).

Equally important differences between the two models concern the lipids. In the mosaic model, the lipid bilayer is not an uninterrupted continuum, and the ionic heads of the lipids are in contact with the aqueous phase rather than buried under, and constrained by, a layer of protein. This latter factor has an interesting and perhaps somewhat surprising consequence. The lipid–globular protein mosaic model suggests that certain characteristics of intact membranes and their isolated lipids in aqueous dispersion *might be more closely similar* than the Davson-Danielli-Robertson model would predict. The characteristics in question would be those which are strongly dependent on the physical state of the phospholipid ionic heads. For example, the response of membranes and aqueous dispersions of phospholipid bilayers to changes in ionic composition would be expected to be more alike if the membrane were organized as a lipid–globular protein mosaic than if it were a Davson-Danielli-Robertson structure, because in the latter case the ion atmosphere at the membrane surface would be vastly different from that at the surface of the lipid bilayer. This point is taken up again in Sections IV,G and V,C.

The major difference between the Benson and the lipid–globular protein mosaic models is that in the latter, the bulk of the lipid exists as a bilayer phase which is physically distinct from the bulk of the globular protein in the plane of the membrane, whereas in the former, the bulk of the lipid and protein are integrated in a single phase. In the mosaic model, the marvelously intricate ways in which single protein chains generally fold up in order to maximize their internal hydrophobic interactions and hydrogen bonds are not interfered with by the intercalation of a large number of chemically bland fatty acid chains.

Up to this point, the consideration of membrane models has been largely theoretical, dealing with the thermodynamic requirements for a system of lipids, proteins, and oligosaccharides organized in an aqueous environment. Consequently, only the grossest features of structure have been considered. Three models of membrane structure have been discussed which differ in these gross features. Other models have been devised and could be discussed at this level of detail, but the three models considered focus on the basic problems of the nature of lipid–protein interactions and structure. Little has been said, however, concerning the multiplicity of membrane proteins, the detailed distribution of lipids in the membrane, or the larger-scale organizations of the membrane, among other important aspects of membrane structure. This is because the thermodynamic considerations so far presented do not contribute insight into these important problems. They must either be explored by suitable experiments on membrane systems or left to speculation for the present. We shall therefore turn to some experimental studies of membranes to assess their contribution to an understanding of the organization of membranes. First, we shall try to ascertain whether they can discriminate among the different gross models of membrane structure that have been discussed and, second, whether they shed light on the enormous problems that remain.

IV. SOME STRUCTURAL STUDIES OF MEMBRANE SYSTEMS

It is not intended that this section be an exhaustive review of all of the relevant experiments that have been carried out with membrane systems. Our object is rather to select a few studies that bear most directly on the discussion in previous sections, particularly those which might, or are purported to, discriminate among the three gross models discussed and those which suggest more detailed structural features than have so far been presented. The permeability characteristics of membranes, for example, are not discussed here because all three models are characterized by a hydrophobic core which can equally well account for the low permeability of membranes to ions and high permeability to lipophilic substances.

Stoeckenius and Engelman (1969) have analyzed a large body of experiments regarding membrane structure, but only in order to discriminate between what they designate as "bilayer" and "subunit" models. This morphological designation is discussed in Section III,A. Here, the only point to be made is that in their analysis, they do not attempt to discriminate effectively between the models in Figs. 3a and c because by their criteria the models are not sufficiently different. Furthermore, there are several important matters of interpretation about which we hold different views. Therefore, the following discussion, although it covers some of the same ground, differs in major respects from that of Stoeckenius and Engelman.

Some of the experiments discussed in this section involve physical techniques such as nuclear magnetic resonance, electron spin resonance, fluorescence spectroscopy, infrared spectroscopy, and differential thermal analysis, which are taken up in considerably greater detail in Chapters 2 and 3.

A. Electron Microscopy

Electron microscopy has exerted a profound influence on the study of membranes. In principle, this is as it should be since with this method the most direct visualization of the molecular details of membrane structure can be achieved. On the other hand, the preparation of specimens for electron microscopy in a manner which maintains their original detailed molecular structural organization, and the visualization and identification of individual molecular species in an electron micrograph, are two problems which have been very difficult to solve successfully. As a consequence, electron microscopy has not yet yielded the kind of unambiguous structural information about membranes which it is potentially capable of yielding.

The standard method of specimen preparation at the present time involves (a) fixation, (b) replacement of the aqueous medium with some suitable monomer, (c) polymerization of the monomer, and (d) thin sectioning of the polymer block containing the specimen. Differential staining of the specimen is produced either in the fixation step or after sectioning by topical application of some suitable reagent. By this procedure, using $KMnO_4$ or OsO_4 as fixatives and any of several polymer embedding systems, a number of investigators in the 1950's obtained the now-classic railroad-track pictures of membranes. These studies were particularly pursued by Robertson (1964), who considered the ubiquitous "unit membrane" pattern as strong evidence for the membrane model shown in Fig. 3a. There are, however, many serious problems with this interpretation. First, fixation with OsO_4, and particularly with $KMnO_4$, markedly alters the conformations of protein molecules, including those of membrane proteins (Lenard and Singer, 1968a). Second, the substitution of an organic monomer such as methacrylate, Epon, or Vestopal for the water in the specimen can grossly change the structures of the membrane proteins and lipids (Singer, 1962; see Section IV,D). For at least these two reasons, the structure of the membranes observed in such thin sections may therefore be substantially different from the native structure at the molecular level.

In addition to this serious ambiguity, the significance of the staining patterns observed after OsO_4 or $KMnO_4$ fixation is not clear at the present time. The detailed chemistry of the staining reactions is only partially understood (Korn, 1966). It is entirely possible, for example, that the staining occurs by reaction with hydrophilic and not with hydrophobic groups, in which case the railroad-track pattern might reflect a three-layer hydrophilic–hydrophobic–hydrophilic sandwich across the thickness of the membrane rather than a protein–lipid–protein sandwich. The fact that membranes from which the lipids are first completely removed show an OsO_4 staining pattern similar to that of intact membranes (see below) strongly suggests this. In this event, such railroad-track micrographs would be compatible with any of the three models in Fig. 3. Our conclusion is, therefore, that these electron micrographs, which apparently lend such striking visual support for the Davson-Danielli-Robertson model, in fact do not do so.

Using these conventional techniques, Fleischer *et al.* (1967) carried out an important set of electron microscopic experiments with mitochondria and submitochondrial particles before and after graded solvent extraction of the lipids of the specimens. After the neutral lipids and 95% of the phospholipids were extracted, the OsO_4 staining pattern looked very similar to that of the original sample, and the apparent thickness of the membrane was not much altered. (The peak to peak distance of density in the extracted membrane was 54 Å compared to the original 49 Å.) Closely similar results have been obtained with myelin (Napolitano *et al.*, 1967) and mycoplasma membranes (Terry, 1966) and therefore appear to be of quite general validity. The significance of these results seems clear. If the Davson-Danielli-Robertson model represents the predominant organization of the membranes, the two unfolded protein layers should have collapsed together on removal of essentially all of the lipid, and the apparent membrane thickness should have markedly decreased. Two questions might be raised concerning the interpretation of these experiments. First, might not the membrane lipids have been extracted from the *untreated* membrane during the embedding process? The appearance of the membrane "before" and "after" organic solvent extraction would then perforce be similar. However, it is probable that the original OsO_4-fixed membranes were not *extensively* delipidated by the dehydration and embedding procedures used to prepare the specimens for electron microscopy (cf. Korn and Weisman, 1966; Buschmann and Taylor, 1968). Second, might not the membrane proteins have been converted from an unfolded to a globular form by the lipid extraction process? In this case, the retention of the apparent thickness of the membrane after delipidation would be fortuitous. On the other hand, if this possibility is considered, the embedding monomer itself should have had a similar globularizing effect on the conformation of the proteins of *both* the original and the delipidated membranes; an apparent collapse of the membrane would still be expected on the basis of the Davson-Danielli-Robertson model.

The *ad hoc* introduction of a few transmembrane protein cross-links into the model cannot explain these results, since the collapse would be expected to occur everywhere but in the immediate vicinity of such hypothetical cross-links. We therefore consider these results to be clearly inconsistent with the Davson-Danielli-Robertson model. On the other hand, these results can at least be rationalized by either the Benson or the lipid–globular protein mosaic models, since the proteins of the membrane play as much if not more of a role than the lipids in determining the membrane thickness in these models. The lipid–globular protein mosaic model might lead to the expectation that some shrinkage in the plane of the membrane would occur on removing the lipid, but it is not clear from the experiments of Fleischer et al. (1967) whether such lateral shrinkage was observed. [On the other hand, in the membrane preparations studied by Finean and Martonosi (1965), in which the lipids were enzymatically altered, such lateral shrinkage did occur.] In analogous experiments with myelin (Napolitano et al., 1967), prior fixation of the myelin with glutaraldehyde was necessary to preserve the structure on delipidation, and therefore the dimension in the plane of the membrane might have been preserved by the cross-links so formed.

The ambiguities in experiments such as these with thin sections of polymer-embedded specimens may be circumvented by recently introduced methods of embedding cells and cell organelles in cross-linked proteins, such as thiolated gelatin (Bernhard and Leduc, 1967) or bovine serum albumin (Farrant and McLean, 1969; McLean and Singer, 1970). In the latter case, for example, a specimen is never exposed to a non-aqueous solvent; only aqueous solutions of bovine serum albumin and glutaraldehyde at physiological pH and temperature are used. As a result, lipids and hydrophobic substances are not extracted from the specimens, and proteins such as hemoglobin retain their capacity to bind to specific antibodies. An electron micrograph of a chloroplast embedded in cross-linked bovine serum albumin and sectioned (Nicolson, 1971) is

shown in Fig. 4. In this preparation, the chlorophyll is completely retained in the specimen; in most conventional embedding procedures, the
chlorophyll is extracted.

A radically different method of electron microscopic specimen preparation and visualization is the technique of freeze-etching. Beginning with
the work of Steere (1957), continued refinements in technique (Moor
et al., 1961) and interpretation (Pinto da Silva and Branton, 1970) have
made this a powerful method for the study of membrane structure. In
this technique, a fresh specimen is rapidly frozen in liquid Freon 22; the
frozen specimen, under vacuum at −100°C, is fractured with a micro-
tome knife; some of the frozen water is sublimed (etched) from the
fractured surface if desired; and the surface is then metal shadowed and
replicated. This procedure avoids the fixation, chemical dehydration, and
embedding artifacts of conventional thin-sectioning procedures. On the
other hand, since a shadowed replica of the surface is examined in the

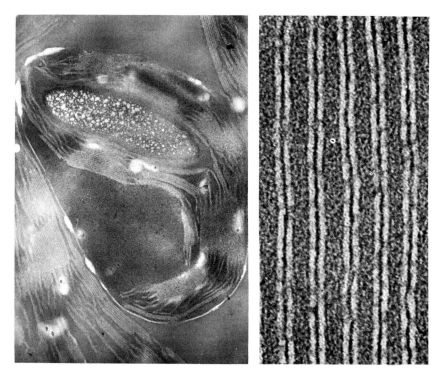

Fig. 4. A spinach chloroplast fixed in glutaraldehyde and embedded in bovine
serum albumin cross-linked with glutaraldehyde. After sectioning, the section was
stained with potassium phosphotungstate at pH 6.8 (Nicolson, 1971). Left, × 26,600;
at right, enlarged cross section of a granum, × 404,000.

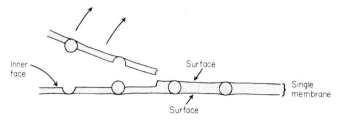

Fig. 5. Illustration of the nature of the fracture process produced during freeze-etching of a membrane, according to Branton (1966) and Pinto da Silva and Branton (1970). [Reproduced from Branton (1966) with permission.]

electron microscope, the method is largely restricted to observations of surface topology, and chemical identification of the morphological structures is difficult to achieve. Partly for this reason, with frozen membranous structures it has been a matter of controversy as to which membrane surfaces have been fractured and are under observation. Branton (1966) has argued that the internal hydrophobic faces of a membrane rather than the outer hydrophilic surfaces (Fig. 5) are preferentially exposed by the fracture process. In the case of the red blood cell membrane, by the use of covalently bound ferritin molecules to mark the outer hydrophilic surfaces, Pinto da Silva and Branton (1970; also Tillack and Marchesi, 1970) have provided strong evidence that the surface shown in Fig. 6 indeed arose from the fracture of the internal hydrophobic region of the membrane.

A characteristic feature of the membrane surface shown in Fig. 6 and of the surfaces of many other types of membrane examined by this technique (Branton, 1969) is a mosaiclike structure consisting of a smooth matrix interrupted by fairly uniform-sized particles. The size and distribution of particles vary with different membrane surfaces. On the other hand, when myelin is fractured, the exposed surfaces are invariably smooth (Fig. 7).

The chemical nature of these particles has not been established. It appears unlikely from the reproducibility of the results that the particles are artifacts of the freeze-etching procedure, although the nagging possibility exists that a surface diffusion of membrane components occurs at the fracture face in spite of the frozen condition of the bulk of the sample. It is certainly tempting to suggest that the particles are integral proteins (or lipoproteins, see Section IV,I) which are sufficiently deeply intercalated into the interior of the membrane and dispersed in the lipid matrix of the mosaic of Fig. 3c. Fracture at the inner hydrophobic face of the membrane model of Fig. 3c would be expected to reveal bumps or cavities where integral protein molecules extended from one of the

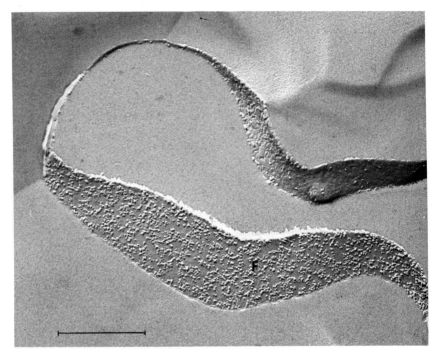

Fig. 6. Electron micrograph of a red blood cell membrane prepared by the freeze-etching procedure (Pinto da Silva and Branton, 1970). The surface labeled F is an internal hydrophobic face of the membrane (see Fig. 5) according to other experiments described in the text. The particles seen on this face average 85 Å in diameter. The bar represents 0.5 μ.

outer hydrophilic surfaces *across* the midline of the membrane, as for the protein molecule schematically represented on the right side of Fig. 3c. On the other hand, for an integral protein molecule which did not extend across the midline of the membrane, as on the left side of Fig. 3c, the corresponding region of the fracture face might be expected to appear smooth. These considerations can explain at least in part the observation (Pinto da Silva and Branton, 1970) that the particles in Fig. 6 can account for only about 30% of the total protein of the membrane. (The case of myelin is discussed further in Section V,A.)

It is evident that the results of freeze-etching experiments with membranes are strikingly consistent with the predictions of the lipid–globular protein mosaic model, provided the particles seen are protein in nature. A direct demonstration (for example, by specific staining) of the chemical nature of the particles is clearly required, but the very nature of the freeze-etch method makes it difficult to see how such direct identi-

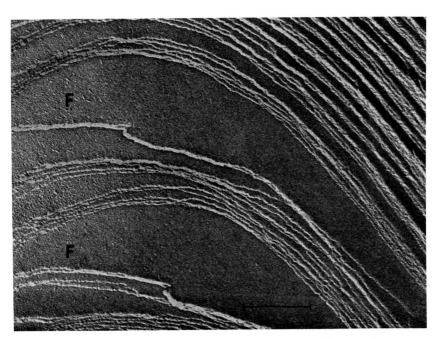

Fig. 7. Electron micrograph of rat myelin prepared by the freeze-etching procedure (Branton, 1967). The surface labeled F is considered to be an internal hydrophobic face of a single myelin layer (see Figs. 5 and 6). The absence of particles such as are observed on the internal faces of red blood cell membranes (Fig. 6) is noteworthy. The bar represents 0.2 μ.

fication can be achieved. Perhaps the use of ferritin-labeled specific antibodies to tag the exterior regions of certain integral proteins (see Section IV,J) of membranes *before* the membranes are frozen and fractured can help in this identification.

B. Myelin and Membranes

Low-angle x-ray diffraction studies of nerve myelin have often been interpreted as providing strong evidence that cell membranes conform to the Davson-Danielli-Robertson model. The argument rests on two pieces of information. (a) The x-ray data on fresh unfixed myelin show that the electron density distribution is periodic in the direction perpendicular to the myelin layers (reflecting the successive repeating layers); within one period, presumably corresponding to one layer thickness, the electron density distribution shows two fairly broad symmetrical maxima separated by a shallow minimum (Finean, 1962; Finean and Burge, 1963). (b) The myelin sheath, as first shown in electron micrographs

by Geren (1954), is continuous with and morphologically indistinguishable from the plasma membrane of a glial cell (for peripheral nerve
myelin, it is a Schwann cell plasma membrane which is so related, while
for central nervous system myelin, it is an oligodendrocyte plasma membrane), from which it appears to be formed during myelinogenesis.

Without any reflection on these x-ray data, which were as good as the
systems and the methodology allowed, it is surprising to realize how
little information they actually convey, especially in view of the unique
interpretation they are often given. The two broad maxima in the electron density distribution across one layer thickness can be satisfied by
any of a large number of different lipid–protein structures. The Davson-
Danielli-Robertson structure, which places the relatively heavier atoms
(P, O, N, etc.) of the myelin layer on the exterior surfaces and the relatively lighter atoms (C and H) in the interior, is one model which
would, qualitatively speaking, lead to such a bimodal electron density
distribution. Obviously, however, in view of the fact that myelin is 80%
lipid, the lipid–globular protein mosaic model in Fig. 3c and probably
also the Benson model, if it were to be applied to myelin, would yield
a similar qualitative result.

The continuity and morphological similarity of myelin and glial cell
plasma membranes is often taken to infer that myelin and membranes
are structurally very similar. It is not clear, however, that this conclusion
is warranted. Chemical studies indicate that myelin has a composition of
lipids and proteins that is very different from that of the usual cell membranes. In the first place, myelin is about 80% lipid and only 20% protein,
whereas isolated cell membranes generally contain significantly more
protein than lipid (see review by Korn, 1969). Second, the fatty acid
content of myelin lipids is unusual. The lipid of the myelin of the central
nervous system is much richer in longer fatty acid chains (19–26 carbon
atoms) than the lipids of gray matter and contains much less ($\frac{1}{5}$) of the
polyunsaturated fatty acids (O'Brien, 1965). Third, the proteins of
myelin, although not yet extensively characterized, appear to be quite
different in character from those of other membrane systems that have
been studied. More than 50% of the protein [the so-called proteolipid protein (Folch-Pi, 1955)] can be extracted into the chloroform-rich phase
of a chloroform–methanol–water two-phase system (cf. Mokrasch, 1969),
and a substantial fraction of the remaining proteins are basic proteins
which include the encephalitogenic antigen (Kies et al., 1965). Therefore,
either of two possibilities exists. (a) The Schwann cell or oligodendrocyte
plasma membrane is essentially similar in lipid and protein composition
to myelin and is therefore a most unusual plasma membrane; or (b) the
plasma membrane of the glial cell is similar to other membranes and very

different from myelin, in which case some profound transformation of the glial cell plasma membrane must occur in the course of myelin formation (see Section V,D). The fact that the lipid composition of myelin is unique and structurally compatible with its role as an electric insulator (O'Brien, 1965), whereas most plasma membranes are chemically and functionally quite different, implies that the second possibility is likely. Therefore, the structure of myelin, while interesting in its own right, is of uncertain relevance to the structure of cell membranes. In the previous section, however, the delipidation experiments of Napolitano et al. (1967) were discussed, and these raise some considerable doubt as to whether the Davson-Danielli-Robertson model is the correct representation even for the structure of myelin (see also Section V,B).

C. Surface Areas Occupied by Lipids

The early experiments of Gorter and Grendel (1925) were intended to determine the state of membrane lipids by comparing the surface area occupied by the extracted lipids as a *monolayer* film at an air–water interface and the area occupied by the same amount of lipid in intact red blood cell membranes. If a continuous lipid bilayer were present in the cell membrane, the surface area occupied by the lipid should be equal to the surface area of the entire cell, and the ratio, r, of lipid surface areas in the monolayer film to the surface area of the cell under the same conditions should ideally be 2.0. There are technical reasons why these classic experiments of Gorter and Grendel and similar ones by Dervichian and Macheboeuf (1938) (which produced different results) were not entirely satisfactory, and a definitive study was later carried out by Bar et al. (1966). One problem with this type of experiment is that the area occupied by the lipid in a monolayer at the air–water interface is not a fixed number but a continuously varying function of the surface film pressure. These authors found that in order for r to equal 2.0, the monolayer film of extracted lipids had to be highly expanded at low surface pressures, a situation very different from that likely to exist in the compact bilayer in the membrane. The numbers calculated from these experiments depend on assumptions about how the cholesterol is packed in the film and in the membrane, but the authors arrived at the large value of 87.5 Å^2 as the surface area occupied by a phospholipid molecule in the monolayer film (see, for example, Gulik-Krzywicki et al., 1967), and the low value of 9 dyn/cm for the surface pressure, at which $r = 2.0$. Some reviewers have attempted to rationalize these numbers away, but they really do not jibe with the predictions of the Davson-Danielli-Robertson model (Bar et al., 1966; Bangham, 1968). This difficulty is resolved, however, if the lipid in the membrane is indeed a bilayer, but the bilayer is discontinuous as in the lipid–globular protein mosaic model, rather than continuous as in the Davson-Danielli-Robertson model. (It is not clear what the Benson model predicts in this case.) In the mosaic model the lipid bilayer occupies only a fraction (roughly half) of the membrane surface area, and protein occupies the remainder. This increases (roughly doubles) the value of r in the calculations of Bar *et al.;* as a result, the calculated area per phospholipid molecule in the monolayer is substantially decreased, and the surface pressure is increased, at which $r = 2.0$. It is difficult to justify the assumptions necessary to make quantitative calculations of surface area and pressure expected for the mosaic model. Qualitatively, however, the revisions suggested by the mosaic model can bring these values into a much more reasonable range as compared to the values required by the Davson-Danielli-Robertson model.

D. Optical Activity and Protein Structure

In the last decade or so, rapid developments in theory and instrumentation have made the study of the optical activity of polypeptides and proteins one of the most powerful sources of information about their conformations in dilute solutions. The application of these measurements to the study of the proteins of intact membranes and membrane fragments has been undertaken recently, and it has yielded some new and useful information, which is discussed in this and the next section.

Two methods of studying the optical activity of polypeptides and proteins are currently in wide use, optical rotatory dispersion (ORD) and circular dichroism (CD). The basic theory and practice of ORD and CD have been reviewed recently (cf. Yang, 1967; Beychok, 1967) and will not be given in detail here. Briefly, when plane polarized light traverses a medium containing an optically asymmetric molecule (one which does not possess a mirror plane of symmetry), the left- and right-handed circularly polarized components of the light are affected differently. If the light is of a wavelength which is absorbed by the asymmetric molecule, the transmitted light will generally be both rotated and ellipticized. The rotation is due to the fact that the refractive indices η_L and η_R for the left- and right-handed circular polarized beams are slightly different, while the elliptical polarization is the result of the two beams being absorbed to different extents. In CD, the molar ellipticity $[\theta]_\lambda$ is directly proportional to the difference in the molar extinction coefficients E_L and E_R for the left- and right-handed circularly polarized beams

$$[\theta]_\lambda = 3300(E_L - E_R) \qquad (10)$$

It is this small difference, $E_L - E_R$, which is measured directly in currently available commercial CD instruments (Velluz et al., 1965).

The relationships among the absorption, CD, and ORD spectra for a single electronic excitation band are schematically shown in Fig. 8. The CD band has, of course, the same width as the absorption band, but it can be positive or negative depending on the magnitudes of E_L and E_R. The ORD spectrum, on the other hand, extends over a much larger wavelength region than the absorption band. The ORD and CD spectra are not independent of one another; one spectrum can be converted into the other by the Kronig-Kramers transform (Moffitt and Moscowitz, 1959). In principle, therefore, CD and ORD spectra yield the same information. In practice, however, the differences between ORD and CD spectra are that (a) ORD measurements, but not CD, can be made at wavelengths appreciably removed from the absorption bands responsible for the optical rotation and (b) CD spectra, being confined to wavelengths of the absorption bands, are therefore easier to interpret in terms of particular

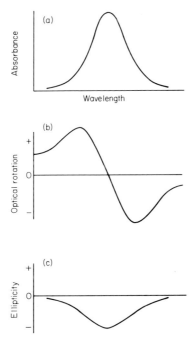

Fig. 8. The relationships, presented schematically, among the adsorption spectrum (a) due to an eletronic transition of a single asymmetric chromophore and the corresponding ORD curve (b) and CD spectrum (c). The maximum in the absorption spectrum, the crossover point in the optical rotatory dispersion curve, and the minimum in the circular dichroism spectrum all occur at the same wavelength.

electronic transitions and, most important for polypeptide studies, where overlapping absorption bands occur, are much better resolved than ORD spectra. For this last reason and the recent development of CD instrumentation capable of measurements down to 185 nm, CD measurements have increasingly become more useful than ORD in polypeptide conformational studies. In the remainder of this chapter, attention is therefore confined to CD.

Although absorption and CD bands occur at the same wavelengths, the CD spectrum has much additional information content. First, a CD band can be positive or negative, as already mentioned; two optical isomers show bands of equal magnitudes but opposite sign. Therefore, the configuration of an optically active compound can be discriminated. Second, the relative magnitudes of different CD bands can be quite different from their counterparts in absorption spectra. A particular electronic transition may be "forbidden" because the electric dipole moment of the transition, **m,** is zero or small. Since the absorption intensity depends on $|m|^2$ the absorption of that band will be weak. On the other hand, the magnitude of the corresponding CD band depends on the vector product $m \cdot \mu$, where μ is the magnetic moment of the same transition. As it may often happen that μ is large when **m** is small, a large CD band may be observed whose corresponding absorption band is so small that it is hard to resolve. Generally, therefore, considerably more information can be obtained from CD than from ordinary absorption spectra.

In the application of CD to the study of the conformations of proteins, attention is focused on the absorption bands of peptide bonds in the wavelength region around 200 nm. The pertinent electronic transitions of the peptide group are fairly well understood (for a brief review, see Schellman and Schellman, 1964). Theory and experiment have shown that the CD spectra of polypeptides are very grossly affected by the conformation which the polypeptide chain adopts, because in different conformations the electronic transitions of neighboring peptide groups are coupled to one another in different ways. Three principal conformations of polypeptide chains have been recognized in protein molecules: the right-handed α-helix; the antiparallel pleated sheet, or β-conformation; and the aperiodic or so-called random coil form. Model polypeptides in these three conformations exhibit markedly different CD spectra (Fig. 9). The CD spectrum for right-handed α-helical polypeptides is especially distinctive. It shows a maximum close to 191 nm and a minimum close to 208 nm due to the split π_1-π^* electronic transition of the peptide bond, and a minimum close to 222 nm due to the n-π^* transition (the latter has a small **m** and a large μ and is difficult to detect in absorption spectra of helical polypeptides). Although certain rigid but nonhelical model peptide struc-

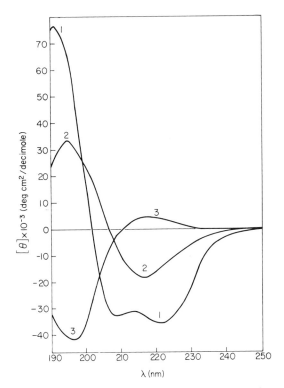

Fig. 9. The CD spectra near 200 nm of model polypeptides in three conformational states. Curve 1, 100% α-helix; curve 2, 100% β-form; curve 3, 100% random coil. [Reproduced from Greenfield and Fasman (1969), with permission.]

tures can also produce splitting of the π_1-π^* transition and a strong n-π^* band (Schellman and Neilsen, 1967), the appearance of two CD minima at 208–209 and 221–222 nm, with magnitudes of $[\theta] > 10^4$, is almost certainly indicative of the presence of a substantial fraction of right-handed α-helix in a high molecular weight polypeptide or protein.

The characteristic α-helical CD spectrum is not greatly affected by the solvent in which the helical polypeptide is dissolved. The wavelengths of the maximum and the two minima appear to be independent of solvent within experimental error; the band intensities are not much affected either, except for the minimum at 208 nm (Carver *et al.*, 1966). The CD spectrum for β-structures, which has been much less extensively studied, however, may be significantly solvent dependent (Iizuka and Yang, 1966).

Procedures for determining the relative amounts of α-helix, β-conformation and random coil in a protein sample have been worked out (cf.

Greenfield and Fasman, 1969) on the assumptions that unique CD spectra characterize each conformation and that these spectra are additive. Since $[\theta]$ at 222 nm is of much larger magnitude for the α-helix than for the other forms, for a protein with more than about 25% helix and little β-structure $[\theta]_{222}$ can be used to obtain a useful first-approximation value for the fraction h of the polypeptide in the α-helical conformation:

$$h = \frac{[\theta]_{222}^{\text{Obs}} - [\theta]_{222}^{\text{RC}}}{[\theta]_{222}^{\text{H}} - [\theta]_{222}^{\text{RC}}} \tag{11}$$

where $[\theta]_{222}^{\text{Obs}}$, $[\theta]_{222}^{\text{RC}}$, and $[\theta]_{222}^{\text{H}}$ are the observed ellipticity for the sample under study, the ellipticity for the completely random coil form, and the ellipticity for the completely α-helical form, respectively. There is some uncertainty about the latter two quantities; in calculations for our membrane experiments we have used -3000 and $-36,000$, respectively (Glaser and Singer, 1971).

An interesting application of CD measurements is to study the effect of the solvent on the conformation of a protein molecule. Early ORD measurements in the visible and near ultraviolet region of the spectrum (Doty, 1959) suggested that many globular proteins which are not extensively α-helical in aqueous solutions become predominantly α-helical in a less polar solvent such as 2-chloroethanol. CD measurements in the regions of the peptide bond absorption band confirm this conclusion (Lenard and Singer, 1966; Hashizume et al., 1967). In Fig. 10, for example, are shown CD spectra taken several years ago for bovine pancreatic ribonuclease (RNase) and bovine serum albumin (BSA) in aqueous buffer and in 2-chloroethanol (Glaser, M., and Singer, S. J., unpublished data). In its native aqueous conformation, RNase has only about 15% of its peptide bonds in α-helical regions, according to x-ray crystallographic studies (Wyckoff et al., 1967) and as is compatible with the CD spectrum (Beychok, 1966) in Fig. 10A. In 2-chloroethanol solution, however, RNase is about 70% helical. Similar but less marked changes occur with BSA with change in solvent. Results such as these provide good evidence (Singer, 1962) for some of the conclusions about the factors that determine protein structure, particularly the role of the solvent, discussed in Section II. They show the following. (a) Globular protein molecules in aqueous solutions can adopt conformations with very little α-helix content in order to be in their lowest free-energy states in water solution. This is clearly not because of any steric constraints on much more extensive α-helix formation, since these proteins become predominantly α-helical in 2-chloroethanol without breaking any covalent bonds. (b) The native aqueous conformation of the protein molecule is stabilized to a large extent by hydrophobic interactions with the solvent.

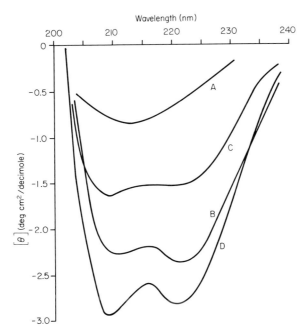

Fig. 10. The CD spectra of bovine pancreatic ribonuclease in 7 mM phosphate buffer, pH 7.4 (A) and in 2-chloroethanol solution (B) and of bovine serum albumin in the buffer (C) and in 2-chloroethanol solution (D), (Glaser, M., and Singer, S. J., unpublished data.)

Replacing the water with 2-chloroethanol changes the conformation largely to the α-helix mainly because hydrophobic interactions (or their equivalent in 2-chloroethanol) are eliminated. (c) The α-helix is more stable in a less polar environment.

E. Circular Dichroism and the Structures of Membrane Proteins

Circular dichroism has certain important advantages over other physical methods for the study of the proteins of intact membranes or membrane fragments. The sensitivity of currently available CD instruments in the wavelength region of the peptide bond absorption band is large enough so that dilute aqueous suspensions of membranes can be examined. On the other hand, infrared spectral measurements, which also yield information about polypeptide conformation, are usually carried out on dry films in the absence of water. Given the important role of the solvent water in determining macromolecular structures, this represents a severe liability for infrared studies. A second advantage of CD is that there is little spectral interference from lipid and oligosaccharide con-

stituents. Even if amide and ester groups of certain lipids and oligosaccharides have absorption bands in the vicinity of 200 nm, their CD spectral contributions particularly at 222 nm appear to be of small magnitude
compared to that of the protein (Lenard and Singer, 1966; Stone, 1969).
[A contrary claim was made for lipids by Urry et al. (1967). These
authors obtained a CD spectrum for a commercial lecithin preparation
which was not chemically analyzed. On the other hand, the lipids extracted from red blood cell membranes, dissolved in 2-chloroethanol, did
not show a CD spectrum of significant magnitude (Lenard and Singer,
1966).] While nucleic acids might interfere with the CD spectra, their
concentrations in most membrane preparations are generally so small as
to present no problem.

There are certain problems with the method, however. The CD measurements of large membrane fragments are averaged over all the proteins
in the preparation. Given the heterogeneity of the proteins of any one
membrane, the spectra may be averaged over proteins of very similar or
very different conformations without any indication of which situation
prevailed. In addition, there are some CD spectral artifacts which arise
with suspensions of particulate materials (Urry and Ji, 1968; Urry and
Krivacic, 1970) because of light-scattering and absorption anomalies,
which are briefly discussed below.

The first ORD and CD studies of membrane systems (Ke, 1965; Lenard
and Singer, 1966; Wallach and Zahler, 1966) in the wavelength region
of the peptide bond absorption band, although of low sensitivity, already
suggested that a substantial fraction of the protein of several different
intact membranes and membrane fragments was in the α-helical conformation. This was soon confirmed with other membrane systems as
well (Urry et al., 1967; Mommaerts, 1967). More recently, better instrumentation has allowed more accurate CD spectra to be obtained, an
example of which is shown in Fig. 11. In this figure a spectrum of intact
human red blood cell membranes in dilute aqueous buffer, as prepared
by the method of Dodge et al. (1963), is compared with that of the
protein hemoglobin. These CD spectra are discussed in detail elsewhere
(Glaser and Singer, 1971). Hemoglobin is a protein which is about 70%
in the α-helical conformation. The two CD spectra have been superimposed by adjusting their relative $[\theta]$ scales in Fig. 11 to show that the
shapes of the two spectra and the wavelengths of their maxima and two
minima are very similar. There are two significant differences, however.
(a) If the spectra are superimposed at about 222–223 nm, then it is clear
that the entire n-π^* band is somewhat shifted to longer wavelengths and
has a minimum at about 223 nm instead of 222 nm for the membrane
sample (the "red shift" observed by Lenard and Singer, 1966; Wallach

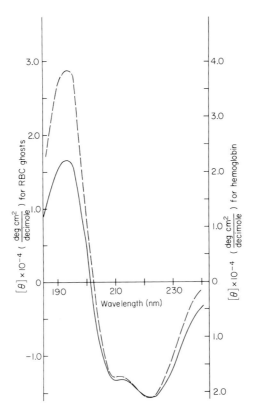

Fig. 11. A comparison of the CD spectra of human hemoglobin (- - -) and intact human red blood cell membranes (—) in 7 mM phosphate buffer, pH 7.4, with respective ellipticity scales adjusted to give similar minima near 222 nm (Glaser and Singer, 1971). This comparison makes evident the "red shift" in the membrane spectrum at wavelengths greater than 222 nm and the reduced ellipticity of the membrane preparation at the maximum near 194 nm (see text for details).

and Zahler, 1966). (b) The value of [θ] at the maximum at 194 nm is only about half as large for the membrane sample as for hemoglobin. A variety of explanations has been given for these differences (Lenard and Singer, 1966; Wallach and Zahler, 1966; Urry et al., 1967). Studies with physically fragmented red blood cell membranes (Glaser and Singer, 1971), however, have shown conclusively that both of these differences are largely or completely attributable to optical artifacts pointed out by Urry and his co-workers (Urry and Ji, 1968; Urry and Krivacic, 1970). The red shift near 222 nm appears to be an artifact due to different extents of light-scattering of the left- and right-handed circularly polarized components, while the reduced value of [θ] at around 190 nm is probably due to a combination of light-scattering and absorption flattening effects. Taking these artifacts into account, the corrected membrane spectrum and the hemoglobin spectrum are even more nearly similar in shape than they appear in Fig. 11.

These CD results provide the following information. It is clear from the characteristic double minimum that a substantial fraction (estimated below) of the protein in the intact red blood cell membrane is in the right-handed α-helical conformation. The close correspondence of the membrane and hemoglobin spectra in the wavelength region from 200–223 nm suggests that no significant amount of β-structure is present in the membrane proteins, since the shape of that region of the spectrum would be markedly affected by β-structure. The fraction of the membrane protein not in the α-helical form is therefore assigned to the random coil conformation. Although the $n-\pi^*$ band is artifactually red shifted, the magnitude of $[\theta]$ at the minimum of this band is not seriously affected (Glaser and Singer, 1971); the absorption flattening artifact for the protein in a thin membrane is negligibly small near 222 nm although it is not negligible near 190 nm. From these considerations, it follows that the observed value of $[\theta]_{222} = -16,000$, can be used to estimate the fractional helicity by Eq. (11); this yields the result that on the average, about 40% of the protein of the intact human red cell membrane is in the right-handed α-helical conformation. If the optical artifacts have been underestimated by us, then the α-helical content is even larger.

The CD spectra of a wide range of membranes and membrane fragments of different types and sources are closely similar (Fig. 12), indicating that little or no β-conformation and 30–40% α-helical conformation are present in the membrane proteins in each case.

The partial helical character of the membrane proteins suggests that the proteins are generally in a more or less globular form within the intact membranes. Like the soluble proteins RNase and BSA (Fig. 10), membrane proteins adopt a conformation which is that of lowest free energy in their particular membrane environment; in the presence of a protein denaturant such as $6 M$ guanidinium salt, they are dissociated and converted to the random coil conformation, and in 2-chloroethanol to essentially completely α-helical form, according to CD spectra (Fig. 2 in Lenard and Singer, 1966). That a major portion of the membrane protein is in the α-helical form is not consistent with the Davson-Danielli-Robertson model (Fig. 3a). In fact, this model would lead to the expectation that most of the membrane protein would be in the antiparallel pleated sheet, or β-conformation, whereas little or no β-conformation is detectable by CD measurements or infrared spectra (Maddy and Malcolm, 1965; Wallach and Zahler, 1966). Maddy (1966) suggests that a

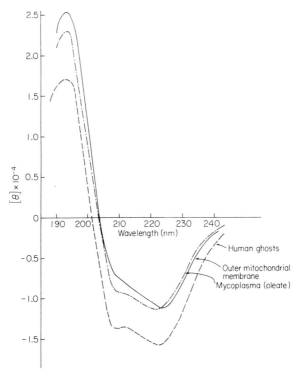

Fig. 12. The CD spectra in neutral aqueous buffers of intact human red blood cell membranes, of the outer membranes of rat liver mitochondria, and of the membranes of *Mycoplasma laidlawii* with the lipid fatty acids enriched in oleic acid. The spectrum was the same for the membranes with the lipids enriched in elaidic acid (Glaser, M., Simpkins, H., and Singer, S. J., unpublished data.)

substantial fraction of the protein in the Davson-Danielli-Robertson model might be permitted to be in the α-helical conformation. However, such protein would have to possess a large fraction of hydrophilic groups in order to expose them both to the water phase and to the polar heads of the lipids for electrostatic bonding. This would require the polypeptide to be highly charged and would not be likely to result in a stable α-helix.

The plasma membranes of red blood cells and Ehrlich ascites carcinoma cells gave no evidence of any β-structure by infrared spectral analyses. There appeared, however, to be some β-structure in lyophilized films of preparations of inner and outer membranes of rat liver mitochondria (Wallach *et al.*, 1969). Whether the preparative procedure, including lyophilization, might have induced the formation of some β-structure in this case is not clear, but in any event the proportion of β-structure was small. Since globular proteins contain significant amounts of β-structure, the presence of some β-structure in the proteins of intact membranes is not inconsistent with the lipid–globular protein mosaic model. The Davson-Danielli-Robertson model, however, would appear to require that β-structure be the predominant conformational feature of membrane proteins.

What might the remarkable similarity of different membrane CD spectra mean? Perhaps the similarity is fortuitous, given the gross heterogeneity of the proteins of any one membrane and the fact that the spectra are mainly determined by the α-helical content. On the other hand, there are reasons to believe that the CD spectral similarities are significant. For the large majority of soluble proteins that have been examined by CD, distinctive individual spectra often quite different from those of Fig. 12 have been observed. In particular, only a relatively small number of soluble proteins have α-helical contents of 30% or greater (Timasheff *et al.*, 1967). If the membrane CD spectra were simply uncorrelated averages over a random set of membrane proteins, therefore, they might be expected to show much less α-helical character and perhaps substantial differences from one membrane to another. We suggest, therefore, that some general structural characteristics of membranes and membrane proteins are reflected in the close similarities of the different CD spectra. This inference is considered further in Section IV,J.

F. The Action of Phospholipases on Membranes

Enzymes capable of catalyzing the hydrolysis or formation of specific bonds in lipids are likely to be increasingly important in studies of membrane structure. Unfortunately, although a variety of enzymatic activities have been described, in only a few cases have relatively pure enzymes been isolated. Furthermore, the rates of some enzymatic hydrolyses have been shown to depend strongly on the local environment of the susceptible bonds within lipid bilayers.

Phospholipases A_1 and A_2 (Van Deenen, 1969) hydrolyze specific ester linkages in phosphoglycerides to form monoacylphosphoglycerides. These products are effective detergents and disrupt membranes when present in low concentrations. Therefore, it is impossible to interpret the results of the phospholipase A digestion of a membrane in terms of its original structural organization.

Phospholipase C enzymes (Macfarlane, 1948) catalyze the removal of the phosphorylamine moieties from certain phosphoglycerides. The residual diglyceride does not have the detergent properties. The enzyme from *Bacillus cereus* has been highly purified (Ottolenghi, 1965), but in several studies cruder enzyme preparations have been utilized. Finean and

Martonosi (1965) showed that although the action of a crude enzyme preparation on muscle microsomes released 60–70% of the phosphorylated amines of the membrane phospholipids, no disruption or gross alteration in the railroad-track appearance of the membrane in thin-section electron microscopy resulted. Lenard and Singer (1968b) demonstrated that a crude enzyme rapidly liberated about 70% of the phosphorous of intact red cell membranes without disrupting the membrane and without significantly altering the protein CD spectrum of the membrane. This study has recently been extended by Glaser et al. (1970), using a purified phospholipase C (Ottolenghi, 1965). These investigators confirmed that no detectable change in the CD spectrum of the red cell membrane followed phospholipase C treatment and further showed by proton magnetic resonance measurements that the fatty acid chains of a large fraction (75%) of membrane lipids were rendered much more mobile than they were in the untreated membrane. Furthermore, the protein in the treated and untreated membranes had about the same conformational stability toward increasing temperature. Gordon et al. (1969) have reported that, on the contrary, the action of a (crude) phospholipase C on membrane preparations did result in CD changes. These changes, however, were almost certainly artifactual, to be attributed to light-scattering changes in the suspension of treated membranes (Glaser et al., 1970; Urry and Krivacic, 1970).

The results of Lenard and Singer (1968b) and Glaser et al. (1970) led them to the following conclusions. (a) The polar heads of the phospholipids in the intact membrane must be accessible to the active site of the phospholipase C molecule in order to form the enzyme–substrate complex readily; (b) electrostatic interactions between the polar heads of the phospholipids and the charged groups of the protein cannot be the predominant source of stability of the membrane or even a primary determinant of the conformation of the membrane proteins, since removal of 70% of the polar heads leaves the membrane intact and the protein conformation not detectably altered by CD spectral criteria; and (c) the bulk of the lipids and proteins are largely independent of one another since the physical state of a substantial fraction of the lipids can be markedly altered without corresponding effects on the physical state of the protein.

These results are therefore inconsistent with both the Davson-Danielli-Robertson and Benson models as representing the predominant mode of molecular organization of red cell membranes. The former model leads to the expectation that the enzymatically catalyzed hydrolysis of the protein-covered phosphoester bonds of the phospholipids would be slow, perhaps immeasurably so, and that if it did occur, it would result in the release of a significant fraction of the membrane protein which was no longer capable of being bound. The Benson model, which involves an intimate interaction of the fatty acid chains of the lipids and the polypeptide chains of the protein, would seem to predict that a profound change in the physical state of most of the fatty acid chains would lead to a profound change in the protein conformation and the membrane stability. The lipid–globular protein mosaic model, however, can rationalize these results. Since it proposes that the ionic heads of the phospholipids are exposed to the bulk aqueous phase, the ready enzymatic catalysis of their removal is predicted. Furthermore, if hydrophobic rather than electrostatic interactions hold the membrane components together, but the phospholipid and protein are largely sequestered from one another, the removal of the bulk of the ionic lipid groups might not lead to any major effects on the stability of the membrane proteins.

G. Phase Changes of Membrane Lipids

Several kinds of physical experiments have been performed which bear on the state of the lipids of membranes, and these are discussed in Chapters 1–3. Suffice it to mention here that an important series of experiments has been carried out in a coordinated manner with intact membrane preparations from the prokaryotic organism, *Mycoplasma laidlawii* (Razin, 1969). These include (a) differential calorimetry (Steim *et al.*, 1969; Melchior *et al.*, 1970), (b) x-ray diffraction (Engelman, 1970), and (c) spin labeling (Tourtellotte *et al.*, 1970). The differential calorimetry experiments indicate that the bulk of the heterogeneous lipids of the intact membrane show a broad phase transition at a mean temperature very similar to that of the isolated lipids in aqueous suspension. Since the isolated lipids are presumably in the bilayer form, these results indicate that at least a large fraction of the lipids in the intact membrane are also arranged in a bilayer form at temperatures that are sufficiently below the "melting point." The spin labeling results support this conclusion. There are certain limits on the interpretation of these results, however, which should be appreciated.

The results are difficult to quantitate. That is, it is not certain whether all or only a large fraction of the phospholipids of the intact membrane are arranged as bilayer. The possibility that some, perhaps as much as 30%, of the phospholipid may be chemically and physically distinguished from the remainder is not excluded by any of the data.

Furthermore, these results do not prove that the bilayer is a continuum. The phase transition of the lipids which is observed in the differential calorimetry and x-ray experiments is not a typical "melting" as is observed with ordinary small-molecule solid–liquid transitions. Because the phospholipids are confined in two dimensions in the bilayer, and particularly because the fatty acid chains are extended structures, the phase transition most probably involves a cooperative increase in the rotational mobility of the individual fatty acid chains around a single axis perpendicular to the plane of the bilayer (Hoffman, 1959). The cooperative unit need not be an essentially infinite bilayer. The size of the cooperative unit which is required to account for the observed differential calorimetry data is in fact small (Melchior *et al.*, 1970). (This, however, may be in part artifactual due to the heterogeneity of the lipids in the mycoplasma membrane.) The point to be made is that a lipid–globular protein mosaic model involving a *discontinuous* lipid bilayer (Fig. 3c) is not inconsistent with the data discussed in this section. Of course, the Davson-Danielli-Robertson model satisfies the data. On the other hand, the data render the Benson model unlikely for mycoplasma membranes.

H. Fluorescent Probes and Membranes

An interesting development in biophysical studies of membranes involves the use of certain fluorescent dyes which fluoresce intensely in nonpolar media but not in water (see Chapter 2). The most widely used such dye is 1-anilinonaphthalene-8-sulfonic acid (ANS) (Weber and Laurence, 1954). These dyes have been particularly used to probe for hydrophobic regions in protein molecules (cf. Stryer, 1965; McClure and Edelman, 1966) and more recently to investigate the properties of membranes (Tasaki *et al.*, 1968; Rubalcava *et al.*, 1969; Vanderkooi and Martonosi, 1969; Azzi *et al.*, 1969). Because ANS has been shown to bind to hydrophobic sites on simple soluble proteins, it has sometimes been implicitly assumed that ANS binding to membranes occurs at protein sites also. However, the studies of Rubalcava *et al.* (1969) and Vanderkooi and Martonosi (1969) strongly indicate that with red cell membranes and muscle microsomal membranes, respectively, the ANS binds mainly, if not exclusively, to sites involving the phospholipid portion of the membrane. The binding to phospholipid and detergent micelles very likely involves the intercalation of the amphipathic ANS molecule in the lipid bilayer or micelle, with the sulfonate ionic group directed to the exterior surface of the bilayer or micelle and the nonpolar part into the hydrophobic interior.

One important result of these studies is particularly relevant. This involves the effects of cation concentration on the binding and fluorescence of ANS in membranes as compared to aqueous dispersions of phospholipids (Vanderkooi and Martonosi, 1969). The effects of cations were remarkably similar in the two types of systems. Increasing concentrations of monovalent cations enhanced the ANS fluorescence in suspensions of microsomal membranes or in micelles of dipalmitoylphosphatidylcholine. Different monovalent cations at the same concentration gave the same fluorescent enhancement, which varied linearly with ionic strength up to some maximum enhancement (Rubalcava et al., 1969). With either Ca^{++} or Mg^{++}, similar effects were observed but at equivalent concentrations of about one-hundredth that of the monovalent ions. With $LaCl_3$, the same enhancement occurred at concentrations about another factor of 100 lower than that for divalent cations. The interpretation these authors placed upon their results is that the effect of an increase in the concentration of a given ion type is to change the ion atmosphere, or double layer, at the surface of the membrane or of the phospholipid bilayer, thereby decreasing the charge repulsions between the ionic heads of the phospholipid molecules. The resultant changes in lipid structure slightly, but significantly, alter the binding properties of the membrane or micelle for ANS, and this enhanced binding is observed as a fluorescence increase. (The quantum yield of fluorescence, however, is not changed.) On the other hand, while the effect of any one electrolyte depended linearly on the ionic strength, the effect of an increase in the charge on the cation was much larger than would be expected from an increase in ionic strength alone, suggesting that binding of the multivalent cations to the ionic heads of the phospholipids might be involved, the phospholipid bilayer acting as an ion exchanger (Abramson et al., 1964).

The most interesting result, however, was that the effects of increasing cation concentration and the unusual effects of increasing cation charge on the ANS fluorescence *were closely parallel* for the membranes and for the dispersions of dipalmitoylphosphatidylcholine, indicating that the ionic heads of the lipids in both cases exhibited closely similar interactions with ion atmospheres. We suggest that this is strong evidence that the bulk of the phospholipids in membranes have their ionic heads exposed to the aqueous environment as they do in dispersions of isolated phospholipids. This is consistent with the lipid-globular protein mosaic model but *not* with the Davson-Danielli-Robertson model, which covers the ionic heads with protein (see Section III,G). The ANS results likewise suggest that it is unlikely that the phospholipids are arranged as the Benson membrane model proposes, since it is so radically different from a bilayer. (Other aspects of ANS experiments with membranes are considered in Section V,C.)

I. Lipid–Protein Interactions in Membranes

The data discussed in the previous two sections demonstrate that the bulk of the phospholipids and the proteins of membranes are not tightly coupled to one another. On the other hand, there is considerable evidence that lipid–protein interactions play a direct role in a variety of membrane functions. These effects may be observed at the whole membrane level, or with individual enzyme or antigen activities isolated from membranes. Two examples of the former are (a) the studies of Fleischer *et al.* (1967) on the controlled partial extractions of the lipids of mitochondria, the effects of such delipidation on the loss of electron transport activity, and the reconstitution of activity on restoration of phospholipids and coenzyme Q; and (b) the demonstration by Schairer and Overath (1969) that the rate of thiomethylgalactoside accumulation into intact *E. coli* cells depends on the fatty acid composition of the lipids of the cell membrane, possibly by exerting a structural effect on the *lac* permease protein (Fox and Kennedy, 1965). A few examples of many individual enzyme activities which are dependent on lipids, often on specific lipids, are the β-hydroxybutyrate dehydrogenase of mitochondria (Jurtshuk *et al.*, 1961), the ion-dependent ATPases of red cell membranes (Ohnishi and Kawamura, 1964; Roelofsen *et al.*, 1966), the galactosyltransferase from the membranes of *Salmonella typhimurium* (Rothfield *et al.*, 1969), and the vectorial phosphorylation of sugar by membranes of *E. coli* (Milner and Kaback, 1970). (The effects of lipids on membrane enzymes are discussed in greater detail in Chapter 6.)

It is possible that these effects of lipids on membrane functions have a trivial explanation, namely that the lipids are required to disaggregate the membrane proteins, perhaps uncovering their active sites in the process. On the other hand, the wide distribution of these lipid effects suggests that a more general explanation is required. This could be the effect of lipid on the conformations of individual membrane proteins. Such effects are to be expected on theoretical (Section II,C) and experimental (Section IV,D) grounds. In the absence of the lipids, or in the presence of altered lipids, the conformations of these enzymes might be significantly different from those that are required for the expression of their native activities.

The apparent conflict between the physical results that suggest that most of the membrane lipid does not interact strongly with the membrane proteins and the chemical evidence for lipid–protein interactions may be reconciled if it is postulated that in the intact membrane some relatively small fraction of the lipids is structurally differentiated from the rest and is more directly involved in interactions with the proteins. At the present time, there is no really satisfactory evidence for such a differentiation of the lipids, but neither is it firmly excluded by any evidence to our knowledge. In the phospholipase C experiments, for example, a maximum of only about 70% of the phospholipids are hydrolyzed from different membranes; this at least allows the possibility that the remaining 30% are structurally differentiated, although other explanations for these results are not ruled out. It has long been known (Parpart and Ballantine, 1952) that the lipids of membranes are not equally extractable by particular organic solvent combinations, and that "weakly" and "strongly" binding lipid fractions appear to exist by these criteria. Those results, however, do not provide good evidence for the structural differentiation of lipids because solvents certainly alter membrane structure and may thereby generate lipid–protein combinations not present in the original membrane.

Such structural differentiation of the lipids, if it exists, might have a number of different chemical and thermodynamic bases; for example, anionic phospholipids such as phosphatidylserine, phosphatidylglycerol, or phosphatidylinositol might be distributed differently from zwitterionic lipids in the vicinity of membrane proteins because of a bias introduced by electrostatic interactions. Such electrostatic interactions are viewed as occurring in the lipid–globular protein mosaic model through the side-by-side juxtaposition of anionic lipids and the cationic residues of protein (Fig. 3c); under such circumstances these interacting ionic groups would *remain* in contact with water, in contrast to the Davson-Danielli-Robertson formulation. It is of some interest in this connection that the restoration of the ion-dependent ATPase activities of red blood cell membranes deprived of lipids was found to be more effective with phosphatidylserine than with other phospholipids (Ohnishi and Kawamura, 1964; Roelofsen et al.; 1966). The fact that dioleylphosphatidylethanolamine was over 600 times more effective than dioleylphosphatidylcholine in restoring galactosyltransferase activity (Rothfield et al., 1969) strongly suggests that the difference in size of the zwitterionic group may play an important role in differentiating otherwise closely similar phospholipids.

The possibility of at least two states for membrane lipids could be incorporated into the mosaic model in Fig. 3c by associating a small fraction of the membrane lipid with the protein moieties, creating a mosaic of lipoproteins interspersed with the remaining lipid in a bilayer. This aspect of lipid–protein interactions in membranes deserves a great deal of experimental attention in the future.

J. The Action of Proteases on Membranes

It is an interesting fact that the action of proteases on membranes is often limited. For example, Ito and Sato (1969) found that tryptic digestion of rabbit liver microsomes stopped when about 22–25% of the membrane-bound protein was released. Further addition of enzyme had no effect, and the same result was obtained with other proteolytic enzymes. These results, however, cannot readily be employed to discriminate among the different gross models of Fig. 3. The enzymes may have access to only one side of the membrane, and the heterogeneity of membrane proteins may make for a heterogeneity of susceptibilities to proteolysis, so the limited digestion of only about one-quarter of the membrane can be rationalized by any of the models.

On the other hand, proteolytic digestion may be useful in the selective release of protein fragments from the surface of a membrane. This may be illustrated by an example. Ito and Sato (1968) studied the properties of cytochrome b_5 from rabbit liver microsomes. If the intact microsomal membranes were treated with trypsin, the cytochrome b_5 spectral activity was solubilized in a nonaggregating protein of molecular weight 12,000 (trypsin–b_5). On the other hand, if intact microsomal membranes were treated with detergent, the b_5 spectrum was found associated with a protein (detergent–b_5) which had a molecular weight of 25,000 but which aggregated in the absence of detergent or urea. If detergent–b_5 was treated with trypsin, the b_5 spectrum was found in a soluble 12,000 molecular weight fragment. These authors concluded that cytochrome b_5 is a single polypeptide chain but that it is cleaved by tryptic hydrolysis into two roughly equal-sized moieties: One, carrying the b_5 activity, is itself quite soluble, but the other, the non-b_5 portion, confers hydrophobicity upon the entire molecule and presumably is responsible for its attachment to the membrane. The b_5-carrying portion can be cleaved off the intact membrane by trypsin. The susceptible bonds are presumably, therefore, accessible to the enzyme, and this moiety is probably exterior to the membrane. The other moiety is probably interior.

It appears that a glycoprotein of molecular weight about 31,000 in human red blood cell membranes is also cleaved by tryptic treatment of the membranes into soluble glycopeptides of about 10,000 molecular weight, while the remaining portions are quite hydrophobic, again suggesting the existence of a single membrane protein differentiated into exterior and interior regions (Winzler, 1969).

In both of these cases, the protein involved is an integral protein of its respective membrane, by the criterion that detergents or organic-solvent extraction are required to release it from the membrane. Although it may be hazardous to attach too much significance to these two sets of experiments, the interesting possibility arises that integral proteins of membranes may often be enzymes or endowed with other metabolic activities (such as antigens, drug or hormone receptors, or permeases) and *consist of two types of structurally distinct regions, one type interior and one exterior to the membrane* (Fig. 13). Each region would be compact and globular. The interior region would be the part of the protein

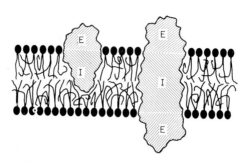

Fig. 13. A schematic representation of a lipid–globular protein mosaic model, with integral proteins of the membrane differentiated into exterior (E) and interior (I) regions (see text for details).

that is bound to the fatty acid chains of the lipids in the interior of the membrane, and the (one or two) exterior regions would be parts of the protein bearing a specific active site or an antigenic determinant that required exposure to the aqueous phase on one or both of the outer surfaces of the membrane. By virtue of their interior regions, such proteins would be critically involved in the maintenance of the structure of the membrane and yet would not be solely "structural proteins" in the sense used by Green and co-workers (Richardson *et al.*, 1963), since by virtue of their exterior regions they could be enzymes, etc. The differentiation of a membrane protein into interior and exterior regions would not require that two regions of equal size exist, or that the amino acid sequences involved be sequestered into two or three distinct segments of a single polypeptide chain. While such a partition of the sequence into two discrete halves apparently occurs with the cytochrome b_5 of microsomes (Ito and Sato, 1968), it is not necessary that this be the case. It is conceivable that the single polypeptide chain could have successive segments weaving in and out of the interior and exterior regions of the protein. In that case, proteolytic digestion would not be expected to cleave off the whole of the exterior regions of the protein, and so the results of Ito and Sato (1968) might not be observable with other membrane proteins even if the existence of interior and exterior regions were a more general phenomenon. On the other hand, it is also possible that the bipartite situation with cytochrome b_5 is not uncommon. The polypeptide chain of an integral membrane protein may have an amino acid sequence divided into linear segments that fold up more or less independently of each other to give three-dimensional interior and exterior regions to the molecule. Other types of proteins are known, e.g., the immunoglobulins and the histones, whose single polypeptide chains appear to be differentiated into two nearly equal-sized linear segments. Amino acid sequence determinations carried out with pure membrane proteins are still some time off, but they should be keenly interesting.

If integral membrane proteins are generally differentiated into interior and exterior regions, the interior regions of different proteins, being adapted to bind to the hydrophobic interior of the membrane, may be structurally related to some extent. In particular, they may all be largely in the α-helical conformation, since in Section IV,D it was demonstrated that many different proteins adopt a 70–80% α-helical conformation in an environment of low polarity. The exterior regions, on the other hand, may exhibit a wide range of conformations, as is observed with ordinary soluble proteins. These speculations may provide an explanation, therefore, for the observations discussed in Section IV,E that widely different membranes exhibit closely similar CD spectra, reflecting the presence of about 30–40% of the protein in the α-helical conformation in the intact membranes. If the α-helicity were mainly due to the interior regions of integral proteins and the interior regions constituted on the average roughly half of the structure of different membrane proteins, these similarities would follow.

Certain experimental predictions follow from these speculations. One is that if an individual membrane-bound enzyme functions as an integral membrane protein and consists of interior and exterior regions, then when it is isolated and examined as a "pure" lipoprotein complex, [e.g., the acetylcholinesterase liberated from red blood cells (Burger *et al.*, 1968)], it should exhibit a protein CD spectrum very similar to that of intact membranes, with roughly similar α-helical content.

K. Summary and Conclusions

The discussion of the selected experimental evidence in this section may be summarized as follows.

1. The railroad-track staining patterns observed with OsO_4- or $KMnO_4$-fixed and polymer-embedded membranes in thin sections do not yield information that can discriminate among the different membrane models.

2. Electron microscopic observations of extensively delipidated membranes, showing an almost unaltered staining pattern, are not consistent with the Davson-Danielli-Robertson model but can be rationalized by the Benson and the lipid–protein mosaic models.

3. The particulate structures revealed by freeze-etching techniques to be in the interior of membranes are predicted to exist and to be protein by the lipid–globular protein mosaic model; they are inconsistent with the Davson-Danielli-Robertson model; and they can be rationalized by the Benson model.

4. Myelin is chemically quite different from most cell membranes that have been adequately studied, and information obtained from structural studies of myelin are therefore not clearly relevant to cell membranes. Even so, the x-ray data for myelin do not discriminate between the Davson-Danielli-Robertson and the lipid–globular protein mosaic models.

5. The results of surface film measurements of membrane lipids at air–water interfaces are more readily rationalized by the lipid–globular protein mosaic model than by the Davson-Danielli-Robertson model.

6. Circular dichroism measurements indicate that on the average, some 30–40% of the protein of intact membranes is in the α-helical conformation. This is inconsistent with structural expectations for the Davson-Danielli-Robertson model but can be rationalized by the Benson and mosaic models.

7. Phospholipase C treatment of red cell membranes cleaves off about 70% of the phosphorylated amines of the phospholipids without disrupting the membrane or changing the protein CD spectrum. These results are inconsistent with the Davson-Danielli-Robertson model, are not easily reconciled with the Benson model, but can be rationalized by the lipid–globular protein mosaic model. These experiments rule out the possibility that membrane proteins and lipids are held together predominantly by electrostatic interactions.

8. Differential calorimetry and x-ray measurements on mycoplasma membranes suggest that at least the major fraction of the lipids in intact membranes undergoes a broad change of phase very similar to that of the isolated membrane lipids dispersed in water. This is expected for the Davson-Danielli-Robertson model, is not inconsistent with the lipid–globular protein mosaic model, but is strong evidence against the Benson model.

9. The effects of ionic composition on the binding of the fluorescent dye ANS to aqueous dispersions of isolated phospholipid bilayers and to membranes are very similar, suggesting that the ionic structure and ion-exchange capacity at both types of lipid surface is similar. The lipid–globular protein mosaic model predicts this, because the ionic heads of the phospholipids of the membrane are exposed to the aqueous environment as they are in isolated phospholipid bilayers; however, the Davson-Danielli-Robertson model, in which the ionic heads are largely blanketed by protein, would suggest quite different properties for the two systems.

As this summary indicates, there are significant pieces of experimental evidence that cannot be reconciled with either the Davson-Danielli-Robertson or the Benson models as representing the predominant pattern of organization of the lipids and proteins of membranes. To the best of our knowledge and objectivity, none of the data presented, nor any other results so far obtained, are clearly inconsistent with the lipid–globular protein mosaic model. Although direct and unambiguous evidence for the mosaic model has not yet been obtained, its consistency with experiment and its thermodynamic feasibility recommend it for further investigation. The lipid–globular protein mosaic model therefore provides a good working hypothesis for the gross organization of a membrane, and in the next section some of the functions of membranes and the possible mechanisms of their biosynthesis and transformation are considered in terms of this model.

V. THE LIPID–GLOBULAR PROTEIN MOSAIC MODEL AND MEMBRANE FUNCTION

The lipid–globular protein mosaic model illustrated in Fig. 3c is clearly a gross schematic model intended to convey a very general picture of the predominant interactions and organization of the proteins and lipids of membranes. There are many directions in which it can be diversified and refined, although there is not much experimental basis for such elaboration at the present time. The possibility exists, for example, as discussed in Section IV,I, that membrane lipids may be structurally differentiated and while the major part is sequestered as the bilayer matrix of the mosaic, some smaller part is in a closer interaction with the integral proteins of the membrane. It has also been considered (Section III,D) that some generally small fraction of membrane proteins might be globular *peripheral* proteins, not intimately associated with lipid, and bound to the membrane in considerable part by electrostatic interactions. Another important possibility discussed in Section IV,J is that a variety of enzymes and other biochemically active proteins may function as integral proteins and that a globular protein molecule may be structurally differentiated into an interior and exterior region (Fig. 13). Some further possible refinements are discussed in the next two sections. In the remaining two sections, the lipid–globular protein mosaic model is used as a basis for speculations about possible mechanisms of membrane function and biosynthesis.

A. Multiple-Subunit Proteins in Membranes

One reason that the lipid–globular protein mosaic model is attractive is that it provides a rational, rather than an *ad hoc*, basis for the penetration of proteins into the matrix of a membrane. Whether a protein molecule completely spans the thickness of a membrane, however, or is only partially intercalated into it might depend in part on the molecular

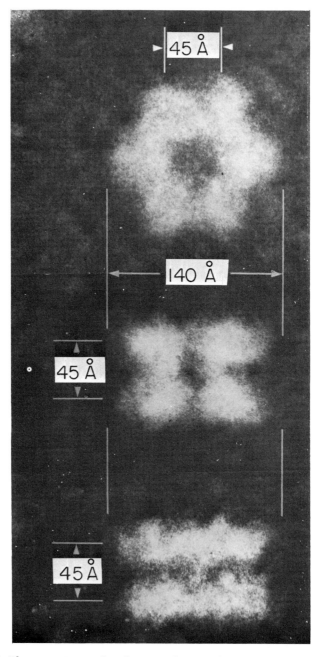

Fig. 14. Electron micrographs of negatively stained molecules of glutamine synthetase of *E. coli* in three views. The molecule consists of 12 subunits stacked in two

weight of its polypeptide chain and also on its specific subunit structure. The molecular weights of molecularly dispersed membrane proteins cover a wide range of values (cf. Kiehn and Holland, 1968; Lenard, 1970); other things being equal, the larger the molecular weight, the larger the linear dimensions of the globular protein and the more likely it will extend completely across the membrane. In addition, however, a particular single protein may exist in the membrane as specific non-covalently bound aggregate of identical or nonidentical subunits. The specific aggregate may then span the membrane, although a single glob-ular subunit would not be extended enough to do so.

There is no adequate evidence at present that such multiple-subunit proteins exist in membranes, but the fact that they are extremely common and functionally very important among the cytoplasmic proteins (cf. Atkinson, 1969) suggests that they may be present and functional in membranes also. By analogy with their soluble counterparts, they might provide the mechanism that imparts two very important characteristics to membranes.

1. Although pores, and even protein pores, have been postulated as ion-transport mediators in membranes, simple proteins do not exhibit pores, their internal structure being quite compact for good thermody-namic reasons (Section II,B). Subunit proteins, however, often contain narrow water-filled channels near the central axis of the aggregate, and similar membrane proteins could thereby provide specific pores across a membrane. The roughly 10 Å water-filled channel down the center of the tetrameric hemoglobin molecule (Perutz, 1969) is the best-known example in a soluble protein. Another more complex example is the multiply allosteric enzyme, glutamine synthetase of E. coli (Valentine et al., 1968) which exists in solution as an aggregate of 12 identical sub-units arranged in two hexagonal rings with a hole running down the central axis of the molecule (Fig. 14). Other examples are discussed by Valentine (1969). If narrow channels were present in at least some multiple-subunit proteins in membranes, they might be expected to provide permanent pores for water molecules, but not for larger mole-cules, to permeate. Such pores would be expected to be lined with some of the ionic and polar residues of the proteins; these residues would be in contact with water, and therefore would exhibit hydrophilic interac-tions with the solvent (Section II,D). Clearly, some of these pores might be structurally specific for the permeability of certain ions through the membrane and might exclude others. Such fixed specific ion "carriers"

layers of six units each. The top view is down the axis of the molecule and reveals the hole which traverses the molecule. [Reproduced from Valentine et al. (1968) with permission.]

might then have structural properties roughly analogous to those of anti-biotic ion carriers such as alamethicin (Mueller and Rudin, 1968). Ala-methecin is a cyclic 19-amino acid oligopeptide which imparts selective cation transport to synthetic bilayer membrane systems, apparently by forming a stacked aggregate of six molecules extending through the membrane. The postulated protein subunit aggregates might contain both a central core capable of binding specific ions and an outer hydro-phobic surface at which the protein is attached to the interior of the membrane. Such subunit proteins in membranes could therefore mediate specific ion transport through membranes.

2. Soluble multiple-subunit proteins are often allosteric enzymes, achieving their effects by *quaternary rearrangements** of their subunits that occur upon binding a specific small-molecule ligand to an active site on one of the subunits. There are many well-studied examples of this phenomenon (for a review, see Atkinson, 1969), including the en-zyme glutamine synthetase mentioned above. [With this protein, a remarkably complex set of allosteric controls exists. Each of at least eight different compounds metabolically derived from glutamine only partially inhibits the activity of the enzyme; collectively they nearly completely inhibit it (Woolfolk and Stadtman, 1967).] The quaternary rearrange-ments which multiple-subunit proteins undergo can both amplify and project some considerable distance away, the effects of specific ligand binding.

Similar subunit proteins embedded in a membrane could provide a plausible general mechanism for many important membrane phenomena that are not adequately understood at present. All of these phenomena involve a change in a membrane function or property that accompanies the binding of some apparently unrelated small molecule; two examples are the effect of acetylcholine binding on the ion permeability of the postsynaptic membrane of a nerve–muscle junction (Podleski and Changeux, 1969) and the effect of specific hormones on the stimulation of the membrane-bound enzyme adenylcyclase (Robison *et al.*, 1969). In the first example, the binding of acetylcholine at a receptor site on

---

* Structural changes that occur in protein molecules on binding a ligand are all generally referred to as "conformational changes." It might be more useful con-ceptually if the latter term were used to denote only those changes in structure which occur within a single polypeptide chain and to refer to those structural changes in which the subunits have their relative orientation altered as "quaternary rearrangements." There is no clear evidence from x-ray data for hemoglobin (Perutz, 1969) at the resolution of 2.8 Å that the individual α- and β-chains undergo con-formational changes on binding oxygen, although they might occur and be too small to be detected; the major observable structural effect is a large reorientation of the α- and β-chains relative to one another.

one subunit could induce a quaternary rearrangement of the different subunits of the same protein molecule and thus change the ion-permeability characteristics of a pore extending down the central axis of the aggregate. In the second example, two different kinds of subunit might be present in the same protein molecule, an allosteric subunit and a catalytic one, as in the soluble enzyme aspartyl transcarbamylase (Gerhart and Schachman, 1965). The allosteric subunit might contain the hormone receptor site on the outer surface of the membrane, while the catalytic site on the other subunit would be on the cytoplasmic side of the membrane. The binding of the hormone could result in a quaternary rearrangement affecting the catalytic site on the other side of the membrane. The allosteric subunit containing the hormone receptor might be different in the plasma membranes of thyroid, adrenal cortical, or pineal gland cells to account for the fact that in each of these cells adenylcyclase is specifically stimulated by a different hormone.

The stability of such hypothetical multiple-subunit membrane proteins would depend, for the reasons given in Sections II and III, on interaction with the membrane lipids and the aqueous environment. The search for and isolation of such proteins must be designed with this in mind.

Complex structures associated with membranes, such as the flagella of bacterial membranes, or the headpieces and stalks of the inner mitochondrial membrane, may also be accommodated in the lipid–globular protein mosaic model. These may be complex protein subunit aggregates whose structures are initiated at and protrude from some specific integral protein unit in the membrane.

It is possible that the structure of myelin is also a mosaic of lipid and protein, containing a larger fraction of lipid than most membranes; the reason that myelin is an electric insulator may be that the proteins it contains have predominantly low molecular weights and do not interact to form large subunit aggregates. Under such circumstances, the protein molecules might also be predominantly globular and embedded in the mosaic, as in Fig. 3c, but unable to span the thickness of a single myelin layer, thereby providing no ion-permeable pores. Consistent with this hypothesis is the result that in freeze-etching experiments with myelin, cleavage within the single myelin layer reveals a perfectly smooth surface, whereas cleavage within a typical membrane shows a particulate structure (Section IV,A, Fig. 7).

B. The Membrane Matrix: Lipoprotein Subunits or Continuum?

Much attention has been given in recent years to the hypothesis that membranes are organized as an assembly of distinct repeating lipoprotein subunits, as contrasted to a continuum such as the Davson-Danielli-Robertson model proposes. (These lipoprotein subunits of a membrane are not to be confused with the subunits of the individual proteins of membranes discussed in the preceding section.) The question not only is of intrinsic interest to membrane structure, but it bears directly on the problem of membrane biosynthesis (see Section V,D).

The experimental evidence which has been adduced in favor of a lipoprotein subunit assemblage as a general mode of organization of membranes has been thoroughly reviewed by Stoeckenius and Engelman (1969) and will not be detailed here. We agree with their conclusion that the evidence at present is not convincing. The great heterogeneity of membrane proteins and lipids requires that if lipoprotein subunits did constitute a membrane, either they would have to be very large to be identical, or they would have to be correspondingly heterogeneous both chemically and structurally. If the latter were the case, it is unclear what would determine how such heterogeneous subunits would be arranged and assembled. Furthermore, the existence of discrete lipoprotein subunits would seem to require that the synthesis of lipid and membrane protein be reasonably synchronized and that the ratio of lipid to protein in a particular membrane be fairly constant. Neither of these expectations has been borne out in recent experiments with mycoplasma (Kahane and Razin, 1969). The membranes of chloramphenicol-treated cells (with protein synthesis stopped) had a significantly lower buoyant density than those of untreated cells, showing that a substantial increase in the ratio of lipid to protein had been produced in the membrane with no apparent effect on membrane function.

The lipid–globular protein mosaic model clearly has structural features of both a continuum and (at least superficially) a lipoprotein subunit assembly. In particular, if the protein of a membrane were of only one or a few kinds [as, for example, in the outer segments of visual receptor rods (cf. Bownds and Gaide-Huguenin, 1970) or in viral membranes], its arrangement in a lipid–globular protein mosaic might exhibit a fairly uniform periodic structure in the plane of the membrane, whereas in a more heterogeneous membrane such as that of the red blood cell, the mosaic arrangement of the many different integral proteins of the membrane, with perhaps variable amounts of intervening lipid, might blur out any appearance of periodicity. On the other hand, there is nothing intrinsic to the lipid–globular protein mosaic model that requires the existence of discrete lipoprotein subunits in the membrane. The fact that treatment of membranes with detergents usually results in the appearance of small lipid–protein aggregates has sometimes been used to suggest that such aggregates are the membrane subunits. One would predict, however, that detergent treatment of a membrane organized as a continuum of a lipid–globular protein mosaic would split the membrane within the lipid portions of the mosaic and produce small aggregates such as are experimentally observed.

If the lipid–globular protein mosaic is indeed organized as a continuum, is it one of proteins embedded in a lipid continuum or of lipids embedded in a protein continuum (i.e., which component provides the matrix of the membrane structure)? The schematic cross-sectional representation of the lipid–globular protein mosaic model in Fig. 3c deliberately avoids the issue, which must be joined when the third dimension of the model is specified. In favor of the view that the lipid provides the matrix of the continuum are the following arguments. (a) It accommodates more easily the results of Kahane and Razin (1969) mentioned above that the lipid–protein ratio of a mycoplasma membrane can be significantly increased without apparent alteration of membrane properties; since most of the lipid in the bilayer portion of the mosaic does not interact strongly with the protein (Section IV,F), extra lipid could be inserted without much effect. If protein–protein contacts provided the matrix of the continuum, however, and the lipids filled the spaces between, there might be greater difficulty in accommodating a significant additional amount of lipid into the membrane, if it were uniformly distributed throughout. (b) It allows many different arrangements of grossly heterogeneous integral membrane proteins in the plane of the membrane.* If protein–protein contacts provided the matrix, these contacts would presumably be specific, and a highly organized and uniform arrangement of the membrane proteins might result. The greater the heterogeneity of the membrane proteins, the more difficult it is to envision a regular repeating protein matrix.

The electron microscopic experiments discussed in Section IV,A, which revealed no marked alteration in the railroad-track appearance of fixed and embedded membranes if their lipids were previously extracted, might appear to be more consistent with a protein than with a lipid matrix for the mosaic structure. However, the possibility that some lateral shrinkage occurred in the specimens, as might be expected if the lipid provided the matrix, has not been ruled out. Also, bilayer-type phase changes observed for the lipids in intact membranes do not necessarily imply that the lipids form the continuum, since, as is discussed in

---

* In fact, a lipid matrix in which only relatively weak interactions among the major portions of the lipids and proteins existed allows the possibility for a *translational* diffusion of membrane lipids and proteins *in the plane of the membrane*. That such diffusion occurs at significant rates is indicated by the recent studies of Frye and Edidin (1970).

Section IV,G, the size of the cooperative unit involved in the particular phase change is not known. It is not inconceivable that a bounded unit containing of the order of 100 lipid molecules might "melt" at a temperature not noticeably different from that of an infinite bilayer, if melting involved a cooperative increase in a uniaxial rotational motion of the long fatty acid chains.

We are inclined to the view that if the membrane is a continuous lipid–globular protein mosaic, its matrix is provided by lipid, but we acknowledge that there is no firm evidence at present to rule out the possibility that the matrix is provided by protein.

C. Structural Changes and Membrane Function

There is a great deal of interest in the possibility that by analogy with the properties of many soluble proteins, some important membrane functions are mediated by conformational changes in membrane proteins and, further, that these changes show long-range cooperativity and are therefore capable of transmitting certain biochemical signals over long distances in the plane of the membrane (Changeux et al., 1967). The long-range cooperativity discussed by Changeux et al. (1967) is based on a repeating lipoprotein subunit model of membrane structure, in which each unit can exist in one of two conformational states. For the reasons given in the preceding section, however, a complex membrane may be an aperiodic continuum of a lipid–globular protein mosaic rather than an assembly of discrete lipoprotein subunits, and the treatment of Changeux et al. (1967) is not relevant to such a model.

Different mechanisms can be envisioned for generating long-range cooperativity with a mosaic continuum model in which the lipid forms the matrix of the continuum. It is possible for structural changes to occur in either the integral protein or the lipid portions of the mosaic, and these changes might or might not interact with one another. The possibility of quaternary rearrangements in hypothetical subunit proteins of membranes, for example, has already been discussed in Section V,A. These changes could also influence the local lipid structure. Furthermore, lipids in bilayers can undergo structural changes of their own. These might be phase changes (Kavanau, 1965; Luzzati et al., 1969), but they need not be. They could, for example, involve compression and decompression ("breathing") motions of the lipid molecules in the plane of the membrane. The work of Rubalcava et al. (1969) and also of Vanderkooi and Martonosi (1969) quite clearly demonstrates that changes in salt composition and concentration can significantly change the structures of both intact membranes and detergent micelles, as is detected by changes in binding of the fluorescent dye ANS (see Section IV,H). These structural changes appear to result from the effects of changes in the ion atmosphere and of ion binding on the electrostatic repulsions of the ionic heads of the phospholipid or detergent molecules. With an increase in salt concentration or cation charge, these electrostatic repulsions are decreased and the lipid bilayer or detergent micelle may be slightly compressed in the plane of the membrane. There is in fact some evidence that the addition of bivalent cations to monolayers of phospholipids can produce a closer packing of the molecules in the film (cf. Bangham and Papahadjopoulos, 1966). In an already compact membrane, this compression might involve only a very small change in the distance separating any two adjacent lipid molecules, but summed over a considerable number of molecules in a bilayer region of a mosaic structure, it could result in significant movements in the plane of the membrane and could be propagated over large distances.

Of particular interest is the observation that Ca^{++} and Mg^{++} could produce the same structural changes in these ANS–lipid systems at about a 100-fold lower concentration than Na^+. Binding of the divalent cations to the ionic heads of the phospholipids, as well as ion-atmosphere effects, appear to be implicated.

On the basis of these experiments, one can conceive of a local change in the state of compression of the lipid in a bilayer being transmitted (i.e., having structural effects) appreciable distances away in the plane of a mosaic membrane. If this local change were a *transient* one, say due to a transient change in the Na^+–Ca^{++} ratio,* a "pressure wave" might be transmitted for considerable distances along the lipid bilayer portion of a mosaic membrane, due to successive compression and decompression of the lipid at a particular local area on the membrane (much as sound travels through air). Such "pressure waves" might also produce small but significant transient structural changes in the proteins embedded in the mosaic at distances well removed from the point of initiation of the wave.

The relatively fluid state of the fatty acid chains of the phospholipids of many functional membranes (see Chapters 1–3) under physiological conditions may be connected with these "breathing" motions of the lipid portions of the membrane. If the fatty acid chains were in a crystalline array, the response of the phospholipids to ion-binding and ion-atmosphere changes would be much smaller.

* For example, Ca^{++} has been shown to destroy the impermeability of bimolecular lipid membranes of phosphatidylserine to simple univalent cations (Papahadjopoulos and Bangham, 1966) and may therefore have a similar effect at certain lipid regions of a lipid–globular protein mosaic membrane (Loewenstein, 1966).

These speculations may provide an explanation for the fascinating effects observed by Tasaki *et al.* (1968, 1969) with excitable membranes of nerve cells. These authors applied ANS or acridine orange to nerves and looked for changes in the fluorescence of the dye and in other optical properties of the nerve on passage of an impulse, or depolarization wave, down the nerve membrane. As the depolarization wave passed a section of the membrane, transient small changes following the same time course as the depolarization wave were observed in light-scattering and birefringence, and in the fluorescence of ANS or acridine orange. The binding sites of the ANS or acridine orange in this complex system were not established, but in view of the work of Rubalcava *et al.* (1969) and Vanderkooi and Martonosi (1969), it is likely that a large part, if not all, of the dye was bound to sites involving the lipid portions of the nerve cell membrane rather than the protein portions. If then the transient state of compression of the lipid were somehow coupled to the transient local state of polarization of the membrane (perhaps through the transient local ionic composition), the passage of a wave of depolarization down the excitable membrane could be accompanied by, or be the result of (see footnote, p. 211), a pressure wave in the lipid. The pressure wave could then have a transient effect on the extent of binding of ANS to sites in the lipid of the membrane, thereby producing the fluorescence and other optical effects observed by Tasaki *et al.* (1968).

It is interesting in this context that there was a marked effect of the Na^+–Ca^{++} ratio on the optical effects observed in the nerve transmission experiments of Tasaki *et al.* (1968) which are strikingly similar to effects observed by Rubalcava *et al.* (1969) and Vanderkooi and Martonosi (1969) with ANS dissolved in dispersions of phospholipids and in membranes. The latter effects were attributed to differences in Na^+ and Ca^{++} binding to the ionic heads of the phospholipids (see Section IV,G).

The speculations in this section have perhaps been too detailed to be warranted by the present state of knowledge of membrane structure. Our object, however, was not so much to defend any specific hypothesis about nerve impulse initiation and conduction, but rather to show that looking at membrane phenomena in terms of the lipid–globular protein mosaic model can lead to different ideas about their underlying mechanisms. In particular, the lipid–globular protein mosaic model requires that one recognize that different functions may be mediated either at the lipid portions or the protein portions of the mosaic; i.e., since these portions are in large part physically segregated, they may function independently to some extent, but they may also interact.

These speculations also demonstrate a point made earlier (Section III,G), namely, that the lipid–globular protein mosaic model leads to the expectation that for functions in which the lipids of the membrane are primarily involved, a membrane and an isolated phospholipid bilayer may exhibit very similar properties. This is because the model places the phospholipids mainly in a bilayer structure with their ionic heads exposed to the aqueous medium, rather than coated with a layer of protein. This would not be expected of the Davson-Danielli-Robertson model nor the Benson model. In terms of the lipid–protein mosaic model, therefore, the very intriguing experiments that have been carried out with synthetic lipid membranes (Bangham, 1968) or lipid "black films" (Henn and Thompson, 1969; Mueller and Rudin, 1968; Del Castillo et al., 1966) are perhaps all the more relevant to an understanding of those membrane functions which directly involve the lipid portions of membranes.

D. The Transformation and Biosynthesis of Membranes

The proposal that the predominant pattern for the structure of membranes is the lipid–globular protein mosaic model raises questions about how the components of such a membrane might be synthesized and assembled. If the mosaic consists of proteins embedded in a lipid continuum, then the synthesis of lipid and protein would not have to be tightly coupled; a loose coupling in fact appears to characterize the mycoplasma membrane (Kahane and Razin, 1969). How then do the integral proteins get into the membrane, particularly if, as nascent chains not yet bound to lipid, they would be very insoluble in the cytoplasm? Do they first bind lipid to become soluble, then diffuse from the site of synthesis through the cytoplasm and attach to the membrane as lipoprotein units? If this were the case, how would any nonrandom arrangement of proteins in membranes be generated? How, for example, would a plasma membrane protein get incorporated exclusively into the plasma membrane and not into the outer membrane of a mitochondrion if diffusion through the cytoplasm were involved?

In speculating about these problems, I have been very much influenced by the studies of Morgan and his co-workers (Howe et al., 1967) on the transformation of cell membranes upon infection with paramyxoviruses. A paramyxovirus particle has a lipoprotein membranelike structure as part of its outer coat, and it has been known for some time from electron microscopic observations that virus particles in the process of synthesis seem to bud out of the plasma membrane of the infected cell; the newly excreted particle appears to have surrounded itself with and excised, a piece of the cell membrane. On the other hand, the cell and viral membranes are not identical. Their surface antigens do not cross-react, and the viral coat proteins appear to be products of the viral, and not the cell, genome. Howe *et al.* (1967) utilizing the ferritin–antibody technique, showed that at the site on the plasma membrane of the infected cell where a viral bud was forming, the region of the bud consisted completely of the viral surface antigen, with a large number of viral coat protein molecules presumably closely packed together (Fig. 15). Further-

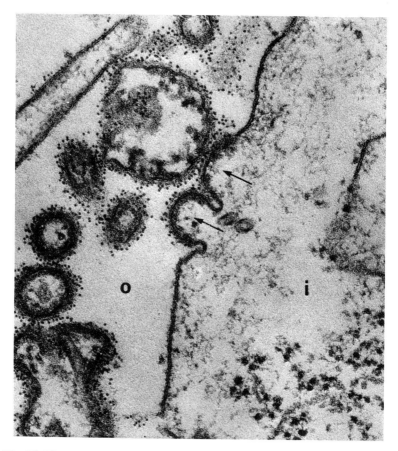

Fig. 15. Electron micrograph of a HeLa cell infected with parainfluenza virus and stained with a ferritin–antibody conjugate specific for the viral coat antigen. The small round black dots are ferritin particles denoting the presence of the viral antigen. The outside and inside of the cell are denoted by o and i, respectively. The arrows point to regions of the cell membrane where viral buds are in the process of formation and where the intense staining by ferritin–antibody shows the sharply defined localization of the viral coat antigen to the bud region. Note also the increased density underlying the membrane in the region of the viral bud. [Reproduced from Howe *et al.* (1967) with permission.]

more, at the region of the viral bud, the cytoplasmic side of the membrane appeared appreciably thicker than the adjacent plasma membrane.

The clustering of viral proteins in the bud region of the membrane certainly does not jibe with a random diffusion process for their incorporation into the plasma membrane. Rather it suggests that many individual subunits (protein or lipoprotein) of the viral coat are *inserted* successively into the plasma membrane to form a coherent tight cluster continuous with the plasma membrane. This could occur, for example, if the machinery for synthesizing coat protein (messenger RNA and ribosomes) is attached to the cytoplasmic side of the plasma membrane (where it appears thicker in electron micrographs); as the viral membrane protein is synthesized, it could become directly incorporated in the membrane without being first liberated into the cytoplasm. That region of the membrane then might enlarge and bow out as successive units of the viral protein are made at that site and inserted. Lipids would probably also be inserted, but not necessarily synchronously or bound to individual protein units. They might diffuse from their sites of synthesis in the cell and dissolve spontaneously in the region of the bud where the recently inserted viral protein had produced a local reduction in the concentration of membrane lipid.

I suggest that a mechanism like this may be a more general one that operates in most membrane biosynthesis. [Upon trying to learn something about membrane biosynthesis, I found that this possibility had already been suggested by Dallner et al. (1966).] A ribosome, or polyribosome, with bound messenger RNA may become attached to a cell membrane. There would have to be specificity to this attachment, so that the membrane proteins would ultimately become associated with the right membrane. There are various ways this might occur: by specific interaction between a protein subunit in a membrane and its duplicate just synthesized on a polyribosome, or in part by a physical sequestration of specific messenger RNA molecules, such as occurs inside a mitochondrion (cf. Truman, 1963; Beattie et al., 1967). As the individual protein molecules are made, they would then be inserted directly into the membrane without first being solubilized and not necessarily synchronously with lipid. By means of such a mechanism, clusters of identical protein molecules or clusters of several different protein molecules translated from the same polycistronic messenger RNA molecule (such as an enzyme complex) could arise in limited areas of a membrane, and different proteins could be incorporated into cell membranes at different stages of cell differentiation as the appropriate messenger RNA molecules became available (Dallner et al., 1966). This last possibility provides a mechanism for transformations of membranes not only during viral infections, but also during ordinary development. For example, the possibility has earlier been discussed (Section IV,B) that the formation of the myelin sheath from a glial cell plasma membrane involves a profound compositional transformation of the plasma cell membrane without disruption of the membrane. This could conceivably occur as follows. At some suitable stage of differentiation, specific polyribosome–messenger RNA complexes might attach to fixed specific sites on the plasma membrane, such that all further growth of the membrane involved the incorporation of only myelin-specific proteins into it. This could lead to a unidirectional growth of the newly synthesized portions of the membrane. Because of their much higher lipid–protein ratio and their specific lipid–protein composition, these newly synthesized portions might then be more stable in a lamellar form rather than as an isolated single membrane layer, and this might then provide part of the stimulus for the formation of the multilayered myelin sheath.

That intracellular polyribosome–messenger RNA complexes do indeed exist to a considerable extent bound to membranes is well documented for many types of cells and membranes (for review, see Hendler, 1968). It has not yet been established to our knowledge that messenger RNA molecules carrying messages for membrane proteins are attached to their specific cell membranes; therefore, the discussion in this section remains hypothetical.

VI. CONCLUDING REMARKS

In this chapter, I have tried to approach the problem of the gross organization of the lipids, proteins, and saccharides of membranes in a rational manner, starting from the body of thermodynamic information and structural insights that have been derived in the past decade or so mainly from the study of proteins. This information, while quantitatively still in a primitive stage of development, nevertheless appears qualitatively to encompass all of the kinds of noncovalent interactions that are most important in macromolecular systems. If there are any major factors peculiar to membranes that have not already been encountered with simpler systems, they are likely to be connected with the unusual electrostatic situation at the surface of a membrane (cf. Mauro, 1962) and with the effects of these surface charges on ion binding (Abramson et al., 1964) and on the structure of water in the region of the membrane (Cerbon, 1967). As has been pointed out many times, the concentration of ionic head groups packed in a phospholipid bilayer are at a two-dimensional concentration equivalent to a 5–10 M solution of a uni-univalent electrolyte.

Consideration of the presently available information, however, and of a body of experimental results obtained with membrane systems has led us to conclude that the lipid–globular protein mosaic model is the most satisfactory of the current models for the gross molecular organization of membranes. Nothing could be more obvious, however, then that we are a very long way from the kind of detailed structural understanding of membranes which is necessary if their functions are to be adequately understood. If the lipid–globular protein mosaic model is accepted as a working model of membrane structure, many questions still remain unanswered. Among these are the following. Is there a fraction of the total lipids which interacts closely and specifically with the integral proteins of membranes? Which proteins function as the integral proteins? Does the lipid or the protein provide the matrix of the mosaic membrane? How are the many different membrane proteins distributed in the mosaic, and how is that distribution determined? It is hoped that the lipid–globular protein mosaic model can serve a useful function by helping to focus more sharply directed experimental approaches on these and other problems of membrane structure and function.

Acknowledgments

This chapter was written mainly while I was on leave of absence at the Battelle Seattle Research Center and the Department of Biochemistry at the University of Washington. It is a pleasure to express my gratitude to Dr. Ronald S. Paul and staff of the Battelle Center and to Dr. Hans Neurath of the University of Washington not only for making my leave a pleasant one, but for making the preparation of this chapter possible. I cannot conceive how I could have written it otherwise. I want also to acknowledge my intellectual debts to my former and present colleagues, particularly Dr. John Lenard and Mr. Michael Glaser. The kindness and cooperation of Drs. D. Branton, G. Fasman, C. Morgan, E. R. Stadtman, and Mr. G. Nicolson in furnishing some of the figures are also gratefully acknowledged. The original work from our laboratories over the past several years which is discussed in this article was supported by grants AI-04255 and GM-15971 from the National Institutes of Health, U. S. Public Health Service, and by grant B6-1466E from the National Science Foundation.

References

Abramson, M. B., Katzmann, R., and Gregor, H. P. (1964). *J. Biol. Chem.* 239, 70.
Atkinson, D. E. (1969). *Annu. Rev. Microbiol.* 23, 47.
Azzi, A., Chance, B., Radda, G. K., and Lee, C. P. (1969). *Proc. Nat. Acad. Sci. U. S.* 62, 612.
Bakerman, S., and Wasemiller, G. (1967). *Biochemistry* 6, 1100.
Bangham, A. D. (1968). *Progr. Biophys. Mol. Biol.* 18, 29.
Bangham, A. D., and Papahadjopoulos, D. (1966). *Biochim. Biophys. Acta* 126, 181.

Bar, K. S., Deamer, D. W., and Cornwell, D. G. (1966). *Science* **153**, 1012.
Beattie, D. S., Basford, R. E., and Koritz, S. B. (1967). *Biochemistry* **6**, 3099.
Benson, A. A. (1966). *J. Amer. Oil Chem. Soc.* **43**, 265.
Benson, A. A., and Singer, S. J. (1965). *Abstr. 150th Nat. Meet. Amer. Chem. Soc.* p. 8c.
Bernhard, W., and Leduc, E. (1967). *J. Cell Biol.* **34**, 757.
Beychok, S. (1966). *Science* **154**, 1288.
Beychok, S. (1967). *In* "Poly-α-Amino Acids" (G. D. Fasman, ed.), p. 293. Marcel Dekker, New York.
Bownds, D., and Gaide-Huguenin, A. C. (1970). *Nature (London)* **225**, 870.
Brandts, J. F. (1969). *In* "Structure and Stability of Biological Macromolecules" (S. N. Timasheff and G. D. Fasman, eds.), p. 213. Marcel Dekker, New York.
Branton, D. (1966). *Proc. Nat. Acad. Sci. U. S.* **55**, 1048.
Branton, D. (1967). *Exp. Cell Res.* **45**, 703.
Branton, D. (1969). *Annu. Rev. Plant Physiol.* **20**, 209.
Burger, S. P., Fujii, T., and Hanahan, D. J. (1968). *Biochemistry* **7**, 3682.
Buschmann, R. J., and Taylor, A. B. (1968). *J. Cell Biol.* **38**, 252.
Carver, J. P., Schechter, E., and Blout, E. R. (1966). *J. Amer. Chem. Soc.* **88**, 2562.
Cerbon, J. (1967). *Biochim. Biophys. Acta* **144**, 1.
Changeux, J.-P., Thiéry, J., Tung, Y., and Kittel, C. (1967). *Proc. Nat. Acad. Sci. U. S.* **57**, 335.
Cheesman, D. F., and Davies, J. T. (1954). *Advan. Protein Chem.* **9**, 439.
Cohn, E. J., and Edsall, J. T. (1943). *In* "Proteins, Amino Acids, and Peptides," p. 201. Reinhold, New York.
Dallner, G., Siekevitz, P., and Palade, G. E. (1966). *J. Cell Biol.* **30**, 97.
Davson, H., and Danielli, J. F. (1952). "The Permeability of Natural Membranes," 2nd ed. Cambridge Univ. Press, London and New York.
Del Castillo, J., Rodriguez, A., Romero, C. A., and Sanchez, V. (1966). *Science* **153**, 185.
Dervichian, D., and Macheboeuf, M. (1938). *C. R. Acad. Sci.* **206**, 1511.
Dodge, J. T., Mitchell, C., and Hanahan, D. J. (1963). *Arch. Biochem. Biophys.* **100**, 119.
Doty, P. (1959). *In* "Biophysical Science" (J. L. Oncley, ed.), p. 112. Wiley, New York.
Edsall, J. T., and Wyman, J., Jr. (1958). "Biophysical Chemistry," Vol. 1, p. 258. Academic Press, New York.
Engelman, D. M. (1970). *J. Mol. Biol.* **47**, 115.
Engelman, D. M., and Morowitz, H. J. (1968). *Biochim. Biophys. Acta* **150**, 385.
Epstein, C. J., Goldberger, R. F., and Anfinsen, C. B. (1963). *Cold Spring Harbor Symp. Quant. Biol.* **28**, 439.
Farrant, J. L., and McLean, J. D. (1969). *Abstr. 27th Meet. Electron Microsc. Soc. Amer.* p. 422.
Finean, J. B. (1962). *Symp. Int. Soc. Cell Biol.* **1**, 89.
Finean, J. B., and Burge, R. E. (1963). *J. Mol. Biol.* **7**, 672.
Finean, J. B., and Martonosi, A. (1965). *Biochim. Biophys. Acta* **98**, 547.
Fleischer, S., Fleischer, B., and Stoeckenius, W. (1967). *J. Cell Biol.* **32**, 193.
Folch-Pi, J. (1955). *In* "Biochemistry of the Developing Nervous System" (H. Waelsch, ed.), pp. 121–136. Academic Press, New York.
Fox, C. F., and Kennedy, E. P. (1965). *Proc. Nat. Acad. Sci. U. S.* **54**, 891.
Frank, H. S., and Evans, M. W. (1945). *J. Chem. Phys.* **13**, 507.

Franks, F. (1968). *In* "Membrane Models and the Formation of Biological Membranes" (L. Bolis and B. A. Pethica, eds.), p. 43. Wiley, New York.

Frye, C. D., and Edidin, M. (1970). *J. Cell Sci.* **7**, 319.

Geren, B. B. (1954). *Exp. Cell Res.* **7**, 558.

Gerhart, J. C., and Schachman, H. K. (1965). *Biochemistry* **4**, 1054.

Glaser, M., and Singer, S. J. (1971). *Biochemistry* **10**, 1780.

Glaser, M., Simpkins, H., Singer, S. J., Sheetz, M., and Chan, S. I. (1970). *Proc. Nat. Acad. Sci. U. S.* **65**, 721.

Goddard, E. D., Hoeve, C. A. J., and Benson, G. C. (1957). *J. Phys. Chem.* **61**, 593.

Gordon, A. S., Wallach, D. F. H., and Straus, J. H. (1969). *Biochim. Biophys. Acta* **183**, 405.

Gorter, E., and Grendel, F. (1925). *J. Exp. Med.* **41**, 439.

Green, D. E., Haard, N. F., Lenaz, G., and Silman, H. I. (1968). *Proc. Nat. Acad. Sci. U. S.* **60**, 277.

Greenfield, N., and Fasman, G. D. (1969). *Biochemistry* **8**, 4108.

Gulik-Krzywicki, T., Rivas, E., and Luzzati, V. (1967). *J. Mol. Biol.* **27**, 303.

Gulik-Krzywicki, T., Schechter, E., Luzzati, V., and Faure, M. (1969). *Nature (London)* **223**, 1116.

Haldar, D., Freeman, K., and Work, T. S. (1966). *Nature (London)* **211**, 9.

Hanahan, D. J. (1969). *In* "Red Cell Membrane" (G. A. Jamieson and T. J. Greenwalt, eds.), p. 83. Lippincott, Philadelphia, Pennsylvania.

Hashizume, H., Shiraki, M., and Imahori, K. (1967). *J. Biochem. (Tokyo)* **62**, 543.

Hendler, R. W. (1968). *Protides Biol. Fluids, Proc. Colloq.* **15**, 37.

Henn, F. A., and Thompson, T. E. (1969). *Annu. Rev. Biochem.* **38**, 241.

Herskovits, T. T., Singer, S. J., and Geiduschek, E. P. (1961). *Arch. Biochem. Biophys.* **94**, 99.

Hoffman, J. D. (1959). *Arch. Sci. Phys. Natur.* **12**, 36.

Howe, C., Morgan, C., St. Cyr, C. de V., Hsu, K. C., and Rose, H. M. (1967). *J. Virol.* **1**, 215.

Iizuka, E., and Yang, J. T. (1966). *Proc. Nat. Acad. Sci. U. S.* **55**, 1175.

Ito, A., and Sato, R. (1968). *J. Biol. Chem.* **243**, 4922.

Ito, A., and Sato, R. (1969). *J. Cell Biol.* **40**, 179.

Jurtshuk, P., Jr., Sezuki, I., and Green, D. E. (1961). *Biochem. Biophys. Res. Commun.* **6**, 76.

Kahane, I., and Razin, S. (1969). *Biochim. Biophys. Acta* **183**, 79.

Katchalski, A., and Oster, G. (1969). *In* "The Molecular Basis of Membrane Function" (D. C. Tosteson, ed.), p. 1. Prentice-Hall, Englewood Cliffs, New Jersey.

Kauzmann, W. (1959). *Advan. Protein Chem.* **14**, 1.

Kavanau, J. (1965). "Structure and Function of Biophysical Membranes," Vol. 1. Holden-Day, San Francisco, California.

Ke, B. (1965). *Arch. Biochem. Biophys.* **112**, 554.

Kiehn, E. D., and Holland, J. J. (1968). *Proc. Nat. Acad. Sci. U. S.* **61**, 1370.

Kies, M. W., Thompson, E. B., and Alvord, E. C. (1965). *Ann. N. Y. Acad. Sci.* **122**, 148–160.

Klotz, I. M., and Farnham, S. B. (1968). *Biochemistry* **7**, 3879.

Klotz, I. M., and Franzen, J. S. (1962). *J. Amer. Chem. Soc.* **84**, 3461.

Korn, E. D. (1966). *Science* **153**, 1491.

Korn, E. D. (1969). *Annu. Rev. Biochem.* **38**, 263.

Korn, E. D., and Weisman, R. A. (1966). *Biochim. Biophys. Acta* **116**, 309.

Lenard, J. (1970). *Biochemistry* **9**, 1129.

Lenard, J., and Singer, S. J. (1966). *Proc. Nat. Acad. Sci. U. S.* **56**, 1828.

Lenard, J., and Singer, S. J. (1968a). *J. Cell Biol.* **37**, 117.

Lenard, J., and Singer, S. J. (1968b). *Science* **159**, 738.

Lenaz, G., Haard, N. F., Silman, H. I., and Green, D. E. (1968). *Arch. Biochem. Biophys.* **128**, 293.

Loewenstein, W. R. (1966). *Ann. N. Y. Acad. Sci.* **137**, 441.

Lumry, R., and Biltonen, R. (1969). *In* "Structure and Stability of Biological Macromolecules" (S. N. Timasheff and G. D. Fasman, eds.), p. 65. Marcel Dekker, New York.

Luzzati, V., Gulik-Krzywicki, T., Tardieu, A., Rivas, E., and Reiss-Husson, F. (1969). *In* "The Molecular Basis of Membrane Function" (D. C. Tosteson, ed.), p. 79. Prentice-Hall, Englewood Cliffs, New Jersey.

McClure, W. O., and Edelman, G. M. (1966). *Biochemistry* **5**, 1908.

Macfarlane, M. G. (1948). *Biochem. J.* **43**, 587.

McLean, J. D., and Singer, S. J. (1970). *Proc. Nat. Acad. Sci. U. S.* **65**, 122.

Maddy, A. H. (1966). *Int. Rev. Cytol.* **20**, 1.

Maddy, A. H., and Malcolm, B. R. (1965). *Science* **150**, 1616.

Marchesi, V. T., and Steers, E., Jr. (1968). *Science* **159**, 203.

Mauro, A. (1962). *Biophys. J.* **2**, 179.

Melchior, D. L., Morowitz, H. J., Sturtevant, J. M., and Tsong, T. Y. (1970). *Biochim. Biophys. Acta* **219**, 114.

Milner, L. S., and Kaback, H. R. (1970). *Proc. Nat. Acad. Sci. U. S.* **65**, 683.

Moffitt, W., and Moscowitz, A. (1959). *J. Chem. Phys.* **30**, 648.

Mokrasch, L. C. (1969). *In* "Handbook of Neurochemistry" (A. Lajtha, ed.), Vol. 1, pp. 171–193. Plenum Press, New York.

Mommaerts, W. F. H. M. (1967). *Proc. Nat. Acad. Sci. U. S.* **58**, 2476.

Moor, H., Mühlethaler, K., Waldner, H., and Frey-Wyssling, A. (1961). *J. Biophys. Biochem. Cytol.* **10**, 1.

Mueller, P., and Rudin, D. O. (1968). *Nature* (*London*) **217**, 713.

Napolitano, L., Lebaron, F., and Scaletti, J. (1967). *J. Cell. Biol.* **34**, 817.

Nicolson, G. (1971). *J. Cell. Biol.* **50** (in press).

O'Brien, J. S. (1965). *Science* **147**, 1099.

Ohnishi, T., and Kawamura, H. (1964). *J. Biochem.* (*Tokyo*) **56**, 377.

Ottolenghi, A. C. (1965). *Biochim. Biophys. Acta* **106**, 510.

Papadhadjopoulos, D., and Bangham, A. D. (1966). *Biochim. Biophys. Acta* **126**, 185.

Parpart, A. K., and Ballantine, R. (1952). *In* "Modern Trends in Physiology and Biochemistry" (E. S. G. Barrón, ed.), p. 135. Academic Press, New York.

Perutz, M. F. (1969). *Proc. Roy. Soc., Ser. B* **173**, 113.

Pinto da Silva, P., and Branton, D. (1970). *J. Cell Biol.* **45**, 598.

Podleski, T. R., and Changeux, J.-P. (1969). *In* "Fundamental Concepts in Drug-Receptor Interactions" (D. J. Triggle, J. F. Danielli, and J. F. Moran, eds.), p. 93. Academic Press, New York.

Ralston, A. W., and Hoerr, C. W. (1946). *J. Amer. Chem. Soc.* **68**, 2460.

Razin, S. (1969). *Annu. Rev. Microbiol.* **23**, 317.

Richardson, S. H., Hultin, H. O., and Green, D. E. (1963). *Proc. Nat. Acad. Sci. U. S.* **50**, 821.

Robertson, J. D. (1964). *In* "Cellular Membranes in Development" (M. Locke, ed.), p. 1. Academic Press, New York.

Robison, G. A., Butcher, R. W., and Sutherland, E. W. (1969). *In* "Fundamental Concepts in Drug-Receptor Interactions" (D. J. Triggle, J. F. Danielli, and J. F. Moran, eds.), p. 59. Academic Press, New York.

Roelofsen, B., Baadenhuyzen, H., and Van Deenen, L. L. M. (1966). *Nature* (*London*) 121, 1379.

Rosenberg, S. A., and Guidotti, G. (1969). *In* "Red Cell Membrane" (G. A. Jamieson and T. J. Greenwalt, eds.), p. 93. Lippincott, Philadelphia, Pennsylvania.

Rothfield, L., Weiser, M., and Endo, A. (1969). *J. Gen. Physiol.* 54, No. 1, Part 2, 27.

Rubalcava, B., Martinez de Munoz, D., and Gitler, C. (1969). *Biochemistry* 8, 2742.

Schairer, H. U., and Overath, P. (1969). *J. Mol. Biol.* 44, 209.

Schellman, J. A. (1955). *C. R. Trav. Lab. Carlsberg., Ser. Chim.* 29, 223.

Schellman, J. A., and Nielsen, E. B. (1967). *In* "Conformation of Biopolymers" (G. N. Ramachandran, ed.), Vol. 1, p. 109. Academic Press, New York.

Schellman, J. A., and Schellman, C. (1964). *In* "The Proteins" (H. Neurath, ed.), 2nd ed., Vol. 2, p. 1. Academic Press, New York.

Singer, S. J. (1962). *Advan. Protein Chem.* 17, 1.

Steere, R. L. (1957). *J. Biophys. Biochem. Cytol.* 3, 45.

Steim, J. M., Tourtelotte, M. E., Reinert, J. C., McElhaney, R. N., and Rader, R. L. (1969). *Proc. Nat. Acad. Sci. U. S.* 63, 104.

Stephen, H., and Stephen, T. (1963). *In* "Solubilites of Inorganic and Organic Compounds," Vol. 1, Part 2, p. 1201. Pergamon Press, Oxford.

Stoeckenius, W. (1962). *Symp. Int. Soc. Cell Biol.* 1, 349.

Stoeckenius, W., and Engelman, D. M. (1969). *J. Cell Biol.* 42, 613.

Stone, A. L. (1969). *In* "Structure and Stability of Biological Macromolecules" (S. N. Timasheff and G. D. Fasman, eds.), p. 353. Marcel Dekker, New York.

Stryer, L. (1965). *J. Mol. Biol.* 13, 482.

Tanford, C. (1962). *J. Amer. Chem. Soc.* 84, 4240.

Tasaki, I., Watanabe, A., Sandlin, R., and Carnay, L. (1968). *Proc. Nat. Acad. Sci. U. S.* 61, 883.

Tasaki, I., Carnay, L., and Sandlin, R. (1969). *Science* 163, 683.

Terry, T. M. (1966). Ph.D. Thesis, Yale University, New Haven, Connecticut.

Tillack, T. W., and Marchesi, V. T. (1970). *J. Cell Biol.* 45, 649.

Timasheff, S. N., Susi, H., Townend, R., Stevens, L., Gorbunoff, M. J., and Kumosinki, T. F. (1967). *In* "Conformation of Biopolymers" (G. N. Ramachandran, ed.), p. 173. Academic Press, New York.

Tourtellotte, M. E., Branton, D., and Keith, A. (1970). *Proc. Nat. Acad. Sci. U. S.* 66, 909.

Truman, D. E. S. (1963). *Exp. Cell Res.* 31, 313.

Urry, D. W., and Ji, T. H. (1968). *Arch. Biochem. Biophys.* 128, 802.

Urry, D. W., and Krivacic, J. (1970). *Proc. Nat. Acad. Sci. U. S.* 65, 845.

Urry, D. W., Medniecks, M., and Bejnarowitz, E. (1967). *Proc. Nat. Acad. Sci. U. S.* 57, 1043.

Valentine, R. C. (1969). *In* "Symmetry and Function of Biological Systems at the Macromolecular Level" (A. Engström and B. Strandberg, eds.), p. 165. Wiley, New York.

Valentine, R. C., Shapiro, B. M., and Stadtman, E. R. (1968). *Biochemistry* 7, 2143.

Van Deenen, L. L. M. (1969). *In* "The Molecular Basis of Membrane Function" (D. C. Tosteson, ed.), p. 47. Prentice-Hall, Englewood Cliffs, New Jersey.

Vandenheuvel, F. A. (1963). *J. Amer. Oil Chem. Soc.* 40, 455.

Vanderkooi, J., and Martonosi, A. (1969). *Arch. Biochem. Biophys.* 133, 153.

Velluz, L., Legrand, M., and Grosjean, M. (1965). "Optical Circular Dichroism."
     Academic Press, New York.
von Hippel, P. H., and Wong, K.-Y. (1965). *J. Biol. Chem.* **240**, 3909.
Wallach, D. F. H., Graham, J. M., and Fernbach, B. R. (1969). *Arch. Biochem.
     Biophys.* **131**, 322.
Wallach, D. F. H., and Zahler, P. H. (1966). *Proc. Nat. Acad. Sci. U. S.* **56**, 1552.
Ward, A. F. (1940). *Proc. Roy. Soc., Ser. A* **176**, 412.
Weber, G., and Laurence, D. J. R. (1954). *Biochem. J.* **56**, 31.
Winzler, R. J. (1969). *In* "Red Cell Membrane" (G. A. Jamieson and T. J. Green-
     walt, eds.), p. 157. Lippincott, Philadelphia, Pennsylvania.
Woodward, D. O., and Munkres, K. D. (1966). *Proc. Nat. Acad. Sci. U. S.* **55**, 827.
Woolfolk, C. A., and Stadtman, E. R. (1967). *Arch. Biochem. Biophys.* **118**, 736.
Wyckoff, H. W., Hardman, K. D., Allewell, N. M., Inagami, T., Johnson, L. N., and
     Richards, F. M. (1967). *J. Biol. Chem.* **242**, 3984.
Yang, J. T. (1967). *In* "Poly-α-Amino Acids" (G. D. Fasman, ed.), p. 239. Marcel
     Dekker, New York.

5

THE CONCEPT OF PERIPLASMIC ENZYMES

LEON A. HEPPEL

ABBREVIATIONS

ADP   Adenosine diphosphate
DNA   Deoxyribonucleic acid
EDTA   Ethylenediaminetetraacetate
RNA   Ribonucleic acid
UDP   Uridine diphosphate

I. INTRODUCTION

The problem of enzyme localization in bacteria is often a difficult one and only tentative conclusions are possible even after considerable research. The problem is not easy when the enzyme is found to be firmly attached to a particulate structure, because this could have taken place by secondary adsorption during the process of cell disruption and isolation of particulate fractions. As an example, ribonuclease I of *Escherichia coli* is found, in a latent state, firmly associated with the 30 S ribosomal subunit when the bacteria are disrupted and examined. However, when the purified ribonuclease is mixed with a suspension of ribosomes *in vitro*, the enzyme specifically associates with the 30 S subunit and becomes latent (Neu and Heppel, 1964b). It is possible that, *in vitro*, the enzyme occurs to some extent bound to ribosomes, but the experiment just cited throws doubt on the available evidence. Poly A polymerase is also found associated with ribosomes, but Hardy and Kurland (1966) showed very clearly that the association is an artifact. Thus, no functional relationship could be established and in fact poly A polymerase could be eluted from ribosomes under some conditions. The enzyme was not specifically bound to one of the two subunits nor was there a well-defined number of binding sites. A similar problem arises when an enzyme is isolated in association with a membrane or cell envelope fraction. Often the enzyme can be eluted, although with difficulty. In such cases it is reasonable to suppose that an association exists *in vivo*, but this is not absolutely certain. It is very helpful if the membrane-bound enzyme differs in some significant property from the enzyme in the free state.

The enzymes being considered in this chapter present even greater difficulties. They are a group of degradative enzymes found in *E. coli* and certain related gram-negative bacteria. When sonic or alumina extracts of the bacteria are examined, these proteins are all found in the supernatant fraction after high-speed centrifugation to remove ribosomes. This is where ordinary cytoplasmic enzymes are found, but other evidence indicates that the present group of enzymes are external to the protoplasmic membrane in the living cell. This evidence is entirely indirect. Thus, the enzymes considered here are selectively released by procedures that do not set free any internal proteins, namely by osmotic shock and the formation of spheroplasts with lysozyme (muramidase) and EDTA. Further evidence for surface localization is provided by the fact that the activity of these enzymes can be measured with intact cells even if the substrate cannot penetrate the organism (Brockman and Heppel, 1968; Torriani, 1968a). A wall barrier between the enzymes and the external medium is believed to exist because for most substrates the enzymes are partially cryptic. For these and other reasons, we speak of periplasmic enzymes, a term coined by Mitchell (1961). This means that they are believed to exist freely or to occupy sites in a region between the cell wall and the cytoplasmic membrane. Most of the studies of periplasmic enzymes are concerned with *E. coli*, but other gram-negative bacteria have also been investigated.

II. METHODS FOR SELECTIVE RELEASE OF PERIPLASMIC ENZYMES

A. Treatment with EDTA and Lysozyme

Repaske (1958) showed clearly that when *E. coli* and other gram-negative bacteria are treated with EDTA, they become susceptible to small concentrations of lysozyme and undergo lysis. This is due to partial removal of wall structure and conversion to osmotically sensitive "spheroplasts." In the presence of 0.5 M sucrose, which penetrates very slowly, spheroplasts are osmotically stabilized and can be maintained for hours. Malamy and Horecker (1961) were the first to show the selective release of a periplasmic enzyme by this procedure. Alkaline phosphatase was set free into the sucrose medium during the formation of spheroplasts whereas three internal enzymes, glucose 6-phosphate dehydrogenase, glutamic dehydrogenase and β-galactosidase, remained within the spheroplast structure.

Spheroplasts are prepared as follows. The bacterial cells are washed with 0.01 M Tris–HCl buffer, pH 8, and suspended at a concentration of about 10^{10} cells per milliliter in a medium consisting of 20% sucrose in 0.03 M Tris–HCl buffer, pH 8. This is accomplished by dispersing 1 g (wet weight) of cells in 80 ml of sucrose–Tris medium at 24°C. The suspension is supplemented with EDTA to give a concentration of 10^{-3} M and, 15 seconds later, lysozyme (10 μg/ml) is added. Formation of spheroplasts is usually complete within 10 minutes. The process is followed by examination in a phase contrast microscope by measuring the decrease in number of surviving organisms on nutrient agar plates and by following the decrease in turbidity when diluted tenfold in distilled water. The procedure has been modified in various ways. Often the order

of addition of EDTA and lysozyme is reversed, but in our hands this
has made no difference in yield of spheroplasts.

It should be emphasized that *E. coli* can be converted to spherical,
osmotically sensitive structures by growth in the presence of penicillin,
but these penicillin protoplasts show no release of periplasmic enzymes
except for what can be accounted for by gross lysis.

B. Osmotic Shock

In this procedure the bacteria are first suspended in a concentrated
solution of sucrose in the presence of EDTA. Then they are suddenly
shifted to a medium of low osmotic strength. The details of the procedure
vary, depending on the age of the culture.

Bacteria in stationary phase are treated as follows. The cells are har-
vested and washed several times with 0.01 M Tris–HCl plus 0.03 M NaCl
solution. The washed cells (1 g, wet weight) are suspended in 80 ml of
20% sucrose plus 0.03 M Tris–HCl, pH 7.3, at 23°C. This corresponds to
10^{10} cells per milliliter. The suspension is treated with EDTA to give a
concentration of 10^{-3} M and mixed in a flask on a rotary shaker (about
180 rpm). After 5–10 minutes the mixture is centrifuged for 10 minutes at
13,000 g, at 4°C. The supernatant fluid is removed and the well-drained
pellet is rapidly mixed with 80 ml of cold water. After 5–10 minutes of
stirring at 3°C, the mixture is centrifuged as before. The supernatant
fluid (shock fluid) is removed.

For bacteria in exponential phase of growth the procedure is modified
by reducing the concentration of EDTA in the sucrose solution to 10^{-4} M.
Also, the sucrose-treated pellet is dispersed, not in cold water, but in cold
0.5 mM MgCl$_2$. Penrose *et al.* (1968) have varied the method somewhat
for cells in late exponential or early stationary phase. The sucrose-treated
pellet is initially dispersed in cold distilled water, but 1 minute later
MgCl$_2$ is added to a concentration of 10^{-3} M. The method for osmotic
shock has also been modified in various ways when it is necessary to
handle large amounts of cells for preparative purposes. Usually this involves increasing the concentration of EDTA and working with denser
cell suspensions, such as 1 g (wet weight) per 20 ml.

The osmotic shock procedure causes selective release of about 4% of the
cell protein and, when properly carried out, does not reduce the viability
of the cell suspension. Various parts of the procedure have been evaluated. Among a number of amine buffers that have been tested, only Tris
was found to be satisfactory (Neu, 1969). The amount of EDTA must not
be too large; this results in breakdown of RNA, loss of viability, and cell
lysis (Nossal and Heppel, 1966; Neu *et al.*, 1967; Leive and Kollin, 1967).

On the other hand, use of a critical level of EDTA is an essential part of osmotic shock. It greatly improves the selective release of certain hydrolytic enzymes and it enables bacteria to withstand osmotic shock more successfully, as judged by reduced leakage of β-galactosidase (an internal enzyme), total protein, and other material absorbing at 260 nm (Anraku and Heppel, 1967). Sucrose may be replaced by another osmotic stabilizer, that is, a substance obtainable in high osmolarity that penetrates the bacterial cell very slowly. Sodium chloride will serve (Nossal and Heppel, 1966; Kundig et al., 1966). The concentration of sucrose has been varied from 10 to 40% with similar results as far as selective release of enzymes is concerned.

III. THE SURFACE STRUCTURE OF NORMAL E. COLI, SPHEROPLASTS, AND OSMOTICALLY SHOCKED CELLS

A. The Surface Structure of E. coli

It is not our purpose to review the anatomy of the bacterial cell wall; this has been done quite recently (Glauert and Thornley, 1969). However, a brief description is appropriate here because we are concerned with enzymes localized near the cell surface and with treatments that alter surface structures. Studies have been carried out with the electron microscope using negative staining, shadow-casting, freeze-etching, and quick-freezing of unfixed cells.

The surface layers of gram-negative bacteria consist basically of an inner or plasma membrane, an outer membrane, and various intermediate layers between the two membranes. It is convenient to consider these structures, starting from the innermost and proceeding in order toward the outside of the cell. As one leaves the cytoplasm one first encounters the plasma membrane, a single dense layer 60–80 Å thick. Next one encounters the rigid glycosaminopeptide (peptidoglycan, mucopeptide, murein) layer, also called the dense intermediate layer, measuring 30–80 Å, depending on the intensity of staining and other factors. This is the layer which is attacked by lysozyme. According to Bayer and Anderson (1965) the plasma membrane protects the rigid layer from attack by lysozyme from below when cells are cut open. However, if the membrane is separated from the rigid layer by plasmolysis, the enzyme can then attack the rigid layer from underneath. In plasmolysis the protoplasmic sac shrinks away from the more rigid wall, creating a space between the plasma membrane and rigid layer. Plasmolysis occurs in the sucrose stage of the osmotic shock procedure.

Proceeding outward from the rigid layer, one next encounters the outer

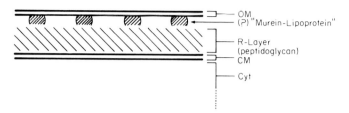

Fig. 1. Diagram of cell envelope of *E. coli*. The location of "murein-lipoprotein" (Braun and Rehn, 1969) is speculative. Abbreviations: OM, outer membrane; CM, cytoplasmic membrane; Cyt, cytoplasm.

membrane, which is discussed in detail in Chapter 6. Channels are found in this layer. In fact, Bayer and Anderson (1965) claim that channels exist which provide communication all the way between the cytoplasmic membrane and the environment.

The surfaces of most gram-negative bacteria appear to be smooth in negatively stained preparations, but newer methods have revealed a more complex structure. Thus, the outer lipoprotein layer is described as occurring in patches, and regular patterns of morphological units have been reported (Glauert and Thornley, 1969). The outermost layers contain antigen sites and bacteriophage receptor sites, and are the source of endotoxic activity.*

B. The Surface Structure of Spheroplasts

A detailed description of EDTA–lysozyme spheroplasts has been provided by Birdsell and Cota-Robles (1967). The cytoplasm assumes a spherical shape. The outer membrane is ruptured by the EDTA and the broken membrane coils on itself, thus exposing the plasma or cytoplasmic membrane to the environment. The ruptured outer membrane, consisting

---

* *Editor's note:* It is likely that additional substructure, presumably composed of protein, is present between the plasma membrane and outer membrane. This is based on the relatively large amount of protein in the gram-negative cell envelope and on the occasional demonstration of globular elements beneath the outer membrane in thin sections of *E. coli*. The studies of Weidel *et al.* (1960) and Braun and Rehn (1969) suggest that these elements may represent a specific protein or proteins covalently linked to the peptidoglycan of the rigid layer. This protein also appears to contain a covalently linked lipid component which may anchor it in the outer membrane, and which has been called "murein-lipoprotein" (Braun and Rehn, 1969). Figure 1 schematically illustrates these structures. In addition, several areas of adhesion or continuity between cytoplasmic and outer membranes can often be seen by electron microscopy. These areas may represent direct bridges between the two membranes.

of a mass of complex coils, remains attached as a kind of cap covering part of the spheroplast surface. A single membrane, the cytoplasmic membrane, is the outermost boundary between the cytoplasm and the environment over large areas of the spheroplast surface. If the term "protoplast" is restricted to those cells lacking all wall components, then Birdsell and Cota-Robles feel that it should not be applied to spheroplasts.

Antibody studies have been carried out with a periplasmic protein, the sulfate-binding protein, which was partially released on spheroplast formation (Pardee and Watanabe, 1968). The residual binding activity on the spheroplast was not inhibited by antiserum which was able to inactivate the purified sulfate-binding protein. It is possible that the residual binding protein was buried in the "cap" region where the spheroplast was not completely denuded of wall components.

C. The Alterations in Surface Structure Caused by Osmotic Shock

The osmotic shock procedure is carried out in two stages. In stage I the bacterial cell is exposed to 0.5 M sucrose in the presence of EDTA. Exposure to sucrose or another osmotic stabilizer causes plasmolysis. In plasmolyzed cells the cytoplasmic membrane limits the more condensed cytoplasm, and the rigid wall retains the shape of the cell as it was before exposure to sucrose (Birdsell and Cota-Robles, 1967). There are areas where the cytoplasmic membrane apparently adheres unusually firmly to the rigid wall layers and these resemble extensions of the cytoplasm. Bayer (1968) has described these extensions of the protoplasmic membrane in detail. He found from 200 to 400 wall–membrane associations per bacterium in *E. coli* B. Sometimes chromosomal material was observed close to the adhesions, as if nuclear material were anchored there. Bayer speculates that these regions of adhesion may be weak areas in the wall or sites of attachment of T1 and T7 phages and that the adhesions between membrane and wall could represent ducts.

The time course of plasmolysis is as follows (Bayer, 1968). Most of the bacterial protoplasts shrink almost at once. After 10–15 minutes the cells begin to swell once more and the former protoplasmic volume is reestablished. This would suggest that stage I should not be prolonged unnecessarily. A part of the shock procedure is the use of EDTA. This does not cause selective release of periplasmic enzymes by itself, but it does increase cell permeability and releases a portion of the lipopolysaccharide of the cell wall into the medium (Leive, 1965a,b; Levy and Leive, 1970). As expected, lipopolysaccharide is also released in stage I of osmotic shock (Anraku and Heppel, 1967).

In stage II of osmotic shock the cells are subjected to an abrupt decrease in the environmental osmotic pressure. According to Bayer (1967), who studied osmotic shock in the *absence* of EDTA, this change in osmotic pressure causes a rapid inflow of water into the protoplasm and a swelling of the protoplast. Only the rigid wall prevents the cell from bursting. In the electron microscope fingerlike extrusions are seen to emerge from the bacterial wall. These extrusions arise from weak areas in the rigid layer, which Bayer feels are sites of wall synthesis. Viability after this treatment is very low. In some manner, use of EDTA modifies the condition of the cell so that it is able to survive these violent osmotic transitions. We do not know if extrusions also occur when shock is done in the presence of EDTA.

IV. ENZYMES THAT APPEAR TO BE PERIPLASMIC IN LOCATION

This section contains a brief description of the individual enzymes which can be called periplasmic by applying an operational definition; they are selectively released by osmotic shock or by converting bacterial cells into spheroplasts. It can be seen that they are, by and large, degradative enzymes. In a number of cases, similar enzymes are found in mammalian lysosomes. In gram-positive bacteria enzymes of similar specificities are actually secreted into the medium.*

A. Alkaline Phosphomonoesterase

Alkaline phosphomonoesterase, or alkaline phosphatase, is produced by *E. coli* and other bacteria in response to low levels of inorganic phosphate in the growth medium. It permits the cell to derive phosphorus from phosphate esters in the medium, even though almost all of these esters are unable to pass the membrane barrier as such.

Malamy and Horecker (1961) observed that alkaline phosphatase is liberated quantitatively into the surrounding medium when the cells are converted into spheroplasts with EDTA and lysozyme. They succeeded in purifying and crystallizing the enzyme from the sucrose medium (Malamy and Horecker, 1964). It is more convenient to purify alkaline phosphatase from the fluid obtained by subjecting *E. coli* to osmotic shock and removing the shocked cells (Torriani, 1968a; Brockman and Heppel, 1968). A single-step purification procedure, beginning with osmotic shock fluid, has been described by Simpson *et al.* (1968).

---

* *Editor's note:* Gram-positive bacteria contain a cytoplasmic membrane and a rigid cell wall but lack the outer membrane found in gram-negative cells (see Fig. 1). This raises the interesting possibility that the outer membrane acts as a barrier to release of periplasmic enzymes in gram-negative bacteria.

Alkaline phosphatase is unusually heat stable, especially in the presence of $0.01\ M$ Mg^{++}. The enzyme consists of two identical subunits, and it contains Zn^{++}. Enzyme containing Co^{++} is also active. At pH 2–3, in the cold, it dissociates and this process can be reversed.

Schlesinger and Barrett (1965) postulated that subunits of alkaline phosphatase are released from the polyribosomal site of synthesis and diffuse to some region within the cell wall matrix. There they are thought to interact with zinc atoms and to dimerize. Torriani (1968b) provided experimental evidence for the existence of a pool of alkaline phosphatase monomers in the particulate fraction of cell extracts. This was determined by finding an increase in enzymatic activity on incubating the particles with $2 \times 10^{-4}\ M$ $ZnCl_2$ for 30 minutes at room temperature, conditions which promote dimerization. This particulate fraction corresponded to membrane and it contained more monomer than was found attached to ribosomes. This indicated that a step subsequent to synthesis on the ribosomal subunit was limiting the formation of active enzyme.

Torriani thereupon carried out pulse labeling of cells with [^{14}C]arginine under conditions of enzyme synthesis. The cells were washed and converted into spheroplasts by lysozyme–EDTA treatment. This treatment released 98% of the active enzyme into the medium. The pool of radioactive monomers was entirely in the endoplasm. Its size was such that from it could be formed 81% as much active enzyme as had been released in the EDTA–lysozyme step. The size of the pool was increased by decreasing Zn^{++} in the growth medium. Lack of this metal cation caused a low level of active enzyme, but the cells contained a full complement of monomers. However, by the time the cultures reached the stationary phase of growth, the free monomers were partially lost or denatured.

Schlesinger (1968) considered the following question. Does transport to the periplasmic space occur before or after dimerization of the subunits? He was able to make use of specific antibodies that discriminate between the subunit and the active dimer to answer this question. He prepared spheroplasts under conditions that enabled the cells to continue protein synthesis. It was found that spheroplasts can synthesize alkaline phosphatase subunits which are secreted into the culture medium. This suggests transport of the subunit across the bacterial cell membrane and subsequent dimerization to active enzyme in the periplasmic space.

B. 5′-Nucleotidase

The enzyme, 5′-nucleotidase, occurs constitutively in *E. coli* and it is released nearly quantitatively by osmotic shock (Neu and Heppel, 1965). Formation of the enzyme is inhibited neither by high concentrations of inorganic phosphate nor by glucose. Also, the activity of the enzyme is not inhibited by high concentrations of phosphate. A specific protein inhibitor for 5'-nucleotidase was discovered by Melo and Glaser (1966) and Dvorak *et al.* (1966). Because of this inhibitor, almost no activity can be detected in cell extracts unless they are first heated to destroy the inhibitor. The inhibitor is not released by osmotic shock and therefore 5'-nucleotidase is present in shock fluid in a fully active state.

Hydrolytic activity against uridine diphosphoglucose was also found in shock fluid and it, too, was sensitive to a protein inhibitor (Melo and Glaser, 1966). Further study (Glaser *et al.*, 1967) showed that UDP-sugar hydrolase and 5'-nucleotidase activity are associated with the same enzyme. Thus, the ratio of 5'-nucleotidase to UDP-sugar hydrolase activity remained constant over a 1000-fold range of purification. The protein inhibitor was purified and found to inhibit both the 5'-nucleotidase and UDP-sugar hydrolase activity. Heat inactivation experiments provided further evidence of identity. Glaser *et al.* (1967) also incubated [^{14}C]-uridine-labeled UDP-D-glucose in the presence of a large pool of non-radioactive 5'-UMP and allowed hydrolysis to proceed for a brief interval. The results were consistent with some hydrolysis of nonradioactive 5'-UMP to form uridine and concomitant hydrolysis of [^{14}C]uridine-labeled UDP-D-glucose *directly* to [^{14}C]*uridine*, without mixing with the pool of "cold" UMP. This would suggest an enzyme-bound complex of 5'-UMP as an intermediate in UDP-D-glucose hydrolysis. The enzyme cleaves UDP-D-glucose, UDP-D-galactose, and UDP-*N*-acetyl-D-galactosamine at essentially the same rate.

Neu (1967a) reported a 5000-fold purification of the 5'-nucleotidase of *E. coli* and measured the ratios of specific activity against uridine diphosphate glucose, bis(*p*-nitrophenyl) phosphate, 5'-AMP, and ATP. These ratios all remained constant throughout the purification, and heat inactivation curves were found to parallel each other. Thus, it would appear that the last three compounds must be added to the list of substrates for "5'-nucleotidase"; the enzyme has a wide range of specificity and its name is hardly appropriate. Neu's preparation was judged to be pure by molecular sieve chromatography, polyacrylamide disc gel electrophoresis, and ultracentrifugation. A molecular weight of 52,000 was found.

The 5'-AMP activity is stimulated 100-fold by $5 \times 10^{-3}\ M$ Co^{++} and 200-fold by the further addition of Ca^{++} (Neu, 1967a). When *E. coli* cells were grown in the presence of ^{65}Zn, this isotope was accumulated. The cells were subjected to osmotic shock and 5'-nucleotidase was purified to homogeneity from the shock fluid. The purification was accompanied by considerable enrichment with respect to ^{65}Zn, which could not be removed by dialysis. Superimposable peaks of enzyme activity and ^{65}Zn were observed (Dvorak and Heppel, 1968). Treatment of 5'-nucleotidase with acid caused loss of activity associated with release of ^{65}Zn. The preparation could be reactivated by low concentrations of each of several divalent metal ions. It was suggested that the enzyme is a metalloprotein, possibly a zinc metalloenzyme. In this connection, it is of interest that *E. coli* cells grown in the presence of EDTA revealed a selective depression in the levels of cyclic phosphodiesterase, 5'-nucleotidase, and alkaline phosphatase. Six other enzymes were unaffected (Dvorak, 1968).

The enzyme 5'-nucleotidase is present in a number of the Enterobacteriaceae and it is released from many of them by osmotic shock or during the formation of EDTA–lysozyme spheroplasts (Neu, 1967b, 1968a; Neu and Chou, 1967). Members of the *Klebsiella–Enterobacter* group consistently release only 50% of the enzyme by osmotic shock. The *Proteus* species contain a 5'-nucleotidase but it cannot be released by osmotic shock. In spite of these differences it would appear that a quite similar enzyme exists in all of these species. In every case there is lack of control by phosphate and no evidence of catabolite repression. Furthermore, partially purified enzyme derived from shock fluids of *Shigella sonnei*, *Aerobacter aerogenes* and *Salmonella heidelberg* and from sonic extracts of *Proteus vulgaris* are quite similar with respect to metal stimulation, heat stability, substrate, and reaction with the protein inhibitor. In fact, the protein inhibitor purified from *E. coli* inhibits the 5'-nucleotidase activity of other members of the Enterobacteriaceae.

Results with a mutant *E. coli* that forms minicells are rather interesting (Dvorak *et al.*, 1971). Minicells are small round bodies, free of DNA, that are formed throughout the growth cycle by polar budding (Adler *et al.*, 1967). The enzyme 5'-nucleotidase is enriched in the minicells but its protein inhibitor is entirely restricted to the rod forms of the organism.

C. Cyclic Phosphodiesterase

Cyclic phosphodiesterase was discovered by Anraku (1964a,b). It catalyzes two reactions: (a) the hydrolysis of ribonucleoside 2',3'-cyclic phosphates to ribonucleoside 3'-phosphates and (b) the hydrolysis of ribonucleoside 3'-phosphate to nucleoside and inorganic phosphate. Over a 900-fold purification the ratio of specific activity of phosphodiesterase to 3'-nucleotidase remains constant. The purified enzyme preparation is free of ribonuclease and deoxyribonuclease activities. Bis-*p*-nitrophenyl phosphate and *p*-nitrophenyl phosphate are also hydrolyzed, but at slower rates. (Enzymes that are otherwise highly specific are frequently found to hydrolyze nitrophenyl esters.) Cyclic phosphodiesterase is activated by Co^{++} and inhibited by Zn^{++}, Cu^{++}, and EDTA.

Cyclic phosphodiesterase is released from *E. coli* in high yield by osmotic shock (Neu and Heppel, 1965) and has been purified from this source (Dvorak *et al.,* 1967; Neu, 1968c). It was also purified from the osmotic shock fluids of *Serratia marcescens, A. aerogenes, S. heidelberg,* and *S. sonnei* (Neu, 1968c). That fraction of the activity not released by shock was recovered in sonic extracts of shocked cells and also purified. In addition, the enzyme was purified from sonic extracts of *P. vulgaris* from which hardly any activity can be released by osmotic shock. Enzyme fractions from all of these sources were similar in many properties. The ratio of activity against 3'-nucleotide and cyclic nucleotide remained constant throughout all of the purifications, and there was parallel loss of cyclic phosphodiesterase and 3'-nucleotidase activities on heating. The pH optima were similar for all of the enzymes (about pH 7.8), and in every case stimulation was greatest with Co^{++}. Acrylamide gel electrophoresis at pH 8.9 gave single protein bands that migrated identical distances. For enzymes purified from all of these sources, molecular sieve chromatography showed identical elution volumes for Sephadex G-100, and comparison with protein standards indicated molecular weights in the neighborhood of 60,000. When cells grown in the presence of $^{65}Zn^{++}$ were used as a source of enzyme, purification of cyclic phosphodiesterase was accompanied by enrichment of $^{65}Zn^{++}$ activity (Dvorak and Heppel, 1968). Further, enzyme preparations could be inactivated by EDTA and, after subsequent dialysis, activity could be partially restored by addition of zinc ions.

D. Acid Hexose Phosphatase and Other Acid Phosphatases

1. Acid Hexose Phosphatase

Several acid phosphatases are released into the medium when *E. coli* are converted into EDTA–lysozyme spheroplasts or when cells are subjected to osmotic shock (Neu and Heppel, 1964a). One of these acid phosphatases was purified by DEAE-cellulose and hydroxylapatite chromatography to a point where it was homogeneous on disc gel electrophoresis (Dvorak *et al.,* 1967). The purified enzyme has no metal requirements and is active in the presence of EDTA, and it is highly specific for certain sugar phosphate esters. These include glucose 6-phosphate, fructose 6-phosphate, fructose 1,6-diphosphate and mannose 6-phosphate. Compounds such as α-D-glucose 1-phosphate and α-D-galactose 1-phosphate are also hydrolyzed, but the corresponding β-esters are *not*. Pentose esters are more slowly hydrolyzed. Many other phosphate esters were tested and found to be resistant.

The pH optimum is between 5.5 and 6.0 and K_m is $3.3 \times 10^{-4} M$ for glucose 6-phosphate. Acid hexose phosphatase is subject to catabolite repression (von Hofsten, 1961; Dvorak et al., 1967), so that the lowest levels of activity are observed with glucose as a carbon source. Several attempts have been made to induce formation of the enzyme in E. coli by having glucose 6-phosphate or glucose 1-phosphate present in the growth medium (von Hofsten, 1961; Dvorak et al., 1967). However, when compared with equivalent concentrations of free glucose and inorganic phosphate, no significant differences in the levels of acid hexose phosphatase are noted at any stage of growth.

A similar sugar phosphate phosphohydrolase has been purified from Neisseria meningitidis (Lee and Sowokinos, 1967, 1969). This enzyme is also released by osmotic shock. The α-D-aldohexose 1-phosphates and D-ketohexose 1-phosphates are the best substrates. Again, if the phosphate group of D-aldohexose 1-phosphates is esterified in the β-configuration at the first carbon atom, enzymatic hydrolysis does not occur.

2. Nonspecific Acid Phosphatase

Another acid phosphatase fraction also released by osmotic shock is not as well characterized and may consist of several enzymes (Dvorak et al., 1967). This fraction is poorly adsorbed to DEAE-cellulose at neutral pH. The best substrate is p-nitrophenyl phosphate and substantial rates of hydrolysis are observed with 3'-AMP, fructose 6-phosphate, glucose 6-phosphate, and β-glycerophosphate. The pH optimum is near 5.0, and these are striking effects of metal ions.

3. Acid Phosphatase Active at pH 2.5

An acid phosphatase is released from E. coli K12 by osmotic shock, and this enzyme has an unusual pH optimum, at 2.5. The enzyme also shows stimulation by chloride ion (Hafkenscheid, 1968).

E. Nucleases

1. Ribonuclease I

This enzyme was originally purified from E. coli ribosomes (Spahr and Hollingworth, 1961) where it occurs in a latent state. This latency means that the activity cannot be measured unless the ribosomes are first incubated with EDTA or with urea. The enzyme is localized in the 30 S ribosomal subunit, but this observation could be an artifact since purified ribonuclease I specifically binds to 30 S ribosomal subunits and becomes latent (Neu and Heppel, 1964b). The enzyme is released by osmotic shock and during spheroplast formation (Neu and Heppel, 1964a). Almost all of the enzyme is set free during exponential growth, but, rather uniformly, only 50% of the activity is released from cells in stationary phase. It has been suggested that part of the enzyme may change in location as the culture ages, from a membranal to a ribosomal site. A fraction of ribonuclease tightly bound to the cell envelope of *E. coli* has also been described (Anraku and Mizuno, 1967). In this connection, it is of interest that a temperature-sensitive *E. coli* mutant exists whose ribosomal RNA becomes degraded when cells are exposed to temperatures of 40°C (Nozawa *et al.*, 1967). It is suggested that the mutant has a fragile cell membrane from which ribonuclease is released at 40°C.

Ribonuclease I has no metal requirements and hydrolyzes ribonucleic acid and various homopolymers, first to nucleotides terminated with a 2',3'-cyclic phosphate group, and finally to nucleoside 3'-phosphates.

2. Deoxyribonuclease I

This is the major deoxyribonuclease of *E. coli*. It hydrolyzes DNA to oligonucleotides terminated with a 5'-phosphomonoester end group, and it is inhibited by extremely small concentrations of RNA. The enzyme is widely used to produce "nicks" in one strand of the DNA duplex. Deoxyribonuclease I is released by osmotic shock and during spheroplast formation (Cordonnier and Bernardi, 1965; Nossal and Heppel, 1966; Obinata and Mizuno, 1968). The enzyme differs from the periplasmic enzymes in that a considerable fraction of the activity is released in stage I of osmotic shock (the treatment with 20% sucrose containing EDTA), rather than in stage II.

It has been speculated that a nuclease compartmentalized near the cell surface could be responsible for "host cell restriction," resulting in abortive infection of certain strains of *E. coli* by various bacteriophages (Smith and Pizer, 1968). No infection of the restrictive hosts occurs and the phage DNA is degraded.

F. Other Enzymes Released by Osmotic Shock

1. ADP-Glucose Pyrophosphatase

This enzyme was discovered by Melo and Glaser (1966). It is specific for ADP-D-glucose and the reaction products are 5'-AMP and α-D-glucose 1-phosphate. The enzyme does not have an associated 5'-nucleotidase activity, as is true for UDP-glucose pyrophosphatase. It is released in high yield both by osmotic shock and during formation of EDTA–lysozyme spheroplasts. The Michaelis constant is $8 \times 10^{-5} M$ and the pH optimum lies between 7 and 8. A protein inhibitor exists which seems to be an internal protein because it is not released by shock; this inhibitor is distinct from the one which reacts with UDP-sugar hydrolase.

In *Salmonella typhimurium* LT2, both a soluble adenosine diphosphate glucose pyrophosphatase and a particle-bound nucleoside diphosphate sugar pyrophosphatase of broader specificity are present.

2. Asparaginase II

Extracts of *E. coli* K12 contain two asparaginases, of which one, called asparaginase II, is active against the growth of mouse lymphomas and is an effective inhibitor of cell-free protein synthesis in microbial extracts. Asparaginase II appears only in cells grown under anaerobic conditions. In contrast to asparaginase I, which is an internal enzyme, it is released by osmotic shock and can be measured with intact cells (Cedar and Schwartz, 1967). It is also released on spheroplast formation.

3. Penicillinase

Datta and Richmond (1966) observed release of penicillinase, which is mediated by an R factor, when *E. coli* were subjected to osmotic shock. The extent of release was only 35%. Neu (1968d) also investigated episomally mediated penicillinases from both *E. coli* and *S. typhimurium*. He observed total release of enzyme by osmotic shock. Neu reported that penicillinases whose synthesis is chromosomally directed are bound to the cell more firmly and may not be periplasmic. Burman *et al.* (1968) derived a somewhat different conclusion from their experiments. Spheroplast formation released almost all of the enzyme, both from strains carrying an episome and from those strains whose penicillinase was entirely under chromosomal control. They felt that the chromosomally and episomally mediated enzymes are both located outside the cell membrane.

4. Enzymes Catabolizing Deoxyribonucleosides

A whole complex of enzymes concerned with the catabolism of deoxyribonucleosides can be released by osmotic shock. Thus, Munch-Petersen (1967) found that thymidine phosphorylase is set free by this treatment, an observation also made by Kammen (1967) and Barth *et al.* (1968). A later report by Munch-Petersen (1968) added deoxyriboaldolase, purine deoxyribonucleoside phosphorylase and deoxyribomutase to the list of enzymes released by osmotic shock. Munch-Petersen observed that in fact cells cleave exogenous thymidine and effectively catabolize the liberated deoxyribosyl group. Toluenization or sonic treatment causes complete inhibition. From this, she concluded that cleavage of thymidine by intact cells is catalyzed by an organized complex of enzymes situated near the cell surface. The deoxyribosyl acceptor group is bound in this complex, and the cleavage is tightly linked to the deoxyribomutase and deoxyriboaldolase reactions.

5. An Adenylating Enzyme That Inactivates Streptomycin

Three groups of investigators have reported that in strains of *E. coli* which are resistant to streptomycin due to an extrachromosomal R factor, enzymatic adenylation of streptomycin occurs (Harwood and Smith, 1969; Umezawa *et al.*, 1968; Yamada *et al.*, 1968). The enzymatic reaction requires ATP or dATP and Mg^{++}. The enzyme is released by osmotic shock. If, indeed, it occurs in the periplasmic space, ATP must become available there is some manner.

G. Other Proteins Released by Osmotic Shock

In addition to the enzymes listed above, another group of proteins is also released by osmotic shock. These proteins are characterized by the fact that they specifically bind small substrate molecules, but so far no enzymatic activity has been found for any of them. Their molecular weights appear to fall in the range 25,000–35,000, and a number of them have been found to be quite heat stable. It is believed that they are involved in transport (for a discussion of their role see Chapter 7). So far, binding proteins have been described for inorganic sulfate (Pardee, 1968); D-galactose (Anraku, 1968; Boos, 1969); L-leucine, L-valine, and L-isoleucine (Piperno and Oxender, 1968); L-leucine, specifically (Furlong and Weiner, 1971); L-arginine proteins (O. H. Wilson and Holden, 1969); L-lysine, L-glutamine, and cystine (Weiner *et al.*, 1970); L-arabinose (Hogg and Englesberg, 1969); phenylalanine (Klein *et al.*, 1970); and histidine, (Rosen and Vasington, 1970).

The proteins released by osmotic shock, and proteins so far observed to remain within the cell during shock procedure, are listed in Table I.

V. EVIDENCE FOR PERIPLASMIC LOCALIZATION OF ENZYMES

A. Measurement of Activity with Intact Cells Using Substrates Believed Not to Penetrate Cells

If a substrate is believed not to penetrate the cell and yet an enzyme activity can be measured using intact cells, this is taken as evidence for surface localization. The enzymes released from *E. coli* and related gram-negative bacteria by osmotic shock are accessible to substrate when intact cells are used as a source of enzyme. However, in general the enzymes are

TABLE I

SENSITIVITY OF VARIOUS PROTEINS TO RELEASE BY OSMOTIC SHOCK

Proteins Released by Osmotic Shock

Alkaline phosphatase	Deoxyribomutase
5'-Nucleotidase	Purine deoxyribonucleoside phosphorylase
Acid hexose phosphatase	Streptomycin-adenylating enzyme
Nonspecific acid phosphatase	Sulfate-binding protein
Acid phosphatase (pH optimum 2.5)	Protein binding leucine, isoleucine, and valine
Cyclic phosphodiesterase	Specific leucine-binding protein
ADP-Glucose pyrophosphatase	Arginine-binding proteins
Ribonuclease I	Lysine-binding protein
Asparaginase II	Arabinose-binding protein
Penicillinase	Glutamine-binding protein
DNA Endonuclease I	Cystine-binding protein
Thymidine phosphorylase	Phenylalanine-binding protein
Deoxyriboaldolase	Histidine-binding protein

Proteins Not Released by Osmotic Shock[a]

Inorganic pyrophosphatase	Ribonuclease II (phosphodiesterase)
β-Galactosidase	Glucose 6-phosphate dehydrogenase
Polynucleotide phosphorylase	Glutamic dehydrogenase
Histidyl-tRNA synthetase	Adenylic acid pyrophosphorylase
Adenosine deaminase	Guanylic acid pyrophosphorylase
Thiogalactoside transacetylase	DNA exonuclease I
Uridine phosphorylase	5'-Nucleotidase inhibitor protein
Lactic dehydrogenase	Glycerol kinase
Leucine aminopeptidase	UDP-Glucose pyrophosphorylase
Certain other dipeptidases	UDP-Galactose 4-epimerase[b]
DNA polymerase	

[a] Taken from Heppel (1967).
[b] D. W. Wilson (1971).

partially cryptic. This means that the specific activity, measured with intact cells, is less than that found for an equivalent cell extract. The percent of activity expressed by intact cells, compared with sonic extracts, varies with the substrate for alkaline phosphatase, acid hexose phosphatase, and cyclic phosphodiesterase (Brockman and Heppel, 1968). This is interpreted to mean that the phosphate esters vary in ease of penetration of the wall barrier. There are striking differences even among related compounds (Table II).

The case of 5'-nucleotidase is especially striking because the activity can be measured with intact cells but not with cell extracts, presumably because the enzyme is periplasmic and its protein inhibitor is internal. The inhibitor is not released by osmotic shock (Melo and Glaser, 1966; Neu, 1968b).

Glucose 6-phosphate is one of the unusual phosphate esters that can

TABLE II

ALKALINE PHOSPHATASE ACTIVITY OF *E. coli* SUSPENSIONS EXPRESSED AS
PERCENT OF ACTIVITY OF EQUIVALENT CELL EXTRACTS[a]

Substrate	Activity of suspension (% of extract)	Substrate	Activity of suspension (% of extract)
5′-UMP	76	Phenolphthalein P	10
2′,3′-UMP	45	Sodium pyrophosphate[b]	19
5′-CMP	89	α-Glycerophosphate	93
2′,3′-CMP	76	β-Glycerophosphate	92
5′-GMP	65	D-Glucose-6-P	92
2′,3′-GMP	25	D-Galactose-6-P	90
5′-AMP	21	D-Mannose-6-P	80
3′-AMP	7	PNPP[b]	45
2′-AMP	8	Ethanolamine P	96
ATP	36	Serine P[b]	99

[a] Strain C90, constitutive for alkaline phosphatase, was used. Cells were grown to stationary phase. The activity of a suspension of washed cells is given as a percent of the activity of an equivalent sonic extract. Assays were at 37°C in Tris–HCl buffer (pH 8.3). Under these conditions on other phosphatase shows appreciable activity in *E. coli*. The concentration of substrate was $2.5 \times 10^{-3} M$ except as noted. [Data are taken from Brockman and Heppel (1968).]

[b] Concentration of substrate was $4 \times 10^{-3} M$.

be transported into *E. coli* cells. With this compound it has been observed that whole cells of a mutant strain, unable to transport this ester through the protoplasmic membrane, nevertheless are able to hydrolyze it (Brockman and Heppel, 1968) by means of their acid hexose phosphatase.

Hydrolysis of nonpenetrating phosphate esters by intact bacterial cells may occur rapidly enough to permit the use of these esters as a source of carbon (Neu, 1967b).

B. Other Lines of Evidence

1. Histochemical Evidence for Localization of the Enzymes Released by Osmotic Shock

Alkaline phosphatase has been studied cytochemically at the electron microscope level in several laboratories (Done *et al.*, 1965; Kushnarev and Smirnova, 1966; Wetzel *et al.*, 1971). These reports generally agree that the reaction product appears at the periphery of the cells, but they differ with regard to the precise localization of these enzymes. The studies of Done *et al.* indicated that alkaline phosphatase is in the periplasmic space, but the two other groups localized the enzyme only to the cell surface. In a later more complete study, Wetzel *et al.* (1971) found a periplasmic location of this enzyme in a large number of specimens. They felt that formalin fixation and the use of calcium as a precipitating cation can interfere with periplasmic staining in specimens prepared for electron microscopy.

The studies of Nisonson *et al.* (1969) indicate that 5'-nucleotidase is bound to the surface of the cell in intact *E. coli*. The enzyme was observed in the periplasmic space only when the permeability barrier of the cell was altered by EDTA. Preliminary studies suggested that cyclic phosphodiesterase is also located at the cell surface. These results are in partial conflict with the results of Wetzel *et al.* (1971), who observed that most of the specimens examined for cyclic phosphodiesterase and fixed in glutaraldehyde showed staining of the periplasmic space. After fixation in formalin, most of the precipitate was restricted to the cell surface. Similar results were obtained by these workers when acid hexose phosphatase was examined, using glucose 6-phosphate as the substrate and reaction conditions such that other activities would not be measured. A further observation made with preparations examined both by light and electron microscopy was that alkaline phosphatase, acid hexose phosphatase, and cyclic phosphodiesterase appear to be concentrated in polar caplike enlargements of the periplasmic space.

All of these investigations agree in locating the enzymes released by osmotic shock somewhere in the cell envelope. Agreement on precise localization has not been reached, and results seem to vary depending on the nature of the fixative and the reagent used to trap inorganic phosphate released by the enzymatic reaction.

2. Use of Mutants with Defective Cell Walls

Mangiarotti *et al.* (1966) obtained mutants with apparent defects in their cell envelopes. These mutants were dependent on sucrose in the medium for growth; in the absence of an osmotic stabilizer such as sucrose, cell lysis occurred. One of these mutants, which also happened to be constitutive for alkaline phosphatase, released 60% of the enzyme into the culture medium during exponential growth in medium containing 20% sucrose. This is rather striking evidence for surface localization since the cells showed good growth with no evidence of gross lysis.

C. Studies with Binding Proteins

Binding proteins are considered in Chapter 7, but it should be pointed out here that similar cytochemical efforts have been made to localize them. Thus, Pardee and Watanabe (1968) made use of a nonpenetrating inhibitory dye to show that sulfate-binding protein is outside the cell membrane of S. *typhimurium*. By means of antibody they showed that the protein does not extend outside the cell wall.

Treatment of the leucine-binding protein with antiserum causes strong inhibition of leucine binding, but treatment of whole cells with antiserum has no effect on leucine transport (Penrose *et al.*, 1968). Histochemical studies indicate that the leucine-binding protein is clearly in the cell envelope, although precise localization is not possible (Nakane *et al.*, 1968).

VI. "SURFACE" ENZYMES IN OTHER MICROORGANISMS

This review is concerned with periplasmic enzymes, almost all of which have been described only for *E. coli* and certain related species of gram-negative bacteria. Space does not allow a review of enzymes considered to be at or near the surface of other microorganisms. Nevertheless, it is useful to make brief mention of a random selection of "surface" enzymes of gram-positive bacteria and yeast.

Several enzymes which resemble the periplasmic enzymes of *E. coli* are actually exoenzymes in *Bacillus subtilis*. The ribonucleases secreted into the medium have been studied in many laboratories. An enzyme similar to *E. coli* alkaline phosphatase is also an exoenzyme in *B. subtilis* (Cashel and Freese, 1964). However, at least three additional enzymes are selectively released from *B. subtilis* when cells are converted into protoplasts with lysozyme. Birnboim (1966) described a nuclease which prefers single-stranded DNA. Momose *et al.* (1964) reported on a 5'-nucleotidase, the major portion of which was released on protoplast formation. Mauck and Glaser (1970) discovered a "periplasmic" nucleoside diphosphate sugar hydrolase that was formed by *B. subtilis* W23 under conditions of limitation of inorganic phosphate in the medium. It is released into the medium during spheroplast formation and the activity can be measured with intact cells.

Cell-bound penicillinase in *Bacillus licheniformis* is selectively released during lysozyme treatment, according to Sargent *et al.* (1968), who also reviewed earlier studies with this enzyme. It is fully available to substrate; up to 98% can be released by trypsin in hypertonic sucrose. It is partially available to antibodies and is found in the membrane fraction of spheroplasts. Antibody inhibition has been especially useful in localizing this and other enzymes in gram-positive bacteria (Kushner and Pollock, 1961).

Surface enzymes of yeast have been extensively studied. An important early investigation is that of Demis *et al.* (1954) in which invertase was localized at the cell surface by the use of uranium salts which inhibit the activity in whole cells. The inhibition was reversed by low concentrations of inorganic phosphate in the medium, much lower than the known intracellular concentration of phosphate. Gascon *et al.* (1968) have found that a strain of yeast contains two different, although immunologically related, invertases, one external and the other internal. The external enzyme happens to be a glycoprotein. Several investigators have shown that surface enzymes can be eluted from living yeast cells by growth in the presence of unusually large concentrations of KCl. For example, Weimberg and Orten (1965, 1966) found that invertase and acid phosphatase could be eluted with KCl. Finally, a study by Trevithick and Metzenberg (1966) showed that fractionation of "light" invertase, the monomer, from "heavy" invertase, the aggregated form, occurs at the cell wall of *Neurospora crassa*.

VII. SPECULATIONS ON THE FUNCTION OF PERIPLASMIC ENZYMES

Most investigators have postulated that the periplasmic enzymes are concerned with the external environment of the bacterial cell. Thus, they make possible the utilization of a wide variety of materials, most of which are unable to pass the protoplasmic membrane as such. The materials include sugar phosphates, nucleotides and polynucleotides, and many other phosphorylated compounds. The fact that many of the same enzymes are periplasmic in gram-negative bacteria and are exoenzymes in gram-positive bacteria argues in favor of this assumption. It has been suggested that the periplasmic enzymes are, in a sense, analogous to lysosomes of mammalian cells. They are compartmentalized and could be released on cell death to autolyze the cell and make it available to survivors in a culture. This possible digestive function is discussed by De Duve and Wattiaux (1966), who also considered the idea that lysosomes owe their distant origin to the infolding of a membrane secreting exoenzymes (and possibly periplasmic enzymes).

Another possible role, also related to the environment, is a protective one; that is, guarding the organism against injurious agents such as phage nucleic acid. It must be admitted, however, that the evidence in favor of either a digestive or a protective function is not entirely convincing.

Glaser and his associates (1967) favor the hypothesis that periplasmic enzymes help maintain a suitable level of nucleotides in the cell. Thus, they would serve as an alternative to feedback inhibition. These workers argue that coenzyme nucleotides occur in the environment of the normal cell in extremely low concentrations. Thus, it makes more sense to assume that UDP-sugar hydrolase, for example, is more concerned with intracellular than extracellular UDP-glucose. This hypothesis, of course, demands that the periplasmic enzymes have access to the intracellular pool of nucleotide. It is known that nucleotides are excreted into the medium under certain abnormal conditions (Neu et al., 1967; Hurwitz et al., 1963), and it is possible that outward movement through the protoplasmic membrane may occur physiologically. It is presumed that nucleotides must become available outside the membrane in some manner to provide precursors for wall constituents during a repair process (as when cells recover from the effects of EDTA). Also, as noted earlier, at least one periplasmic enzyme, the streptomycin adenylating enzyme, requires the presence of ATP as a cosubstrate.

Glaser et al. presented the following arguments in favor of their theory. The nucleoside diphosphate sugar hydrolases, in conjunction with 5′-nucleotidase and α-D-sugar-1-P phosphatase, allow for the complete degradation of UDP-glucose that has come from inside, to form nucleoside and sugar. Both of these are easily transported back into the cell. The specificity of the periplasmic enzymes is such that they act on nucleotides that yield easily metabolizable compounds. These workers considered, as an alternative to nucleotide movement outward through the protoplasmic membrane, the possibility that periplasmic enzymes act on both sides of the membrane. They suggested that the enzymes might be controlled by their protein inhibitors or an unfavorable K_m. The present reviewer does not consider this to be a very attractive idea.

If the intracellular sugar nucleotide pool is accessible to the hydrolases, which serve as control agents, then the nucleoside diphosphate sugar pool would turn over at a rather rapid rate even in the absence of glycosyl transfer reactions. Ward and Glaser (1968, 1969) tested this prediction. They discovered an E. coli mutant with cryptic UDP-sugar hydrolase; the enzyme was no longer available to external substrate although it could still be released by osmotic shock. The authors assumed that the mutation resulted in a change in the specific attachment site for the enzyme; that is, it faced inward and was totally accessible to the intracellular pool. Ward and Glaser concluded that in the mutant most of the turnover of the UDP-glucose and UDP-galactose pool cannot be accounted for by polymer synthesis and, presumably, is accounted for by the hydrolysis of these nucleotides. The conclusions are quite interesting, although they do rest on a number of assumptions and postulates.

VIII. GENERAL CONCLUSIONS

The field of periplasmic proteins has developed considerably since the reviewer's previous treatment of the subject (Heppel, 1967). The evidence is nearly conclusive that a family of proteins is located "at or near the surface of the cell" in *E. coli* and certain other members of the Enterobacteriaceae. However, their location is not known with certainty. Also, we do not know if they have definite attachment sites or if they exist in a free state, sandwiched between membrane and wall layers. For the present, it is wise to consider the designation of these proteins as periplasmic to be operational, emphasizing their selective release, and to await further developments. This is an active field of investigation in which exciting results can be anticipated.

References

Adler, H. I., Fisher, W. D., Cohen, H., and Hardigree, A. (1967). *Proc. Nat. Acad. Sci. U. S.* **57**, 321.
Anraku, Y. (1964a). *J. Biol. Chem.* **239**, 3412.
Anraku, Y. (1964b). *J. Biol. Chem.* **239**, 3420.
Anraku, Y. (1968). *J. Biol. Chem.* **243**, 3116, 3123, and 3128.
Anraku, Y., and Heppel, L. A. (1967). *J. Biol. Chem.* **242**, 2561.
Anraku, Y., and Mizuno, D. (1967). *J. Biochem. (Tokyo)* **61**, 70.
Barth, P. T., Beacham, I. R., Ahmed, S. I., and Pritchard, R. H. (1968). *Biochim. Biophys. Acta* **161**, 544.
Bayer, M. E. (1967). *J. Bacteriol.* **93**, 1104.
Bayer, M. E. (1968). *J. Gen. Microbiol.* **53**, 395.
Bayer, M. E., and Anderson, T. F. (1965). *Proc. Nat. Acad. Sci. U. S.* **54**, 1592.
Birdsell, D. C., and Cota-Robles, E. H. (1967). *J. Bacteriol.* **93**, 427.
Birnboim, H. C. (1966). *J. Bacteriol.* **91**, 1004.
Boos, W. (1969). *Eur. J. Biochem.* **10**, 66.
Braun, V., and Rehn, K. (1969). *Eur. J. Biochem.* **10**, 426.
Brockman, R. W., and Heppel, L. A. (1968). *Biochemistry* **7**, 2554.
Burman, L. G., Nordstrom, K., and Boman, H. G. (1968). *J. Bacteriol.* **96**, 438.
Cashel, M., and Freese, E. (1964). *Biochem. Biophys. Res. Commun.* **16**, 541.
Cedar, H., and Schwartz, J. H. (1967). *J. Biol. Chem.* **242**, 3755.
Cordonnier, C., and Bernardi, G. (1965). *Biochem. Biophys. Res. Commun.* **20**, 555.
Datta, N., and Richmond, M. H. (1966). *Biochem. J.* **98**, 204.
De Duve, C., and Wattiaux, R. (1966). *Annu. Rev. Physiol.* **28**, 435.
Demis, D. J., Rothstein, A., and Meier, R. (1954). *Arch. Biochem. Biophys.* **48**, 55.
Done, J., Shorey, C. D., Loke, J. P., and Pollak, J. K. (1965). *Biochem. J.* **96**, 27C.
Dvorak, H. F. (1968). *J Biol. Chem.* **243**, 2640.
Dvorak, H. F., and Heppel, L. A. (1968). *J. Biol. Chem.* **243**, 2647.
Dvorak, H. F., Anraku, Y., and Heppel, L. A. (1966). *Biochem. Biophys. Res. Commun.* **24**, 628.
Dvorak, H. F., Brockman, R. W., and Heppel, L. A. (1967). *Biochemistry* **6**, 1743.

246   LEON A. HEPPEL

Dvorak, H. F., Wetzel, B. K., and Heppel, L. A. (1970). *J. Bacteriol.* **104**, 543.
Furlong, C. E., and Weiner, J. H. (1970). *Biochem. Biophys. Res. Commun.* **38**, 1076.
Gascon, S., Neumann, N. P., and Lampen, J. O. (1968). *J. Biol. Chem.* **243**, 1573.
Glaser, L., Melo, A., and Paul, R. J. (1967). *J. Biol. Chem.* **242**, 1944.
Glauert, A. M., and Thornley, M. J. (1969). *Annu. Rev. Microbiol.* **23**, 159–198.
Hafkenscheid, J. C. M. (1968). *Biochim. Biophys. Acta* **167**, 582.
Hardy, S. J. S., and Kurland, C. G. (1966). *Biochemistry* **5**, 3676.
Harwood, J. H., and Smith, D. H. (1969). *J. Bacteriol.* **97**, 1262.
Heppel, L. A. (1967). *Science* **156**, 1451.
Hogg, R., and Englesberg, E. (1969). *J. Bacteriol.* **100**, 423.
Hurwitz, C., Rosano, C. L., and Peabody, R. A. (1963). *Biochim. Biophys. Acta* **72**, 80.
Kammen, H. O. (1967). *Biochim. Biophys. Acta* **134**, 301.
Klein, W. L., Dahms, A. S., and Boyer, P. D. (1970). *Fed. Proc., Fed. Amer. Soc. Exp. Biol.* **29**, 341.
Kundig, W., Kundig, F. D., Anderson, B., and Roseman, S. (1966). *J. Biol. Chem.* **241**, 3243.
Kushnarev, V. M., and Smirnova, T. A. (1966). *Can. J. Microbiol.* **12**, 605.
Kushner, D. J., and Pollock, M. R. (1961). *J. Gen. Microbiol.* **26**, 255.
Lee, Y., and Sowokinos, J. R. (1967). *J. Biol. Chem.* **242**, 2264.
Lee, Y., and Sowokinos, J. R. (1969). *J. Biol. Chem.* **244**, 1711.
Leive, L. (1965a). *Proc. Nat. Acad. Sci. U. S.* **53**, 745.
Leive, L. (1965b). *Biochem. Biophys. Res. Commun.* **21**, 290.
Leive, L., and Kollin, V. (1967). *Biochem. Biophys. Res. Commun.* **28**, 229.
Levy, S. B., and Leive, L. (1970). *J. Biol. Chem.* **245**, 585.
Malamy, M., and Horecker, B. L. (1961). *Biochem. Biophys. Res. Commun.* **5**, 104.
Malamy, M., and Horecker, B. L. (1964). *Biochemistry* **3**, 1893.
Mangiarotti, G., Apirion, D., and Schlessinger, D. (1966). *Science* **153**, 892.
Mauck, J., and Glaser, L. (1970). *Biochemistry* **9**, 1140.
Melo, A., and Glaser, L. (1966). *Biochem. Biophys. Res. Commun.* **22**, 524.
Metzenberg, R. L. (1963). *Biochim. Biophys. Acta* **77**, 451.
Mitchell, P. (1961). In "Biological Structure and Function" (T. W. Goodwin and O. Lindberg, eds.), Vol. 2, p. 581. Academic Press, New York.
Momose, H., Nishikawa, H., and Katsuya, N. (1964). *J. Gen. Appl. Microbiol.* **10**, 343.
Munch-Petersen, A. (1967). *Biochim. Biophys. Acta* **142**, 228.
Munch-Petersen, A. (1968). *Eur. J. Biochem.* **6**, 432.
Nakane, P. K., Nichoalds, G. E., and Oxender, D. L. (1968). *Science* **161**, 182.
Neu, H. C. (1967a). *J. Biol. Chem.* **242**, 3896.
Neu, H. C. (1967b). *J. Biol. Chem.* **242**, 3905.
Neu, H. C. (1968a). *J. Bacteriol.* **95**, 1732.
Neu, H. C. (1968b). *Biochemistry* **7**, 3766.
Neu, H. C. (1968c). *Biochemistry* **7**, 3774.
Neu, H. C. (1968d). *Biochem. Biophys. Res. Commun.* **32**, 258.
Neu, H. C. (1969). *J. Gen. Microbiol.* **57**, 215.
Neu, H. C., and Chou, J. (1967). *J. Bacteriol.* **94**, 1934.
Neu, H. C., and Heppel, L. A. (1964a). *J. Biol. Chem.* **239**, 3893.
Neu, H. C., and Heppel, L. A. (1964b). *Proc. Nat. Acad. Sci. U. S.* **51**, 1267.
Neu, H. C., and Heppel, L. A. (1965). *J. Biol. Chem.* **240**, 3685.

Neu, H. C., Ashman, D. F., and Price, T. D. (1967). *J. Bacteriol.* **93**, 1360.
Nisonson, I., Tannenbaum, M., and Neu, H. C. (1969). *J. Bacteriol.* **100**, 1083.
Nossal, N. G., and Heppel, L. A. (1966). *J. Biol. Chem.* **241**, 3055.
Nozawa, R., Horiuchi, T., and Mizuno, D. (1967). *Arch. Biochem. Biophys.* **118**, 402.
Obinata, M., and Mizuno, D. (1968). *Biochim. Biophys. Acta* **155**, 98.
Pardee, A. B. (1968). *Science* **162**, 632.
Pardee, A. B., and Watanabe, K. (1968). *J. Bacteriol.* **96**, 1049.
Penrose, W. R., Nichoalds, G. E., Piperno, J. R., and Oxender, D. L. (1968). *J. Biol. Chem.* **243**, 5921.
Piperno, J. R., and Oxender, D. L. (1968). *J. Biol. Chem.* **243**, 5914.
Repaske, R. (1958). *Biochim. Biophys. Acta* **30**, 225.
Rosen, B. P., and Vasington, F. D. (1970). *Fed. Proc., Fed. Amer. Soc. Exp. Biol.* **29**, 342.
Sargent, M. G., Ghosh, B. K., and Lampen, J. O. (1968). *J. Bacteriol.* **96**, 1329.
Schlesinger, M. J. (1968). *J. Bacteriol.* **96**, 727.
Schlesinger, M. J., and Barrett, K. (1965). *J. Biol. Chem.* **240**, 4284.
Sheimin, R. (1959). *J. Gen. Microbiol.* **21**, 124.
Simpson, R. T., Vallee, B. L., and Tait, G. H. (1968). *Biochemistry* **7**, 4336.
Smith, H. S., and Pizer, L. I. (1968). *J. Mol. Biol.* **37**, 137.
Spahr, P. F., and Hollingworth, B. R. (1961). *J. Biol. Chem.* **236**, 823.
Torriani, A. (1968a). *Methods Enzymol.* **12**, Part B, 212–218.
Torriani, A. (1968b). *J. Bacteriol.* **96**, 1200.
Trevithick, J. R., and Metzenberg, R. L. (1966). *J. Bacteriol.* **92**, 1010.
Umezawa, H., Takasawa, S., Okanishi, M., and Utahara, R. (1968). *J. Antibiot., Ser. A* **21**, 81.
von Hofsten, B. (1961). *Biochim. Biophys. Acta* **48**, 171.
Ward, J. B., and Glaser, L. (1968). *Biochem. Biophys. Res. Commun.* **31**, 671.
Ward, J. B., and Glaser, L. (1969). *Arch. Biochem. Biophys.* **134**, 612.
Weidel, W., Frank, H., and Martin, H. H. (1960). *J. Gen. Microbiol.* **22**, 158.
Weimberg, R., and Orton, W. L. (1965). *J. Bacteriol.* **90**, 82.
Weimberg, R., and Orton, W. L. (1966). *J. Bacteriol.* **91**, 1.
Weiner, J. H., Berger, E. A., Hamilton, M. N., and Heppel, L. A. (1970). *Fed. Proc., Fed. Amer. Soc. Exp. Biol.* **29**, 341.
Wetzel, B. K., Spicer, S. S., Dvorak, H. F., and Heppel, L. A. (1970). *J. Bacteriol.* **104**, 529.
Wilson, D. W. (1971). Unpublished data.
Wilson, O. H., and Holden, J. T. (1969). *J. Biol. Chem.* **244**, 2743.
Yamada, T., Tipper, D., and Davies, J. (1968). *Nature (London)* **219**, 288.

Structure–Function Relationships in Biological Membranes

6

ENZYME REACTIONS IN BIOLOGICAL MEMBRANES

L. ROTHFIELD and D. ROMEO

ABBREVIATIONS

PE   Phosphatidylethanolamine
LPS   Lipopolysaccharide

I. INTRODUCTION

A large number of cellular enzymes are located in membranes. Indeed, it is likely that most "particulate" enzymes are in truth membrane-bound enzymes whose particulate nature reflects their association with the lipid matrix of biological membranes. Included in this category are enzymes located in plasma membranes, mitochondria, microsomes, and in other subcellular organelles. In fact, if one is willing to extend the definition to proteins which catalyze the physical translocation of substrates, the class of membrane enzymes can be extended to include proteins responsible for transmembrane transport, such as bacterial permeases.

The existence of membrane-bound enzymes has been known for many years, but until recently these enzymes were thought to be refractory to solubilization and hence were considered unsuitable for purification and detailed study. In the past ten years, however, it has become clear that many membrane-bound enzymes require the lipid components of membranes for activity and that previous difficulties in "solubilization" were due to the failure to recognize this fact.

Fleischer and his co-workers (Fleischer and Klouwen, 1961; Fleischer et al., 1962) were the first to show clearly a lipid requirement in enzyme reactions by demonstrating that lipid extraction resulted in a marked decrease in activity of several reactions of the mitochondrial electron transport system and that activity was restored by the addition of phospholipids. Since then a large number of membrane-bound enzymes from mitochondria, microsomes, and bacteria have been studied and have been shown to be affected by phospholipids.

These observations suggest that it should be possible to dissociate membrane-bound enzyme systems into phospholipid and protein components and to restore the biological activity by recombining the purified components, thus achieving the reconstitution of a functional portion of the original membrane. Thus far this has been convincingly achieved in only a few cases, as discussed below.

The significance of this type of experiment is severalfold. First, it permits study of a new parameter of enzyme catalysis, namely the role of lipids as cofactors in enzyme reactions. Second, it provides a model system for study of the physical and chemical aspects of lipid–protein interactions of biological significance. Third, it permits detailed study of perhaps the most important unanswered question of membrane structure—the nature of the interaction of lipids and proteins in membranes. Therefore, a discussion of enzyme reactions in membranes must focus chiefly on the several roles of membrane lipids in these reactions. They can be divided into two general categories: (a) lipids as physical cofactors in membrane enzyme reactions and (b) lipids as covalently linked intermediates in membrane enzyme reactions.

II. PARTICIPATION OF LIPIDS IN MEMBRANE ENZYME REACTIONS

A. Lipids as Physical Cofactors

A requirement for phospholipids in the activity of many membrane-bound enzymes has now been shown. In all cases the basic experimental scheme is similar—enzyme activity is lost when lipid is removed from an enzyme system and is restored when lipid is added back. In most cases the lipids do not form covalently linked intermediates in the reaction sequence, and they therefore play the role of physical cofactors, activating the enzyme system but not themselves participating in the reaction.

Several possible mechanisms can account for this type of effect. (a) The lipid can activate the substrate. This has been demonstrated clearly with the enzymes that glycosylate bacterial lipopolysaccharides (see below). (b) The lipid can directly activate the enzyme, perhaps by inducing a conformational change in the protein. It is well established that nonpolar compounds can induce reversible changes in protein structure (Wettlaufer and Lovrien, 1964) and it is likely that similar phenomena are of importance in biological systems. Thus, Scanu and Hirz (1968) have shown that the lipids of human serum lipoproteins stabilize the predominantly α-helical configuration of the peptide chains, presumably through hydrophobic interactions (see Chapter 4 for a more detailed theoretical discussion of this subject). This general mechanism, i.e., direct activation of the enzyme, is assumed to operate in many cases, although with little evidence except in the case of mitochondrial β-hydroxybutyrate dehydrogenase (see below). (c) The lipid can act as an organizer in multienzyme systems, directing the sequential arrangement of enzyme proteins within the membrane structure in systems such as the electron transport chain. Direct evidence of this mechanism is lacking at present.

B. Lipids as Covalently Linked Intermediates ("Carrier Lipids")

Membrane lipids can also participate in enzyme-catalyzed reactions by forming covalently linked intermediates in the reaction sequence. Thus far this has been shown to occur in the synthesis of macromolecules in bacteria, and in all cases examined the "carrier lipids" are polyisoprenoid compounds. This subject is discussed in detail in Chapter 8.

III. STUDIES OF INDIVIDUAL ENZYMES

A. Introduction

Detailed studies of the role of lipids in the activity of membrane enzymes require that the enzyme proteins be available in purified form. Unfortunately, only a few membrane enzymes have been isolated free of lipids and even fewer have been highly purified. Most of the enzymes that have been studied are of bacterial origin or are from mitochondria or microsomes of higher organisms. The best-studied individual enzymes are the purified transferase enzymes involved in biosynthesis of bacterial lipopolysaccharides and β-hydroxybutyrate dehydrogenase and cytochrome oxidase of beef heart mitochondria, but some important information has also been obtained from studies of other membrane enzymes which have, of necessity, been examined under less clearly defined conditions.

B. Methods of Isolation

1. General Considerations

In general, solubilization of membrane proteins requires the disruption of the organized membrane structure. Nonpolar bonds (lipid–lipid, lipid–protein, protein–protein) are believed to play a major role in stabilizing the structure of membranes (see Chapter 4) and it is therefore not surprising that agents which disrupt such bonds are often effective in solubilizing membrane-bound enzymes. Much of this work has involved mitochondrial enzymes [for further details see King and Howard (1967)].

2. Detergents

Detergents were among the first compounds to be used to solubilize membrane enzymes. Bile salts and other detergentlike molecules were used to solubilize components of the mitochondrial electron transfer chain and segments of the chain were then fractionally precipitated by the addition of salt to the soluble extracts. Several electron transfer complexes have been isolated by this general approach, although problems arise in removing the detergent and in further fractionating the protein components of the complexes. The presence of multiple proteins in these preparations has been shown by examining the fractions by gel electrophoresis in the presence of urea (Takayama et al., 1964).

More recently, a wide variety of newer synthetic detergents has become available. Enzyme activity is frequently retained in the presence of low concentrations of these detergents, and preliminary evidence indicates that membrane proteins may be differentially extracted by some of them. They may therefore prove to be of use in both extracting and fractionating some membrane enzymes.

3. Organic Solvents

Organic solvents have also been used to extract membrane proteins, but enzymologists have often been reluctant to use this approach because they fear that the enzymes will be irreversibly denatured. This fear may be ill founded in the case of proteins which are normally located in nonaqueous environments such as the lipid-rich structure of biological membranes. Indeed, for some membrane enzymes the nonpolar milieu of organic solvents may prove a happier home than the usual aqueous buffer systems used for classical water-soluble enzymes. There is now evidence that many membrane enzymes are quite stable in the presence of organic solvents. These include many components of the mitochondrial electron transfer chain (Fleischer et al., 1962) and several membrane-bound bacterial enzymes (Endo and Rothfield, 1969; Higashi et al., 1970). In addition, combinations of organic solvents and HCl have been successfully used to extract membrane proteins (Zahler and Wallach, 1967), although thus far no enzymatic studies have been performed on the extracted proteins.

4. Lipolytic Enzymes

Treatment with phospholipases results in the solubilization of certain membrane enzymes. These agents should be ideal for this purpose because their action is highly specific for the lipid components of the membrane and some of the possible artifacts of chemical treatments are thereby avoided. Snake venoms have generally been used as sources of the lipolytic activities, but the presence of a variety of other hydrolytic activities in the crude venom preparations is a potential source of difficulty. Release of cytochrome b_5 by gentle disruption of microsomes with pancreatic lipase and of cytochrome b_5 reductase by treatment of microsomes with snake venom has been described by Strittmatter and Velick (1956). A variety of mitochondrial enzymes has also been released by treatment with phospholipases, including β-hydroxybutyrate dehydrogenase (Fleischer et al., 1966) and NADH-dehydrogenase (King and Howard, 1967). These are described in greater detail below.

5. Chaotropic Agents

A particularly useful class of compounds for the extraction of membrane enzymes are the so-called chaotropic agents. These inorganic ions have in common the property of favoring the transfer of nonpolar residues into an aqueous environment, presumably due to the ability of the agents to disorder the structure of water. Because of this property they have been used to alter the secondary and tertiary structure of proteins and nucleic acids and to increase the water solubility of small molecules. Their relative effectiveness follows the order

$$SCN^- > ClO_4^- > NO_3^- > I^- > Br^- > Cl^- > SO_4^=, CH_3COO^-, F^-$$

Certain other small molecules, such as guanidine and urea, may work in a similar manner although the effects of the organic molecules on the ordered structure of water are a matter of controversy (Glasel, 1970).

Hatefi and his co-workers (Hatefi and Stempel, 1969; Davis and Hatefi, 1969) have used chaotropic agents to resolve complex I of the mitochondrial electron transport chain (DPNH-CoQ reductase). Treatment of the complex with NaSCN, NaClO₄, guanidinium-HCl, or urea resulted in solubilization of several proteins and permitted detailed studies of the DPNH-dehydrogenase component of the system. Guanidinium-SCN has also been used successfully to solubilize several bacterial membrane proteins (Robertson *et al.*, 1971), and it is likely that the use of this class of agents will have general applicability to many membrane enzyme systems.

C. Bacterial Enzymes

1. *Enzymes Involved in Lipopolysaccharide Biosynthesis: Dissociation and Reconstitution of a Membrane Enzyme System*

a. Membrane-Bound Glycosyltransferases. It was shown in 1964 that several of the enzymes involved in biosynthesis of the lipopolysaccharides of gram-negative bacteria required phospholipids for activity (Rothfield and Horecker, 1964), and these enzymes have since been the subject of intensive investigation. These studies defined one role of phospholipid in membrane-bound enzyme reactions, namely "substrate activation." In this role, lipids interact physically with the substrate, thereby converting it into an active participant in the reaction.

The substrates in these reactions are lipopolysaccharides located in the cell envelope of gram-negative bacteria (see Chapters 5 and 8 for further discussion of the bacterial cell envelope). The lipopolysaccharides are large molecules containing a lipid portion ("Lipid A") covalently linked to a complex polysaccharide (Fig. 1), and they are closely associated with phospholipids and proteins within the membrane (Rothfield and Pearlman-Kothencz, 1969). As discussed below, phospholipid and protein are both involved in the enzyme reactions leading to biosynthesis of the lipopolysaccharide molecule.

Biosynthesis of the core region of the molecule occurs by the stepwise transfer of sugars from nucleotide sugars to the growing polysaccharide chain. A series of membrane-bound transferase enzymes has been defined which catalyzes this sequential transfer of sugar residues. Beginning with the initial glucose residue of the core, the following sequence of reactions has been demonstrated in *Salmonella typhimurium* by Osborn and her co-workers (Osborn and D'Ari, 1964) and in *Escherichia coli* by Heath and his collaborators (Edstrom and Heath, 1964; also seen Chapter 8):

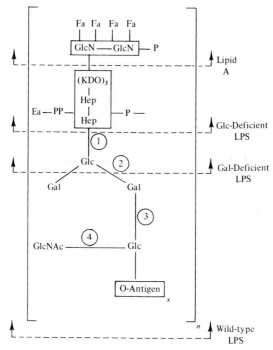

Fig. 1. Structure of the lipopolysaccharide of *Salmonella typhimurium*. The Lipid A region and the structure of several mutant lipopolysaccharides are shown (Osborn *et al.*, 1964). In each case the polysaccharide terminates at the point indicated by the dashed lines. The numbers indicate the site of action of several of the transferase enzymes involved in biosynthesis of the core region of the polysaccharide. Abbreviations: Fa, fatty acid; KDO, 2-keto-3-deoxyoctonyl; Hep, heptosyl (L-glycero-D-mannoheptosyl); Ea, ethanolamine; GlcN, glucosaminyl; Glc, glucosyl; Gal, galactosyl; GlcNAc, N-acetylglucosaminyl.

$$\text{UDP-Glc} + \text{glucose-deficient LPS} \rightarrow \text{Glc-LPS} \tag{1}$$

$$\text{UDP-Gal} + \text{Glc-LPS} \rightarrow \text{Gal-Glc-LPS} \tag{2}$$

$$\text{UDP-Glc} + \text{Gal-Glc-LPS} \rightarrow \text{Glc-[Gal-Glc-LPS]} \tag{3}$$

$$\text{UDP-GlcNAc} + \text{Glc-Gal-Glc-LPS} \rightarrow \text{GlcNAc-[Glc-Gal-Glc-LPS]} \tag{4}$$

To study the enzymes of lipopolysaccharide biosynthesis, mutant strains are used which contain incomplete lipopolysaccharides (Fig. 1) since these incomplete lipopolysaccharides can act as acceptors for sugar transfer in the presence of the other components of the enzyme system. In studies of the galactosyltransferase reaction, for example, the galactose-deficient acceptor lipopolysaccharide is obtained from mutant strains which are unable to incorporate galactose into lipopolysaccharides. In some cases the defect is due to an inability of the mutant to synthesize UDP-galactose, whereas in other mutants the defect is in the transferase enzyme itself.

b. Studies of Purified Glycosyltransferases. All of the transferase enzyme activities are largely particulate, being located in the insoluble membrane structure together with the lipopolysaccharides which are the substrates in the reactions. After sonication of the cells, a portion of the activity of several of these enzymes is also present in the soluble fraction. The enzyme activities are identified by their ability to catalyze the transfer of glucose or galactose from the appropriate nucleotide sugar to the endogenous lipopolysaccharide of the mutant cell envelope preparations (Rothfield et al., 1964). Two of the enzymes (UDP-glucose:lipopolysaccharide glucosyltransferase I and UDP-galactose:lipopolysaccharide α,3-galactosyltransferase, enzymes 1 and 2 in Fig. 1) have been purified. The availability of these enzymes in purified form permitted detailed study of the mechanism of the reaction and led to the discovery of the role of membrane phospholipids in this membrane enzyme system.

A role for membrane lipids in the glucosyl- and galactosyltransferase reactions [reactions (1) and (2), above] is now firmly established. When lipids are extracted from the cell envelope fraction by treatment with organic solvents, the residual cell envelope still contains the galactose-deficient or glucose-deficient lipopolysaccharide but has lost its ability to act as acceptor in the transferase reaction (Rothfield and Horecker, 1964). In contrast, the glycosyl acceptor activity of the lipid-depleted cell envelope can be completely restored by adding back the lipid extract (Fig. 2).

Isolated lipopolysaccharides can also be tested for their ability to act as acceptors in the transferase reaction by measuring the transfer of [^{14}C]galactose or [^{14}C]glucose from UDP-[^{14}C]galactose or UDP-[^{14}C]-glucose to the lipopolysaccharide in the presence of the transferase enzyme. In this case there is also a requirement for phospholipid in the reaction since acceptor activity is not seen unless phospholipid is added.

$$\text{UDP-Glc} + \text{glucose-deficient LPS} \xrightarrow[\text{phospholipid}]{\text{enzyme}} \text{Glc-LPS} + \text{UDP} \qquad (1a)$$

$$\text{UDP-Gal} + \text{galactose-deficient LPS} \xrightarrow[\text{phospholipid}]{\text{enzyme}} \text{Gal-LPS} + \text{UDP} \qquad (2a)$$

The enzymatic parameters of the reconstituted system (i.e., V_{max}, K_m, yield) are almost identical with those of the intact cell envelope, suggesting that the molecular organization of the reconstituted material is similar to that of the native system within the membrane.

Phosphatidylethanolamine was shown to be the active component of

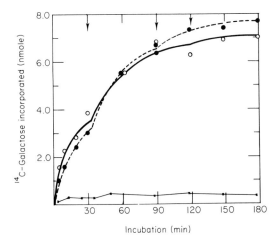

Fig. 2. Effect of lipid extraction on the galactosyltransferase reaction. Incorporation of [^{14}C]galactose was determined by incubating UDP-[^{14}C]galactose, soluble enzyme, and acceptor and measuring incorporation of radioactivity into acid-insoluble material. In the experiments using extracted cell envelopes, the cell envelope acceptor was first extracted by treatment with ethanol–ether (1:1). The acceptors in each experiment are indicated in the figure. Additional enzyme and UDP-[^{14}C]galactose were added at the times indicated by the arrows. Legend: O—O, extracted cell envelope plus lipid extract; ●—●, unextracted cell envelope; x—x, extracted cell envelope.

the lipids extracted from the bacterial cell envelope, but several other synthetic and natural phospholipids were also active in the system (Rothfield and Pearlman, 1966). To be active in the transferase system, lipids must contain an intact diacylglycerophosphate backbone (see Fig. 3). In keeping with this requirement, no activity is seen with lysophosphatidylethanolamine, sphingolipids, triglycerides, fatty acids, etc.

In addition to the requirement for a diacylglycerophosphate backbone, the nature of the fatty acid residues (R^1 and R^2) and of the polar head group (R^3) also markedly affect the ability of phospholipids to participate in the reaction. Activity requires that the phospholipid contain *cis*-unsaturated or cyclopropane fatty acyl groups (Fig. 4). Since cyclopro-

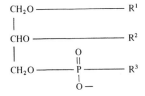

Fig. 3. Structure of glycerophosphatides.

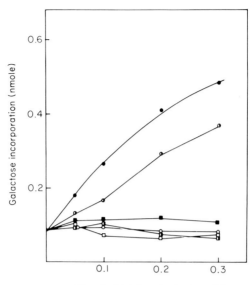

Fig. 4. Effect of different fatty acid components on the activity of phosphatidylethanolamines in the galactosyltransferase reaction. The *E. coli* phosphatidylethanolamine contained 54% cyclopropane fatty acids, 43% saturated fatty acids, and
3% *cis*-unsaturated fatty acids. The *A. agilis* phosphatidylethanolamine (*A. agilis* PE[a])
contained 65% *cis*-unsaturated and 35% saturated fatty acids. The *A. agilis* PE[b]
refers to hydrogenated *A. agilis* phosphatidylethanolamine (i.e., 100% saturated fatty
acids). Legend: ●, *A. agilis* PE[a]; □, *A. agilis* PE[b]; ◑, *E. coli* PE; ○, dipalmitoyl
PE; ■, didecanoyl PE; ◧, dihexanoyl PE. (Endo and Rothfield, 1969.)

pane fatty acids differ from unsaturated fatty acids by the presence of a
cyclopropane ring in place of a double bond, this eliminates the possibility that the double bonds of unsaturated phospholipids participate
directly in the reaction by virtue of their electronic configuration or
chemical reactivity. Since *cis*-cyclopropane and unsaturated fatty acid
residues are similar in shape (Fig. 5) and in physical properties (i.e.,
both are highly mobile when compared to straight-chain saturated fatty
acids), it is likely that their similar activity is related to spatial and
physical characteristics required in the interaction of the phospholipids
with lipopolysaccharide molecules.

Specificity also resides in the polar position of the phospholipid. A
variety of polar groups can be substituted for ethanolamine without loss
of activity, but the presence of a choline residue in the R[3] position results in loss of activity of the phospholipid in the galactosyltransferase
reaction.

The major role of phospholipid in these reactions is to interact with

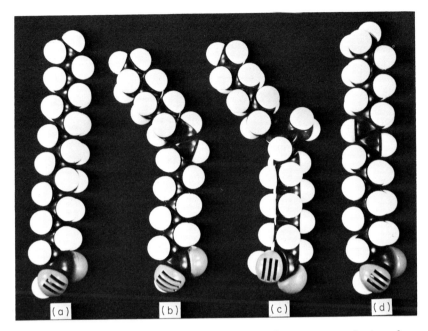

Fig. 5. Comparison of molecular models of saturated, *cis*-unsaturated, *cis*-cyclopropane, and *trans*-unsaturated fatty acids. (a) Hexadecanoic acid (C-16 saturated); (b) *cis*-9,10-hexadecenoic acid (C-16 *cis*-unsaturated); (c) *cis*-9,10-methylenehexadecanoic acid (C-17 *cis*-cyclopropane); (d) *trans*-9,10-hexadecenoic acid (C-16 *trans*-unsaturated).

the lipopolysaccharide substrate. Evidence for this was obtained by demonstrating the following sequence of reactions in the galactosyltransferase system:

$$PE + LPS \rightarrow LPS \cdot PE \tag{5}$$

$$LPS \cdot PE + Enz \rightarrow Enz \cdot LPS \cdot PE \tag{6}$$

$$Enz \cdot LPS \cdot PE + UDP\text{-}Gal \rightarrow Gal\text{-}LPS \cdot PE + UDP(+Enz) \tag{7}$$

In the first step [Eq. (5)] a physical interaction occurs, leading to formation of a multimolecular LPS·PE complex (Rothfield and Takeshita, 1965), but in order to form the active complex it is necessary to heat and slowly cool the mixture of lipopolysaccharide and phospholipid. The heating procedure is thought necessary to promote the reshuffling of individual molecules from the originally separated micelles of lipopolysaccharide and phosphatidylethanolamine, resulting in formation of the more stable binary complex of the two components. The phospholipid and lipopolysaccharide molecules are not covalently bound to each other since the binary complex can be dissociated by extracting the phospha-

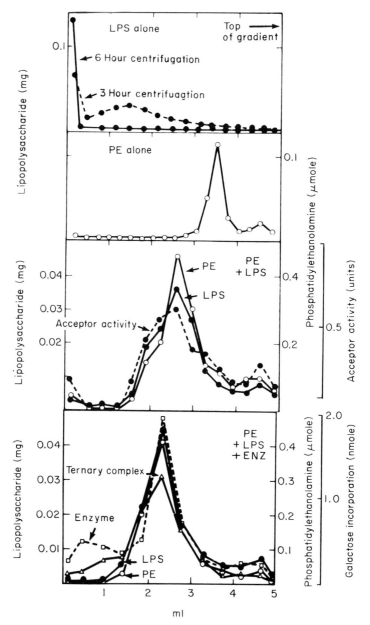

Fig. 6. Isolaton of binary and ternary complexes by gradient centrifugation to equilibrium. Acceptor activity was measured by adding UDP-[¹⁴C]galactose and enzyme to each fraction; ternary complex was measured by adding only UDP-[¹⁴C]galactose to each fraction. (From Weiser and Rothfield, 1968.)

tidylethanolamine with organic solvents. In the second step [Eq. (6)] the transferase enzyme binds to the LPS·PE complex with formation of a ternary complex (Enz·LPS·PE). Sucrose gradient centrifugation to equilibrium has been used to isolate the binary and ternary complexes (Fig. 6), and the isolated complexes are active in the enzyme system when the additional required components are added, demonstrating that the functional enzyme system can indeed be reconstituted from the purified components (Weiser and Rothfield, 1968).

Recent interest has focused on the question of the molecular architecture of the reconstituted complexes, since a similar molecular organization probably exists within the cell envelope. This question has not yet been definitively answered, but a model has been proposed which is consistent with the available evidence (Rothfield et al., 1966). In this model the basic membrane framework consists of a planar array of phospholipid and lipopolysaccharide molecules intermixed in a bimolecular leaflet structure (Fig. 7), which corresponds to the binary complex described above [LPS·PE, Eq. (5)]. Protein molecules, including the transferase enzyme, are then viewed as interpolated into this basic lipid–lipopolysaccharide framework in a still-undefined manner. Several lines of evidence are consistent with the model. (a) Lipopolysaccharides and phospholipids are amphipathic molecules, having a polar region (the phosphorylethanolamine portion of phosphatidylethanolamine and the

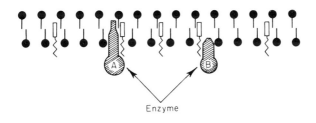

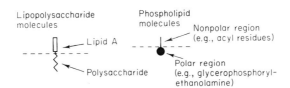

Fig. 7. A speculative model of the portion of the membrane containing the galactosyltransferase enzyme system. Two possible locations of the enzyme are shown. A, A portion of the enzyme penetrates into the nonpolar interior of the membrane. B, Enzyme is located only in the polar portion of the membrane. Present evidence does not permit a choice between the two possibilities.

polysaccharide portion of lipopolysaccharide) and a nonpolar region (the acyl residues of phosphatidylethanolamine and lipopolysaccharide). As a result, when placed in aqueous solvents both compounds tend to form ordered liquid crystalline structures with dimensions charactersitic of bimolecular leaflets (Rothfield and Horne, 1967). In addition, x-ray diffraction studies of phospholipids and of lipopolysaccharides in aqueous suspension (Luzzati *et al.*, 1962; Finean, 1953; Burge and Draper, 1967) indicate that the fatty acid chains are oriented perpendicular to the surface of the flat sheets as required by a bimolecular leaflet arrangement. The physicochemical characteristics of the molecules therefore tend to favor formation of a bimolecular leaflet. (b) Electron microscopy indicates that the dimensions of the outer membrane of the cell envelope and of reconstituted lipopolysaccharide–phosphatidylethanolamine complexes are compatible with the bimolecular leaflet structure (DePetris, 1967; Rothfield *et al.*, 1966). (c) The functional transferase system has been successfully reconstituted in a monomolecular film.

c. Reconstitution of the Galactosyltransferase System in a Monomolecular Film. It has been known for many years that lipid molecules arrange themselves into a molecular monolayer when spread on the surface of a water-filled vessel. In such films the molecules are arranged in a planar array and are oriented with the nonpolar hydrocarbon residues directed upward into the air phase while the polar portions of the lipid molecules remain in the aqueous phase (see Fig. 8). The monolayer therefore is analogous to half of a bimolecular leaflet and can be used as a model system for studying molecular interactions that may normally occur in bimolecular leaflets. Penetration of additional molecules into the film is indicated by changes in surface pressure, and the film can be removed for direct analysis. Support for the model shown in Fig. 7 has emerged from studies using this technique.

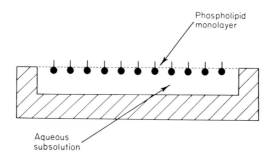

Fig. 8. Diagrammatic representation of monolayer device.

It has been shown that molecules of lipopolysaccharide, phosphatidyl-ethanolamine, and enzyme protein will form a mixed monolayer which possesses the full catalytic activity normally seen within the intact cell envelope (Romeo et al., 1970a,b). The experiment was performed in four steps. (1) A monomolecular film of phosphatidylethanolamine was formed at the air–water interface. (2) Lipopolysaccharide was injected into the aqueous subphase, and penetration of lipopolysaccharide mole-cules into the phospholipid monolayer was indicated by an increase in the surface pressure of the film (Fig. 9). When the resulting mixed LPS·PE monolayer was removed from the surface and assayed by addi-tion of enzyme and UDP-[^{14}C]galactose, it was an active acceptor in the transferase reaction without preliminary heating. In contrast, simple mixtures of the two components are inactive as acceptors in the enzyme system when the heating step is omitted. This indicates that the lipopoly-saccharide and phosphatidylethanolamine molecules in the mixed mono-layer were in the proper configuration to participate in the enzyme reac-tion. (3) Enzyme was then injected into the subphase below the mixed LPS·PE monolayer; penetration of enzyme molecules into the mono-layer was indicated by an increase in the surface pressure of the film (Fig. 9) and was confirmed by direct analysis of the film. (4) In the final

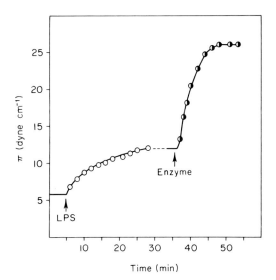

Fig. 9. Sequential penetration of lipopolysaccharide and enzyme protein into a monomolecular film of phosphatidylethanolamine; π indicates change in surface pressure. (From Romeo et al., 1970b.)

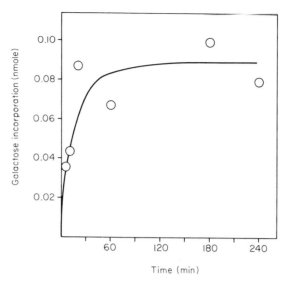

Fig. 10. Enzyme activity within a monomolecular film containing phosphatidyl-ethanolamine, lipopolysaccharide, and transferase enzyme; UDP-[³H]galactose was introduced into the subsolution below a mixed monolayer (Enz·LPS·PE) and incorporation of [³H]galactose into the film was measured by acid precipitation of the removed film at intervals. Each point represents a separate experiment. (From Romeo et al., 1970b.)

step, UDP-[³H]galactose was injected into the subphase below the ternary monolayer (Enz·LPS·PE). After a suitable incubation period, the monolayer was removed and incorporation of [³H]galactose into lipopolysaccharide was determined (Fig. 10), demonstrating that the enzyme reaction took place within the monomolecular film.

These experiments indicated that enzyme, lipopolysaccharide, and phosphatidylethanolamine molecules were organized within the monolayer in the manner required for activity of the transferase system. Since the monolayer is equivalent to half of a molecular bilayer, this lends support for the model in which the components are part of a mixed bimolecular leaflet within the membrane.

The orientation of the transferase enzyme molecules within the structure has not yet been established. A portion of the enzyme molecule is surely present in the polar region (Fig. 7, model B) since the galactosyl donor was present in the subsolution and since the polysaccharide chains which act as acceptors in the reaction face the aqueous subphase. At present it is not known whether a portion of the enzyme also extends into the hydrophobic interior of the membrane structure (Fig. 7, model A).

2. Membrane ATPase: Allotopy in Membrane Enzymes

Membrane-bound ATPase's are ubiquitous in nature. In animal cells the best studied have been the mitochondrial ATPase which appears to be involved in oxidative phosphorylation (see Chapter 9) and the Na^+–K^+-activated ATPases of the plasma membrane of erythrocytes and other cells, which are a part of the active transport system for cations (see Chapter 7). In addition to these ATPases of mammalian cells, ATPase activity is also present in the plasma membranes of bacteria. This is not surprising since the bacterial plasma membrane is the site of the active transport of cations; in many organisms it also contains the electron transport system and is therefore analogous to the inner membrane of the mitochondria of eucaryotic cells (Salton, 1967).

In 1965, Abrams succeeded in solubilizing the membrane ATPase of *Streptococcus fecalis*, and the availability of the soluble enzyme made it possible to study the factors involved in the association of the enzyme with the membrane structure. It soon became apparent that cations were required for binding the ATPase to the membrane, as shown by the observation that most of the membrane-bound ATPase was released when the membranes were washed in cation-free solutions (Abrams and Baron, 1968). A variety of cations were effective in preventing release of the bound ATPase, including Mg^{++}, Mn^{++}, spermidine^{+++}, and Ca^{++}.

The binding was easily reversible. Although a loose reassociation of ATPase with the ATPase-depleted membrane was observed even in the absence of Mg^{++}, a stable ATPase–membrane complex could be formed only in the presence of the cation. In addition, the binding experiments indicated that the membrane could accommodate only a limited number of ATPase molecules, implying a limited number of sites in the membrane that are available to the enzyme. On the basis of this type of evidence, Abrams and Baron (1968) suggested that Mg^{++} ions (and presumably other multivalent cations) act as ligands binding the ATPase to specific binding sites in the membrane (Fig. 11). On the other hand, the evidence does not exclude an alternate possibility, namely that the cation induces a conformational change in the enzyme or in components of the membrane, thereby permitting the protein to bind to the membrane without the cation acting directly as a ligand.

Several analogies exist between the bacterial membrane ATPase and the mitochondrial ATPase studied by Racker and his co-workers (see Chapter 9). In each case the membrane-bound enzyme differs from the solubilized enzyme in several properties. In the most striking parallel, the bacterial and mitochondrial enzymes are both inactivated by N,N'-dicyclohexylcarbodiimide (DCCD) when the enzymes are membrane

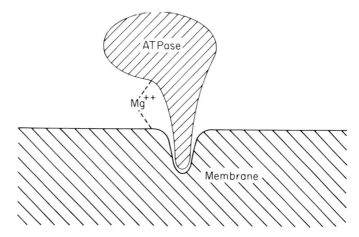

Fig. 11. Suggested role of Mg^{++} in facilitating the binding of ATPase to the plasma membrane of *S. fecalis*. (Freely adapted from Abrams and Baron, 1968.)

bound (Hanold *et al.*, 1969; Bulos and Racker, 1968). In contrast, the sensitivity to DCCD disappears when the enzymes are solubilized. Thus the presence of the enzyme within the native membrane or in the reconstituted enzyme–membrane complex is associated with changes in the properties of the protein. Racker has called this property "allotopy."

Allotypy appears to be a general characteristic of membrane enzymes. In most of the examples discussed elsewhere in this chapter, a catalytic function was markedly altered when lipids were removed from the membrane, as shown by the loss of an enzyme activity. In the case of the bacterial and mitochondrial ATPases, the alteration is less extreme but also involves regulation of a catalytic activity, as manifested by an altered sensitivity to inhibitors (DCCD for the bacterial enzyme; DCCD or oligomycin for the mitochondrial enzyme).

3. The PEP–Phosphotransferase System: A Membrane Enzyme System Involved in Transport

The cell envelopes of many bacteria contain an enzyme system catalyzing the transfer of phosphate from phosphoenolpyruvate to sugars (Fig. 12), and the studies of Roseman and his co-workers have indicated that the enzyme system is involved in the transport of sugars into the cell (for further details see Chapter 7). In *E. coli* and *Salmonella typhimurium*, at least four proteins are required for the PEP-dependent phosphorylating activity. Two of the protein components (II-A and II-B) appear to be located in the membrane, while the other two (HPr and enzyme I) are recovered in the soluble fraction of cell extracts. In 1969,

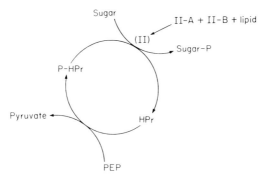

Fig. 12. The PEP–phosphotransferase system.

Roseman demonstrated that lipid was required for activity of the membrane-bound components of the system and subsequent studies revealed that the active system can be reconstituted *in vitro* from the protein components plus phospholipid (Roseman, 1969).

The phosphotransferase system is one of the very few membrane enzyme systems in which the individual proteins have been successfully obtained in soluble form. Therefore, in addition to the great inherent interest of its relation to sugar transport, the system offers another excellent opportunity to study the details of protein–lipid interactions in membranes.

Phosphatidylglycerol is the active phospholipid in the system, whereas other phospholipids show much less activity. This was the first demonstrated involvement of phosphatidylglycerol in membrane enzyme reactions. It emphasizes the point that a wide variety of phospholipids are involved in the activation of membrane enzyme systems and that a relatively high degree of specificity exists. Thus even minor phospholipids can play a major role in regulating the catalytic activity of important protein components of membranes. The overall phospholipid composition of the membrane is therefore less important than the ability or inability of the cell to synthesize specific phospholipids and to ensure that they are localized in the proper portion of the membrane (i.e., adjacent to the appropriate catalytic proteins).

4. Isoprenoid Alcohol Phosphokinase: A Membrane Enzyme Soluble in Organic Solvents

The membrane fraction of *Staphylococcus aureus* contains an enzyme (isoprenoid alcohol phosphokinase) which catalyzes the phosphorylation of isoprenoid alcohols. The studies of Higashi and Strominger (1970) have shown that phospholipid is required for activity of the enzyme, thereby placing it in the general category of lipid-activated membrane enzymes. *In vivo* the product of the enzyme-catalyzed reaction is presumably the C_{55}-isoprenoid alcohol phosphate involved in biosynthesis of the peptidoglycan of the cell wall (see Chapter 8).

The enzyme has the unusual property of being solubilized when the membrane is treated with acidic *n*-butanol at room temperature and of being quite stable in several organic solvents (Higashi *et al.*, 1970). The acidic butanol extract contains all three components of the system—enzyme, phospholipid, and C_{55}-isoprenoid acceptor. On lowering the temperature to $-15°C$ the enzyme and phospholipid are precipitated while the acceptor remains in the supernatant fraction.

Evidence that the enzyme system requires phospholipid was obtained by further fractionation of the enzyme–phospholipid mixture on DEAE-cellulose under conditions usually used to fractionate lipids. Two fractions were obtained. Fraction B contained the enzyme protein, whereas fraction A consisted of a mixture of phosphatidylglycerol and cardiolipin. Enzyme activity was not seen unless both fractions were present in the assay mixture.

There is a marked specificity for phosphatidylglycerol or cardiolipin of bacterial origin in this enzyme system. Phosphatidylglycerol and cardiolipin are equally active, while a variety of other phospholipids from bacteria or from higher organisms are much less effective. A part of this specificity appears to reside in the fatty acid portion of the phospholipid molecule since bacterial cardiolipins are active while cardiolipin from beef heart is much less active. Since the inactive mammalian phospholipids contain high percentages of polyunsaturated fatty acids, detailed studies of the role of the fatty acids residues will be of interest. Moreover, in common with most other lipid-activated membrane enzymes, the polar portion of the phospholipid is also of importance. Thus bacterial phosphatidylethanolamine is considerably less active than phosphatidylglycerol from the same organism. An unusual feature of the system is the considerable activity seen with lysophatidylglycerol and lysocardiolipin.

5. The Nitrate Reductase System: Assembly of a Membrane Multienzyme System

The cell envelope fraction of *E. coli* contains an inducible electron transport chain which is capable of transferring electrons to nitrate. Recently, mutants of the nitrate reductase system have been isolated which appear to be defective in assembly of this membrane-bound multienzyme system. Since studies of membrane assembly have not previously been possible, the system is of great potential interest.

The toxic effect of chlorate on *E. coli* provides the basis for mutant selection, since chlorate toxicity requires an intact nitrate reductase system. Mutants of the nitrate reductase system can therefore be selected on the basis of their chlorate resistance. Mutations in any of several distinct genetic loci (*chl A*, *chl B*, *chl C*, and *chl E*) result in loss of nitrate reductase activity. In 1967, Puig and Azoulay observed that mutations in the *chl A* or *chl B* genes are pleiotropic, resulting in the loss of several normal membrane components (Puig and Azoulay, 1967; Pischaud *et al.*, 1967). These include nitrate reductase and formate hydrogen lyase activities (Azoulay *et al.*, 1967) and a cytochrome component of the normal membrane (Ruig-Herrera and DeMoss, 1969). This led to the suggestion that the products of the *chl A* and *chl B* genes are required for proper assembly of the portion of the membrane containing the components of the nitrate reductase system. This idea was supported by the altered sedimentation properties of a crude cell envelope fraction from a *chl A* mutant (Azoulay *et al.*, 1969), suggesting that the overall structure of the membrane differed from that of the wild-type strain.

Evidence that assembly required the products of both the *chl A* and *chl B* genes emerged from *in vitro* complementation studies (Azoulay *et al.*, 1969). Nitrate reductase activity was restored to approximately 10% of the wild-type level when supernatant fractions from both mutants were mixed, whereas the separate supernatant fractions showed no activity. At the same time there was formation of a new high molecular weight particle. The new particle contained about one-third of the reconstituted nitrate reductase activity and a similar proportion of the cytochrome b_1 of the original supernatant fractions. However, the reconstituted particle also contained 15–20% of the total soluble protein of the original supernatant fraction, and it has not been established that all of this very large amount of protein represents membrane proteins. Further studies will be required to determine whether the reconstituted particle actually represents membrane reassembly.

It has been suggested by Azoulay and his co-workers that the normal products of the *chl A* and *chl B* genes are a "structural protein" and an "assembly factor," respectively. In the absence of either of these proteins, it is thought that the other components of the nitrate reductase system are unable to assemble themselves into a membranous structure and therefore remain in the soluble fraction. In this view, the soluble fractions from *chl A* and *chl B* mutants are thought to contain all the normal constituents of the membrane-bound system except for the mutant gene product. The ability to reconstitute nitrate reductase activity together with formation of a particulate structure when soluble fractions from *chl A* and *chl B* mutants are mixed is consistent with this view. It should be pointed out that several alternate explanations, not involving mem-

brane reassembly, are also possible and proof of this interesting hypothesis must await further experimentation. However, the system may
provide one of the few opportunities to study the key question of membrane biogenesis.

Direct evidence that a single membrane protein is required for several
of the activities of the nitrate reductase system has been obtained by
Schnaitman (1969). A chlorate-resistant mutant was studied and found
to be deficient in the activities of both nitrate reductase and formate
dehydrogenase. Since these two enzyme activities are regulated independently in the wild-type organism, it is likely that at least two proteins
are responsible for the two enzyme activities.

Gel electrophoretic analysis of the cell envelope fraction revealed that
the mutant lacked one of the protein peaks of the wild-type organism
(Fig. 13). This protein does not seem to be either the nitrate reductase
or the formate dehydrogenase enzyme since its concentration does not
correlate with changes in the activities of these enzymes when they are
differentially induced in wild-type organisms. It was suggested that the
protein may play a structural role in the organization of the multienzyme
system. If so, it may prove to be identical to the postulated products of
the *chl* A or *chl* B genes (see above). It should be emphasized that proof

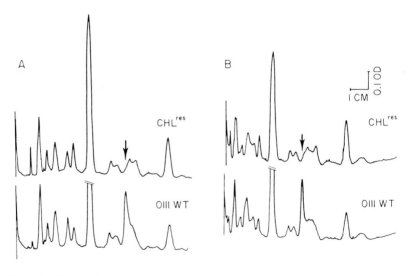

Fig. 13. Electrophoretic patterns of membrane proteins from wild-type (0111
WT) and a chlorate-resistant mutant (*CHL*^res) of *E. coli* subjected to gel electrophoresis in the presence of sodium dodecyl sulfate. The arrows indicate the position
of the protein missing in the mutant strain. (A) 7.5% acrylamide gels, and (B) 10%
acrylamide gels. (From Schnaitman, 1969.)

of these hypotheses will require the purification of the individual components of the system, followed by their use in *in vitro* reconstitution experiments. Thus far, this has not been achieved.

D. Microsomal Enzymes: Lysophosphatides and Membrane Enzymes

Lipids are required for full activity of a number of microsomal enzymes. These enzymes have not yet been obtained in soluble form, and the experiments have therefore been limited to studying enzyme activities after lipid extraction or degradation in intact microsomes.

The DPNH-cytochrome *c* reductase of hen liver microsomes has been studied by Jones and Wakil (1967), who showed an approximately 80% decrease in enzyme activity when most of the microsomal lipids were removed by treatment with acetone. Enzyme activity was completely restored by readdition of the extracted lipids. Studies with purified lipids showed that approximately 1:1 mixtures of phosphatidylcholine and lysophosphatidylcholine were most effective in restoring activity. The effectiveness of the lysophosphatides contrasts with the results of similar studies in most other enzyme systems in mitochondria and bacteria. In these other systems lysophosphatides are quite ineffective, as might be predicted from their detergentlike activity, which causes disruption of most organized lipid structures. Interpretation is made difficult by the presence of many components in the particulate microsomal preparation and by the known tendency of the substrate, cytochrome *c*, to interact directly with the phospholipids. Therefore, it is not now possible to ascribe the role of lipids to their interaction with the enzyme protein, with the substrate, or with other components of the microsomal preparation.

In parallel studies, Jones, Wakil and their co-workers (Jones *et al.*, 1969) discovered a similar role for lipids in the stearyl-CoA desaturase activity of hen liver microsomes. Extraction of intact microsomes with aqueous acetone resulted in loss of the desaturase activity, and most of the enzyme activity was restored when appropriate mixtures of lipids were added back to the system. Mixtures of phosphatidylcholine, triglycerides, and free fatty acids were much more effective than individual phospholipids. Similarities were noted between the lipid mixtures that were effective in the stearyl-CoA desaturase system and in the DPNH-cytochrome *c* reductase system (see above), although triglyceride was not required in the latter system. In both systems the unusual requirement for lysophosphatides or free fatty acids suggests that a detergentlike action may be involved.

The palmityl-CoA desaturase activity of microsomes is similar in many respects to the stearyl-CoA desaturase in its requirement for lipid (Jones et al., 1969).

Glucose-6-phosphatase is another microsomal enzyme which appears to require phospholipid for activity (Dutter et al., 1968). Treatment of intact microsomes with phospholipase C caused a marked loss of glucose-6-phosphatase activity under conditions leading to hydrolysis of approximately two-thirds of the total microsomal phospholipid. Activity was completely restored by the addition of a variety of phospholipids, but lysophosphatidylcholine was effective in restoring activity at much lower concentrations than were the other lipids tested. As in studies of several other microsomal enzymes (see above) this suggests that the added lipid may be acting as a detergent, perhaps disrupting the aggregated structure of the lipid-depleted microsomes, thereby making the enzyme more available to molecules of substrate.

Treatment with phospholipase C also inhibits both the ATPase activity and Ca^{++} transport of intact microsomes, indicating the importance of lipid–protein interactions in these processes (Martonosi et al., 1968). Both functions can be fully restored by the addition of micellar dispersions of phosphatidylcholine, lysophosphatidylcholine, and phosphatidic acid. The degree of unsaturation of the phospholipids does not play an important role in this interaction since synthetic dipalmitylphosphatidylcholine was at least as effective in restoring ATPase activity and calcium transport as was a phosphatidylcholine with a high degree of unsaturation.

A similar requirement for phospholipids has been reported for a Na^+–K^+-activated ATPase activity prepared from a "membrane fraction" of cerebral cortex (Tanaka and Strickland, 1965) and kidney cortex (Fenster and Copenhaver, 1965). The enzyme activities in the two membranes are activated by different phospholipids. The brain enzyme displays full activity with phosphatidylcholine, whereas lysophosphatidylcholine and phosphatidic acid are less active, and the acidic phospholipid cardiolipin is inactive. On the other hand, activation of the kidney enzyme is best accomplished by the negatively charged phosphatidylserine.

There are also indications that microsomal phospholipids may be involved in fatty acid saturation (Pande and Mead, 1963), in fatty acid synthesis (Foster and McWhorten, 1969), in an enzyme activity which hydrolyzes the alkenyl ether bond of plasmalogens (Ellingson and Lands, 1968), and in aromatic hydroxylase (Imai and Sato, 1960) and phosphatidic acid phosphatase activities (Coleman and Hubscher, 1963). These enzyme systems all act on lipid-soluble substrates; the effect of the phospholipids on the micellar state of the substrates has not yet been defined and may explain the observed results.

E. Mitochondrial Enzymes

1. Electron Transfer Chain

a. *Effect of Lipids on Electron Transfer Activity.* The idea that lipids play an active role in the activity of membrane-bound enzymes of the mitochondrion emerged from the studies of Fleischer and his co-workers, who first clearly demonstrated a requirement for phospholipids in the reactions of the mitochondrial electron transfer chain (Fleischer and Klouwen, 1961; Fleischer et al., 1962). It has now been shown that lipids affect the activity of all segments of the chain, although the degree of involvement varies depending on the enzyme activities being studied (Brierley et al., 1962). Since purified enzymes have not been available, most experiments have been performed on whole mitochondria or on submitochondrial fractions which probably contain more than one enzymatic component.

In the initial experiments mitochondria were studied before and after phospholipid had been removed by extraction with 90% acetone, and the lipid extraction procedure caused a decrease in all of the enzyme activities (Table I). The most marked decrease was seen in succinoxidase and in succinate–cytochrome c reductase activities, which fell to less than 5% of the value seen in untreated mitochondrial electron transfer particles, with restoration to the native level by the addition of phospholipids. Since both activities require most of the steps from succinate to O_2 (see Fig. 3 in Chapter 9), the lipid requirement may represent the summation of minor effects on several of the enzymes or may reflect a major requirement for phospholipid in only one step of the sequence.

Several major subsegments of the electron transport chain have been isolated (see Section VI,A of Chapter 9) and have also been tested for lipid requirements in a similar manner (Brierley et al., 1962; Brierley and Merola, 1962). These include succinate-CoQ reductase, CoQ-cytochrome c reductase, and cytochrome c oxidase. Different phospholipids were not equally effective in stimulating activity of the lipid-depleted enzyme preparations, but with the most active phospholipids the amount of lipid required for maximum stimulation was usually similar to the amount present in the original preparation before lipid extraction. Since the crude enzyme preparations contained many components, detailed studies of the mechanism of the lipid effects were not possible.

The phospholipid requirement for each of the complexes of the electron transfer chain was also demonstrated when the lipids were first removed by exposing the complexes to phospholipase A. In these experiments, the membrane-bound phospholipids were first degraded by

TABLE I

EFFECT OF LIPIDS ON MITOCHONDRIAL ENZYME ACTIVITIES

Enzyme	Source	Activity (% of original)		Reference
		After extraction of lipid	After readdition of lipid	
Succinate–cyt c reductase	Mitochondria	6	100	Fleischer et al. (1962)
Cyt c oxidase	Mitochondria	6	50	Fleischer et al. (1962)
Succinoxidase	Mitochondria	$(0.1)^a$	$(0.7)^a$	Brierley et al. (1962)
Cyt c oxidase	Mitochondria	2.5	25	Brierley et al. (1962)
Succinate–cyt c reductase	Mitochondria	$(0.2)^a$	$(1.3)^a$	Brierley et al. (1962)
CoQ-cyt c reductase	Mitochondria	15	30	Brierley et al. (1962)
Succinate-CoQ reductase	Mitochondria	$(0.8)^a$	$(1.4)^a$	Brierley et al. (1962)
Succinate-CoQ reductase	Submitochondrial complexes	17	43	Brierley et al. (1962)
CoQ-cyt c reductase	Submitochondrial complexes	8	50	Brierley et al. (1962)
Cyt c oxidase	Submitochondrial complexes	2	50	Brierley and Merola (1962)
Cyt c oxidase	Submitochondrial complexes	10	100	McConnell et al. (1966)
NADH-CoQ$_1$ reductase	Submitochondrial complexes	40^b	100	Machinist and Singer (1965)

[a] Units of enzyme activity.
[b] Treatment of electron transport particles with phospholipase A.

phospholipase A treatment and the breakdown products (i.e., lysophospholipids and fatty acids) were removed by a wash procedure. This treatment resulted in loss of the electron transfer activities described above, as well as NADH-oxidase and Mg^{++}-stimulated ATPase. In all cases the enzyme activities reappeared when phospholipids were added back to the lipase-treated system (Casu et al., 1966; Fleischer and Fleischer, 1967). This general method is one of the mildest procedures available for the removal of phospholipids from membranes and permits the removal of controlled amounts of phospholipid since the enzymatic degradation can be halted at any time. By the use of this technique of controlled phospholipid degradation, it was shown that a significant amount of phospholipid can be removed before any loss of enzyme activity is seen. Thus, only a portion of the phospholipid of the membrane is required for the activity of any one enzyme (Fleischer et al., 1970).

In somewhat similar studies, Pesch and Peterson (1965) suggested that phospholipid–protein interactions also determine the substrate specificity of mitochondrial NAD(P) transhydrogenase. In these studies treatment with phospholipase A or extraction with organic solvents resulted in a preparation capable of transferring hydrogen to the acceptor only from NADH and not from NADPH, presumably due to disruption of the phospholipid–protein association of the native membrane.

The degree of lipid specificity was very striking in all of these studies of mitochondrial enzymes. In all cases the most active phospholipid was cardiolipin, which was approximately as active as the total mitochondrial phospholipid extract. On the other hand, comparable concentrations of phosphatidylcholine showed negligible activity despite the fact that phosphatidylcholine is the major phospholipid component of native mitochondria. It seems reasonable to conclude that phospholipids are not randomly distributed throughout the membrane and that molecules of cardiolipin are probably located in close apposition to those components of the electron transfer chain that require cardiolipin for activity.

b. Cytochrome c–Phospholipid Interactions. Of all the portions of the mitochondrial electron transfer chain, the terminal segment has been best studied. The isolation of cytochrome oxidase (cytochrome $c:O_2$ oxidoreductase) soon led to the realization that active preparations always contained significant amounts of phospholipids (Griffiths and Wharton, 1961; Yonetani, 1961; Greenlees and Wainio, 1959; Horie and Morrison, 1963; Fleischer et al., 1961). Soon thereafter, evidence that the phospholipids were actually required for enzyme activity came from several laboratories, using the techniques of lipid depletion and readdition described above (see preceding section).

More recently, attention has focused on the details of the interactions of cytochrome c with phospholipids. Studies in Crane's laboratory (Widmer and Crane, 1958; Das et al., 1962; Das and Crane, 1964) demonstrated that cytochrome c–phospholipid complexes can be extracted from electron transfer particles by treatment with isooctane and that similar complexes are formed when purified cytochrome c and phospholipids are mixed in vitro. Several different phospholipids have been effective in forming these complexes. Cardiolipin and phosphatidylethanolamine were most effective, whereas formation of complexes between cytochrome c and phosphatidylcholine were difficult to demonstrate unless very large amounts of the phospholipid were used. In most studies (Das et al., 1965; Kimelberg and Lee, 1970) the cytochrome c–phospholipid interaction in the reconstituted complexes was thought to be primarily electrostatic, based largely on the observation that initial binding was inhibited by increased ionic strength and changes in pH.

It was therefore concluded that the polar groups of the phospholipids play a major role in the interaction with cytochrome c. On the other hand, the fatty acid residues of the phospholipid also appear to affect the interaction since a saturated phospholipid (dimyristoyl phosphatidyl-ethanolamine) was much less effective than were phosphatidylethanol-amines containing longer-chain unsaturated fatty acids.

Studies of the reconstituted cytochrome c–phospholipid complexes by low-angle x-ray diffraction (Gulik-Krzywicki et al., 1969) suggested that cytochrome c was located on the polar surfaces of phospholipid leaflets. Together with the suggestive evidence for electrostatic bonding (see above) this prompted Kimelberg and Lee (1970) to suggest that the cytochrome c–phospholipid interaction in vitro represents electrostatic bonding of cytochrome c to the polar head groups of bimolecular leaflets of phospholipid. The relevance of this to the location of cytochrome c in the architecture of the intact mitochondrial membrane is still conjectural.

2. Other Mitochondrial Enzymes

a. β-Hydroxybutyrate Dehydrogenase. One of the first demonstrations of a lipid requirement for activity of an isolated enzyme was the report that the β-hydroxybutyrate dehydrogenase of beef heart mitochondria required lecithin (phosphatidylcholine) for activity. The requirement for lipid was discovered when it was observed that procedures leading to solubilization of the enzyme from mitochondrial particles resulted in the parallel release of a heat-stable factor that was required for activity. The heat-stable factor proved to be phospholipid (Sekuzu et al., 1961; Jurtshuk et al., 1961). The complete enzyme is now viewed as a lipid–protein complex, and the lipid-free apoenzyme has been partially purified (Sekuzu et al., 1963). This system appears to illustrate the second pos-sible role of phospholipid in membrane-bound enzyme reactions, namely direct "enzyme activation," although the experimental evidence for this mechanism is still somewhat equivocal.

Phosphatidylcholine was the only purified lipid able to activate the enzyme, whereas a wide variety of other lipids were inactive, including phosphatidylethanolamine, phosphatidylinositol, cardiolipin, sphingo-myelin, lysophosphatidylcholine, and neutral lipids (Sekuzu et al., 1963; Fleischer et al., 1966). In this respect the enzyme differs from the enzyme complexes of the mitochondrial electron transfer chain (see above) which require cardiolipin or phosphatidylethanolamine for full activity and which show much less activity with phosphatidylcholine.

In addition to the requirement for choline as the polar residue, specificity also resides in the fatty acid portion of the phospholipid molecule. Phosphatidylcholines containing unsaturated fatty acids were active in the system, while little activity was seen when the fatty acid residues were saturated. In this requirement for unsaturated fatty acyl groups, the enzyme is similar to almost all other lipid-activated enzyme systems which have been studied.

Preincubation studies provided presumptive evidence that a major role of phospholipid may be to directly activate the enzyme. Prompt initiation of the reaction occurred if enzyme, phosphatidylcholine, and DPN were mixed and preincubated for 15 minutes before adding β-hydroxybutyrate. In contrast, there was a significant lag period when the enzyme was omitted from the preincubation mixture. Although the situation is complex, the evidence is consistent with the idea that direct interactions of phospholipid with both enzyme and DPN are required for activity.

b. Acyl-CoA Synthetase. A GTP-dependent acyl-CoA synthetase has been purified from liver mitochondria and appears also to fall in the class of lipid-activated enzymes (Galzigna *et al.*, 1967). The purified enzyme contains both phosphatidylcholine and phosphopantothenine and is active on both short- and long-chain fatty acid substrates. Its activity is largely lost after extraction with acetone and can be restored by preincubation with phosphatidylcholine (Sartorelli *et al.*, 1967). It was suggested that the phospholipid is linked to the protein by electrostatic bonds (Galzigna *et al.*, 1969). This hypothesis is based partly on the finding that phosphatidylethanolamine cannot replace phosphatidylcholine in the system, suggesting that the polar group of the phospholipid is of importance in the interaction with the enzyme protein (Rossi, 1970) although alternative explanations are obviously possible.

IV. REGULATION OF THE ACTIVITY OF MEMBRANE ENZYMES: SOME SPECULATIONS

Membrane-bound enzymes can be viewed as one structural element in an organized array of other molecules. Since the behavior of many membrane enzymes is markedly affected by their insertion into the membrane structure, neighboring molecules (e.g., phospholipids, proteins) may play a major role in regulating the catalytic activity of these enzymes.

Membrane lipids are known to be in a dynamic state, with significant rates of turnover of whole molecules and parts of individual phospholipid molecules (White and Tucker, 1969). Changes in membrane phospholipids may therefore serve a major regulatory function. For example, many membrane enzymes (see above) require phospholipids for activity and activity is lost if the fatty acid residues are changed from unsaturated to saturated or if a single acyl residue is removed, resulting in formation of a lysophosphatide. Similarly, the activity of many membrane enzymes is markedly affected by the nature of the polar portion of the phospholipid. Thus changes in phospholipid structure have a major effect on enzyme function. Since the phospholipids of interest to an enzyme are those in its immediate vicinity, such regulatory changes in structure need only affect phospholipid molecules in relatively localized regions of the membrane, and changes in overall membrane lipid composition might be minor.

This implies a mosaic structure in which lipid molecules are not randomly distributed throughout the membrane. It is profitable to think in terms of functional units, each unit consisting of an individual enzyme and the phospholipid molecules in its immediate vicinity. In the case of enzymes which require the less major phospholipids, such as phosphatidylglycerol in the phosphotransferase system of Roseman and his collaborators (Roseman, 1969), it seems reasonable to predict a considerable enrichment for the required lipid in the immediate neighborhood of the protein components of the enzyme system. It should be emphasized that this does *not* imply structurally discontinuous lipoprotein subunits, since the concept of a functional unit consisting of an enzyme and its phospholipid neighbors is consistent with a variety of models of membrane structures.

In most known biological membranes most of the acyl residues of the membrane phospholipids are *cis*-unsaturated, and this is reflected in the fact that most membrane enzymes require unsaturated phospholipids for activity.* Mutants of *E. coli* unable to synthesize unsaturated fatty acids are unable to grow in the absence of unsaturated fatty acid supplements (Silbert and Vagelos, 1967). Under these conditions the membrane phospholipids contain a much lower than normal percentage of unsaturated residues, resulting in disturbances of many cellular functions including membrane transport (Fox, 1969). Proudlock *et al.* (1969) have also

---

* The membrane phospholipids of some bacteria contain cyclopropane fatty acids (e.g., *E. coli* in the stationary phase of growth) or branched fatty acids (e.g., many gram-positive bacteria) in place of unsaturated acyl groups. These unusual residues are very similar to *cis*-unsaturated fatty acids in their physical properties (e.g., the high degree of mobility of the hydrocarbon chains), in their location in the phospholipid molecule (i.e., primarily in the β position), and in their ability to activate membrane enzymes (e.g., UDP-galactose:lipopolysaccharide α,3-galactosyltransferase). They differ markedly from straight-chain saturated acyl groups in all of these properties. Therefore, the conclusions drawn for unsaturated fatty acids in phospholipids probably also apply to cyclopropane and branched-chain acyl residues.

described a mutant of *Saccharomyces cerevisiae* which is unable to grow aerobically in the absence of unsaturated fatty acids. In the case of the *S. cerevisiae* mutant, depletion of unsaturated fatty acids does not affect the oxidative ability of the mutant mitochondria, but the energy of respiration is not available for growth apparently due to lack of respiratory control and to a very low level of phosphorylation. The fall in these two mitochondrial functions closely parallels the decrease in percentage of unsaturated fatty acids in the mitochondria. The important role of unsaturated fatty acids in many membrane functions both *in vitro* and *in vivo* is therefore well established.

On the other hand, it is striking that certain organisms can tolerate quite major changes in composition of their membrane phospholipids including situations in which unsaturated residues have been largely replaced by saturated groups. In the case of *Mycoplasma laidlawii* the organisms will tolerate conditions (i.e., growth in the presence of stearate) leading to replacement of nearly 90% of the membrane acyl residues by saturated groups (Steim *et al.*, 1969). Since essentially all lipid-activated membrane enzymes require phospholipids with a much higher percentage of unsaturated or branched fatty acids, it follows that if phospholipids in stearate-grown *M. laidlawii* were distributed randomly throughout the membrane, this would result in loss of activity of most membrane enzymes including those whose functions are essential for survival of the organism. On the other hand, survival would be possible if phospholipids in the immediate vicinity of critical membrane enzymes were protected from the overall perturbation of fatty acid composition. This implies the concentration of the few remaining unsaturated phospholipids in these critical regions.

All of these facts suggest that consideration of the molecular organization of membranes should include the strong likelihood that lipid molecules are distributed in a nonrandom manner within the membrane structure and that this microheterogeneity may play an important role in regulating the function of specific membrane enzymes. Essentially nothing is now known about the regulation of the synthesis of specific phospholipids *in vivo* and this would seem a fruitful field for investigation. Similarly, the mechanism whereby individual phospholipids become localized in close proximity to specific enzyme proteins is also unknown.

A second mechanism of control of membrane enzyme function also suggests itself. Since a membrane consists of an ordered and continuous array of interacting molecules, a change in a single molecule could be transmitted through the membrane by the sequential perturbation of adjacent molecules. The result would be a cascade effect, affecting the activity of enzymes at many sites within the membrane. In operational terms, this is an application of the general theory proposed by Changeux and his collaborators (1967), who suggested that many membrane functions can be explained on the basis of cooperative molecular interactions throughout the lattice structure of the membrane. One might expect that this mechanism would produce generalized effects on a large number of catalytic membrane proteins.

At the time this chapter was written there were essentially no reports of experiments bearing directly on the regulation of the activity of membrane enzymes *in situ*. However the genetic tools are now at hand to begin a more direct attack on this problem in bacterial systems. Several groups have now isolated mutants defective in various stages of lipid biosynthesis. This permits the controlled alteration of the lipid composition of membranes *in vivo* and makes it possible to investigate the effects of these changes on the function of specific enzymes in the intact membrane. Major advances in our knowledge of this important area can be expected within the next few years.

It should be kept in mind that all of the above considerations are highly speculative at this time, and they fail to consider physical effects of changes in lipid composition (i.e., changes in fluidity of the hydrocarbon residues, phase changes, etc.) which might affect nonenzymatic membrane functions such as permeability. The mechanisms whereby the physical characteristics of membrane lipids affect membrane function remains a largely unexplored field.

References

Abrams, A. (1965). *J. Biol. Chem.* **240,** 3675.

Abrams, A., and Baron, C. (1968). *Biochemistry* **7,** 501.

Azoulay, E., Puig, J., and Pichimoty, F. (1967). *Biochem. Biophys. Res. Commun.* **27,** 270.

Azoulay, E., Puig, J., and Couchoud-Beaumont, P. (1969). *Biochim. Biophys. Acta* **171,** 238.

Brierley, G., and Merola, A. J. (1962). *Biochim. Biophys. Acta* **64,** 205.

Brierley, G., Merola, A. J., and Fleischer, S. (1962). *Biochim. Biophys. Acta* **64,** 218.

Bulos, B., and Racker, E. (1968). *J. Biol. Chem.* **243,** 3891.

Burge, R. E., and Draper, J. C. (1967). *J. Mol. Biol.* **28,** 205.

Casu, A., Fleischer, B., and Fleischer, S. (1966). *Fed. Proc. Fed. Amer. Soc. Exp. Biol.* **25,** 413.

Changeux, J.-P., Thiéry, J., Tung, Y., and Kittel, C. (1967). *Proc. Nat. Acad. Sci. U. S.* **57,** 335.

Coleman, R., and Hubscher, G. (1963). *Biochim. Biophys. Acta* **73,** 257.

Das, M. L., and Crane, F. L. (1964). *Biochemistry* **3,** 696.

Das, M. L., Hiratsuka, H., Machinist, J. M., and Crane, F. L. (1962). *Biochim. Biophys. Acta* **60,** 433.

Das, M. L., Haak, E. D., and Crane, F. L. (1965). *Biochemistry* **4,** 859.

Davis, K. A., and Hatefi, Y. (1969). *Biochemistry* **8,** 3355.

DePetris, A. (1967). *J. Ultrastruc. Res.* **19**, 45.
Dutter, S. M., Byrne, W. L., and Ganoza, M. C. (1968). *J. Biol. Chem.* **243**, 2216.
Edstrom, R., and Heath, E. (1964). *Biochem. Biophys. Res. Commun.* **16**, 576.
Ellingson, J. S., and Lands, W. E. M. (1968). *Lipids* **3**, 111.
Endo, A., and Rothfield, L. (1969). *Biochemistry* **8**, 3500, 3508.
Fenster, L. J., and Copenhaver, J. H., Jr. (1967). *Biochim. Biophys. Acta* **137**, 406.
Finean, J. B. (1953). *Biochim. Biophys. Acta* **10**, 371.
Fleischer, B., Casu, A., and Fleischer, S. (1966). *Biochem. Biophys. Res. Commun.* **24**, 189.
Fleischer, S., and Fleischer, B. (1967). *Methods Enzymol.* **10**, 355.
Fleischer, S., and Klouwen, H. (1961). *Biochem. Biophys. Res. Commun.* **5**, 378.
Fleischer, S., Klouwen, H., and Brierley, G. (1961). *J. Biol. Chem.* **236**, 2936.
Fleischer, S., Brierly, G., Klouwen, H., and Slautterback, D. B. (1962). *J. Biol. Chem.* **237**, 3264.
Fleischer, S., Casu, A., and Fleischer, B. (1970). Personal communication.
Foster, D. W., and McWhorten, W. P. (1969). *J. Biol. Chem.* **244**, 260.
Fox, C. F. (1969). *Proc. Nat. Acad. Sci. U. S.* **63**, 850.
Galzigna, L., Rossi, C. R., Sartorelli, L., and Gibson, D. M. (1967). *J. Biol. Chem.* **242**, 2111.
Galzigna, L., Sartorelli, L., Rossi, C. R., and Gibson, D. M. (1969). *Lipids* **4**, 1.
Glasel, J. A. (1970). *J. Amer. Chem. Soc.* **92**, 372.
Greenlees, J., and Wainio, W. W. (1959). *J. Biol. Chem.* **234**, 658.
Griffiths, D. E., and Wharton, D. C. (1961). *J. Biol. Chem.* **236**, 1850.
Gulik-Krzywicki, T., Schechter, E., Luzzatti, V., and Faure, M. (1969). *Nature* (*London*) **223**, 1116.
Hanold, F. M., Baarda, J. R., Baron, C., and Abrams, A. (1969). *J. Biol. Chem.* **244**, 2261.
Hart, C. J., Leslie, R. B., Davis, M. A. F., and Lawrence, G. A. (1969). *Biochim. Biophys. Acta* **193**, 308.
Hatefi, Y., and Hanstein, W. G. (1969). *Proc. Nat. Acad. Sci. U. S.* **62**, 1129.
Hatefi, Y., and Stempel, K. E. (1969). *J. Biol. Chem.* **244**, 2350.
Higashi, Y., and Strominger, J. L. (1970). *J. Biol. Chem.* **245**, 3697.
Higashi, Y., Siewert, G., and Strominger, J. L. (1970). *J. Biol. Chem.* **249**, 3691.
Horie, S., and Morrison, M. (1963). *J. Biol. Chem.* **238**, 1855.
Imai, Y., and Sato, R. (1960). *Biochim. Biophys. Acta* **42**, 164.
Jones, P. D., and Wakil, S. J. (1967). *J. Biol. Chem.* **242**, 5267.
Jones, P. D., Holloway, P. W., Petuffo, R. O., and Wakil, S. J. (1969). *J. Biol. Chem.* **244**, 755.
Jurtshuk, P. J., Sekuzu, I., and Green, D. E. (1961). *Biochem. Biophys. Res. Commun.* **6**, 76.
Kimelberg, H. C., and Lee, C. P. (1970). *J. Membrane Biol.* **2**, 252.
King, T. E., and Howard, R. L. (1967). *Methods Enzymol.* **10**, 275.
Luzzati, V., Mustacchi, H., Skoulios, A., and Husson, F. (1962). *Acta Crystallogr.* **13**, 660.
McConnell, D. G., Tzagoloff, A., MacLennan, D., and Green, D. E. (1965). *J. Biol. Chem.* **241**, 2373.
Machinist, J. M., and Singer, T. R. (1965). *J. Biol. Chem.* **240**, 3182.
Martonosi, A., Donley, J., and Halpin, R. A. (1968). *J. Biol. Chem.* **243**, 61.
Osborn, M. J., and D'Ari, L. (1964). *Biochem. Biophys. Res. Commun.* **16**, 568.

Osborn, M. J., Rosen, S. M., Rothfield, L., Zcleznick, L., and Horecker, B. L. (1964). *Science* **145**, 783.

Pande, S. V., and Mead, J. F. (1968). *J. Biol. Chem.* **243**, 352.

Pesch, L. A., and Peterson, J. (1965). *Biochim. Biophys. Acta* **96**, 390.

Pischaud, M., Puig, J., Pichimoty, F., Azoulay, E., and LeMinor, L. (1967). *Ann. Inst. Pasteur, Paris* **112**, 24.

Proudlock, J. W., Hoslam, J. M., and Linnane, A. W. (1969). *Biochem. Biophys. Res. Commun.* **37**, 847.

Puig, J., and Azoulay, E. (1967). *Compt. Rend.* **264**, 1507.

Robertson, J., Moldow, C., and Rothfield, L. (1971). *Fed. Proc. Fed. Amer. Soc. Exp. Biol.* **30**, 1119.

Romeo, D., Girard, A., and Rothfield, L. (1970a). *J. Mol. Biol.* **53**, 475.

Romeo, D., Hinckley, A., and Rothfield, L. (1970b). *J. Mol. Biol.* **53**, 491.

Roseman, S. (1969). *J. Gen. Physiol.* **54**, 138s.

Rossi, R. C. (1970). Personal communication.

Rothfield, L., and Horecker, B. L. (1964). *Proc. Nat. Acad. Sci. U. S.* **52**, 939.

Rothfield, L., and Horne, R. W. (1967). *J. Bacteriol.* **93**, 1705.

Rothfield, L., and Pearlman, M. (1966). *J. Biol. Chem.* **241**, 1386.

Rothfield, L., and Pearlman-Kothencz, M. (1969). *J. Mol. Biol.* **44**, 477.

Rothfield, L., and Takeshita, M. (1965). *Biochem. Biophys. Res. Commun.* **20**, 521.

Rothfield, L., Osborn, M. J., and Horecker, B. L. (1964). *J. Biol. Chem.* **239**, 2788.

Rothfield, L., Takeshita, M., Pearlman, M., and Horne, R. W. (1966). *Fed. Proc. Fed. Amer. Soc. Exp. Biol.* **25**, 1495.

Ruig-Herrera, J., and DeMoss, J. A. (1969). *J. Bacteriol.* **99**, 720.

Salton, M. R. J. (1967). *Annu. Rev. Microbiol.* **21**, 424.

Sartorelli, L., Galzigna, L., Rossi, C. R., and Gibson, D. M. (1967). *Biochem. Biophys. Res. Commun.* **26**, 90.

Scanu, A., and Hirz, R. (1968). *Proc. Nat. Acad. Sci.* **59**, 890.

Schnaitman, C. (1969). *Biochem. Biophys. Res. Commun.* **37**, 1.

Sekuzu, I., Jurtshuk, P. J., and Green, D. E. (1961). *Biochem. Biophys. Res. Commun.* **6**, 71.

Sekuzu, I., Jurtshuk, P. J., and Green, D. E. (1963). *J. Biol. Chem.* **238**, 975.

Silbert, D., and Vagelos, P. R. (1967). *Proc. Nat. Acad. Sci. U. S.* **58**, 1579.

Steim, J., Reinert, J. C., Tourtellotte, M. E., McElhaney, R. N., and Roder, R. L. (1969). *Proc. Nat. Acad. Sci. U. S.* **63**, 104.

Strittmatter, P., and Velick, S. F. (1956). *J. Biol. Chem.* **221**, 253.

Takayama, K., MacLennan, D., Tzagoloff, A., and Stoner, C. D. (1964). *Arch. Biochem. Biophys.* **114**, 223.

Tanaka, R., and Strickland, K. P. (1965). *Arch. Biochem. Biophys.* **111**, 583.

Weiser, M., and Rothfield, L. (1968). *J. Biol. Chem.* **243**, 1320.

Wettlaufer, D. B., and Lovrien, R. (1964). *J. Biol. Chem.* **239**, 596.

White, D. C., and Tucker, A. N. (1969). *J. Lipid Res.* **10**, 220.

Widmer, C., and Crane, F. L. (1958). *Biochim. Biophys. Acta* **27**, 203.

Yonetani, T. (1961). *J. Biol. Chem.* **236**, 1680.

Zahler, P., and Wallach, D. F. H. (1967). *Biochim. Biophys. Acta* **135**, 371.

7

THE MOLECULAR BASIS
OF MEMBRANE TRANSPORT SYSTEMS

E. C. C. LIN

ABBREVIATIONS

ADP	Adenosine 5′-diphosphate
AMP	Adenosine 5′-monophosphate
ATP	Adenosine 5′-triphosphate
CoA	Coenzyme A
DNP	2,4-Dinitrophenol
EDTA	Ethylenediaminetetraacetate
GTP	Guanosine 5′-triphosphate
G-1-P	Glucose 1-phosphate
G-6-P	Glucose 6-phosphate
L-α-GP	L-α-Glycerophosphate
HPr	Histidine-containing protein of the phosphotransferase system
IPTG	Isopropyl-1-thio-β-D-galactopyranoside
α-MG	Methyl-α-glucopyranoside
M protein	Membrane-associated protein coded by the y gene of the *lac* operon
NAD	Nicotinamide adenine dinucleotide
NEM	N-Ethylmaleimide
ONPG	o-Nitrophenyl-β-D-galactopyranoside
PCMB	p-Chloromercuribenzoate
PEP	Phosphoenolpyruvate
RNA	Ribonucleic acid
TDG	β-D-Galactosyl-1-thio-β-D-galactopyranoside or thiodigalactoside
TMG	Methyl-1-thio-β-D-galactopyranoside
TPG	Phenyl-1-thio-β-D-galactopyranoside

I. INTRODUCTION

It is now well known that ions and lipid-insoluble molecules can cross cell membranes at significant rates only through the intervention of specific membrane-associated proteins. The basic mechanisms by which these proteins mediate transport can be divided into several classes.

The simplest one is known as *facilitated diffusion,* whereby substrate molecules are merely equilibrated across the cell membrane by protein carriers (Danielli, 1954); no metabolic energy is consumed. Glucose and glycerol, for example, traverse the membrane by such a mechanism in red blood cells (Jacobs, 1950; LeFevre and LeFevre, 1952; Jacobs, 1954; Widdas, 1954; Rosenberg and Wilbrandt, 1957; LeFevre and McGinniss, 1960).

A more complicated mechanism is known as *active transport,* whereby substrate molecules are delivered across the cell membrane against electrochemical potentials. This kind of process was recognized in early studies of intestinal (Bárány and Sperber, 1939) and renal (Smith, 1951) functions. Three modes of translocation can be distinguished among the currently known examples of active transport: (a) *simple active transport*, as represented by the lactose system of *Escherichia coli*, in which a substrate is moved without necessitating the migration of another substance; (b) *linked transport*, as exemplified by the uptake processes for certain amino acids (Christensen, 1970) and sugars (Crane, 1968b) by intestinal epithelial cells, in which the movement of the organic molecules is apparently accompanied by the migration of sodium ions in the same direction; and (c) *exchange transport*, as illustrated by the ubiquitous sodium pump, in which the migration of Na^+ is geared to the countermigration of K^+ ions (Glynn, 1957). (The translocation of both substrates against concentration gradients distinguishes this energy-dependent process from the counterflow phenomenon; see Section II,B,3.)

More recently, a kind of transport mechanism has been described in bacteria, whereby carbohydrates are captured and delivered into the cytoplasm not as free sugars, but as their phosphorylated derivatives (Roseman, 1969). Since this process combines the uptake process with the first step of dissimilation, it might be termed *phosphorylating transport*. The expenditure of a high-energy phosphate for substrate activation distinguishes this from the familiar cases of active transport in which a high-energy compound is utilized for the activation of a membrane protein.

Approaches to the understanding of transport phenomena have been diverse; the emphasis has ranged from genetic and physiological controls to membrane electrophysiology. In this chapter an attempt is made to review on a molecular basis the current state of knowledge of several specific transport systems. In the main, the subjects treated are only those in which at least one protein component has been identified and isolated (for discussions of other aspects of biological transport systems see Albers, 1967; Christensen, 1969; Kaback, 1970a,b; Kennedy, 1970; Lin, 1970; Whittam and Wheeler, 1970).

II. THE LACTOSE SYSTEM

A. Introduction

1. An Inducible Accumulation System in E. coli

It has been long known that β-galactosides are hydrolyzed by intact cells of *E. coli* considerably more slowly than by disrupted cells (Deere *et al.*, 1939; Lederberg, 1950), thus suggesting the existence of a permeation barrier in the cell membrane. In 1955, Cohen and Rickenberg reported that the uptake of radioactively labeled TMG (a nonmetabolizable analog) by cells of E. coli was controlled by an inducible process (Cohen and Rickenberg, 1955). A subsequent report by Rickenberg and collaborators (1956) greatly clarified the nature of this accumulation. These investigators showed that cells that were induced to metabolize lactose could accumulate the ^{35}S-labeled β-galactoside, TMG, up to several percent of their dry weight against a high concentration gradient. The labeled material in the cells could be readily extracted by boiling water and was chromatographically indistinguishable from the original [^{35}S]TMG. The induction of this transport mechanism was blocked by any treatment that prevented normal protein synthesis, e.g., the addition of the antibiotic chloramphenicol or amino acid analogs, or the deprivation of the required amino acid in the case of an auxotrophic mutant. The investigators therefore concluded that the induction of this transport system entailed the *de novo* formation of a protein which can specifically recognize the substrate and assure its catalytic transfer across the cellular osmotic barrier. To this protein they gave the name "permease," adding that this does not require that the protein catalyze "active transport." They were also able to show that among mutants defective in the fermentation of lactose, some lacked β-galactoside permease (cryptic), while others lacked β-galactosidase.

Although the apparent energy dependence of the uptake process and the extraordinary amount of the substrate which could be retained by the cell argued against adsorption to specific binding sites as the mechanism of β-galactoside accumulation by induced cells, it remained to be demonstrated more directly that the intracellular substrate was mostly in solution. This was achieved by Sistrom (1958), who converted intact E. coli cells to osmotically sensitive spheroplasts by Tris–lysozyme–EDTA treatment and showed that only when these were prepared from cells induced in the lactose system would exposure to TMG or lactose cause swelling. The joint genetic and biochemical attack in the study of this transport system was to open up a fruitful approach which paved the way for the elucidation of numerous transport systems in cell membranes at the molecular level.

2. Acetylation of Accumulated Substrates

Thiogalactosides, such as TMG, although immune to hydrolysis by β-galactosidase, can be slowly acetylated by the action of an acetyl-CoA-dependent transacetylase (Rickenberg et al., 1956; Zabin et al., 1959, 1962; Herzenberg, 1961; Zabin, 1963; Alpers et al., 1965). Acetylated

TMG is gradually extruded into the medium and is not recaptured by any specific transport systems (Herzenberg, 1961; Wilson and Kashket, 1969). Wilson and Kashket (1969) have cautioned that the occurrence of this process may jeopardize accurate kinetic measurements and that for this reason it is safer to employ transacetylase-negative mutants which can be readily isolated by a simple autoradiographic technique (Zwaig and Lin, 1966a).

B. Kinetics

1. Substrate Specificity and Kinetics

The lactose permease has affinity for a number of α- and β-galactosides including TMG, TDG, TPG, ONPG, and melibiose (6-O-α-D-galacto-pyranosyl-D-glucose) (Kepes and Cohen, 1962; Prestidge and Pardee, 1965). The kinetics of accumulation of various substrates can be described by the following Michaelis-Menten relationships (Rickenberg et al., 1956; Kepes, 1960; Winkler and Wilson, 1966):

$$V_{influx} = V_{max}^{influx} \frac{[S]_{ex}}{K_m^{influx} + [S]_{ex}} \tag{1}$$

$$V_{efflux} = V_{max}^{efflux} \frac{[S]_{in}}{K_m^{efflux} + [S]_{in}} \tag{2}$$

where $[S]_{ex}$ and $[S]_{in}$ represent, respectively, the extracellular and the intracellular concentration of a galactoside, V represents the velocity of transfer across the membrane, and K_m is the Michaelis constant for transport. Since at equilibrium

$$V_{influx} = V_{efflux} \tag{3}$$

it follows that

$$V_{max}^{influx} \frac{[S]_{ex}}{K_m^{influx} + [S]_{ex}} = V_{max}^{efflux} \frac{[S]_{in}}{K_m^{efflux} + [S]_{in}} \tag{4}$$

The entry rate can be measured simply by the initial uptake rate of a nonmetabolizable substrate. The exit rate, first shown by Koch (1964) to be dependent upon induction as well, can be studied by preloading the cells with a substrate and measuring its rate of loss from the cells in a substrate-free medium. For accurate determination of the exit rate, it is often important to block recapture of the substrate by adding to the suspension medium either competitive inhibitors or metabolic inhibitors such as azide and DNP.

With the chromogenic substrate, ONPG, the entry rate can be conveniently followed by the rate of o-nitrophenol liberation in β-galactosidase-positive cells, since the relative activity of the hydrolase has been shown to be much greater than that of the permease (Rickenberg *et al.*, 1956; Koch, 1964). The exit rate of ONPG, however, can be measured only with β-galactosidase-negative cells preloaded with isotopically labeled substrate.

Both the catalyzed entry and the exit are greatly curtailed by the sulf-hydryl inhibitor, PCMB (Kepes, 1969; Wong and Wilson, 1970). The rate of noncarrier-mediated exit of lactose from such poisoned cells has been studied by Winkler and Wilson (1966) and was found to be of the order of 2 nmole/g wet weight of cells per minute per 1 mM concentration gradient at 10°C, or 4 orders of magnitude lower than the V_{max} of catalyzed influx.

More elaborate kinetic treatments of the transport system have been developed by Koch (1967) and Kepes (1969). However, in any kinetic study, it is well to bear in mind the following. (a) The chemical environment inside the cell may vary critically during the course of the experiment; the change in concentration of the transport substrate itself may be an important factor in regulating the behavior of the carrier protein. (b) The growth condition may affect the chemical composition of the membrane as well as the patterns of metabolites in the cytoplasm. (c) There may be more than one route of exit, especially when the intracellular concentration of the accumulated compound becomes high. (d) Peculiarities of specific bacterial strains may be involved; even stocks from the same original strain have been found to behave differently as a result of genetic drift.

2. The Steady State

It has been found that when the intracellular concentration of an accumulated β-galactoside ceases to rise, an equilibrium has been attained during which the rate of uptake is undiminished but is exactly balanced by the rate of efflux. The existence of such a dynamic state was demonstrated in the following way (see Fig. 1). Cells constitutive in the lactose system suspended in mineral medium with an energy source were divided into three equal portions. At zero time, nonradioactive TMG was added to each; labeled TMG was added only to the first sample and the accumulation of radioactivity by the cells was monitored. To the two other samples, radioactive TMG was added at different times well after steady-state accumulation of labeled TMG was achieved in the first sample. The rate of uptake of labeled TMG in the latter two cases was similar to that observed in the first case. In all three cases, the same height of intracellular concentration of labeled TMG was achieved. Thus, even when

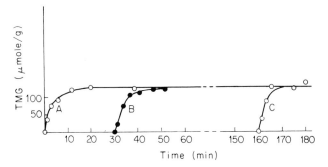

Fig. 1. Kinetics and steady-state of TMG uptake. Cells of *E. coli* ML308 were suspended in a mineral medium containing maltose and incubated at 15°C. A, [^{14}C] TMG and cold TMG (1 mM) were introduced together at zero time. B, Cold TMG added at zero time and labeled TMG introduced at 30 minutes. C, Cold TMG added at zero time and labeled TMG introduced at 160 minutes. (From Kepes, 1960.)

the net concentration of TMG inside the cells ceased to change, the influx rate was undiminished (Kepes and Monod, 1957; Kepes, 1960).

The total amount of a substrate which could be accumulated varied according to the compound and was not correlated with the influx K_m of the compound. Thus, the capacity for retention of TPG was only about one-fifth that of TMG, even though the K_m of uptake of the former was about one-half that of the latter (Kepes, 1960). Presumably the maximal concentration gradient achievable by a compound reflected largely the ratio of the entry K_m to the exit K_m [see Eq. (5) below].

3. Counterflow

Koch (1964) showed that when cells poisoned by azide and preloaded with cold TMG were resuspended in a medium containing radioactive TMG, rapid transient uptake of the labeled TMG occurred against a concentration gradient which was followed by a gradual release of the intracellular radioactivity to a baseline expected of simple equilibrium across the cell membrane. Cells not preloaded showed only a small transient rise of radioactivity followed by a return to the baseline. The effect of "preloading" was not found in cells lacking the permease (Winkler and Wilson, 1966; Wong and Wilson, 1970). The phenomenon of counterflow, predicted by Widdas (1952) and verified by Park and associates (1956) and Rosenberg and Wilbrandt (1957) in their studies of sugar movement across erythrocyte membranes, may be explained as follows. Suppose that at the beginning of the measurement, labeled molecules are present only in the medium and unlabeled molecules are present only inside the cells. The entry of labeled substrates would encounter no competition. On the other hand, the initial presence of a high internal concentration of the unlabeled molecules would competitively block efflux of the labeled molecules once they entered the cell. Thus during the initial period, the rate of entry of the labeled substrate should exceed that of its exit. With time, the unlabeled substrate would leave the cell and eventually the external and internal chemical concentrations of the nonradioactive substrate would approach the same value. With the disappearance of this gradient, the internal to external concentration of the labeled substrate would also level off. The net transient movement of the labeled substrate into an energy-deprived cell is thus driven at the expense of the high internal concentration of the unlabeled substrate. Conversely, if energy-

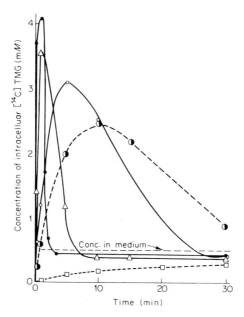

Fig. 2. Counterflow of TMG in mutants possessing different quantities of membrane carrier. Mutants, grown into exponential phase on casein hydrolyzate, were centrifuged, washed in mineral medium, resuspended in 4 ml of mineral medium containing sodium azide (30 mM) and TMG (20 mM) and incubated for 30 minutes at 25°C. Cells were then centrifuged, and the supernatant fluid poured off. The small pellet at the bottom of the tube was broken up by rapid squirting with 4 ml of mineral medium containing sodium azide and [^{14}C]TMG (0.5 mM, 0.7 μCi/ml) followed by brief agitation on a Vortex mixer. Samples of 0.5 ml were removed at 15 seconds, 1, 5, 10, 15, and 30 minutes. Legend: ●, 100% carriers (*E. coli* ML308); △, 25% carriers (*E. coli* ML308-1320); ○, 8% carriers (*E. coli* ML308-1710); ◑, 4% carriers (*E. coli* ML308-1337); □, 0% carriers (*E. coli* ML35). (From Wong and Wilson, 1970.)

uncoupled cells are first equilibrated with labeled substrate, the addition of cold substrate to the medium would cause transient depletion of the intracellular labeled substrate, since the external cold substrate initially blocks only the influx of the labeled substrate.

More detailed kinetic analyses of counterflow were carried out by Kepes (1969) and Wong and Wilson (1970). In both studies the initial rate of influx seemed to vary in direct proportion to the number of permease molecules per cell, and with increasing numbers of these protein molecules, the height of the peak of overshoot increased and the maximum was achieved sooner (Fig. 2). It is noteworthy that Kepes controlled the number of carriers (or permeases) per cell by manipulating the degree of induction, whereas Wong and Wilson accomplished this by controlled partial inactivation with NEM or by the use of mutants with different degrees of leakiness. The results with these various approaches were all in agreement.

4. Exchange Diffusion

If the diffusion coefficient of a carrier–substrate complex in the membrane is greater than that of the free carrier itself, then preloading a cell with substrate A is expected to accelerate the entry of substrate B because the complexing of the carrier by substrate A from within would allow it to return faster to the external surface where, upon the dissociation of A, another molecule of B can be picked up. This phenomenon, called "exchange diffusion," and not to be confused with counterflow, had been previously studied by Mawe and Hempling (1965) and by Levine et al. (1965) in the case of glucose transport in red cells and by Heinz and Walsh (1958) in the case of glycine transport in Ehrlich carcinoma cells. More recently Robbie and Wilson were able to demonstrate exchange diffusion in the transport of lactose in E. coli. Their success depended on the recognition that the presence of a cell envelope impeded the study of unidirectional flux because substrate molecules about to leave the cell might be recaptured by the membrane transport system before they could cross the cell wall layer to escape into the medium. Once the recapture phenomenon was controlled or taken into account, the true unidirectional flow of a substrate, measured by initial rates, could indeed be shown to be influenced by the presence of another compound on the opposite side of the membrane. In the case of most of the substrates, positive exchange diffusion was observed (i.e., the substrate–carrier complex appeared to have a higher diffusion constant in the membrane than the free carrier), but in the case of TDG, the effect was negative (i.e., the substrate–carrier complex appeared to have a lower diffusion constant in the membrane than the free carrier). On the other hand, ONPG appeared to have little or no effect on the diffusion rate of the carrier (Robbie and Wilson, 1969).

C. Energy Requirement

Inhibitors that affect energy metabolism abolish the accumulation of galactosides against concentration gradients without preventing facilitated diffusion (Rickenberg et al., 1956; Koch, 1963, 1964; Winkler and Wilson, 1966; Carter et al., 1968). Further evidence for the coupling of metabolic energy to substrate accumulation came from an experiment in which it was shown that oxygen consumption of resting cells constitutive in the lactose transport system increased in the presence of TMG or TPG (Kepes and Monod, 1957; Kepes, 1960, 1964). Although steady-state TMG accumulation was reached within 2–3 minutes, the higher rate of oxygen consumption persisted for 30–40 minutes, indicating that the maintenance of the high concentration gradient required sustained expenditure of metabolic energy.

If resting cells pregrown on [^{14}C]fructose were exposed to TMG or TPG, the release of $^{14}CO_2$ was stimulated. It was estimated that for each additional atom of oxygen consumed by the cells, 3.8 molecules of TMG were accumulated. Assuming a P/O ratio of 3, the cost of the net transport of one molecule of TMG would be one high-energy phosphate bond. However, as subsequently pointed out by Kennedy (1970), there is no evidence for tight coupling of respiration to the utilization of high-energy bonds in bacteria. Indeed, the available information points otherwise. Moreover, the purity of the substrates used seemed not to be established; the presence of a trace of metabolizable sugar may also be expected to stimulate the release of $^{14}CO_2$ from the prelabeled cells.

The isolation of an E. coli mutant which has only about 10% of the active transport activity of its parent as measured by TMG accumulation, but which has a slightly higher capacity to carry out facilitated diffusion as measured by in vivo hydrolysis of ONPG, provides further evidence that entry and accumulation against a concentration gradient can be functionally dissociated (Wong et al., 1970).

Clues to the manner in which the energy input modifies the carrier protein in the transport system are beginning to emerge. Winkler and Wilson (1966) stressed the fact that the affinity of the binding site for the substrate should be different on the two surfaces of the membrane if the compound were to be concentrated across the cell membrane against a gradient and pointed out that under conditions of substrate saturation, Eq. (4) discussed above simplifies to

$$\frac{[S]_{\text{ex}}}{K_m{}^{\text{influx}} + [S]_{\text{ex}}} = \frac{[S]_{\text{in}}}{K_m{}^{\text{efflux}} + [S]_{\text{in}}} \tag{5}$$

Since under conditions of active transport one expected the $K_m{}^{\text{influx}}$ to be smaller than $K_m{}^{\text{efflux}}$, the question arose as to whether energy input was required to generate the low or the high affinity state of the transport protein. The investigators found that under physiological conditions, the $K_m{}^{\text{influx}}$ was in fact lower than the $K_m{}^{\text{efflux}}$, but under conditions of energy uncoupling (treatment with azide or iodoacetate*) the influx constant remained unchanged, whereas the efflux constant was reduced to the value of the influx constant.

Schachter and Mindlin (1969) reached the same conclusions from less direct experiments. These experiments were based on the observations by previous investigators (see below) that β-galactosides protect the lactose transport system against inactivation by NEM. Schachter and Mindlin found that energy uncouplers such as cyanide and DNP enhanced the protection of the transport system by β-galactosides against NEM inactivation and argued that in the absence of adequate metabolic energy, a larger proportion of the carriers became converted to the state with high affinity for the substrate.

The experiments based on NEM inactivation of the transport system, however, have to be interpreted with some caution. First, Schachter and Mindlin found that IPTG, like TDG, offered effective protection against NEM, whereas Kennedy and co-workers found it ineffective (Carter et al., 1968), although both groups employed E. coli ML. Secondly, earlier experiments of Schachter and Mindlin employing longer NEM treatment gave altogether opposite results with respect to the effect of energy uncoupling on substrate protection against inactivation by NEM (Schachter et al., 1966). In a study by Manno and Schachter (1970) in which the cells were separated from the incubation medium by centrifugation of the suspension through silicone, it was concluded that energy uncoupling increased the K_m and reduced the V_{max} for TMG in the entry process. Unfortunately, the investigators were unable to present initial rates because the centrifugation process required about 1 minute. Neither is it clear whether inactivation of the sulfhydryl group of the transport system in metabolically poisoned cells was prevented (see Section II,D,5 below).

The identification of the compound which serves as the direct energy donor for lactose accumulation promises to be one of the most exciting

---

* Iodoacetate is presumed to block glycolysis at the level of 3-phosphoglyceraldehyde oxidation, preventing the glycolytic generation of ATP. Kennedy (1970) reported, however, that this reagent under certain conditions also inactivated the specific lactose transport protein. Thus the use of the compound requires careful control.

aspects in the study of membrane transport. Kennedy and his collaborators showed that when cells of *E. coli* were treated with cold Tris–HCl in such a way as to cause the loss of ions and metabolites, the addition of ATP in combination with magnesium increased the ability of azide-poisoned cells to accumulate TMG against a concentration gradient. Interestingly, the rate of hydrolysis of ONPG under similar conditions could also be increased severalfold by ATP and magnesium. Only purine nucleoside triphosphates stimulated this reaction, and ATP was the most effective. Phosphoenolpyruvate (PEP) had no effect. Unexpectedly the effect of ATP on ONPG hydrolysis is primarily through increasing the V_{max} of the transport reaction (Scarborough *et al.*, 1968). This effect is particularly noteworthy because it suggests that the turnover rate of the carrier can be changed by an energy donor even when uphill transport is not required. In retrospect, results from this experiment are consistent with the finding that the rate of ONPG hydrolysis was reduced by 25–50% in cells simply poisoned by azide (Carter *et al.*, 1968).

The ATP stimulation of ONPG hydrolysis by cold Tris-treated cells was confirmed by West (1969), who carried out further experiments to explore whether the activation occurred intracellularly or extracellularly. His results indicated that ATP was more effective when the cells were diluted with a hypotonic solution containing this compound than when it was added during the 10 minute assay of ONPG hydrolysis. When the activation procedure was repeated with [14C]ATP, more radioactivity was retained by the cells when the labeled ATP was present during the dilution than when it was added afterward to the assay mixture. Hence it seemed that activation of the transport system took place inside by the ATP which leaked into the cell as a result of a sudden change in external osmotic pressure during cell dilution.

Evidence suggesting an indirect role of ATP was first obtained by Pavlasova and Harold (1969), who found that even under anaerobic conditions, the accumulation of TMG was abolished by DNP, tetrachlorosalicylanilide, and carbonylcyanide-*m*-chlorophenylhydrazone, compounds known to uncouple oxidative phosphorylation. Yet the presence of these uncouplers did not seem to impair the generation of high-energy phosphate compounds from glycolysis, as attested by the resistance of the uptake process for glycerol and α-MG to treatment with the three uncouplers.

Barnes and Kaback (1970) reported that the accumulation of lactose and TMG by isolated membrane preparations was greatly stimulated by the provision of D-lactate, which they found to be converted to pyruvate during the transport process. Succinate acted in a similar but less effective fashion and was converted to fumarate. These workers also found that the stimulation of lactose transport by D-lactate required oxygen and was effectively counteracted by the uncouplers of oxidative phosphorylation, 2-heptyl-4-hydroxyquinoline-N-oxide, carbonylcyanide-m-chlorophenylhydrazone, and DNP.

It therefore appeared likely that an intermediate in or connected with the chain of oxidative phosphorylation served as the obligatory energy donor for lactose transport and that the oxidation of D-lactate and succinate by the flavodehydrogenases resulted in the formation of this intermediate. Under anaerobic conditions, high-energy compounds such as PEP or ATP probably generated this intermediate by reversal of a pathway associated with the oxidative process, but this transformation was likewise inhibited by compounds that affect oxidative phosphorylation. Kennedy (1970) recently found that the accumulation of galactosides by aerobically grown cells oxidizing succinate was not inhibited by 30 mM sodium fluoride. Since fluoride rapidly penetrated intact cells at this concentration and was expected to be an effective inhibitor of various enzymes, especially enolase, it was concluded that the accumulation process did not depend upon continuing glycolysis.

D. Alteration of Transport Activity

1. Interference of Galactoside Accumulation by Substrates of Other Transport Systems

The displacement of one compound, A, from the cell by the presence of another compound, B, in the medium can conceivably occur in two ways. (a) Both A and B have affinity for the binding site for transport, in which case the addition of B to a system already equilibrated with A will cause an abrupt displacement of A, because B at first competes for the entry reaction and not for the exit reaction. This is another way of describing counterflow, discussed above. The phenomenon was illustrated in its simplest form in the pioneer study of Rickenberg and associates (1956) when it was shown that radioactive TMG accumulated by the cells could be rapidly displaced by the addition of unlabeled TPG to the medium. (b) The presence of B may activate a control mechanism of the transport system for A (for example, by uncoupling the energy supply), thus preventing retention of the substrate against a concentration gradient. The first model cannot account for the interference of substrates of the glucose uptake system with the accumulation of β-galactosides observed in several studies. For example, both glucose and its analog α-MG reduced the rate of entry of ONPG and accelerated the exit of TMG in cells grown on glucose but not on a number of other substrates (Kessler and Rickenberg, 1963; Koch, 1964; Boniface and Koch, 1967; Winkler and Wilson, 1967). Growth on glucose, it should be mentioned, increased the α-MG uptake system severalfold both in *E. coli* K12 (3100) (Hagihira *et al.*, 1963) and ML308 (Winkler and Wilson, 1967). The effect of α-MG on the lactose system, therefore, might have been dependent upon the achievement of high intracellular concentrations of the analog and its phosphorylated derivative. However, Winkler and Wilson (1967) were able to raise the concentration of α-MG and α-MG phosphate (see Section III) in succinate-grown cells by bathing them in an incubation medium with a high concentration of α-MG without being able to inhibit ONPG hydrolysis. Interestingly, Kepes (1960) found that glucose could expel preloaded TMG even from cells lacking β-galactoside permease, suggesting that TMG was also concentrated in these cells by another permeation system whose retentive capacity could also be affected by glucose (see Lin, 1970).

There does not seem to be a reciprocal interference with the glucose transport system by substrates of the lactose permease. Thiogalactosides, for instance, have no inhibitory effects on the uptake of α-MG even in cells constitutive in the lactose system. Furthermore, preloading of metabolically poisoned glucose-grown cells with α-MG did not promote counterflow of TMG (Winkler and Wilson, 1967).

Boniface and Koch (1967) observed that when cells constitutive in the lactose system were grown on succinate, the *in vivo* hydrolysis of ONPG was not appreciably sensitive to inhibition by α-MG. If such cells were transferred to a glucose medium and allowed three doublings, there was a threefold increase in the capacity of the total culture to hydrolyze ONPG and this permeation process became sensitive to inhibition by α-MG. However, during the course of growth on glucose, the ONPG hydrolysis which persisted when measured in the presence of α-MG remained unchanged despite the severalfold increase in cell number. The authors, therefore, interpreted the results to mean that all of the galactoside permease units newly laid down during growth on glucose were of a different nature, distinguishable from the old units by their sensitivity to inhibition by α-MG. Although this interpretation is concordant with the discovery of Fox that lipid elements synthesized at the time of galactoside permease formation appear to be stably associated with the transport units (see Section II,F,2), the possibility is not excluded that growth on glucose results in the induction of another element in limited quantities, which mediates the interaction between α-MG (or α-MG phosphate) and ONPG, without distinguishing old galactoside permease molecules from new ones. For instance, Kaback (1971) observed that metabolites of glucose inhibited the oxidation of

D-lactate by membrane vesicles and suggested that some of the effects described above were related to the interference of energy generation for β-galactoside transport.

2. Irreversible Inhibitors

Compounds that react with sulfhydryl groups, such as PCMB and NEM, block irreversibly both the entry and exit of galactosides through this transport system, whether or not the process is coupled with energy (Kepes, 1960; Fox and Kennedy, 1965; Schachter and Mindlin, 1969). The inhibition can be antagonized by substrates such as TDG (Kepes, 1960).

3. Effect of K Ions

Schachter and Mindlin (1969) showed that the steady-state level of TMG accumulation decreased progressively as the K^+ concentration in the incubation medium increased from 0.05 to 0.35 M. Within the same range of K^+ concentration, there was no effect on the rate of ONPG hydrolysis, although the growth rate of the cells began to decline above 0.2 M K^+. This interesting ionic effect merits further study.

4. Effects of Temperature

The entry process, as measured by ONPG hydrolysis, was found to be depressed 70–90-fold by lowering the incubation temperature from 28° to 0°C. The exit process, as measured by the half-time of escape of pre-accumulated TMG, decreased by a factor of 200 within the same temperature range (Koch, 1964). This differential temperature effect would account for the observation that the steady-state level of substrate accumulation achieved at 14°C was several times higher than that at 34°C (Kepes, 1960, 1962b). A more systematic study later showed that the efflux rate of TMG increased at least 40-fold from 5° to 35°C, whereas the influx rate increased about 13-fold (Schachter and Mindlin, 1969). The observation that the maximal level of accumulation of another substrate, TDG, did not show any striking difference between 14° and 34°C (Kepes, 1962b), however, would suggest that nonspecific membrane leakage was not an important factor in the efflux rates. This discrepancy in the behavior between the two substrates of the lactose system makes it worthwhile to reinvestigate the temperature dependence of the efflux rates of various substrates from preloaded cells of the same culture, with their β-galactoside permeases poisoned or not poisoned by an inhibitor such as NEM.

5. Changes in Stability of Transport Activity

β-Galactoside transport activity, whether measured by TMG accumulation or ONPG hydrolysis, was found to be less stable than β-galactosidase in cells deprived of a carbon and energy source (Rickenberg et al., 1956). The inactivation of the transport activity was retarded by chloramphenicol and largely prevented by supplying the cells with a metabolizable compound such as succinate (Koch, 1963, 1964). The lability of this transport system seemed to be related to autooxidation of sulfhydryl groups of the specific transport protein since the presence of TDG protected against this spontaneous inactivation, and the addition of β-mercaptoethanol restored the lost activity provided that the inactivation by aeration was not carried out for long periods (Carter et al., 1968). A prolonged state of inactivation may explain why an earlier attempt to rescue the transport activity by β-mercaptoethanol did not meet with success (Koch, 1963). It would be interesting to establish that the inactivation process involved solely the M protein, and not sulfhydryl groups in other membrane proteins as well. A comparative study of different transport systems might be helpful.

E. Specific Transport Protein

1. Isolation of the M Protein, the y Gene Product

Although the participation of a specific protein in the transport of β-galactosides was surmised immediately after the discovery of this inducible process, isolation of the protein defied experimental attempts for almost a decade. In 1965, Fox and Kennedy (1965), exploiting the facts that a sulfhydryl group is essential for transport and that this functional group can be effectively protected from sulfhydryl inhibitors by TDG (Kepes, 1960), were able to label selectively the protein associated with lactose transport. To achieve this, one batch of cells was induced in the lactose system by IPTG and exposed to unlabeled NEM in the presence of TDG. Unreacted NEM was then destroyed by the addition of mercaptoethanol. When the cells were removed and incubated with [¹⁴C]NEM without TDG, the sulfhydryl groups formerly eclipsed by TDG became available to attack by the labeled reagent. A parallel batch of cells grown in the absence of the inducer for the lactose system was processed in an identical fashion, except that [³H]NEM was employed at the last step. Since the uninduced cells should contain negligible quantities of the protein in question, any molecule which became attached to the [³H]NEM after elimination of TDG would be irrelevant to the specific lactose transport system. The two batches of cells were then mixed.

In this mixture all the proteins except the one which was inducible by IPTG and protected by TDG should be doubly labeled. The cells were sonically disrupted and the particulate fraction containing membrane fragments was extracted to release the proteins which were then purified to maximize the ^{14}C to ^{3}H ratio (Fig. 3). To exclude any artifact resulting from nonequivalent properties of the [^{14}C]NEM and [^{3}H]NEM, the labeling procedure was reversed; the induced cells were treated with [^{3}H]NEM and the noninduced cells with [^{14}C]NEM (Fox and Kennedy, 1965). When the NEM-labeled specific membrane component was hydrolyzed in HCl, the radioactive NEM was recovered in association with cysteine as expected (Carter *et al.*, 1968).

This transport protein, or M protein, was found to represent about 3% of the total membrane protein. Relative mobility during electrophoresis and Sephadex gel filtration of a highly purified preparation gave a molecular weight of about 30,000 (Jones and Kennedy, 1969). The fact that this membrane protein represented the product of the *y* gene of the *lac* operon was established by several lines of evidence. (a) It was constitutively produced in cells constitutive in the lactose system; (b) amber mutations in the *y* gene abolished the protein; and (c) a temperature-sensitive revertent of a point mutant in the *y* gene produced a thermolabile protein. The temperature-sensitive character of the permeation

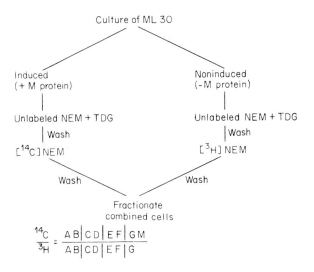

Fig. 3. Plan of the "double-label" procedure for the isolation of the M protein. Induced cells contained the M protein in addition to the other proteins represented by A to G. Only the M protein should have a high ^{14}C to ^{3}H ratio. (From Kennedy *et al.*, 1966.)

system and the M protein was shown by genetic recombination to be controlled by a locus in the *lac* operon (Fox *et al.*, 1967).

Kolber and Stein (1966, 1967) also reported the isolation of a membrane-bound protein believed to be involved in the specific transport of lactose, but their results could not be substantiated (Stein, 1969; Kennedy, 1970).

2. Binding Sites on the M Protein

The dissociation constant for the TDG–M protein complex, estimated by protection of the protein in whole cells or extracts by thiodigalactoside against NEM inactivation, varied from 5 to $7.5 \times 10^{-5} M$ (Carter *et al.*, 1968), which is close to $2 \times 10^{-5} M$, the K_m for TDG transport (Kepes, 1960). Surprisingly, other substrates of this transport system (lactose, ONPG, IPTG, and TMG) could not protect the M protein even at 5 mM, nor did they interfere with the protection of the M-protein by TDG even when introduced at ten times the concentration of the digalactoside. On the other hand, TDG was found to be very effective in blocking the hydrolysis of ONPG by whole cells. It was therefore suggested that the M protein has two binding sites, one for TDG (and melibiose), and the other for lactose, ONPG, TMG, IPTG, and TDG (Carter *et al.*, 1968).

F. Unsaturated Fatty Acids and the Transport System

1. Effects on Transport

Schairer and Overath (1969) investigated the rates of TMG accumulation in cells of *E. coli* containing different unsaturated fatty acids. Analysis of the data according to the Arrhenius equation revealed that membranes containing fatty acids with *cis*-double bonds gave activation energies in the range of 8–9 kcal, whereas those containing a *trans*-unsaturated fatty acid gave a value of 40 kcal. Fox and his collaborators (Wilson *et al.*, 1970) made a similar study in which they measured *in vivo* hydrolysis of both ONPG and *p*-nitrophenyl-β-D-glucoside. (The latter compound is taken up through the action of the PEP-phosphotransferase system, which is discussed in Section III.) In both cases the transport but not the hydrolysis of the substrate was rate limiting. The Arrhenius plots showed clear breaks in the slope at 13°–14°C for cells supplemented with oleate and at 7°–8°C for cells supplemented with linoleate. Both transport systems behaved strikingly alike within the temperature range studied, although the two substrates were transported by systems that are mechanistically different, one dependent upon the

PEP system and the other not. The Arrhenius constants obtained in the 15°–25°C decade for cells containing oleate and linoleate were about three times as high as the corresponding values obtained by Schairer and Overath. It was suggested that the different results might be ascribed to the fact that in the earlier study, transport was measured by the accumulation of the substrate against a concentration gradient under a condition in which the efflux rate might have been considerable. The latter process has been shown to have a higher temperature coefficient than that for influx (Kepes, 1960; Koch, 1964; Schachter and Mindlin, 1969).

2. Importance to the Maturation of the Transport System

Using an *E. coli* mutant defective in unsaturated fatty acid synthesis isolated by Silbert and collaborators (Silbert *et al.*, 1968), Fox (1969) observed that induction of the ability to accumulate [^{14}C]TMG or to hydrolyze ONPG required the addition of an unsaturated fatty acid such as oleate, *cis*-monoenoic, *cis*-dienoic, *cis*-trienoic, or cyclopropane fatty acids. In contrast, the induced synthesis of β-galactosidase and transacetylase during the first 30 minute period was unimpaired in the absence of such fatty acids. However, once the lactose transport activity was induced in the presence of a requisite fatty acid, function persisted after the critical agent was removed; the half-life was about one and a half hours at 37°C. Chloramphenicol, β-mercaptoethanol, and TDG, which could effectively retard or prevent the inactivation of the transport activity in normal cells (Koch, 1964; Carter *et al.*, 1968), were ineffectual in preventing the loss of the transport activity in the mutant.

G. The Model for Lactose Transport

A simple model that accommodates the presently available data (Kennedy *et al.*, 1966) is presented in Fig. 4. The M protein is envisaged as the carrier molecule which undergoes translational or rotational motion in the lipid membrane by thermal diffusion, exposing its substrate site alternately to the extracellular and intracellular surfaces. While it is internally exposed and in the presence of an energy supply, the protein may be converted from form M to form M_i; in this latter form the protein has very little affinity for the substrate. The protein can be reconverted from the M_i form to form M when it is externally oriented. The predominance of the conformation with high substrate affinity when the M protein is externally exposed and of the conformation with low substrate affinity when it is internally exposed would account for the uphill flow of the substrate into the cell. When the energy input is absent, the

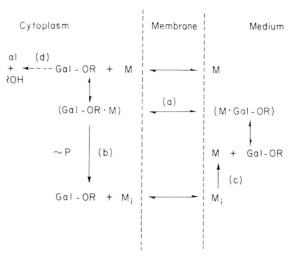

Fig. 4. A model of the β-galactoside transport system. The substrate, Gal-OR, may combine reversibly with the M protein and cross the cell membrane by process (a). If an energy source is available, the carrier protein may be converted by reaction (b) to a form M_i with little affinity for the substrate. Form M may be regenerated from M_i by process (c). Under physiological conditions, entered substrate molecules would be rapidly hydrolyzed by process (d), catalyzed by β-galactosidase. (Adapted from Kennedy, 1970.)

M protein exists solely in the high-affinity state and can serve only to facilitate the equilibration of the substrate across the cell membrane.

The existence of M proteins in two distinguishable states, one with the essential sulfhydryl group readily accessible from external environment and the other not, was demonstrated by Yariv and collaborators (1969). It was found that the lactose permeases inactivated by a bulky mercurial compound, N-(3-mercuri-2-methoxypropyl)poly-DL-alanylamide (P-Hg), could be completely reactivated by either mercaptoethanol or poly-DL-alanylcysteine (P-SH), whereas the lactose permeases inactivated by a small mercurial compound, p-chloromercuribenzene sulfonate (PMBS), could be completely reactivated by mercaptoethanol but only partially reactivated by P-SH. The investigators suggested that blocking the essential sulfhydryl with either mercurial agent froze the permeases in the state in which they reacted. Being small, PMBS readily attacked the permeases in either of the two states, whereas P-Hg, being large, attacked the permeases only when they were more exposed and hence trapped all of them in the external state. This would explain why the bulky P-SH was able to regenerate all of the permeases blocked by P-Hg but only a portion (about 50%) of those blocked by PMBS, and the

EEC Lion Biolog. Machines

~~Constants of diseases — Mon dia~~

Active transport

- single – no other molecules transported
- linked – intestine
- exchange – e.g. Na^+ K^+ pump

LACTOSE SYSTEM

$$V_{influx} = V_{max}^{influx} \frac{[S]_{ex}}{K_m^{influx} + [S]_{ex}}$$

↓ or efflux ↓ efflux — or intracellular — in

high internal conc.

$K_m^{influx} < K_m^{ex}$

Counterflow — > intracellular conc. nevertheless labeled extra will flow in (since permease), but will (at first) not flow out on acct of > conc. of unlabeled. Result – gradual equilibration.

Exchange diffusion — if EA diffuses faster than E alone, adding intracellular

ATP stimulates TMG (may repress) ± OMPG (full TK + offense) kinetip.

Glucose removes apparr to interfere with glucose, or cause ... in glucose. Outside the cell WEM + PCB14 on interaction ... - Kinetic ... rates ... to phenom. High K_t values, low T + ... units.

Galactose permease - 'M' perm, contains sulphhydryl groups; repressed by grow? low green of 2 low-J rates - one TDG, oth TDG + OMPG, TMG, IPTG
On-site converted to M_i with little acetate present.

Lack of crisp functional during transport disulf. of ... & TDG arises are not large. Read ... and ...

Phosphoenolpyruvate phosphotransferase system & uptake of carbohydrates

$$PEP + HPr \xrightarrow[Mg^{++}]{enzyme I} P\text{-}HPr + pyruvate$$

$$Sugar + P\text{-}HPr \xrightarrow[Mg^{++}]{enzyme II} sugar\text{-}P + HPr$$

smaller mercaptoethanol was able to restore fully the inactivated permeases irrespective of how they were poisoned.

III. THE PHOSPHOENOLPYRUVATE PHOSPHOTRANSFERASE SYSTEM AND THE UPTAKE OF CARBOHYDRATES

A. Biochemistry

Kundig et al. (1964) described a bacterial PEP-dependent phosphotransferase system catalyzing the following reactions:

$$\text{PEP} + \text{HPr} \xrightarrow[\text{Mg}^{++}]{\text{enzyme I}} \text{P-HPr} + \text{pyruvate} \tag{6}$$

$$\text{P-HPr} + \text{sugar} \xrightarrow[\text{Mg}^{++}]{\text{enzyme II complex}} \text{sugar-P} + \text{HPr} \tag{7}$$

The phosphoryl group of PEP is transferred to a nitrogen atom of the histidine imidazole ring of a small protein under the action of enzyme I to yield a phosphorylated protein, P-HPr. The phosphoryl group of P-HPr in turn can be transferred to a hydroxyl group of a sugar under the influence of an enzyme II complex. Later it was shown by Anderson (1968) that it is the N-1 of the histidine imidazole of HPr which accepts the phosphoryl group, thus forming an "energy-rich" phosphoryl–N bond which is acid labile.

The histidine-containing protein, HPr, has been purified to homogeneity from E. coli. It has a molecular weight of 9637 and contains two residues of histidine, but none of cysteine, tyrosine, and tryptophan (Anderson, 1968). The protein withstood heat treatment at 100°C for at least 20 minutes in neutral pH and was precipitated at pH 1 without inactivation (Kundig et al., 1964; Kundig and Roseman, 1966).

A single enzyme I catalyzes reaction (6). This enzyme is found in the soluble (cytoplasmic) fraction of bacterial extract and has been extensively purified (Kundig et al., 1964; Kundig and Roseman, 1966; Roseman, 1969). Both HPr and enzyme I appear to be partially inducible by carbohydrates (Tanaka and Lin, 1967).

A family of enzyme II complexes catalyzes reaction (7); each member of this family is specific for one substrate (Kundig et al., 1964). Enzyme II complexes are associated with cell membranes but can be solubilized (Kundig and Roseman, 1966, 1969; Roseman, 1969). As will be more fully discussed below, these complexes accept substrate from the outside surface of the cell and release the phosphorylated products in the cytoplasm.

The enzyme II complex inducible by and active on galactose and the enzyme II complex inducible by and active on mannitol have each been purified from extracts of *Staphylococcus aureus* into two protein components. Each protein component was found necessary for catalysis (Simoni *et al.*, 1968).

Hanson and Anderson (1968) reported two components for the enzyme II system of *Aerobacter aerogenes* for the phosphorylation of fructose. They found in uninduced cells a constitutive fructose phosphorylating enzyme activity associated with the membrane fraction displaying a high K_m for fructose. In induced cells a soluble protein was found, which when added to the membrane fraction, catalyzed fructose phosphorylation with a low K_m. Hence, the name K_m factor was given to this soluble protein. However, the possibility seems to exist that induced cells actually contained two additional proteins: a soluble one and a membrane-associated one. Each alone was inactive, but the two together catalyzed fructose phosphorylation with a low K_m. The constitutive enzyme activity with the high K_m might be another such complex (with no readily dissociable component) which physiologically acted on a sugar other than fructose. In the presence of the inducible enzyme complex for fructose with low K_m, the contribution of the constitutive enzyme may have been masked.

The enzyme II complex of *E. coli* active on α-methylglucoside (physiologically for glucose utilization) has been resolved into a two-protein component (proteins II-A and II-B) plus a lipid fraction (II-lipid) (see also Chapter 6). The latter could be substituted by phosphatidylglycerol (Kundig and Roseman, 1969). The purified protein II-B migrated as a single band in the presence of sodium dodecylsulfate during polyacrylamide gel electrophoresis at a rate that suggested a molecular weight of approximately 35,000 (Shapiro *et al.*, 1967). Recently it was found that one of the two protein components acted as an intermediate acceptor for the phosphoryl group from HPr (Kundig, 1971).

B. Studies of the PEP-Phosphotransferase System

1. Studies With Osmotically Shocked Cells

The possibility that the phosphotransferase system is involved in the uptake of certain sugars by bacteria was first suggested by observations on *E. coli* cells subjected to osmotic shock in the cold. Cells treated in this manner became impaired in their ability to accumulate radioactive TMG and α-MG from an incubation medium. Concomitantly, HPr was released. The addition of HPr to osmotically shocked cells sometimes stimulated the uptake of TMG and α-MG (Kundig *et al.*, 1966). The effect on the former compound, however, was unexpected since normally the accumulation of TMG was not known to involve a phosphorylated intermediate. Kennedy (1970) suggested that the accumulation of TMG as TMG-phosphate in osmotically shocked cells might have been an artifact resulting from nonspecific phosphorylation by an enzyme II complex when an excessively high concentration of the sugar analog was introduced.

2. Genetic Studies

Unambiguous evidence for the role of the phosphotransferase system in the uptake and metabolism of sugars was obtained in a study of a mutant of A. aerogenes which specifically failed to grow on mannitol. Analysis of cell-free extracts of this mutant revealed the absence of an enzyme II activity inducible by mannitol (Tanaka et al., 1967b). During a search for additional mutants defective in mannitol metabolism two other phenotypes were found. These were pleiotropically defective in the metabolism of several other carbohydrates including sorbitol, glucose, mannose, and fructose. Upon analysis of extracts from these mutants, one was found to lack enzyme I activity and the other HPr activity. Moreover, when cells of these pleiotropic mutants were incubated with labeled mannitol, they failed to accumulate radioactive material even though the enzyme II for mannitol was fully induced. Thus, mannitol was not concentrated unaltered against a gradient as in the case of a classical active transport system but was probably converted into a phosphorylated derivative during translocation across the cell membrane or immediately upon delivery at the intracellular side of the membrane. The phosphorylated product was then retained inside the cell. Since mannitol 1-phosphate could be acted on directly by an NAD-dependent dehydrogenase, no hydrolysis and subsequent rephosphorylation have to be postulated. Indeed, the virtue of this mechanism is that the capture of the substrate and its activation for metabolism are combined in one chemical reaction at the expense of a single energy-rich phosphoryl ester, PEP (Tanaka and Lin, 1967).

Pleiotropic defects of sugar uptake and metabolism as a result of a defect in the phosphotransferase system have also been demonstrated in E. coli (Tanaka et al., 1967a), Salmonella typhimurium (Simoni et al., 1967), and S. aureus (Hengstenberg et al., 1967). Interestingly, in S. aureus lactose uptake is dependent upon the phosphorylation system, which explains why earlier studies failed to reveal any β-galactosidase activity in the extracts of cells grown on lactose (Creaser, 1955; McClatchy and Rosenblum, 1963); lactose phosphate, and not free lactose, is the true substrate of the *S. aureus* β-galactosidase (Kennedy and Scarborough, 1967; Hengstenberg *et al.*, 1967).

C. Exit and Displacement

Despite the fact that substrates entering the cell through the action of the phosphotransferase system are converted and trapped as the phosphorylated derivative, it is possible to expel the accumulated product by the introduction of certain metabolizable compounds. Thus, when cells of *S. typhimurium* pregrown on glycerol and malate were incubated with [^{14}C]α-MG and the accumulation of radioactive material was followed at 37°C until a plateau was reached, addition of glucose caused almost complete displacement of the radioactivity in less than 5 minutes. The effect of glucose did not seem to be primarily dependent upon the fact that the compound is the normal substrate for the α-MG uptake system, since even if the cells were supplemented with metabolizable substrates which enter by an independent route of transport, the amount of [^{14}C]α-MG retainable was greatly reduced. For example, glycerol-grown cells, in the presence of glycerol, accumulated finally only $\frac{1}{2}$–$\frac{1}{3}$ of the amount of [^{14}C]α-MG which could be taken up by resting cells.

The effects of introducing metabolites on the α-MG uptake process was also illustrated by the positive effects of agents or conditions expected to interfere with the generation of metabolic energy. Thus, when cells were incubated with [^{14}C]α-MG and allowed to accumulate radioactive material until a steady-state level was reached, the addition of azide or DNP resulted in a severalfold increase of this level. When cells grown aerobically on fructose were incubated with [^{14}C]α-MG and allowed to accumulate radioactive material maximally in the presence of fructose under aerobic conditions, subsequent withdrawal of oxygen caused an 18-fold increase in the accumulation (Englesberg *et al.*, 1961; Hoffee and Englesberg, 1962).

Studies of *E. coli* gave similar results. For example, when a mutant lacking L-arabinose isomerase (the first enzyme in the catabolic pathway for the pentose) was induced in the L-arabinose permease and incubated in the presence of [^{14}C]α-MG until an equilibrium was reached, the introduction of L-arabinose did not cause any net loss of the accumulated radioactive material. In contrast, the addition of the pentose to wild-type cells induced in the L-arabinose system under similar conditions caused rapid and extensive release of α-MG into the medium. Hence, the competition for a hypothetical common element in the membrane during transport could not explain the release of [^{14}C]α-MG by the cells (Hoffee *et al.*, 1964). In another study, it was found that the transport of TMG, a nonmetabolizable substrate for the lactose permease, did not interfere with the final total uptake of α-MG (Kessler and Rickenberg, 1963), in confirmation of the results of the L-arabinose experiment.

In a study in which the displacement of preaccumulated [^{14}C]α-MG was measured more directly by following the exit rate of radioactive material in the presence or absence of a test substance (Hagihira et al., 1963), the addition of glycerol or L-α-GP was found to accelerate the exit only if the metabolic enzymes for the glycerol system were preinduced. In the case of mutant cells lacking glycerol kinase, but induced in the L-α-GP transport system and L-α-GP dehydrogenase, rapid release of [^{14}C]α-MG occurred upon exposure to the phosphorylated compound but not to glycerol. In addition, it was shown that lactose, but not the nonmetabolizable TMG, accelerated the release of α-MG from cells, and then only if they were induced in the *lac* operon. The belief that the acceleration of the release of α-MG is not directly related to competition for common elements of transport was further strengthened by the observation that the provision of casein hydrolyzate also stimulated exit of the glucoside from cells grown previously on the amino acid mixture. Similar conclusions were made in another study in which it was shown that among the metabolizable carbon sources, glucose was the most potent inhibitor of α-MG uptake in *E. coli* (Halpern and Lupo, 1966). Presumably the special efficacy of glucose reflected the superimposition of the direct competition for entry upon the indirect metabolic effect of the hexose.

The interpretation of the release process is made especially difficult by the fact that most of the α-MG-P formed inside the cell (Rogers and Yu, 1962; Hagihira et al., 1963; Winkler and Wilson, 1967) could be displaced again as free α-MG, even if the chasing agent itself was a phosphorylated compound, such as G-6-P (Winkler, 1966). It is not known whether hydrolysis or transphosphorylation was responsible for the conversion of α-MG-P to α-MG.* Neither is it clear which compounds could act as phosphoryl group acceptors, if the latter occurred. Even more important is the question of whether α-MG exit is catalyzed by the specific enzyme II complex responsible for the uptake and phosphorylation of glucose. A mutant with a thermolabile component of the enzyme II complex may provide the answer.

The release of one substrate by another has also been observed in

---

* The fact that α-MG-P, the immediate product of uptake, is gradually converted intracellularly to the free sugar and that the concentration of the free sugar maintained inside the cell can readily exceed its external concentration serves as a warning that the existence of a chemical gradient is not sufficient evidence for classical active transport. It is necessary to rule out intracellular intermediates.

S. *aureus* (Egan and Morse, 1966). A systematic study to reveal a hierarchical relationship in the displacement reactions may be useful for the understanding of the mechanism and significance of the exit reactions. In particular, the possible involvement of a regulatory mechanism should not be overlooked.

D. Bacterial Membrane Vesicles

Kaback and Stadtman (1966) and Kaback (1968, 1969b, 1970b) developed a method for isolating bacterial membrane vesicles from spheroplasts (produced by penicillin or lysozyme–EDTA treatment) followed by osmotic lysis in the presence of DNase. Electron microscopic examination revealed sacs bound by membranes 65–70 Å thick with diameters varying from 0.15 to 1.5 μ. Vesicles prepared from *E. coli* ML 308-225 were bound by a single trilaminar layer, whereas those prepared from *E. coli* strains W, K12, W2244, and $K_2$1t and from *S. typhimurium* contained up to six trilaminar layers. It was suggested that the paucity of lipopolysaccharide in the membrane of *E. coli* ML may be the reason for the simpler structure. Chemical and enzymatic analyses showed that the membrane fraction retained less than 5% of the RNA and DNA; 15–20% of the protein; less than 1% of the activities of glutamine synthetase, fatty acid synthetase, or leucine-activating enzyme; less than 2% of the enzymes in the periplasmic space, e.g., alkaline phosphatase and 5′-nucleotidase; but about 70% of the phospholipids found in whole cells. The sacs contained less than 3% lipopolysaccharide by dry weight and carried less than 10% of the diaminopimelic acid present originally in the penicillin-induced spheroplasts (strain ML). Disc gel electrophoresis of the sonicated membrane sacs revealed two major bands of protein in contrast to the multibanded pattern given by sonicated whole cells. The composition of the membranes by dry weight was 60–70% protein, 30–40% phospholipids, and about 1% carbohydrate.

The existence of membrane as sealed vesicles was shown by the fact that hypertonicity caused them to shrink and to show increased light-scattering. When the membrane vesicles from wild-type *S. typhimurium* were incubated with [^{14}C]α-MG, more than 90% of the labeled material taken up was identified as [^{14}C]α-MG phosphate. Addition of PEP to the incubation mixture markedly stimulated the uptake of [^{14}C]α-MG. In contrast, membranes prepared from mutants defective in enzyme I or HPr were inactive in substrate uptake.

The membrane preparations differed from intact cells in certain functional features. For example, whereas whole cells accumulated both free α-MG and α-MG-P when incubated with α-MG, membrane sacs accumulated only the phosphorylated compound. Moreover, the phosphorylated sugar in the sacs was not displaceable by carrier α-MG added to the external medium. It was suggested that the absence of a transphosphorylating enzyme system (see Schaefler and Schenkein, 1968) was responsible for these particular properties of the membrane preparation. The kinetic characteristics of α-MG uptake by the vesicles also differed from those of intact cells. (a) The K_m of 4×10^{-6} M for α-MG was low in comparison to the value range of 5×10^{-5} to 1×10^{-4} M observed with intact cells of E. coli (Kepes, 1962a; Hagihira et al., 1963; Kessler and Rickenberg, 1963; Winkler and Wilson, 1967) and 1×10^{-4} M observed with intact cells of S. typhimurium (Hoffee et al., 1964). (b) The V_{max} for α-MG uptake by membrane vesicles was several orders of magnitude lower than that of intact cells, but the addition of HPr and enzyme I enhanced the accumulating activity of the vesicle preparations manyfold.

A number of novel findings were obtained from studies of the membrane vesicles. Substrate molecules presented externally seemed to be preferentially phosphorylated over those presented internally. To show this, Kaback incubated the vesicles first in a high concentration of [^{14}C]glucose under nonphosphorylating conditions, i.e., at 0°C in the presence of NaF for 2–3 hours. Afterwards the vesicles were collected on Millipore filters and washed. A solution of [^{3}H]glucose and PEP at 40°C was then layered on top of the vesicles, and at various times membrane samples were taken for the analysis of their contents. The amount of [^{3}H]glucose phosphate increased progressively over a period of 7 minutes, while that of [^{14}C]glucose phosphate remained at a low constant level, close to that of the free [^{14}C]glucose presumed to be contained in the vesicles. The validity of the conclusion, however, depends upon the assumptions that the [^{14}C]glucose remaining after the washing was essentially all inside the phosphorylating sacs and was present at concentrations expected to be effective.

The function of the phosphotransferase system in membrane vesicles could be modified by agents which affect membrane composition. Thus, digestion of the vesicles with phospholipase D of cabbage abolished completely the capacity for α-MG phosphorylation. Vesicles preloaded with α-MG-P, however, could still retain the phosphorylated compound following treatment with the lipase. Analysis indicated a progressive and selective loss of phosphatidylglycerol in the membranes, with concomitant increases in phosphatidic acid during treatment with the lipase. Surprisingly, despite the removal of the bulk of soluble proteins during the process of preparing the vesicles, sufficient enzyme activities still remained to permit the vesicles to recover spontaneously from the phospholipase D treatment: Within 5 minutes at 46°C, their ability to carry out "vectorial phosphorylation" (uptake of [^{14}C]α-MG on one side of the membrane and the delivery of the [^{14}C]α-MG-P on the other side) was almost fully restored. (Possibly the phosphatidylglycerol at the transport site was replaced not by *de novo* synthesis, but by diffusion from other membrane sites.) A control experiment showed that active transport of proline was only slightly affected by the enzymatic digestion (Milner and Kaback, 1970).

Nutritional alteration of membrane lipid composition could also modify certain transport properties of the PEP-dependent systems. For example, the optimal temperature for phosphorylation during a 5 minute period was 46°C for membranes prepared from cells grown on glucose or succinate. However, sacs from glucose-grown cells began to leak α-MG-P at 35°C, whereas those from succinate-grown cells began to leak at 46°C (Kaback, 1970b). In this connection, it might be noted that Hoffee *et al.* (1964) found the optimal temperature for the rate of entrance of α-MG into intact cells of *S. typhimurium* to be 44°C and the optimal temperature for maximal total uptake to be 24°C, values which agree closely with the membrane preparations from *E. coli*.

The question of whether the membrane-associated enzyme II catalyzes unidirectional phosphorylation because P-HPr is available only on the interior side of the membrane has been of great interest because it bears on the problem of whether the substrate is translocated and phosphorylated at the same combining site of the enzyme II complex. In this regard, it is interesting that the addition of PEP, HPr, and enzyme I to NaF-poisoned vesicle preparations greatly stimulated the formation of α-MG-P which was recovered in the incubation medium instead of in the sacs, suggesting that the active site of enzyme II was available to the external surface (Kaback, 1970b). It would be desirable to confirm these results with spheroplasts themselves to exclude the possibility that some of the vesicular sacs were formed with inverted membranes or were poorly sealed. [In the case of membranes prepared by the same procedure from *Bacillus megaterium* it was possible to determine the orientation of the membrane surface by phase contrast microscopy. No inverted sacs were detected (Kaback, 1971).] It is also pertinent to note here that Simoni and co-workers (1967) observed that an enzyme I-negative mutant of *S. typhimurium* could grow, albeit slowly, on glucose only if preinduced with galactose, which would suggest that the enzyme II complex for glucose (a constitutive system) by itself could not catalyze the entry of glucose at a significant rate and that the induction of the methylgalactoside permease by galactose was necessary for this entry process. Once inside the cell, glucose could be phosphorylated by the classical ATP-dependent kinase, a cytoplasmic enzyme. If the results from both the phosphorylation experiment with membrane vesicles and the growth properties of the enzyme II-negative mutant were correctly interpreted, this would suggest that the same protein is responsible for the translocation and the phosphorylation of the substrate and that in the absence of P-HPr the protein is not sufficiently mobile to facilitate the diffusion of the unphosphorylated substrate. On the other hand, the observation by Pastan and Perlman (1969) that glucose and α-MG could still repress β-galactosidase formation in enzyme I (5% leakiness) and HPr-defective mutants, but not in mutants defective in the enzyme II complex for glucose (4% leakiness), indicated that the enzyme II complex *could* serve to facilitate the diffusion of the substrate in the absence of the phosphorylation reaction. However, it should be borne in mind that enzyme I- and HPr-defective mutants are much more sensitive to catabolite repression than enzyme II mutants and that the difference in the behavior of the two cases might have reflected differential sensitivity to repression rather than differential permeation. The question must be considered to be unsettled.

In a study of the kinetics of substrate uptake by the vesicles prepared from *E. coli*, Kaback (1969a) found that the initial rate of uptake of α-MG was inhibitable by externally added glucose-6-P (G-6-P) or glucose-1-P (G-1-P). Fifty percent inhibition by G-6-P was observed at 5×10^{-4}, 1.5×10^{-3}, and 3×10^{-3} M with membranes prepared from cells grown in glucose, glycerol, and succinate, respectively. Fifty percent inhibition by G-1-P was observed at 5×10^{-5} and 8×10^{-5} M with membranes prepared from cells grown in glycerol and succinate, respectively. Curiously, membranes from glucose-grown cells could be only slightly inhibited by G-1-P. The concentrations of G-1-P employed in these studies were close to the value of 2×10^{-5} M estimated to be in intact cells of *E. coli*.

E. A Model

A simple scheme compatible with the currently known features of phosphorylating transport by the PEP-phosphotransferase system is given in Fig. 5. An enzyme II complex is depicted receiving the substrate approaching from the medium. The enzyme–substrate complex then combines with P-HPr in the membrane (presumably a milieu not accessible to small molecules of high polarity, such as ATP). The transfer of the phosphoryl group from P-HPr to the substrate is then achieved and the products are released into the cytoplasm (reaction A). The discharged enzyme II complex is then returned to the external surface of the membrane to initiate a new cycle of transport.

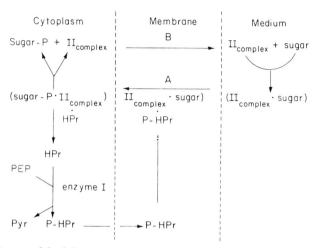

Fig. 5. A model of the phosphotransferase system for the uptake of sugars.

IV. THE BINDING PROTEINS ASSOCIATED WITH BACTERIAL TRANSPORT SYSTEMS

When cells of gram-negative bacteria are subjected to an osmotic shock treatment (Neu and Heppel, 1965; Nossal and Heppel, 1966; Heppel, 1967, 1969), they release certain proteins which bind specific substrates (see Chapter 5 for a further discussion of this subject). Concomitant with the loss of these proteins, certain specific transport systems become impaired. In some cases treatment of the shocked cells with a binding protein can partially restore the related transport system. No binding proteins are released in cases where the transport systems are resistant to the shock treatment, and in general the substrate specificity of the binding protein reflects fairly closely that of the corresponding transport system. The intimate relationship between the binding proteins and the active transport processes is revealed further by the facts that their biosynthesis is jointly controlled and that certain mutations abolish both the protein and the transport process.

Two possible roles can be envisaged for these proteins. (a) They are integral parts of the active transport system in the cell membrane that happen to be readily dissociated from the lipid phase by osmotic shock treatment. (b) They normally reside in the space between the outer limits of the plasma membrane and the inner limits of the cell wall and catalyze the diffusion of the substrate across this space in a manner similar to the facilitation of the diffusion of oxygen through agar membranes (Wittenberg, 1959, 1970), or across a Millipore impregnated with hemoglobin (Wittenberg, 1959, 1970; Scholander 1960). On the basis of available knowledge, the second hypothesis seems more likely inasmuch as a number of active transport systems seems to function without such a binding protein and the absence of the binding protein has not been shown to be associated with an absolute block of transport, especially in the presence of moderate concentrations of the substrate. However, residual transport activities in some cases may be due to the presence of other transport systems with similar specificities. More systematic screening for the existence of external binding proteins and the genetic removal of interfering transport systems will be necessary to settle this important question.

A. The Sulfate-Binding Protein

The sulfate transport system in S. typhimurium has been studied by Pardee, Dreyfuss and their collaborators. Several gene products have been implicated in the concentrative uptake of this compound (Dreyfuss and Monty, 1963). The synthesis of the sulfate transport system was found to be repressible by cysteine (Dreyfuss, 1964), and the operation of the transport appeared to be under kinetic feedback control (Dreyfuss and Pardee, 1966). A sulfate-binding protein was readily released from the cells by osmotic treatment (Dreyfuss and Pardee, 1965; Pardee and Prestidge, 1966; Pardee et al., 1966). Pardee and Watanabe (1968) were able to show that the sulfate-binding protein was located internally to the cell wall, but externally to the cell membrane, by two procedures of general applicability. The first technique involved the inactivation of the protein by a reagent, diazo-7-amino-1,3-naphthalene disulfonate, capable of penetrating the outer bacterial coat but not its plasma membrane. The second procedure involved the use of specific antibodies capable of in-activating the binding function of the protein when isolated, but not when associated with the cells. They found that the transport and sulfate-binding capacity of intact cells were abolished by the chemical reagent but were not affected by the specific antiserum (see also Nakane et al., 1968, cited in Section IV,C). This protein has been purified and crystal-lized and was shown to have a molecular weight of 32,000. Interestingly, it lacked sulfur-containing amino acids. The synthesis of the protein was repressed by cysteine. Mutants have been obtained which are defective both in the transport of sulfate and in the production of the binding pro-tein. The affinity of the binding protein, however, appeared not to be identical with that of the transport process. Whereas the K_m for binding is about $10^{-7} M$, the value for transport is $3.6 \times 10^{-5} M$. Moreover, the transport system displayed a higher apparent affinity for thiosulfate than did the binding protein (Pardee, 1966, 1967, 1968; Langridge *et al.*, 1970).

B. The Phosphate-Binding Protein

A protein that binds phosphate with a K_m of about $1 \times 10^{-6} M$ has been isolated and purified from *E. coli*. It has a molecular weight of 42,000 and has a single binding site. Incubation of wild-type cells shocked with cold water in the presence of this protein stimulated their phosphate uptake. Similar treatment with this protein also stimulated phosphate uptake by a mutant defective in transport and lacking the binding protein (Medveczky and Rosenberg, 1969, 1970). The possible relationship between this protein and alkaline phosphatase or other enzymes associated with phosphate metabolism has apparently not yet been investigated. Several pathways exist for the uptake of phosphate into cells of *E. coli* (Bennett and Malamy, 1970, 1971). The elucidation of the exact role of the binding protein may require more detailed understanding of these several systems.

C. Sugar-Binding Proteins

1. An L-Arabinose-Binding Protein

Hogg and Englesberg (1969) and Schleif (1969) independently isolated an L-arabinose-binding protein from *E. coli*. The molecular weight of this protein is between 32,000 and 35,000. Schleif found that the dissociation constant for binding protein–L-arabinose complex varied markedly with temperature, being $2 \times 10^{-7} M$ at 4°C and $2 \times 10^{-6} M$ at 37°C. Like the active transport system, the binding protein has affinities for sugars such as D-fucose and D-xylose (Novotny and Englesberg, 1966). Treatment of the cells by osmotic shock both released the binding protein and reduced the ability of the cells to concentrate L-arabinose (Hogg and Englesberg, 1969). A mutant has been isolated which could synthesize only 10% of the expected amount of the binding protein and which exhibited no detectable L-arabinose transport activity (Schleif, 1969).

2. The Galactose-Binding Protein

An *E. coli* protein that binds both galactose ($K_m = 0.86 \times 10^{-6} M$) and glucose ($K_m = 0.63 \times 10^{-6} M$) was purified by Anraku (1968a,b). The protein has a molecular weight of 35,000. It has no binding capacity for ATP, and ATP has no effect on the ability of the protein to bind galactose. Cells which were subjected to osmotic shock exhibited reduced capacity to accumulate galactose, and incubation of the shocked cells with the binding protein partly restored the transport process (Anraku, 1967,

1968c). The study has been extended by Boos and Sarvas; their data indicate that the binding protein is intimately associated with methylgalactoside permease (Boos, 1969; Boos and Sarvas, 1970) which has a K_m of about $0.5 \times 10^{-6} M$ for galactose (Rotman and Radojkovic, 1964). Among the mutants isolated as cells defective in the production of methylgalactoside permease, about one-third had reduced or undetectable amounts of the binding protein. Mutations exist that affect the transport activity only, and there are others that affect both the transport activity and the binding protein. These mutational sites are closely linked. Mutations affecting the binding protein and not the transport activity have not been reported.

Recent studies by Hazelbauer and Adler (1971) and by Boos and Kalckar (see Kalckar, 1971) provided evidence that the galactose-binding protein also plays a role as a signal receptor for chemotaxis. It would be interesting to see if binding proteins play a role in all chemotactic responses in bacteria.

D. Amino Acid-Binding Proteins

1. Arginine-Binding Proteins

Two arginine-binding proteins separated by DEAE-cellulose column chromatography have been purified from the shock fluid of suspensions of E. coli cells. The reduced capacity of shocked cells to transport arginine could be partially restored by treatment with these proteins. Both of the proteins have dissociation constants for arginine around 0.5–$1 \times 10^{-6} M$, close to the K_m of the uptake process, which is $2 \times 10^{-6} M$ (Wilson and Holden, 1969).

2. Histidine-Binding Proteins

The specific or high-affinity transport system for histidine in S. typhimurium has been resolved by Ames and Lever (1970) into three components. Two binding proteins, J and K, were released by osmotic shock; about 95% of the binding activity in the shock fluid was accounted for by the J protein and the rest by the K protein. The existence of a third protein, P, presumed to be associated more intimately with the membrane, was inferred from the fact that amber mutations in the hisP locus greatly impaired the transport process. The J and K proteins apparently work in parallel, both in conjunction with the hisP gene product.

3. The LIV-Binding Protein

A protein with a molecular weight of 36,000 and whose formation is repressible by leucine, but not by isoleucine or isoleucine plus valine, was crystallized from *E. coli* by Oxender and his co-workers (Piperno and Oxender, 1966; Penrose *et al.*, 1968) and by Anraku (1968a,b). This protein bound leucine, isoleucine, and valine (LIV) with dissociation constants close to the corresponding K_m's of the transport processes. The isolated protein showed no detectable binding activity for ATP, nor did ATP affect the binding of leucine. By an immunocytochemical technique, it was shown that the leucine-binding protein was localized in the cell envelope of .*E. coli* and not in its cytoplasm (Nakane *et al.*, 1968). The ability of cells to transport leucine was reduced severalfold by osmotic shock treatment. Partial restoration of the transport capacity could be achieved by incubation of the shocked cells with the pure binding protein. The restoration process was enhanced when a protein fraction from the shock fluid, precipitated by ammonium sulfate at 65% saturation, was added. This fraction itself contained no demonstrable binding activity (Anraku, 1968c); its physiological role remains to be revealed.

4. A Leucine-Specific Binding Protein

Furlong and Weiner (1970) isolated and crystallized a binding protein from *E. coli* that is specific for leucine. Like the LIV-binding protein discussed above, this protein has a molecular weight of about 36,000. Its dissociation constant for leucine is 7×10^{-7} M; no affinity for any of the other common amino acids could be demonstrated. The discovery of a second leucine-binding protein led to the search for a second permease for leucine. Indeed, it was found that unlabeled isoleucine did not completely inhibit the initial rate of transport of labeled leucine; the isoleucine-insensitive transport activity could be inhibited by trifluoroleucine. On the other hand, trifluoroleucine failed to inhibit completely the transport of leucine; the residual activity in this case could be inhibited by isoleucine. Thus, it appears that each leucine-binding protein is associated with a corresponding transport system. The existence of a specific leucine-binding protein raises the question as to why leucine alone should be the repressor for the synthesis of the LIV-binding protein. The control of both proteins deserves detailed study.

V. THE TRANSPORT SYSTEM FOR SODIUM AND POTASSIUM IONS

A. Coupling of Na⁺ and K⁺ Fluxes in Red Cells

Studies of the biochemical basis of the transport of Na⁺ and K⁺ have largely centered on red cells because of the ready supply, the relative simplicity of their morphology and metabolism, and the possibility of manipulating their intracellular contents by an osmotic treatment known as "reversible hemolysis." This treatment allows temporary free flow of solutes in and out of the cells while the membrane was distended under a hypotonic condition.

The *transport* of Na$^+$ and K$^+$ ions occurs in opposite directions across the cell membrane and the flow of the two kinds of ions is coupled (Harris, 1954; Glynn, 1956). For the net entry of every two K$^+$, there is a net exit of three Na$^+$ (Post and Jolly, 1957; Post et al., 1967; Post, 1968). In experiments in which erythrocytes were first hemolyzed in a hypotonic solution containing ATP and varying concentrations of Na$^+$ and K$^+$, and the ghosts were then allowed to reseal under isotonic conditions, it was shown that the production of orthophosphate from intracellular ATP was most rapid if the cells contained high concentrations of Na$^+$ and the medium contained high concentrations of K$^+$ (Glynn, 1962b; Whittam, 1962).

B. Inhibition of Na$^+$ and K$^+$ Transport

The discovery by Schatzmann (1953) of the specific inhibition of the transport of Na$^+$ and K$^+$ by the cardiac glycosides and their aglycones provided an important tool for the analysis of this transport phenomenon at both the cellular and the enzymatic level (Glynn, 1964). In the case of the giant axon of the squid, it was found that the introduction of 10^{-5} M ouabain in the external medium caused a precipitous drop in the rate of ^{22}Na$^+$ efflux to one tenth of normal value, whereas injection of a solution containing 10^{-3} M of this compound into the axon had an insignificant effect on Na$^+$ efflux. The drug thus acted only on the external surface. What remained of the Na$^+$ efflux under the inhibition of this compound was no longer dependent on external K$^+$. The residual Na$^+$ efflux was also insensitive to the blockage of aerobic metabolism by cyanide or to the injection of ATP (Caldwell and Keynes, 1959). Ouabain was also found to inhibit Na$^+$–Na$^+$ exchange across the membrane (Glynn, 1964).

Whittam and co-workers (1964) observed that oligomycin was a potent inhibitor of Na$^+$ transport in red cells. Garrahan and Glynn (1967a) showed in preloaded red cells that the exchange of ^{24}Na$^+$ with unlabeled Na$^+$ or K$^+$ in the medium could be partially inhibited by oligomycin (1–10 mg/ml).

C. Energy Requirement

Caldwell and co-workers (1960) were able to produce a transient increase in the efflux of Na$^+$ from cyanide-poisoned giant axons of squid

by intraaxonal injections of ATP, GTP, arginine phosphate, phosphoenolpyruvate, but not creatine phosphate. In a study by Brinley and Mullins
(1968) in which intermediates necessary for oxidative phosphorylation,
glycolysis, or transphosphorylation were removed from the axoplasm, it
was possible to show that among the nine naturally occurring high-energy
phosphate compounds (including acetyl phosphate, PEP, phospho-
arginine, etc.), only ATP was effective in driving the Na^+ pump. Similarly,
the transport of K^+ was shown to be dependent specifically on ATP
(Mullins and Brinley, 1969). Gárdos (1954) showed that erythrocytes
metabolically poisoned by arsenate recovered their ion transport ability
when replenished with exogenous ATP by reversible hemolysis. Garrahan
and Glynn (1965) extended the study by showing that unless ATP or
ADP (which can be converted to ATP by adenylate kinase present in the
cells) was present, no glycoside-sensitive efflux of Na^+ occurred. These
investigators later demonstrated that ATP formation occurred under conditions in which the direction of the net fluxes of Na^+ and K^+ was expected
to be reversed. To accomplish this, the cells were charged by reversible
hemolysis with [^{32}P]orthophosphate and ATP to give an intracellular
ratio of orthophosphate, ADP, and ATP of 15:1:2. The loaded ghosts
were then incubated in a Na^+-rich medium free of K^+. It was found that
the amount of ^{32}P label incorporated into ATP under the above conditions was significantly higher than in the control experiments in which
the ghosts were incubated in a similar medium with ouabain, or in a
medium high in K^+ but low in Na^+ (Garrahan and Glynn, 1966, 1967b).

The stoichiometric relationship between ATP hydrolysis and the translocations of Na^+ and K^+ was carefully studied by Sen and Post (1964),
who charged the red cells with ATP and Na^+ by a mild procedure of reversible hemolysis that permitted half of the cells to retain 95% of their
original hemoglobin while the other half was converted to ghosts. The
cells retaining their hemoglobin were separated for study. Taking into
account that high energy phosphates were generated from the catabolism
of ribose derived from AMP, these workers estimated that the extrusion
of three Na^+, coupled with the uptake of two K^+, costs the cell one high
energy phosphate. Similar values were arrived at by Gárdos (1964), by
Whittam and Ager (1965), and by Garrahan and Glynn (1967a) who also
showed that in a K^+-free medium, the $^{24}Na^+:^{23}Na^+$ exchange was not
accompanied by any ouabain-sensitive hydrolysis of ATP.

Data obtained from the red cells confirmed an earlier study by Leaf
and Renshaw (1957) who also arrived at a figure of one high-energy
phosphate consumed for every three Na^+ transported. Their conclusion
was based on the stimulation of Na^+ transport and oxygen consumption
by isolated frog skin under the influence of neurohypophyseal hormones.

A similar conclusion was drawn from studies on the increased oxygen consumption by the isolated urinary bladder of the toad, *Bufo marinus*, brought about by the introduction of Na^+ in the medium, and on the stimulation of both oxygen consumption and Na^+ transport by neurohypophyseal hormones (Leaf *et al.*, 1959; Leaf and Dempsey, 1960; Frazier *et al.*, 1962).

D. The $(Na^+ + K^+)$-Dependent ATPase

The discovery of a $(Na^+$ and $K^+)$-dependent ATPase in the leg nerves of the shore crab *Carcinus maenas* by Skou (1957) spurred intensive research efforts on the Na^+ and K^+ transport systems. The ATPase in the homogenate of the crab nerves was found to be associated with submicroscopic particles readily sedimentable at 20,000 g. The activity of the enzyme was strictly dependent upon Mg^{++} and was increased by Na^+. In the presence of both Mg^{++} and Na^+, the activity was increased by K^+ up to a certain concentration, after which progressive inhibition set in, presumably due to the interference of K^+ with the sites for Na^+ (Fig. 6). Hess and Pope (1957) found a Mg^{++}-dependent ATPase in rat brain which was also stimulated by Na^+ and K^+ in a synergistic manner. Similar activities were observed in numerous other tissues ranging from the cerebral cortex of the guinea pig (Deul and McIlwain, 1961) and the parotid gland of the dog (Schwartz *et al.*, 1963) to the salt gland of gulls

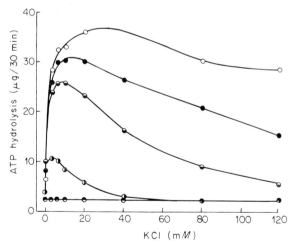

Fig. 6. The activity of ATPase from *Carcinus maenas* in relation to the concentration of K^+ in the presence of Mg^{++} (6 mM) and different concentrations of Na^+. Legend: ○, NaCl 40 mM; ●, NaCl 20 mM; ◐, NaCl 10 mM; ◑, NaCl 3 mM; ◒, 0 mM (Skou, 1957).

(Hokin, 1963), the toe muscle of the frog, and the urinary bladder of the toad (Bonting and Caravaggio, 1963). The electric organ of the eel *Electrophorus electricus* was found to be a good source for this enzyme because little of the ATPase activity in the tissue was independent of Na^+ and K^+ (Albers and Koval, 1962; Glynn, 1963). However, purification of the enzyme from this source proved to be difficult, apparently because of instability.

Post and his collaborators (1960) purified the ATPase from human red cells about 30-fold utilizing osmotic shock and Tris–glycylglycine treatments. Two types of enzyme activity were found, one dependent upon added Na^+ and K^+, and the other not. The concentration of Na^+ which gave half-maximal stimulation of the activity was 24 mM and that of K^+ was 3 mM. Transport of the ions by intact cells exhibited K_m's of about 20 mM for Na^+ and 2 mM for K^+. Thus, the *in vitro* and *in vivo* activities correspond closely. In both processes, NH_4^+ substituted for K^+ (but not for Na^+) with a K_m of 8 mM for the enzymatic reaction and 7–16 mM for transport. The ATPase activity was half-maximally inhibited by $10^{-7}\,M$ ouabain and the transport activity by $3–7 \times 10^{-8}\,M$ of the compound.

A number of procedures have been devised for the purification of the $(Na^+ + K^+)$-dependent ATPase activity. The elimination of the activity not dependent on these two cations was generally achieved in the crude preparation. Dispersion of the microsomal particles by sodium iodide and the solubilization of the enzyme complex by deoxycholate or Lubrol were commonly employed treatments (see, for example, Matsui and Schwartz, 1966; Post and Sen, 1967a; Schoner *et al.*, 1967; Medzihradsky *et al.*, 1967; Kahlenberg *et al.*, 1969; Jorgensen and Skou, 1969; Banerjee *et al.*, 1970).

The molecular weight of the ATPase from human erythrocytes and guinea pig cortex was found to be 2.5×10^5 by *in vacuo* radiation inactivation (Kepner and Macey, 1968). The molecular weights of the ATPase complex from guinea pig and beef brains were estimated to be 6.7×10^5 by gel chromatography on agarose (Medzihradsky *et al.*, 1967; Uesugi *et al.*, 1969). The molecular weight of the enzyme complex from pig brain was estimated to be 5×10^5 by radiation inactivation (not specified whether the operation was carried out *in vacuo*), Sepharose–deoxycholate column chromatography, and sedimentation coefficient in sucrose gradient (Mizuno *et al.*, 1968; Nakao *et al.*, 1969).

It should be kept in mind that the enzymes which are solubilized with detergent and partially purified by gel chromatography or salt precipitation may still be associated with fragments of cell membrane as indicated by their phospholipid contents. Molecular aggregation of the enzyme protein is also likely. Hence the true molecular weight of the ATPase may be considerably lower than 2.5×10^5 (Post, 1971; see also Kepner and Macey, 1968).

1. Inhibitors of $(Na^+ + K^+)$ ATPase

All $(Na^+ + K^+)$-activated ATPases studied are inhibitable by cardiac glycosides (Glynn, 1964; Skou, 1965). Dunham and Glynn (1961) observed that inhibition of erythrocyte ATPase by low concentrations of cardiac glycosides could be reversed by raising the concentration of K^+, just as the inhibition of transport by whole cells could be reversed by high external K^+ concentration. Recently hellebrigenin (a relative of strophanthidin) and its 3-haloacetates have been discovered to be highly potent inhibitors of brain $(Na^+ + K^+)$-activated ATPase because of high affinity (Ruoho et al., 1968).

Oligomycin, an inhibitor of oxidative phosphorylation, was observed also to inhibit the ATPases from the electric organ of E. electricus (Glynn, 1962a; Fahn et al., 1966a), the brain of rat (Järnefelt, 1962), the brain of rabbit (Jöbsis and Vremen, 1963), and calf heart muscle (Matsui and Schwartz, 1966). The concentration of the compound required for half-maximal inhibition of the $(Na^+ + K^+)$-dependent ATPase (rabbit brain), however, is more than an order of magnitude higher than that for mitochondrial ATPase (van Groningen and Slater, 1963). With the enzyme from the electric organ, it was shown that oligomycin inhibited ATP hydrolysis but slightly enhanced the Na^+-dependent ATP–ADP exchange (Fahn et al., 1966a). The antibiotic did not affect the phosphorylation of the enzyme by ATP (see Section V,D,2) but inhibited the breakdown of the phosphoenzyme (Whittam et al., 1964).

Sulfhydryl reagents are known to be inhibitors of the ATPases, but the sensitivity to inhibition appears to depend upon the source of the enzyme and the pH of the incubation medium. In general, PCMB is an effective inhibitor at $10^{-5} M$ and NEM at $5 \times 10^{-3} M$ (Schwartz and Laseter, 1964; Skou and Hilberg, 1965; Tanaka and Strickland, 1965; Fahn et al., 1966a,b; Matsui and Schwartz, 1966).

Hokin and Yoda (1965) found that the ATPase from beef kidney was inhibited by diisopropyl fluorophosphate. The inhibition was prevented by ATP, and the protective effect of ATP was antagonized by K^+. The effect of K^+ was in turn reversed by strophanthidin. Since it was known that cardiac glycosides inhibited the sodium pump only from outside the squid axon (Caldwell and Keynes, 1959), that ATP acted only from the inside (Caldwell et al., 1960), and that the inhibition of the transport in red cells by the glycosides could be reversed by high K^+ concentration, it was concluded that K^+ can influence the ATP site from the opposite side of the membrane, i.e., at the K^+ binding site for transport.

2. The Phosphorylated Enzyme Intermediate

Skou (1960) postulated that ATP reacts with the enzyme to form a phosphorylated enzyme intermediate on the basis of the observation that a preparation of crab nerve ATPase, devoid of adenylate kinase activity, catalyzed a [^{32}P]ADP and ATP exchange, but not a ^{32}P$_i$ and ATP exchange.

Post and co-workers (Post and Rosenthal, 1962; Charnock et al., 1963; Charnock and Post, 1963; Post et al., 1965; Post and Sen, 1967b) found that when [γ-^{32}P]ATP was incubated with guinea pig kidney ATPase, a ^{32}P-labeled intermediate was formed which could be precipitated by trichloroacetic acid. This intermediate was considered to be most likely a protein, since most of the label was not extractable with solvents for phospholipids. The presence of Na$^+$ alone increased the amount of the labeled material, and the presence of K$^+$ alone produced the opposite effect. Both ions together accelerated the hydrolysis rate of ATP, but reduced the amount of incorporation of the label into the intermediate. Ouabain, which had no effect upon the formation of this intermediate, abolished the effect of K$^+$. As in the case observed earlier with brain microsomal ATPase by Järnefelt (1961), the kidney enzyme preparation also bound Na$^+$ in the presence of ATP.* Accordingly, the following reaction scheme was proposed (Charnock and Post, 1963):

$$\text{Enzyme} + \text{ATP} + x\text{Na}^+ \longrightarrow [\text{enzyme-P}]\text{Na}_x^+ + \text{ADP} \qquad (8)$$

$$[\text{Enzyme-P}]\text{Na}_x^+ + y\text{K}^+ \xrightarrow[\text{ouabain}]{} \text{enzyme} + \text{P}_i + x\text{Na}^+ + y\text{K}^+ \qquad (9)$$

Similar findings were reported for the enzymes from the electric organ of *E. electricus* (Albers et al., 1963; Siegel and Albers, 1967), the brain of guinea pig (Rodnight and Lavin, 1964), the brain of rat (Ahmed and Judah, 1965; Nagano et al., 1965), the brain of calf (Lindenmayer et al., 1968), and human erythrocytes (Blostein, 1970). Fahn et al. (1968) noted that N-ethylmaleimide or oligomycin did not prevent the Na$^+$-stimulated phosphorylation (reaction 8) of the enzyme from the electric organ but did abolish the effect of K$^+$ in reducing the amount of the phosphorylated intermediate (reaction 9). Stahl (1968) reported that the ATP–ADP exchange reaction catalyzed by ATPase preparation from rat brain was stimulated by oligomycin and inhibited by ouabain. Blostein (1970) found that oligomycin stimulated the ATP–ADP exchange cata-

---

* The amount of Na$^+$ bound seemed in retrospect too large to represent cation–carrier complexes. In view of the recent report on the existence of inverted red cell membrane vesicles by Steck et al. (1970), it is probable that some of the vesicles of the kidney preparation were turned inside out and that these vesicles accumulated Na$^+$ as a product of the reaction (Post, 1971).

lyzed by the human erythrocyte enzyme without affecting the level of the phosphorylated intermediate.

Hokin and co-workers (1965; Kahlenberg et al., 1968; Hokin, 1969) characterized the phosphorylated intermediate further by showing that pretreatment of a guinea pig brain preparation with diisopropyl fluorophosphate, which inhibited ATPase activity irreversibly, markedly reduced the formation of the ^{32}P-labeled intermediate. The phosphorylated intermediate was found to be more labile to alkali than to acid and liberated [^{32}P]orthophosphate after treatment with hydroxylamine or acyl phosphatase. Identification of the side chain of the polypeptide which participated in the phosphorylation reaction was performed on a preparation in which a large fraction of the enzyme was expected to be in the phosphorylated form (incubation with ATP in the presence of Na$^+$, but not K$^+$) and compared with another preparation in which only a small fraction of the enzyme was expected to be in the phosphorylated form as the control (incubation with ATP in the presence of both Na$^+$ and K$^+$). Each enzyme preparation was then precipitated with perchloric acid and hydrolyzed with pepsin, and the peptides were treated in parallel with [2,3-^{3}H]N-(n-propyl)hydroxylamine to replace the phosphoryl groups. A peptide that was more heavily labeled when derived from the first preparation was separated and subjected to digestion by pronase. The labeled material was found to cochromatograph with carrier L-glutamyl-γ-N-(n-propyl)hydroxamate. Nagano and co-workers (1965) independently found that the [^{32}P]orthophosphate was released by hydroxylaminolysis from the ^{32}P-labeled intermediate in an ATPase preparation from rabbit brain. The molecular weight of the phosphorylated protein in dodecylsulfate was found to be considerably smaller than that of the solubilized enzyme preparations discussed earlier (Nakao et al., 1969; Uesugi et al. 1971).

The role of the phosphorylated protein as an integral part of ATPase was strengthened by the finding of Bader and co-workers (1968) that among many different tissues examined, the variation of the amount of the phosphorylated intermediate relative to the activity of the enzyme remained within a factor of two, even though the specific activity of the enzyme varied over 400-fold.

A number of observations showed that the function of ATPase depends upon its proper association with phospholipids. Skou (1961) reported that treatment of the crab nerve ATPase preparation with phospholipase A caused the disappearance of the (Na$^+$ + K$^+$)-dependent activity, but slightly increased the activity observed with Mg^{++} alone. Similar observations were made by Portius and Repke (1963) on the enzyme from heart muscle. Phospholipase C (lecithinase from *Clos-*

tridium welchii) treatment inactivated the enzyme from red cell ghosts (Schatzmann, 1962). With a variety of tissues, it has been shown that the solubilization of the enzyme by deoxycholate treatment during purification led to a loss of ($Na^+ + K^+$)-dependent activity paralleled by a loss of phospholipid content. Addition of a number of phospholipids partially restored the lost activity, but lecithin and phosphatidylserine were found to be most effective, permitting up to tenfold increase of the ouabain-sensitive hydrolytic activity. Phospholipids had no effect on either the ATP–ADP exchange reaction or the labeling of the proteins by [γ-^{32}P]ATP (Tanaka and Abood, 1964; Tanaka and Strickland, 1965; Fenster and Copenhaver, 1967; Post and Sen, 1967a; Tanaka, 1969; Wheeler and Whittam, 1970). It is interesting that during a process of purification of the enzyme from beef brain microsomes, the specific activity of ATPase increased over 30-fold while the phospholipid/protein ratio remained essentially constant (Uesugi *et al.*, 1971).

E. Allosteric Interactions

The requirement that Na^+ and K^+ interact with the transport system from opposite sides of the membrane and the likelihood that more than one site exists for Na^+ and K^+ promise to make the transport ATPase a challenging candidate for the study of allosterism. A number of kinetic investigations have been published, and the results may be briefly summarized as follows. (a) The data are compatible with the existence of homotropic cooperative effects among the Na^+ sites and among the K^+ sites and heterotropic effects between these two kinds of sites. (b) Both the Na^+ and the K^+ sites probably interact heterotropically with the ATP site. (c) The binding of cardiac glycosides is influenced by Na^+ and K^+ in a manner depending upon the presence or absence of ATP and Mg^{++} (Green and Taylor, 1964; Squires, 1965; Robinson, 1967, 1968; Schwartz *et al.*, 1968; Priestland and Whittam, 1968; Garrahan, 1969; Tobin and Sen, 1970; Tobin *et al.*, 1970). The effect of Na^+ on the binding of cardiac glycosides probably constitutes the strongest evidence for allosteric interactions, although it is not clear whether the binding site for cardiac glycosides plays a real role in the normal physiological function of the enzyme. Conformational changes affecting the active center of the enzyme based on its behavior toward inhibitors have recently been discussed by Post and associates (1969) and by Skou (1969). Recent studies of DeWeer (1970a,b) which showed that Na^+ efflux from the squid axon could be stimulated by either external K^+ or Na^+, depending on the ratio of ATP/ADP in the axoplasm, also point to allosteric effects. The availability of homogeneous enzyme preparations, more detailed knowl-

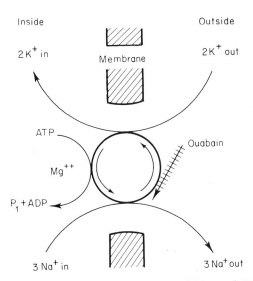

Fig. 7. A model of the coupled transport system of Na+ and K+ (Post, 1968).

edge of the partial reactions, and the establishment of the number of each kind of binding sites on the enzyme seem important for further advances in kinetic studies.

A schematic representation of the sodium pump is given in Fig. 7. The exchange of 3 Na+ for 2 K+ would explain why the operation of this pump builds up an electrical potential across the cell membrane, i.e., is electrogenic. The change in potential in turn would cause the movement of other ions through other pathways in the membrane, resulting in an overall electroneutrality. The coupled transport of Na+ and K+ also accounts for the current produced during the transmission of nerve impulses. Studies of this topic have been made by Rang and Ritchie (1968) and by Thomas (1969).

VI. Na+-DEPENDENT TRANSPORT OF AMINO ACIDS AND SUGARS

The transport of a large number of compounds, mostly amino acids and sugars, depends on the presence of Na+ in the medium. Christensen et al. (1952) observed that the uptake of glycine or alanine by duck red cells decreased when Na+ in the medium was replaced by K+. Specific dependence on Na+ has since been observed in the active transport of tyrosine and phenylalanine by the small intestine of *Rana pipiens* (Csáky, 1961), the transport of alanine into nuclei of calf thymus (Allfrey et al., 1961), the accumulation of glycine by Ehrlich ascites tumor cells (Heinz, 1962; Kromphardt *et al.*, 1963), and the uptake of amino acids by bacteria (Drapeau and MacLeod, 1963; Drapeau *et al.*, 1966; Stevenson, 1966; Frank and Hopkins, 1969; Wong *et al.*, 1969). The apparent K_m for glycine transport into pigeon red cells was found to be lowered as the concentration of Na^+ in the medium was increased (Vidaver, 1964a). A similar effect was found in the uptake of alanine by rabbit red cells (Wheeler and Christensen, 1967a) and rabbit intestinal mucosal cells (Curran *et al.*, 1967) and in the uptake of α-amino-isobutyrate by a marine pseudomonad (Wong *et al.*, 1969). A number of experiments indicated that Na^+ crossed the cell membrane concomitantly with the amino acids (Vidaver, 1964c; Esposito *et al.*, 1964; Schultz and Zalusky, 1964, 1965; Eddy *et al.*, 1967; Wheeler and Christensen, 1967b).

The dependence on Na^+ for the transport of sugars was observed in the intestine of guinea pig (Riklis and Quastel, 1958), rat (Csáky and Zolli-coffer, 1960), and hamster (Bihler and Crane, 1962). The transport of sugars, including the nonmetabolizable 3-O-methylglucose, into mucosal cells of rabbit ileum was associated with a proportional increment in the influx of Na^+ (Capraro *et al.*, 1963; Schultz and Zalusky, 1963, 1964; Goldner *et al.*, 1969).

Crane and co-workers (1961; Crane, 1967; 1968a,b) proposed that Na^+ forms a ternary complex with the carrier–sugar. Since the concentration of Na^+ is usually much higher in the surrounding medium than in the cells, the complex tends to be formed externally and dissociated internally, resulting in the influx of both the sugar and Na^+. The movement of the sugar against a concentration gradient is thus driven by the electrochemical gradient of Na^+. Similar models were proposed for the active transport of amino acids (Riggs *et al.*, 1958; Vidaver, 1964b,c; Curran, 1965; Kipnis and Parrish, 1965; Eddy, 1968a,b; Inui and Christensen, 1966; Christensen, 1970). Such models would account for the inhibitability of both amino acid and sugar transport by ouabain (Csáky *et al.*, 1961; Csáky, 1963; Crane *et al.*, 1961; Rosenberg *et al.*, 1965; Schultz *et al.*, 1966), the reverse flow of glycine out of pigeon red cells against its own concentration gradient when the external concentration of Na^+ was made lower than the internal concentration (Vidaver, 1964b), and the reverse uphill flow of 6-deoxyglucose out of intestinal mucosal cells in a similar situation (Crane, 1964).

Curran (1968) and Goldner and co-workers (1969) pointed out that the Na^+-dependent transport for sugars and amino acids differed in that the deprivation of Na^+ reduced the flux in the former case and increased the K_m in the latter. They suggested that the ternary sugar–Na^+–carrier complex traverses the membrane much more readily than the binary sugar–carrier complex. In contrast, the ternary Na^+–amino acid–carrier complex traverses the membrane at about the same rate as the binary amino acid–carrier complex. Thus, for sugar transport, the lack of Na^+ is tantamount to the presence of a noncompetitive inhibitor, whereas for amino acid transport, the lack of Na^+ is equivalent to the presence of a competitive inhibitor. Crane, however, observed that the absence of Na^+ did cause the apparent K_m of the sugar to increase (Crane et al., 1965; Lyon and Crane, 1966). The discrepancy of the results may be attributable to the fact that Crane and co-workers measured sugar movement across the full thickness of everted intestinal sacs or the concentration of the substrate into whole intestinal segments, processes which depended on events at both the mucosal and serosal sides of the cell, whereas Curran and co-workers carried out direct measurements of unidirectional fluxes across the brush border from the incubation medium to the mucosal epithelial cells.

Figure 8 presents a schematic representation of the Na^+-dependent transport system for amino acids. A similar picture would hold for the sugar transport system. The validity of the sodium gradient model, however, has been questioned by Kimmich (1970), who was able to demonstrate the accumulation of galactose against a concentration gradient in isolated cells of chicken small intestine even when the intracellular concentration of Na^+ was higher than that of the medium. He suggested that a common energized form of the $(Na^+ + K^+)$-dependent ATPase is required for the active transport of both the monovalent cations and the

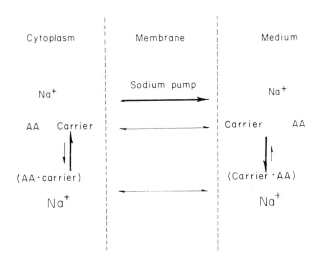

Fig. 8. A model of the Na^+-linked amino acid transport system.

sugars (and amino acids), which would account for the fact that ouabain and oligomycin inhibited both kinds of transport processes.

VII. DISCUSSION

Even from the few systems reviewed here, it is evident that a rich variety of mechanisms has evolved for the transfer of substances across cell membranes. The establishment of a particular mechanism for the transport of a substrate can often be understood on a functional basis, although in most cases it would have been difficult to predict. Thus the desirability of facilitated diffusion for the economical transport of glucose into erythrocytes and other internal tissues of vertebrates can be appreciated in the light of the high concentrations of this sugar that exist in the plasma. However, it cannot be generalized that this mechanism is always employed in situations of substrate abundance; effective scavenging can be achieved with this process if its role is coupled with that of an intracellular enzyme which has an unusually high affinity for the substrate and which catalyzes its conversion to a metabolite that cannot readily escape from the cell. Glycerol is trapped as L-α-glycerophosphate (L-α-GP) in this manner by bacterial cells through the combined action of facilitated diffusion and an ATP-dependent kinase which has a K_m of 10^{-6} M for glycerol (Hayashi and Lin, 1965; Sanno et al., 1968).

Economy of operation is also a prominent feature of the PEP-phosphotransferase system, since no additional expenditure of energy is necessary beyond what is required for the phosphorylation of the substrate, a cost which is unavoidable at some step in the pathway for dissimilation. In this respect, it is analogous to the glycerol kinase system just discussed. The energy saved in the transport reaction, however, is probably offset partially by the cost of the production of the more complex transport machinery, since in view of the virtual irreversibility of the phosphorylation step one would expect an elegant mechanism for kinetic feedback control in addition to the several proteins which mediate the several transphorylation reactions. In the analogous case of the trapping of L-α-GP by glycerol kinase, it is known that metabolic flooding is prevented by the noncompetitive inhibition by fructose diphosphate (Zwaig and Lin, 1966b). It is interesting to note that the PEP-dependent system has been found in facultative anaerobes (presumably also in obligatory anaerobes), but not in strict aerobes in which PEP is not a principal product of energy metabolism (Phibbs and Eagon, 1970; Romano et al., 1970).

The advantage of a simple active transport, such as that for lactose in *E. coli*, is not obvious. Perhaps a broader specificity, and therefore a more versatile role, is permitted by the mere requirement for physical binding as a condition for such active transport, in contrast to binding plus activation for a chemical reaction as is required in phosphorylating transport. The reversibility of active transport systems may also be an advantage which counterbalances the higher cost of operation. In this connection, it is noteworthy that gram-negative bacteria, like *E. coli*, take up lactose by simple active transport whereas gram-positive bacteria, like *S. aureus*, take up the disaccharide by phosphorylating transport. If the hypothesis about the advantages of broad specificity is correct, then one should find that *E. coli* cells can utilize a variety of β-galactosides which are nutritionally inert to *S. aureus*. Although the specificities of the catalytic proteins and the induction mechanisms of the β-galactoside systems have been extensively studied, not much information has been collected on the specificities of growth.

If it is true that the Na^+-linked transport of sugars and amino acids in higher organisms is driven by the Na^+ gradient, the advantage of the mechanism would again be one of economy, since each ATP is able to drive the transport of 3 Na^+. It might be illuminating to see if the Na^+-linked transport of organic molecules is exclusively and always the method of choice for the absorption of sugars and amino acids in organisms with the $(Na^+ + K^+)$-dependent ATPase.

The Na^+–K^+ exchange transport itself is strikingly efficient. The expenditure of one high-energy phosphate bond results in the movement of at least five ions: 3 Na^+ and 2 K^+. The evolutionary emergence of this transport process not only provided a way to control the specific concentrations of Na^+ and K^+ in the cytoplasm, but also, and more importantly, endowed the cell with a means of regulating its volume through adjustment of the internal osmotic pressure (Wilson, 1954; Leaf, 1956), thus freeing the cell from its dependence on a rigid wall structure for the prevention of osmotic lysis.* Once the problem of osmoregulation had been solved by this dynamic membrane mechanism which involves the movement of ions, the foundation was laid for the evolution of an excitable cell (Tosteson, 1964).

---

* In this context it is probably significant that $(Na^+ + K^+)$-dependent ATPase has not been reported for plant cells which are protected by walls. An ouabain-sensitive and K^+-activated ATPase component has been observed in crude homogenates of *E. coli* cells but the activity of this component was less than one tenth of the total ATPase activity (Hafkenscheid and Bonting, 1968). Further substantiation of the true nature of the ouabain-sensitive activity seems necessary.

A class of mechanisms which has not yet received wide attention involves the secretion of a binding agent which is subsequently reabsorbed as a complex with the nutrient or growth factor. For example, various phenolic compounds released into the medium by different species of bacteria are recaptured as iron chelates (Neilands, 1961; Wang and Newton, 1969). A more elaborate adsorptive process has evolved in higher animals for the uptake of vitamin B_{12}. In this case, a protein intrinsic factor, which binds the growth factor, is produced and extruded by certain cells in the stomach and recovered as a complex with the vitamin by the epithelial cells of the small intestine (see Wilson, 1962). Processes of this nature, which may be viewed as *sequestering transport*, are perhaps employed by organisms for securing nutrients that are available only at low concentrations, either because of their low solubility (as in the instance of ferrous or ferric ions) or because of scarcity (as in the case of vitamin B_{12}). Numerous additional mechanisms will undoubtedly be uncovered as details of the as yet obscure means of transfer of proteins, nucleic acids, and other macromolecules become better understood. The entry of DNA into bacterial cells in transformation, the absorption of antibodies by the intestinal cells of newborn animals, and the penetration of steroid and protein hormones into target cells are only a few of the specific examples that readily come to mind.

With respect to detailed molecular mechanisms of transport, the characterization of component units has already been well launched; however the exact role of each component and the interactions of the implicated proteins with the lipid matrix remain difficult challenges. It is encouraging in this regard that investigators have succeeded in introducing solubilized ATPase from *Streptococcus faecalis* into artificial lipid bilayer membranes (Meuller *et al.*, 1962; Hanai *et al.*, 1964; Redwood *et al.*, 1969) and that it has been possible to study changes in the electrical conductance of the bilayer upon the introduction of the enzyme and the modulation of this effect by Mg^{++}, Na^+, K^+, and ATP.

Genetic approaches to the analysis of specific protein and lipid interactions have also been opened up. Mutants of *E. coli* with altered membrane structure have been isolated as clones that grow on lactose and maltose, despite the deletion of the respective specific permeation systems (Ricard *et al.*, 1970). Mutants lacking certain membrane proteins have been isolated as clones that are resistant to colicins (Onodera *et al.*, 1970). The manipulation of lipid composition through the use of fatty acid or glycerol auxotrophs has also been achieved (Fox, 1969). These genetic approaches should effectively supplement the growing number of biochemical and biophysical methods for studies of membrane function.

Acknowledgments

The author is indebted to Joseph B. Alpers, H. R. Kaback, Lowell E. Hokin, Eugene P. Kennedy, Lawrence Rothfield, Thomas H. Wilson and Henry C. P. Wu for helpful comments, and to Sarah Monosson for assistance in the preparation of the manuscript. Work in the laboratory of the author is supported by the National Science Foundation Grant GB-8509 and the U. S. Public Health Service Grants GM11983 and 3 KO3 GM17925.

References

Ahmed, K., and Judah, J. D. (1965). *Biochim. Biophys. Acta* **104**, 112.
Albers, R. W. (1967). *Annu. Rev. Biochem.* **36**, 727.
Albers, R. W., and Koval, G. J. (1962). *Life Sci.* **5**, 219.
Albers, R. W., Fahn, S., and Koval, G. J. (1963). *Proc. Nat. Acad. Sci. U. S.* **50**, 474.
Allfrey, V. G., Meudt, R., Hopkins, J. W., and Mirsky, A. E. (1961). *Proc. Nat. Acad. Sci. U. S.* **47**, 907.
Alpers, D. H., Appel, S. H., and Tomkins, G. M. (1965). *J. Biol. Chem.* **240**, 10.
Ames, G. F., and Lever, J. (1970). *Proc. Nat. Acad. Sci. U. S.* **66**, 1096.
Anderson, B. (1968). Ph.D. Thesis. University of Michigan, Ann Arbor, Michigan.
Anraku, Y. (1967). *J. Biol. Chem.* **242**, 793.
Anraku, Y. (1968a). *J. Biol. Chem.* **243**, 3116.
Anraku, Y. (1968b). *J. Biol. Chem.* **243**, 3123.
Anraku, Y. (1968c). *J. Biol. Chem.* **243**, 3128.
Bader, H., Post, R. L., and Bond, G. H. (1968). *Biochim. Biophys. Acta* **150**, 41.
Banerjee, S. P., Dwosh, I. L., Khanna, V. K., and Sen, A. K. (1970). *Biochim. Biophys. Acta* **211**, 345.
Barnes, E. M., Jr., and Kaback, H. R. (1970). *Proc. Nat. Acad. Sci. U. S.* **66**, 1090.
Bárány, E., and Sperber, E. (1939). *Skand. Arch. Physiol.* **81**, 290.
Bennett, R. L., and Malamy, M. H. (1970). *Biochem. Biophys. Res. Common.* **40**, 496.
Bihler, I., and Crane, R. K. (1962). *Biochim. Biophys. Acta* **59**, 78.
Blostein, R. (1970). *J. Biol. Chem.* **245**, 270.
Boniface, J., and Koch, A. L. (1967). *Biochim. Biophys. Acta* **135**, 756.
Bonting, S. L., and Caravaggio, L. L. (1963). *Arch. Biochem. Biophys.* **101**, 37.
Boos, W. (1969). *Eur. J. Biochem.* **10**, 66.
Boos, W., and Sarvas, M. (1970). *Eur. J. Biochem.* **13**, 526.
Brinley, F. J., Jr., and Mullins, L. J. (1968). *J. Gen. Physiol.* **52**, 181.
Caldwell, P. C., and Keynes, R. D. (1959). *J. Physiol.* **148**, 8P.
Caldwell, P. C., Hodgkin, A. L., Keynes, R. D., and Shaw, T. I. (1960). *J. Physiol.* **152**, 561.
Capraro, Di V., Bianchi, A., and Lippe, C. (1963). *Arch. Sci. Biol. Bologna* **47**, 238.
Carter, J. R., Fox, C. F., and Kennedy, E. P. (1968). *Proc. Nat. Acad. Sci. U. S.* **60**, 725.
Charnock, J. S., and Post, R. L. (1963). *Nature (London)* **199**, 910.
Charnock, J. S., Rosenthal, A. S., and Post, R. L. (1963). *Aust. J. Exp. Biol. Med. Sci.* **41**, 675.

334

E. C. C. LIN

Christensen, H. N. (1969). *Advan. Enzymol.* **32**, 1.
Christensen, H. N. (1970). In "Membrane Metabolism and Ion Transport" (E. E. Bittar, ed.), Vol. I, pp. 393–422. Wiley (Interscience), New York.
Christensen, H. N., Riggs, T. R., and Ray, N. E. (1952). *J. Biol. Chem.* **194**, 41.
Cohen, G. N., and Rickenberg, H. V. (1955). *C. R. Acad. Sci.* **240**, 466.
Crane, R. K. (1964). *Biochem. Biophys. Res. Commun.* **17**, 481.
Crane, R. K. (1967). In "Protides of the Biological Fluids: Proceedings of the 15th Colloquium" (H. Peeters, ed.), p. 227. Elsevier, New York.
Crane, R. K. (1968a). In "Carbohydrate Metabolism and Its Disorders" (F. Dickens, W. S. Whelan, and P. J. Randle, eds.), Vol. I, pp. 25–51. Academic Press, New York.
Crane, R. K. (1968b). In "Handbook of Physiology" (C. F. Code, section ed.), Section 6, Vol. III, pp. 1323–1351. American Physiological Society, Washington, D.C.
Crane, R. K., Miller, D., and Bihler, I. (1961). In "Membrane Transport and Metabolism" (A. Kleinzeller and A. Kotyk, eds.), pp. 439–449. Academic Press, New York.
Crane, R. K., Forstner, G., and Eicholz, A. (1965). *Biochim. Biophys. Acta* **109**, 467.
Creaser, E. H. (1955). *J. Gen. Microbiol.* **12**, 288.
Csáky, T. Z. (1961). *Amer. J. Physiol.* **201**, 999.
Csáky, T. Z. (1963). *Biochim. Biophys. Acta* **74**, 160.
Csáky, T. Z., and Zollicoffer, L. (1960). *Amer. J. Physiol.* **198**, 1056.
Csáky, T. Z., Hartzog, G., III, and Fernald, G. W. (1961). *Amer. J. Physiol.* **200**, 459.
Curran, P. F. (1965). *Fed. Proc. Fed. Amer. Soc. Exp. Biol.* **24**, 993.
Curran, P. F. (1968). *Physiologist* **11**, 3.
Curran, P. F., Schultz, S. G., Chez, R. A., Fuisz, R. E. (1967). *J. Gen. Physiol.* **56**, 1261.
Danielli, J. F. (1954). In "Recent Developments in Cell Physiology" (J. A. Kitching, ed.), pp. 1–14. Academic Press, New York.
Deere, C. J., Dulaney, A. D., and Michelson, I. D. (1939). *J. Bacteriol.* **37**, 355.
Deul, D. H., and McIlwain, H. (1961). *J. Neurochem.* **8**, 246.
DeWeer, P. (1970a). *J. Gen. Physiol.* **50**, 583.
DeWeer, P. (1970b). *Nature (London)* **226**, 1251.
Drapeau, G. R., and MacLeod, R. A. (1963). *Biochem. Biophys. Res. Commun.* **12**, 111.
Drapeau, G. R., Matula, T. I., and MacLeod, R. A. (1966). *J. Bacteriol.* **92**, 63.
Dreyfuss, J. (1964). *J. Biol. Chem.* **239**, 2292.
Dreyfuss, J., and Monty, K. J. (1963). *J. Biol. Chem.* **238**, 1019.
Dreyfuss, J., and Pardee, A. B. (1965). *Biochim. Biophys. Acta* **104**, 308.
Dreyfuss, J., and Pardee, A. B. (1966). *J. Bacteriol.* **91**, 2275.
Dunham, E. T., and Glynn, I. M. (1961). *J. Physiol.* **156**, 274.
Eddy, A. A. (1968a). *Biochem. J.* **108**, 195.
Eddy, A. A. (1968b). *Biochem. J.* **108**, 489.
Eddy, A. A., Mulcahy, M. F., and Thomson, P. J. (1967). *Biochem. J.* **103**, 863.
Egan, J. B., and Morse, M. L. (1966). *Biochim. Biophys. Acta* **112**, 63.
Englesberg, E., Watson, J. A., and Hoffee, P. A. (1961). *Cold Spring Harbor Symp. Quant. Biol.* **26**, 261.
Esposito, G., Faelli, A., and Capraro, V. (1964). *Experentia* **20**, 122.
Fahn, S., Koval, G. J., and Albers, R. W. (1966a). *J. Biol. Chem.* **241**, 1882.

Fahn, S., Hurley, M. R., Koval, G. J., and Albers, R. W. (1966b). *J. Biol. Chem.* **241**, 1890.
Fahn, S., Koval, G. J., and Albers, R. W. (1968). *J. Biol. Chem.* **243**, 1993.
Fenster, L. J., and Copenhaver, J. H., Jr. (1967). *Biochim. Biophys. Acta* **137**, 406.
Fox, C. F. (1969). *Proc. Nat. Acad. Sci. U. S.* **63**, 850.
Fox, C. F., and Kennedy, E. P. (1965). *Proc. Nat. Acad. Sci. U. S.* **54**, 891.
Fox, C. F., Carter, J. R., and Kennedy, E. P. (1967). *Proc. Nat. Acad. Sci. U. S.* **57**, 698.
Frank, L., and Hopkins, I. (1969). *J. Bacteriol.* **100**, 329.
Frazier, H. S., Dempsey, E. F., and Leaf, A. (1962). *J. Gen. Physiol.* **45**, 529.
Furlong, C. E., and Weiner, J. H. (1970). *Biochem. Biophys. Res. Commun.* **38**, 1076.
Gárdos, G. (1954). *Acta Physiol. Acad. Sci. Hung.* **6**, 191.
Gárdos, G. (1964). *Experientia* **20**, 387.
Garrahan, P. J. (1969). *Nature (London)* **222**, 1000.
Garrahan, P. J., and Glynn, I. M. (1965). *Nature (London)* **207**, 1098.
Garrahan, P. J., and Glynn, I. M. (1966). *Nature (London)* **211**, 1414.
Garrahan, P. J., and Glynn, I. M. (1967a). *J. Physiol.* **192**, 217.
Garrahan, P. J., and Glynn, I. M. (1967b). *J. Physiol.* **192**, 237.
Glynn, I. M. (1956). *J. Physiol.* **134**, 278.
Glynn, I. M. (1957). *Progr. Biophys. Biophys. Chem.* **8**, 242.
Glynn, I. M. (1962a). *Biochem. J.* **84**, 75P.
Glynn, I. M. (1962b). *J. Physiol.* **160**, 18P.
Glynn, I. M. (1963). *J. Physiol.* **169**, 452.
Glynn, I. M. (1964). *Pharmacol. Rev.* **16**, 381.
Goldner, A. M., Schultz, S. G., and Curran, P. F. (1969). *J. Gen. Physiol.* **53**, 362.
Green, A. L., and Taylor, C. B. (1964). *Biochem. Biophys. Res. Commun.* **14**, 118.
Hafkenscheid, J. C. M., and Bonting, S. L. (1968). *Biochim. Biophys. Acta* **151**, 204.
Hagihira, H., Wilson, T. H., and Lin, E. C. C. (1963). *Biochim. Biophys. Acta* **78**, 505.
Halpern, Y. S., and Lupo, M. (1966). *Biochim. Biophys. Acta* **126**, 163.
Hanai, T., Haydon, D. A., and Taylor, J. (1964). *Proc. Roy. Soc. London, A* **281**, 377.
Hanson, T. E., and Anderson, R. L. (1968). *Proc. Nat. Acad. Sci. U. S.* **61**, 269.
Harris, E. J. (1954). *In* "Active Transport and Secretion: Symposium No. 8, Society for Experimental Biology," pp. 228–241. Cambridge Univ. Press, London.
Hayashi, S., and Lin, E. C. C. (1965). *Biochim. Biophys. Acta* **94**, 479.
Hazelbauer, G. L., and Adler, J. (1971). *Nature (London)* **230**, 101.
Heinz, E. (1962). *In* "Amino Acid Pools" (J. T. Holden, ed.), pp. 539–544. Elsevier, New York.
Heinz, E., and Walsh, P. M. (1958). *J. Biol. Chem.* **233**, 1488.
Hengstenberg, W., Egan, J. B., and Morse, M. L. (1967). *Proc. Nat. Acad. Sci. U. S.* **58**, 274.
Heppel, L. A. (1967). *Science* **156**, 1451.
Heppel, L. A. (1969). *J. Gen. Physiol.* **54**, 95S.
Herzenberg, L. A. (1961). *Arch. Biochem. Biophys.* **93**, 314.
Hess, H. H., and Pope, A. (1957). *Fed. Proc. Fed. Amer. Soc. Exp. Biol.* **16**, 196.
Hoffee, P., and Englesberg, E. (1962). *Proc. Nat. Acad. Sci. U. S.* **48**, 1759.
Hoffee, P., Englesberg, E., and Lamy, F. (1964). *Biochim. Biophys. Acta* **79**, 337.
Hogg, R. W., and Englesberg, E. (1969). *J. Bacteriol.* **100**, 423.

336   E. C. C. LIN

Hokin, L. E. (1969). *J. Gen. Physiol.* **54**, 327S.
Hokin, L. E., and Yoda, A. (1965). *Biochim. Biophys. Acta* **97**, 594.
Hokin, L. E., Sastry, P. S., Galsworthy, P. R., and Yoda, A. (1965). *Proc. Nat. Acad. Sci. U. S.* **54**, 177.
Hokin, M. R. (1963). *Biochim. Biophys. Acta* **77**, 108.
Inui, Y., and Christensen, H. N. (1966). *J. Gen. Physiol.* **50**, 203.
Ito, I., and Neilands, J. B. (1958). *J. Amer. Chem. Soc.* **80**, 4645.
Jacobs, M. H. (1950). *Ann. N. Y. Acad. Sci.* **50**, 824.
Jacobs, M. H. (1954). *Biol. Bull.* **107**, 314.
Järnefelt, J. (1961). *Biochem. Biophys. Res. Commun.* **6**, 285.
Järnefelt, J. (1962). *Biochim. Biophys. Acta* **59**, 643.
Jöbsis, F. F., and Vreman, H. J. (1963). *Biochim. Biophys. Acta* **73**, 346.
Jones, T. H. D., and Kennedy, E. P. (1969). *J. Biol. Chem.* **244**, 5981.
Jorgensen, P. L., and Skou, J. C. (1969). *Biochem. Biophys. Res. Commun.* **37**, 39.
Kaback, H. R. (1971). Personal communication.
Kaback, H. R. (1968). *J. Biol. Chem.* **243**, 3711.
Kaback, H. R. (1969a). *Proc. Nat. Acad. Sci. U. S.* **63**, 724.
Kaback, H. R. (1969b). *In* "The Molecular Basis of Membrane Function" (D. C. Tosteson, ed.), pp. 421–444. Prentice-Hall, Englewood Cliffs, New Jersey.
Kaback, H. R. (1970a). *Annu. Rev. Biochem.* 561.
Kaback, H. R. (1970b). *In* "Current Topics in Membranes and Transport" (A. Kleinzeller and F. Bronner, eds.), Vol. 1. Academic Press, New York.
Kaback, H. R., Stadtman, E. R. (1966). *Proc. Nat. Acad. Sci. U. S.* **55**, 920.
Kahlenberg, A., Galsworthy, P. R., and Hokin, L. E. (1968). *Arch. Biochem. Biophys.* **126**, 331.
Kahlenberg, A., Dulak, N. C., Dixon, J. F., Galsworthy, P. R., and Hokin, L. E (1969). *Arch. Biochem. Biophys.* **131**, 253.
Kalckar, H. M. (1971). Jean Weigle Memorial Lecture, Calif. Inst. Technol., in preparation.
Kennedy, E. P. (1969). *J. Gen. Physiol.* **54**, 91S.
Kennedy, E. P. (1970). *In* "The Lac Operon" (J. Beckwith and D. Zipser, eds.), p. 49. Cold Spring Harbor, New York.
Kennedy, E. P., and Scarborough, G. A. (1967). *Proc. Nat. Acad. Sci. U. S.* **58**, 225.
Kennedy, E. P., Fox, C. F., and Carter, J. R. (1966). *J. Gen. Physiol.* **49**, 347.
Kepes, A. (1960). *Biochim. Biophys. Acta* **40**, 70.
Kepes, A. (1962a). *In* "Conference on Permeability, Wageningen," p. 97. N. V. Uitgevers-Maatschappij, Tjeenk Willink-Zwolle, Netherlands.
Kepes, A. (1962b). *In* "Recent Progress in Microbiology: 8th International Congress for Microbiology, Montreal" (N. E. Gibbons, ed.), p. 38. Univ. of Toronto Press, Toronto.
Kepes, A. (1964). *In* "The Cellular Functions of Membrane Transport" (J. F. Hoffman, ed.), pp. 155–169. Prentice-Hall, Englewood Cliffs, New Jersey.
Kepes, A. (1969). *In* "The Molecular Basis of Membrane Function" (D. C. Tosteson, ed.), pp. 353–389. Prentice-Hall, Englewood Cliffs, New Jersey.
Kepes, A., and Cohen, G. N. (1962). *In* "The Bacteria" (I. C. Gunsalue and R. Y. Stanier, eds.), Vol. IV, pp. 179–221. Academic Press, New York.
Kepes, A., and Monod, J. (1957). *C. R. Acad. Sci.* **244**, 809.
Kepner, G. R., and Macey, R. I. (1968). *Biochem. Biophys. Res. Commun.* **30**, 582.
Kepner, G. R., and Macey, R. I. (1969). *Biochim. Biophys. Acta* **183**, 241.
Kessler, D. P., and Rickenberg, H. V. (1963). *Biochem. Biophys. Res. Commun.* **10**, 482.

Kimmich, G. A. (1970). *Biochemistry* **9**, 3669.
Kipnis, D. M., and Parrish, J. E. (1965). *Fed. Proc. Fed. Amer. Soc. Exp. Biol.* **24**, 1051.
Koch, A. L. (1963). *Ann. N. Y. Acad. Sci.* **102**, 602.
Koch, A. L. (1964). *Biochim. Biophys. Acta* **79**, 177.
Koch, A. L. (1967). *J. Theor. Biol.* **14**, 103.
Kolber, A. R., and Stein, W. D. (1966). *Nature (London)* **209**, 691.
Kolber, A. R., and Stein, W. D. (1967). *Curr. Mod. Biol.* **1**, 244.
Kromphardt, H., Grobecker, H., Ring, K., and Heinz, E. (1963). *Biochim. Biophys. Acta* **74**, 549.
Kundig, W. (1971). Personal communication.
Kundig, W., and Roseman, S. (1966). *Methods Enzymol.* **9**, 396.
Kundig, W., and Roseman, S. (1969). *Fed. Proc. Fed. Amer. Soc. Exp. Biol.* **28**, 463.
Kundig, W., Ghosh, S., and Roseman, S. (1964). *Proc. Nat. Acad. Sci. U. S.* **52**, 1067.
Kundig, W., Kundig, F. D., Anderson, B., and Roseman, S. (1966). *J. Biol. Chem.* **241**, 3243.
Langridge, R., Shinagawa, H., and Pardee, A. B. (1970). *Science* **169**, 59.
Leaf, A. (1956). *Biochem. J.* **62**, 241.
Leaf, A., and Dempsey, E. (1960). *J. Biol. Chem.* **235**, 2160.
Leaf, A., and Renshaw, A. (1957). *Biochem. J.* **65**, 82.
Leaf, A., Page, L. B., and Anderson, J. (1959). *J. Biol. Chem.* **234**, 1625.
Lederberg, J. (1950). *J. Bacteriol.* **60**, 381.
LeFevre, P. G., and LeFevre, M. E., (1952). *J. Gen. Physiol.* **35**, 891.
LeFevre, P. G., and McGinniss, G. F. (1960). *J. Gen. Physiol.* **44**, 87.
Levi, H., and Ussing, H. H. (1948). *Acta Physiol. Scand.* **16**, 232.
Levine, M., Oxender, D. L., and Stein, W. D. (1965). *Biochim. Biophys. Acta* **109**, 151.
Lin, E. C. C. (1970). *Annu. Rev. Genet.* **4**, 225.
Lindenmayer, G. E., Laughter, A. H., and Schwartz, A. (1968). *Arch. Biochem. Biophys.* **127**, 187.
Lyon, I., and Crane, R. K. (1966). *Biochim. Biophys. Acta* **112**, 278.
McClatchy, J. K., and Rosenblum, E. D. (1963). *J. Bacteriol.* **86**, 1211.
Manno, J. A., and Schachter, D. (1970). *J. Biol. Chem.* **245**, 1217.
Malamy, M. H., and Bennett, L. R. (1971). Personal communication.
Matsui, H., and Schwartz, A. (1966). *Biochim. Biophys. Acta* **128**, 380.
Mawe, R. C., and Hempling, H. G. (1965). *J. Cell. Comp. Physiol.* **66**, 95.
Medveczky, N., and Rosenberg, H. (1969). *Biochim. Biophys. Acta* **192**, 369.
Medveczky, N., and Rosenberg, H. (1970). *Biochim. Biophys. Acta* **211**, 158.
Medzihradsky, F., Kline, M. H., and Hokin, L. E. (1967). *Arch. Biochem. Biophys.* **121**, 311.
Meuller, P., Rudin, D. O., Tien, H. T., and Wescott, W. C. (1962). *Circulation* **26**, 1167.
Milner, L. S., and Kaback, H. R. (1970). *Proc. Nat. Acad. Sci. U. S.* **65**, 683.
Mizuno, N., Nagano, K., Nakao, T., Tashima, Y., Fujita, M., and Nakao, M. (1968). *Biochim. Biophys. Acta* **168**, 311.
Mullins, L. J., and Brinley, F. J., Jr. (1969). *J. Gen. Physiol.* **53**, 704.
Nagano, K., Kanazawa, T., Mizuno, N., Tashima, Y., Nakao, T., and Nakao, M. (1965). *Biochem. Biophys. Res. Commun.* **19**, 759.
Nakane, P. K., Nichoalds, G. E., and Oxender, D. L. (1968). *Science* **161**, 182.

Nakao, M., Nagano, K., Matsui, H., Mizuno, N., Nakao, T., and Tashima, Y. (1969). *In* "Molecular Basis of Membrane Function" (D. C. Tosteson, ed.), pp. 539–544. Prentice-Hall, Englewood Cliffs, New Jersey.

Neilands, J. B. (1961). *In* "Haematin Enzymes" (J. E. Falk, R. Lemberg, and R. K. Morton, eds.), p. 194. Macmillan (Pergamon), Oxford.

Neu, H. C., and Heppel, L. A. (1965). *J. Biol. Chem.* **240**, 3685.

Nossal, N. G., and Heppel, L. A. (1966). *J. Biol. Chem.* **241**, 3055.

Novotny, C. P., and Englesberg, E. (1966). *Biochim. Biophys. Acta* **117**, 217.

Onodera, K., Rolfe, B., and Bernstein, A. (1970). *Biochem. Biophys. Res. Commun.* **39**, 969.

Pardee, A. B. (1966). *J. Biol. Chem.* **241**, 5886.

Pardee, A. B. (1967). *Science* **156**, 1627.

Pardee, A. B. (1968). *J. Gen. Physiol.* **52**, 279S.

Pardee, A. B., and Prestidge, L. S. (1966). *Proc. Nat. Acad. Sci. U. S.* **55**, 189.

Pardee, A. B., and Watanabe, K. (1968). *J. Bacteriol.* **96**, 1049.

Pardee, A. B., Prestidge, L. S., Whipple, M. B., and Dreyfuss, J. (1966). *J. Biol. Chem.* **241**, 3962.

Park, C. R., Post, R. L., Kalman, C. F., Wright, J. H., Jr., Johnson, L. H., and Morgan, H. E. (1956). *Ciba Found. Colloq. Endocrinol.* [*Proc.*] **9**, 240.

Pastan, I., and Perlman, R. L. (1969). *J. Biol. Chem.* **244**, 5836.

Pavlasova, E., and Harold, F. M. (1969). *J. Bacteriol.* **98**, 198.

Penrose, W. R., Nichoalds, G. E., Piperno, J. R., and Oxender, D. L. (1968). *J. Biol. Chem.* **243**, 5921.

Phibbs, P. V., Jr., and Eagon, R. G. (1970). *Arch. Biochem. Biophys.* **138**, 470.

Piperno, J. R., and Oxender, D. L. (1966). *J. Biol. Chem.* **241**, 5732.

Portius, H. J., and Repke, K. (1963). *Naunyn-Schmidebergs Arch. Exp. Pathol. Pharmakol.* **245**, 62.

Post, R. L. (1971). Personal communication.

Post, R. L. (1968). *In* "Regulatory Functions of Biological Membranes" (J. Järnefelt, ed.), pp. 163–176. Elsevier, Amsterdam.

Post, R. L., and Jolly, P. C. (1957). *Biochim. Biophys. Acta* **25**, 118.

Post, R. L., and Rosenthal, A. S. (1962). *J. Gen. Physiol.* **45**, 614A.

Post, R. L., and Sen, A. K. (1967a). *Methods Enzymol.* **10**, 762.

Post, R. L., and Sen, A. K. (1967b). *Methods Enzymol.* **10**, 773.

Post, R. L., Merritt, C. R., Kinsolving, C. R., and Albright, C. D. (1960). *J. Biol. Chem.* **235**, 1796.

Post, R. L., Sen, A. K., and Rosenthal, A. S. (1965). *J. Biol. Chem.* **240**, 1437.

Post, R. L., Albright, C. D., and Dayani, K. (1967). *J. Gen. Physiol.* **50**, 1201.

Post, R. L., Kume, S., Tobin, T., Orcutt, B., and Sen, A. K. (1969). *J. Gen. Physiol.* **54**, 306S.

Prestidge, L. S., and Pardee, A. B. (1965). *Biochim. Biophys. Acta* **100**, 591.

Priestland, R. N., and Whittam, R. (1968). *Biochem. J.* **109**, 369.

Rang, H. P., and Ritchie, J. M. (1968). *J. Physiol.* **196**, 183.

Redwood, W. R., Müldner, H., and Thompson, T. E. (1969). *Proc. Nat. Acad. Sci. U. S.* **64**, 989.

Ricard, M., Hirota, Y., and Jacob, F. (1970). *C. R. Acad. Sci.* **270**, 2591.

Rickenberg, H. V., Cohen, G. N., Buttin, G., and Monod, J. (1956). *Ann. Inst. Pasteur, Paris* **91**, 829.

Riggs, T. R., Walker, L. M., and Christensen, H. N. (1958). *J. Biol. Chem.* **233**, 1479.

Riklis, E., and Quastel, J. H. (1958). *Can. J. Biochem. Physiol.* **36**, 347.

Robbie, J. P., and Wilson, T. H. (1969). *Biochim. Biophys. Acta* **173**, 234.

Robinson, J. D. (1967). *Biochemistry* **6**, 3250.

Robinson, J. D. (1968). *Nature* (*London*) **220**, 1325.

Rodnight, R., and Lavin, B. E. (1964). *Biochem. J.* **91**, 24P.

Rogers, D., and Yu, S. H. (1962). *J. Bacteriol.* **84**, 877.

Romano, A. H., Eberhard, S. J., Dingle, S. L., and McDowell, T. D. (1970). *J. Bacteriol.* **104**, 808.

Roseman, S. (1969). *J. Gen. Physiol.* **54**, 138S.

Rosenberg, I. H., Coleman, A. L., and Rosenberg, L. E. (1965). *Biochim. Biophys. Acta* **102**, 161.

Rosenberg, T., and Wilbrandt, W. (1957). *J. Gen. Physiol.* **41**, 289.

Rotman, B., and Radojkovic, J. (1964). *J. Biol. Chem.* **239**, 3153.

Ruoho, A. E., Hokin, L. E., Hemingway, R. J., and Kupchan, S. M. (1968). *Science* **159**, 1354.

Sanno, Y., Wilson, T. H., and Lin, E. C. C. (1968). *Biochem. Biophys. Res. Commun.* **32**, 344.

Scarborough, G. A., Rumley, M. K., and Kennedy, E. P. (1968). *Proc. Nat. Acad. Sci. U. S.* **60**, 951.

Schachter, D., and Mindlin, A. J. (1969). *J. Biol. Chem.* **244**, 1808.

Schachter, D., Johnson, N., and Kirkpatrick, M. A. (1966). *Biochem. Biophys. Res. Commun.* **25**, 603.

Schaefler, S., and Schenkein, I. (1968). *Proc. Nat. Acad. Sci. U. S.* **59**, 285.

Schairer, H. U., and Overath, P. (1969). *J. Mol. Biol.* **44**, 209.

Schatzmann, H. J. (1953). *Helv. Physiol. Pharmacol. Acta* **11**, 346.

Schatzmann, H. J. (1962). *Nature* (*London*) **196**, 677.

Schleif, R. (1969). *J. Mol. Biol.* **46**, 185.

Scholander, P. F. (1960). *Science* **131**, 585.

Schoner, W., von Ilberg, C., Kramer, R., and Seubert, W. (1967). *Eur. J. Biochem.* **1**, 334.

Schultz, S. G., and Zalusky, R. (1963). *Biochim. Biophys. Acta* **71**, 503.

Schultz, S. G., and Zalusky, R. (1964). *J. Gen. Physiol.* **47**, 1043.

Schultz, S. G., and Zalusky, R. (1965). *Nature* (*London*) **205**, 292.

Schultz, S. G., Fuisz, R. E., and Curran, P. F. (1966). *J. Gen. Physiol.* **49**, 849.

Schwartz, A., and Laseter, A. H. (1964). *Biochem. Pharmacol.* **13**, 337.

Schwartz, A., Laseter, A. H., and Kraintz, L. (1963). *J. Cell. Comp. Physiol.* **62**, 193.

Schwartz, A., Matsui, H., and Laughter, A. H. (1968). *Science* **160**, 323.

Sen, A. K., and Post, R. L. (1964). *J. Biol. Chem.* **239**, 345.

Shapiro, A. L., Vinuela, E., Maizel, J. F. (1967). *Biochem. Biophys. Res. Commun.* **28**, 815.

Siegel, G. J., and Albers, R. W. (1967). *J. Biol. Chem.* **242**, 4972.

Silbert, D. F., Ruch, F., and Vagelos, P. R. (1968). *J. Bacteriol.* **95**, 1658.

Simoni, R. D., Levinthal, M., Kundig, F. D., Kundig, W., Anderson, B., Hartman, P. E., and Roseman, S. (1967). *Proc. Nat. Acad. Sci. U. S.* **58**, 1963.

Simoni, R. D., Smith, M. F., and Roseman, S. (1968). *Biochem. Biophys. Res. Commun.* **31**, 804.

Sistrom, W. R. (1958). *Biochim. Biophys. Acta* **29**, 579.

Skou, J. C. (1957). *Biochim. Biophys. Acta* **23**, 394.

Skou, J. C. (1960). *Biochim. Biophys. Acta* **42**, 6.

Skou, J. C. (1961). *In* "Membrane Transport and Metabolism" (A. Kleinzeller and A. Kotyk, eds.), pp. 228–236. Academic Press, New York.

Skou, J. C. (1965). *Physiol. Rev.* **45**, 596.

Skou, J. C. (1969). *In* "Molecular Basis of Membrane Function" (D. C. Tosteson, ed.), pp. 455–482. Prentice-Hall, Englewood Cliffs, New Jersey.

Skou, J. C., and Hilberg, C. (1965). *Biochim. Biophys. Acta* **110**, 359.

Smith, H. W. (1951). "The Kidney." Oxford Univ. Press, New York.

Squires, R. F. (1965). *Biochem. Biophys. Res. Commun.* **19**, 27.

Stahl, W. L. (1968). *J. Neurochem.* **15**, 511.

Steck, T. L., Weinstein, R. S., Straus, J. H., and Wallach, D. F. H. (1970). *Science* **168**, 255.

Stein, W. D. (1969). *J. Gen. Physiol.* **54**, 935.

Stevenson, J. (1966). *Biochem. J.* **99**, 257.

Takacs, F. P., Matula, T. I., and MacLeod, R. A. (1964). *J. Bacteriol.* **87**, 510.

Tanaka, R. (1969). *J. Neurochem.* **16**, 1301.

Tanaka, R., and Abood, L. G. (1964). *Arch. Biochem. Biophys.* **108**, 47.

Tanaka, R., and Strickland, K. P. (1965). *Arch. Biochem. Biophys.* **111**, 583.

Tanaka, S., and Lin, E. C. C. (1967). *Proc. Nat. Acad. Sci. U. S.* **57**, 913.

Tanaka, S., Fraenkel, D. G., and Lin, E. C. C. (1967a). *Biochem. Biophys. Res. Commun.* **27**, 63.

Tanaka, S., Lerner, S. A., and Lin, E. C. C. (1967b). *J. Bacteriol.* **93**, 642.

Thomas, R. C. (1969). *J. Physiol.* **201**, 495.

Tobin, T., and Sen, A. K. (1970). *Biochim. Biophys. Acta* **198**, 120.

Tobin, T., Banerjee, S. P., and Sen, A. K. (1970). *Nature (London)* **225**, 745.

Tosteson, D. C. (1964). *In* "The Cellular Functions of Membrane Transport" (J. F. Hoffman, ed.), p. 3. Prentice-Hall, Englewood Cliffs, New Jersey.

Uesugi, S., Kahlenberg, A., Medzihradsky, F., and Hokin, L. E. (1969). *Arch. Biochem. Biophys.* **130**, 156.

Uesugi, S., Dulak, N. C., Dixon, J. F., Hexum, T. D., Dahl, J. L., Perdue, J. F., and Hokin, L. E. (1971). *J. Biol. Chem.* **246**, 531.

van Groningen, H. E. M., and Slater, E. C. (1963). *Biochim. Biophys. Acta* **73**, 527.

Vidaver, G. A. (1964a). *Biochemistry* **3**, 662.

Vidaver, G. A. (1964b). *Biochemistry* **3**, 795.

Vidaver, G. A. (1964c). *Biochemistry* **3**, 803.

Wang, C. C., and Newton, A. (1969). *J. Bacteriol.* **98**, 1142.

West, I. C. (1969). *FEBS Lett.* **4**, 69.

Wheeler, K. P., and Christensen, H. N. (1967a). *J. Biol. Chem.* **242**, 1450.

Wheeler, K. P., and Christensen, H. N. (1967b). *J. Biol. Chem.* **242**, 3782.

Wheeler, K. P., and Whittam, R. (1970). *Nature (London)* **225**, 449.

Whittam, R. (1962). *Biochem. J.* **84**, 110.

Whittam, R., and Ager, M. E. (1965). *Biochem. J.* **97**, 214.

Whittam, R., and Wheeler, K. P. (1970). *Annu. Rev. Physiol.* **32**, 21.

Whittam, R., Wheeler, K. P., and Blake, A. (1964). *Nature (London)* **203**, 720.

Widdas, W. F. (1952). *J. Physiol.* **118**, 23.

Widdas, W. F. (1954). *J. Physiol.* **125**, 163.

Wilson, G., Rose, S. P., and Fox, C. F. (1970). *Biochem. Biophys. Res. Commun.* **38**, 617.

Wilson, O. H., and Holden, J. T. (1969). *J. Biol. Chem.* **244**, 2743.

Wilson, T. H. (1954). *Science* **120**, 104.

Wilson, T. H. (1962). "Intestinal Absorption." Saunders, Philadelphia, Pennsylvania.

Wilson, T. H., and Kashket, E. R. (1969). *Biochim. Biophys. Acta* **173**, 501.

Winkler, H. H. (1966). *Biochim. Biophys. Acta* **117**, 231.

Winkler, H. II., and Wilson, T. H. (1966). *J. Biol. Chem.* **241**, 2200.
Winkler, H. H., and Wilson, T. H. (1967). *Biochim. Biophys. Acta* **135**, 1030.
Wittenberg, J. B. (1959). *Biol. Bull.* **117**, 402.
Wittenberg, J. B. (1970). *Physiol. Rev.* **50**, 559.
Wong, P. T. S., and Wilson, T. H. (1970). *Biochim. Biophys. Acta* **196**, 336.
Wong, P. T. S., Thompson, J., and MacLeod, R. A. (1969). *J. Biol. Chem.* **244**, 1016.
Wong, P. T. S., Kashket, E. R., and Wilson, T. H. (1970). *Proc. Nat. Acad. Sci. U. S.* **65**, 63.
Yariv, J., Kalb, A. J., Katchalski, E., Goldman, R., and Thomas, E. W. (1969). *FEBS Lett.* **5**, 173.
Zabin, I. (1963). *Fed. Proc. Fed. Amer. Soc. Exp. Biol.* **22**, 27.
Zabin, I., Kepes, A., and Monod, J. (1959). *Biochem. Biophys. Res. Commun.* **1**, 289.
Zabin, I., Kepes, A., and Monod, J. (1962). *J. Biol. Chem.* **237**, 253.
Zwaig, N., and Lin, E. C. C. (1966a). *Biochem. Biophys. Res. Commun.* **22**, 414.
Zwaig, N., and Lin, E. C. C. (1966b). *Science* **153**, 755.

8

THE ROLE OF MEMBRANES IN THE SYNTHESIS OF MACROMOLECULES

M. J. OSBORN

ABBREVIATIONS*

| Abe | Abequose (3,6-dideoxy-D-galactose) |
| BSM | Bovine submaxillary mucin |

* All sugars are of the D-configuration unless otherwise specified.

Fuc	L-Fucose
Gal	Galactose
GalNAc	N-Acetylgalactosamine
GCL	Glycosyl carrier lipid
Glc	Glucose
GlcNAc	N-Acetylglucosamine
GlcU	Glucuronic acid
HO-GCL	Free alcohol of glycosyl carrier lipid
LPS	Lipopolysaccharide
Man	Mannose
m-DAP	meso-Diaminopimelic acid
MurNAc	N-Acetylmuramic acid
NAN	N-Acetylneuraminic acid
NGN	N-Glycolylneuraminic acid
OSM	Ovine submaxillary mucin
P-GCL	Phosphoryl glycosyl carrier lipid
PP-GCL	Pyrophosphoryl glycosyl carrier lipid
PSM	Porcine submaxillary mucin
Rha	L-Rhamnose

I. INTRODUCTION

While the function of membranes in energy metabolism and ATP synthesis has been a subject of intensive investigation for many years, appreciation of the role of the membrane in macromolecular synthesis has emerged only within the past decade. Present concepts have evolved primarily from studies on the mechanisms of biosynthesis of the complex polysaccharides of bacterial cell walls and glycoproteins of animal tissues and from investigations of the mechanism of chromosome replication in bacteria. The role of membranes in biosynthesis of cell wall polymers and glycoproteins is now well established and considerable insight has been gained into the molecular mechanisms involved in assembly of these complex macromolecules. At the same time, evidence is accumulating rapidly for direct participation of membrane in the replication of DNA, originally postulated by Jacob et al. (1963) on the basis of genetic considerations. This chapter will focus on these three aspects of membrane-directed biosynthesis.

Although the specific biochemical problems posed in synthesis of these diverse types of macromolecule obviously differ, certain generalizations about the role of the membrane in these processes are becoming evident. The first relates to the function of specific membrane lipids. These have been shown to function in two distinct ways: as physical cofactors essential to the activity of membrane-bound enzymes and as true coenzymes participating directly in group transfer reactions. Involvement of phospholipids as physical cofactors appears to be a rather general property of enzymes which normally function in the lipid-rich milieu of the membrane and is discussed in detail in Chapter 6. Participation of lipid coenzymes was initially demonstrated in synthesis of specialized polysaccharides of bacterial cell walls. However, similar coenzymes have now been implicated in heterosaccharide synthesis in higher forms and appear to provide a mechanism of general importance for membrane-associated glycosylation reactions.

Another generalization about the role of the membrane is suggested by architectural considerations. The structural framework of the membrane provides an organized matrix facilitating spatial and temporal coordination of the complex series of events required in assembly of cell walls or orderly replication and separation of chromosomes. This organizational function in synthesis of membrane-associated macromolecules is thus comparable to the function of the ribosome in protein synthesis.

II. SYNTHESIS OF CELL WALL POLYMERS

Membrane-bound enzyme systems have been shown to participate in biosynthesis of a wide variety of complex polysaccharides associated with the cell wall or envelope of bacteria, fungi, and higher plants. The synthesis of polymers at the external surface of the cell poses in acute form a problem of compartmentalization not encountered in biosynthesis of intracellular macromolecules; namely, that the final polymeric products of biosynthesis lie external to the permeability barrier of the cytoplasmic membrane, while the necessary precursor molecules (nucleotide sugars, etc.) are believed to be formed in the cytoplasm on the other side of the membrane barrier. The problems of membrane function in the synthesis of these polymers is therefore twofold. The first is the role of specific membrane constituents and overall membrane structure in the catalytic activity of membrane-bound enzyme proteins. This aspect of membrane function is relevant to all membrane-associated enzyme systems and has generally been approached by investigation of the effects of disrupting membrane structure and the reconstitution of function from purified components (for a detailed discussion, see Chapter 6). The second problem, related more specifically to assembly of cell wall, is that of vectorial synthesis. The mechanism whereby this is accomplished remains one of the most intriguing facets of membrane-directed biosynthesis.

A. Structure of the Bacterial Cell Envelope

Bacteria are surrounded by a rigid cell wall or envelope which is responsible for both the specific shape of the cell and its resistance to mechanical and osmotic injury. These properties are attributable to the presence of a unique macromolecule, the peptidoglycan (also called muco-peptide, glycopeptide, or murein). Peptidoglycan consists of linear poly-saccharide strands cross-linked by oligopeptide units to form a single giant "bag-shaped macromolecule" (Weidel and Pelzer, 1964) the size and shape of the bacterial cell. The overall organization of the envelope structure is of two general types (Fig. 1) corresponding to the sub-division of bacteria into two broad classes, gram-positive and gram-negative. Gram-positive bacteria are typically enveloped by a rather thick (20–80 nm) wall (Fig. 1a) composed of peptidoglycan plus one or more additional heteropolysaccharides. These polysaccharides are covalently attached to glycan strands of peptidoglycan and commonly include poly-mers of the teichoic acid type. Teichoic acids are polymers of polyol–phosphate (or sugar–phosphate) units linked through phosphodiester bonds. The surface organization in gram-negative bacteria (Fig. 1b) is considerably more complex than in gram-positive cells and is more prop-erly described by the term cell envelope rather than cell wall. In these organisms a thin peptidoglycan layer (2–3 nm thick) is sandwiched be-tween the cytoplasmic membrane and an outer membranous structure which contains protein, phospholipid, and lipopolysaccharide. The lipo-polysaccharides comprise a closely related group of polymers which are unique to the cell envelopes of gram-negative bacteria and which are re-sponsible for the major immunological specificities (O-antigen) of these organisms. They are made up of complex polysaccharide chains covalently attached to an unusual glucosamine-containing lipid (lipid A).

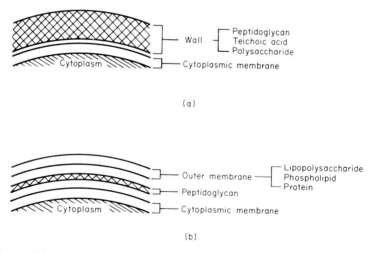

Fig. 1. Schematic representation of the bacterial cell wall. (a) Wall of gram-positive bacteria; (b) envelope of gram-negative bacteria.

B. Role of Polyisoprenol Phosphate Coenzymes (Glycosyl Carrier Lipids)

1. Identification of the Carrier Lipid

Insights of general significance to problems of membrane-directed synthesis emerged from the discovery of a novel type of membrane-bound lipid coenzyme which functions as intermediate carrier of glycosyl groups in biosynthesis of a variety of surface polysaccharides in bacteria. The occurrence of lipid-linked glycosyl intermediates was first described by Strominger and his colleagues (Anderson *et al.*, 1965) in biosynthesis of peptidoglycan, and analogous intermediates are now known to participate in synthesis of the O-antigen of *Salmonella* (Weiner *et al.*, 1965; Wright *et al.*, 1965), the mannan of *Micrococcus lysodeikticus* (Scher *et al.*, 1968), the capsular polysaccharide of *Aerobacter aerogenes* (Frerman *et al.*, 1971), and teichoic acidlike polysaccharides (Douglas and Baddiley, 1968; Brooks and Baddiley, 1969b). The nature of the lipid coenzymes was clarified by Higashi *et al.* (1967) and Wright *et al.* (1967), who carried out mass spectroscopy of the lipids derived from purified oligosaccharide intermediates of peptidoglycan and O-antigen synthesis. Both lipids proved to be phosphomonoesters of a C_{55}-polyisoprenoid alcohol:

$$^{=}O_3P\!-\!O\!-\!CH_2\!-\!CH\!=\!\underset{\underset{CH_3}{|}}{C}\!-\!CH_2\!-\!(CH_2\!-\!CH\!=\!\underset{\underset{CH_3}{|}}{C}\!-\!CH_2)_9\!-\!CH_2\!-\!CH\!=\!\underset{\underset{CH_3}{|}}{C}\!-\!CH_3$$

A similar structure has since been established for the lipid coenzyme in other bacterial systems (Scher *et al.*, 1968; Frerman *et al.*, 1971).

More recently, lipid-linked intermediates have also been implicated in biosynthesis of cell wall polysaccharides in fungi (Tanner, 1969) and higher plants (Kauss, 1969; Villemez and Clark, 1969) and in glycoprotein synthesis in mammalian tissues (Caccam *et al.*, 1969; Behrens and Leloir, 1970). It now seems likely that polyisoprenol phosphate coenzymes are widely distributed in nature and provide a major mechanism (though not the sole mechanism) for glycosyl transfer reactions catalyzed by membrane-bound enzyme systems. In all known cases the lipid functions as intermediate carrier of glycosyl groups between an initial nucleotide sugar donor and the final polysaccharide or protein acceptor, and the term glycosyl carrier lipid (GCL) is therefore suggested to designate this general class of coenzymes.

2. Biosynthesis of Peptidoglycan

a. Enzymatic Mechanism. The peptidoglycans of bacterial cell walls are basically polymers of the disaccharide–peptide repeating unit, *N*-

acetylglucosaminyl-*N*-acetylmuramyl-L-Ala-γ-D-Glu-L-Lys(or DAP)-D-Ala.
In the polymer, GlcNAc-MurNAc disaccharide units are joined glycosidically to form linear polysaccharide chains, and the peptide units are at
least partially cross-linked to each other either directly or through peptide
bridges, as illustrated in Fig. 2. The pathway of biosynthesis of peptidoglycan as described by Strominger and his colleagues for *Staphylococcus
aureus* (Anderson *et al.*, 1965, 1966, 1967; Anderson and Strominger, 1966;
Matsuhashi *et al.*, 1967; Bumsted *et al.*, 1968; Siewert and Strominger,
1968) is summarized in Fig. 3. The process can be divided into the following major stages. (a) Assembly of the complete disaccharide–peptide
repeating unit [reactions (1)–(4), Fig. 3] via a series of intermediates
in which the reducing group of *N*-acetylmuramic acid is in pyrophosphate
linkage to the glycosyl carrier lipid; (b) incorporation of disaccharide–
peptide units into linear polysaccharide strands with liberation of GCL
as the polyisoprenol pyrophosphate [reaction (5), Fig. 3]; (c) formation
of cross-links in nascent polysaccharide chains by transpeptidation between peptide units on adjacent glycan strands; and (d) regeneration of
the active monophosphate form of GCL [reaction (6), Fig. 3]. The complete sequence of reactions is carried out by a membrane fraction which

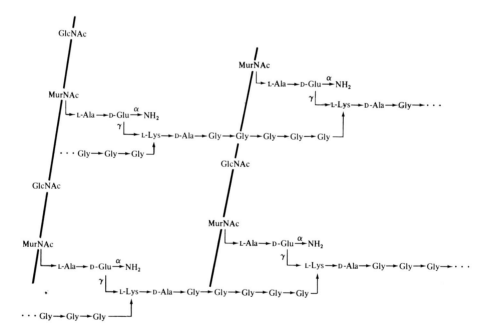

Fig. 2. Structure of the peptidoglycan of *Staphylococcus aureus*.

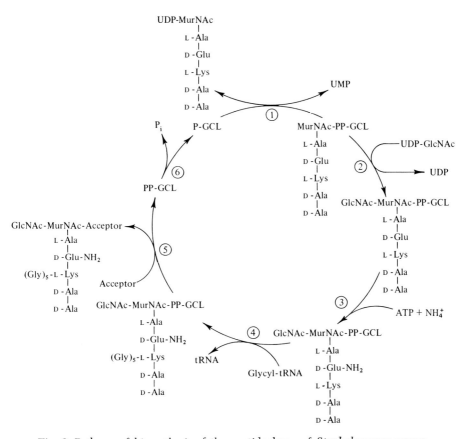

Fig. 3. Pathway of biosynthesis of the peptidoglycan of *Staphylococcus aureus*.

contains not only the enzyme proteins but also glycosyl carrier lipid and endogenous cell wall acceptor.

i. Assembly of the Disaccharide–Peptide Repeating Unit. The initial step, formation of the monosaccharide–lipid intermediate, MurNAc-(pentapeptide)-PP-GCL, occurs by reversible transfer of MurNAc-(pentapeptide)-1-phosphate from UDP-MurNAc-pentapeptide to the polyisoprenol phosphate carrier lipid (P-GCL), with release of UMP [reaction (1), Fig. 3]. The initial clues to the nature of this unusual reaction were derived from early experiments which showed that UMP, rather than the expected UDP, was the nucleotide product (Meadow *et al.*, 1964) and that the membrane-bound MurNAc-pentapeptide intermediate was rapidly discharged by UMP but not by UDP (Struve and Neuhaus, 1965). Transfer of the entire MurNAc-1-phosphate portion of the nucleotide sugar to the acceptor lipid was demonstrated by use of ^{32}P-labeled UDP-MurNAc (Anderson *et al.*, 1965), and the presence of a pyrophosphate linkage in the product was ultimately established by chemical degradation of purified intermediates (Higashi *et al.*, 1967). The allylic phosphate ester linkage of the polyisoprenol phosphate is extraordinary acid labile and is selectively cleaved by brief hydrolysis at pH 4. Degradation of the purified disaccharide intermediate, GlcNAc-MurNAc-(pentapeptide)-PP-GCL, under these conditions permitted recovery of disaccharide-(pentapeptide)-pyrophosphate, which was identified by release of PP_i and disaccharide following hydrolysis in 1 N acetic acid. The reaction is readily reversible, with an apparent K_{eq} of approximately 0.25 (Struve *et al.*, 1966). The high-energy character of the glycosyl–pyrophosphate bond of the nucleotide sugar is therefore fully conserved in the lipid-linked intermediates which act as glycosyl donors in the subsequent polymerization of the polysaccharide chain.

The mechanism of formation of MurNAc-(pentapeptide)-PP-GCL has been investigated in some detail by Neuhaus and co-workers, who have proposed the name, phospho-N-acetylmuramyl-pentapeptide translocase (UMP), for the enzyme (Struve *et al.*, 1966). In addition to the overall transfer of MurNAc-(pentapeptide)-1-phosphate to P-GCL, the membrane-bound enzyme catalyzes rapid and extensive exchange of UMP into UDP-MurNAc-pentapeptide (Struve *et al.*, 1966). On the basis of kinetic studies of the transfer and exchange reactions and differential effects of surfactants, Heydanek *et al.* (1969) have postulated a two-step reaction mechanism involving initial formation of an enzyme-phospho-N-acetyl-muramyl-pentapeptide intermediate:

$$\text{UDP-MurNAc-pentapeptide} + \text{Enz} \rightleftharpoons$$
$$\text{MurNAc-(pentapeptide)-P-Enz} + \text{UMP} \quad (1)$$

$$\text{MurNAc-(pentapeptide)-P-Enz} + \text{P-GCL} \rightarrow$$
$$\text{MurNAc-(pentapeptide)-PP-GCL} + \text{Enz} \quad (2)$$

Definitive analysis of the mechanism of the enzyme awaits isolation and purification of the protein in soluble form. Some progress toward this goal has been reported by Heydanek and Neuhaus (1969), who were able to obtain partial solubilization of the enzyme from *S. aureus* by treatment of membrane fragments with sodium lauroyl sarcosinate, urea, or KOH.

In the second step in assembly of the peptidoglycan repeating unit, MurNAc-(pentapeptide)-PP-GCL acts as acceptor of GlcNAc from UDP-GlcNAc to form the disaccharide–pentapeptide intermediate [reaction (2), Fig. 3]. Addition of GlcNAc appears to occur by a "conventional" glycosyl transfer reaction. The nucleotide product of the reaction is UDP

(Meadow et al., 1964; Anderson et al., 1965), the reaction is not readily reversible, and the available evidence is consistent with direct transfer of the GlcNAc residue from the nucleotide sugar donor to the mono-saccharide–lipid acceptor without intervention of isolable intermediates.

Membrane fractions of S. aureus and M. lysodeikticus are able to utilize the disaccharide–pentapeptide intermediate directly for synthesis of polymeric peptidoglycan. However, in these and other gram-positive organisms in which the peptide subunits contain additional bridge amino acids and/or substituted D-glutamyl residues, those modifications are introduced prior to polymerization, at the level of disaccharide-(penta-peptide)-PP-GCL. The α-carboxyl group of the D-glutamate residue is substituted in S. aureus by an amide group (Muñoz et al., 1966; Tipper et al., 1967) and in M. lysodeikticus by a single glycine residue (Mirelman and Sharon, 1966, 1967; Tipper et al., 1967). Formation of the isogluta-mine amide linkage by membrane fractions of S. aureus was established by Siewert and Strominger (1968) [reaction (3), Fig. 3]. The system re-quired either NH_4^+ or glutamine and ATP; both the monosaccharide- and disaccharide-(pentapeptide)-PP-GCL intermediates were effective as substrates for amidation, but UDP-MurNAc-pentapeptide was totally in-active. Enzymatic addition of the glycine residue by membrane fractions of M. lysodeikticus has also been shown (Katz et al., 1967) to occur at the level of the lipid intermediates, according to the reaction

Glycine + ATP +

$$\text{GlcNAc-MurNAc-PP-GCL} \rightarrow \text{GlcNAc-MurNAc-PP-GCL} + \text{ADP} + P_i \quad (3)$$

L-Ala
γ|
D-Glu$^\alpha$ · COOH
L-Lys
D-Ala
D-Ala

L-Ala
γ|
D-Glu$^\alpha$ · CO-NHCH$_2$COOH
L-Lys
D-Ala
D-Ala

The reaction was independent of tRNA and resulted in formation of ADP and P_i rather than AMP and PP_i. The mechanism of peptide bond forma-tion is therefore quite different from that in tRNA-dependent synthesis of bridge peptide units (see below) and resembles rather the synthesis of glutathione and the pentapeptide subunit itself.

The role of tRNA in synthesis of interpeptide bridge units was sug-gested by the initial observation of Chatterjee and Park (1964) that incorporation of glycine (but not of MurNAc-pentapeptide) into pepti-doglycan in cell-free preparations from S. aureus was abolished by RNase. Subsequent studies by Strominger and co-workers established the general mechanism whereby aminoacyl-tRNA's act as obligatory donors of amino acid residues in assembly of bridge peptide units. Matsuhashi *et al.* (1967) showed unequivocally that glycyl-tRNA was the immediate precursor of the pentaglycine bridge unit in S. *aureus* and that the addition of glycine occurred exclusively at the level of the lipid-linked intermediates [reaction (4), Fig. 3]. Disaccharide-(pentapeptide)-PP-GCL and, to a lesser extent, the monosaccharide intermediate, were effective as acceptors of glycine, while preformed glycine-deficient polymeric peptidoglycan and UDP-MurNAc-pentapeptide were totally inactive. Attachment of the oligoglycine chains to the ε-amino group of the lysine residue was established by dinitrophenylation and Edman degradation of the enzymatic products. Elongation of the oligoglycine chain appears to occur by successive transfer of single glycine residues from glycyl-tRNA to the free amino terminus of the peptide. No evidence for intermediate formation of peptidyl-tRNA could be obtained, and recent experiments of Thorndike and Park (1969) and Kamiryo and Matsuhashi (1969) on the mechanism chain growth have provided strong evidence for stepwise addition of single glycine residues.

Investigations in several other organisms having different bridge structures have confirmed the essential role of tRNA. Thus, the appropriate aminoacyl-tRNA's were found to serve as donors of L-serine and glycine in synthesis of the mixed glycyl–seryl pentapeptide bridge in *Staphylococcus epidermidis* (Petit *et al.*, 1968) of L-threonine and L-alanine in formation of the tetrapeptide bridge of *Micrococcus roseus* (Roberts *et al.*, 1968a). In each case, addition of bridge units to the ε-NH$_2$ group of lysine occurred prior to polymerization, at the level of mono- and disaccharide-(pentapeptide)-PP-GCL intermediates. An interesting variation in the above mechanism has emerged from recent studies of Plapp and Strominger (1970a,b) on synthesis of the L-seryl–L-alanyl bridge unit of *Lactobacillus viridescens*. L-Alanyl- and L-seryl-tRNA were again required as aminoacyl donors; however, the first amino acid was added to UDP-MurNAc-pentapeptide rather than to membrane-bound lipid intermediates. The resulting UDP-MurNAc-hexapeptide product was efficiently incorporated into lipid intermediates, but subsequent polymerization steps depended on addition of the second bridge amino acid at the level of the lipid intermediates.

The important question of whether the tRNA species which participate in peptidoglycan synthesis are different from those involved in protein synthesis has been clarified in part by purification of the relevant tRNA's. Bumsted *et al.* (1968) separated three glycyl-tRNA fractions from S. *aureus,* all of which were approximately equally active in the peptidoglycan system. Two of the three were also active in polypeptide synthesis and were identified by triplet binding as corresponding to known glycine codons. The third fraction however failed to support polypeptide synthesis with natural or synthetic messengers and could not be identified with any of the known glycine anticodons by triplet binding. Similar results were obtained (Petit, *et al.*, 1968) on purification of the L-seryl-tRNA's of *S. epidermidis*. Only three of the four species isolated were active in polypeptide synthesis or triplet binding, although all four were effective as donors of serine in synthesis of the bridge peptide of peptidoglycan. In contrast, purification of the threonyl-tRNA's of *M. roseus* (Roberts *et al.*, 1968b) yielded no evidence of a peptidoglycan-specific fraction. The occurrence in staphylococci of tRNA species apparently unique to peptidoglycan synthesis is of considerable interest, but the general occurrence of these and the specificity relationships *in vivo* remain to be established. The degree of specificity of the bridge-synthesizing enzymes toward their aminoacyl-tRNA substrates may also vary widely. Addition of the single L-alanyl residue in *A. crystallopoietes* was strictly dependent on alanine-specific tRNA; L-alanyl-tRNA[Cys] prepared by catalytic reduction of L-cysteinyl-tRNA was totally inactive as donor of the bridge alanine although still functional in polypeptide synthesis (Roberts *et al.*, 1968a). However, the soluble enzyme from *L. viridescens* responsible for addition of the first bridge amino acid to UDP-MurNAc-pentapeptide exhibited a much broader specificity (Plapp and Strominger, 1970b). Enzyme purified to apparent homogeneity catalyzed transfer of L-alanine, L-serine, L-cysteine, and to a lesser extent glycine from their respective tRNA's to the uridine nucleotide, and L-alanyl-tRNA[Cys] was almost as effective as L-alanyl-tRNA[Ala].

ii. **Polymerization of Disaccharide Units.** Two stages have been detected in conversion of the completed disaccharide–peptide repeating units to mature peptidoglycan: polymerization of disaccharide units to form linear polysaccharide strands, followed by cross-linking of peptide units on adjacent strands to yield the final three-dimensional network. It is thought (Anderson *et al.*, 1967; Ghuysen *et al.*, 1968) that polymerization occurs by successive transfer of disaccharide–peptide units from GCL to endogenous cell wall acceptor, according to reaction (5) (Fig. 3). It should be emphasized that only the glycosyl moiety of the intermediate is transferred and that GCL is released in the form of a pyrophosphate derivative (Siewert and Strominger, 1967) in which the terminal phosphate is derived from UDP-MurNAc-pentapeptide. The nature of the endogenous acceptor or primer has not yet been investigated in detail, but it is thought to be nonreducing termini of preexisting peptidoglycan chains. Thus, membrane fractions obtained by dissolution of cell walls with muramidases or hexosaminidases which degrade the endogenous glycan lose the ability to incorporate lipid-linked intermediates into polymer (Siewert and Strominger, 1968). In addition, incorporation into polymer is inhibited by the antibiotics, vancomycin and ristocetin, which have been shown (Best and Durham, 1965; Sinha and Neuhaus, 1968; Best *et al.*, 1970) to bind tightly to cell wall material. Stepwise addition of disaccharide–peptide units to end groups of glycan chains would fit well with current hypotheses of the mechanism of growth of cell wall, according to which controlled autolysis introduces nicks into peptidoglycan strands at defined growing points and provides new sites for chain elongation.

iii. **Cross-Linking of Peptide Units.** It had been recognized for many years that the nucleotide sugar precursor of peptidoglycan, UDP-MurNAc-pentapeptide, contains two D-alanine residues at the carboxyl terminus of the peptide while the subunit of peptidoglycan contains only a single D-alanine residue. It was originally postulated by Martin (1966) that the cross-linking reaction might involve transpeptidation between the D-alanyl-D-alanine terminus of one peptide unit and the free amino group of the basic amino acid or bridge unit of a second peptide, with release of free D-alanine (Fig. 4). This hypothesis was particularly attractive in that it provided a mechanism for isoenergetic formation of the cross-linking peptide bond in the wall itself, external to the cytoplasmic membrane and independent of energy donors such as ATP. *In vivo* studies of Wise and Park (1965) and Tipper and Strominger (1965, 1968) in *S. aureus* provided support not only for the existence of such a trans-peptidation reaction, but also for identification of this reaction as the penicillin-sensitive step in cell wall synthesis. Direct demonstration of transpeptidation *in vitro* has thus far been achieved only in cell envelope fractions from *E. coli* (Araki *et al.*, 1966a,b; Izaki *et al.*, 1966; Izaki and Strominger, 1968). The evidence strongly supports the reaction shown in Fig. 4. Incorporation of MurNAc-pentapeptide into peptidoglycan was accompanied by liberation of free D-alanine, the ratio of D-alanine to D-glutamate in the polymeric product approached unity and the expected bistetrapeptide–disaccharide dimer could be isolated following digestion of the enzymatic product with lysozyme. The transpeptidation reaction was specifically and irreversibly inhibited by penicillin. Incorporation of MurNAc-pentapeptide into a polymeric product was unaffected by the antibiotic, but the release of D-alanine was inhibited, and this inhibition was paralleled by the appearance of an uncross-linked product which retained the terminal D-alanyl-D-alanine structure of the precursor.

Efforts to detect a similar transpeptidase activity in cell-free preparations from gram-positive bacteria have been unsuccessful for reasons

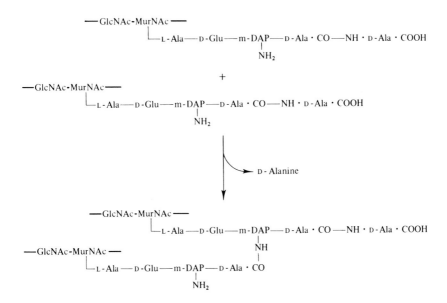

Fig. 4. Mechanism of formation of peptidoglycan cross-links in *Escherichia coli* (m-DAP denotes *meso*-diaminopimelic acid). In organisms in which the peptidoglycan contains interpeptide bridge units (e.g., S. *aureus*), transpeptidation would occur between the D-alanyl-D-alanine terminus of one subunit and the free NH₂-terminus of the bridge peptide of the adjacent subunit.

which are not entirely clear. Lawrence and Strominger (1970a,b) have suggested that a particulate penicillin-sensitive D-alanine carboxypeptidase activity observed in *Bacillus subtilis* may actually represent an "uncoupled" transpeptidase which reacts preferentially with water rather than with the normal amino acceptor.

b. Architectural Relationships in Peptidoglycan Synthesis. At this juncture it is perhaps worthwhile to speculate on the relationship between the events in peptidoglycan synthesis and the molecular architecture of the cytoplasmic membrane and cell wall. The precursors of peptidoglycan (nucleotide sugars, aminoacyl-tRNA's, etc.) presumably originate in the cytoplasm, and it is reasonable to suppose that assembly of the complete disaccharide–peptide repeating unit takes place at or near the inner cytoplasmic surface of the membrane. The peptidoglycan, however, lies outside the membrane and it seems probable that the final steps in synthesis, chain elongation and peptide cross-linking, occur at the outer surface of the membrane or possibly in the wall matrix itself. This implies, first, an asymmetric distribution of enzyme activities at the two surfaces of the membrane and, second, a translocation of disaccharide-(peptide)-PP-GCL from its site of synthesis at the inner surface to its site of utilization at the outer surface of the membrane. Indeed the possible transport function of the lipid coenzyme was emphasized in the initial studies of Anderson *et al.* (1965), and the term "carrier lipid" is appropriate not only to the coenzyme function as intermediate carrier of the activated glycosyl group, but also to cross-membranal transport of the completed oligosaccharide–peptide repeating unit. The mechanism of the postulated translocation through the membrane and the question of whether additional membrane functions are required for the transport of GCL and its derivatives are entirely unknown.

3. Biosynthesis of O-Antigen

a. Enzymatic Mechanism. Shortly after the discovery of the role of lipid-linked intermediates in peptidoglycan biosynthesis, it became obvious that biosynthesis of O-antigen in *Salmonella* occurred by a similar mechanism. The O-antigen specificities of *Salmonella* and other gram-negative enteric bacteria are determined by polysaccharide side chains which comprise the outer portion of the lipopolysaccharide of the cell envelope (Fig. 5). The O-specific chains differ widely in structure from species to species but appear always to be composed of oligosaccharide repeating units (Lüderitz *et al.*, 1966). The mechanism of biosynthesis of O-antigen has been most extensively studied in *S. typhimurium,* in which the basic repeating unit (see Fig. 5) is a branched tetrasaccharide

$$
\begin{array}{c}
\text{Abe} \\
| \\
\cdots \text{Man-Rha-Gal} \cdots
\end{array}
$$

and in *S. newington* and related species which contain a similar trisaccharide unit, $\cdots$Man-Rha-Gal$\cdots$. The pathway of biosynthesis which has emerged from these studies (reviewed in Osborn, 1969; Robbins and Wright, 1971) is summarized in Fig. 6. The entire reaction sequence is catalyzed by the membranous cell envelope fraction which contains both GCL and the final lipopolysaccharide acceptor of the O-antigen chains in addition to all the necessary enzyme proteins. The pathway can be divided into three major phases: (a) assembly of the oligosaccharide repeating unit via a series of glycosyl-PP-GCL intermediates [reactions (1)–(4), Fig. 6], (b) polymerization of oligosaccharide units to form a polysaccharide-PP-GCL intermediate, and (c) transfer of the completed polysaccharide chain to the lipopolysaccharide core.

i. **Assembly of the Oligosaccharide Repeating Unit.** The reactions leading to synthesis of the repeating unit of O-antigen appear to be closely

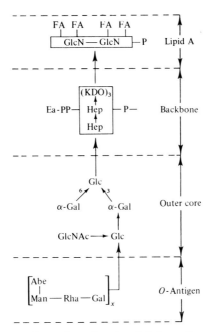

Fig. 5. Structure of the lipopolysaccharide of *Salmonella typhimurium*. Abbreviations: FA, fatty acid; GlcN, glucosaminyl; KDO, 2-keto-3-deoxyoctonyl (3-deoxy-D-mannooctulosonyl); Hep, L-glycero-D-mannoheptosyl; EaPP, pyrophosphorylethanolamine; Glc, glucosyl; Gal, galactosyl; GlcNAc, N-acetylglucosaminyl; Rha, L-rhamnosyl; Man, mannosyl; Abe, abequosyl (3,6-dideoxy-D-galactose). All sugars are of the D-configuration unless otherwise specified.

similar to those occurring in peptidoglycan synthesis. The initial step [reaction (1), Fig. 6] is transfer of galactose 1-phosphate from UDP-galactose to P-GCL to form the monosaccharide–lipid intermediate, Gal-PP-GCL, with stoichiometric release of UMP (Osborn and Yuan Tze-Yuen, 1968). The reaction is again readily reversible, with an apparent equilibrium constant of approximately 0.5. Indeed the ease of reversibility and the presence of other enzyme systems in the cell envelope preparations which compete for the substrate, UDP-galactose, considerably hampered initial attempts to characterize the reaction, and accumulation of Gal-PP-GCL was observed only under conditions in which competing systems were minimized. In the presence of TDP-rhamnose, however, Gal-PP-GCL is efficiently converted to the disaccharide intermediate, Rha-Gal-PP-GCL [reaction (2), Fig. 6] (Weiner *et al.*, 1965; Wright *et al.*, 1965; Dankert *et al.*, 1966), and sequential transfer of mannose from GDP-mannose [reaction (3), Fig. 6] and, in S. *typhimurium*, of abequose from CDP-abequose [reaction (4), Fig. 6] (Osborn

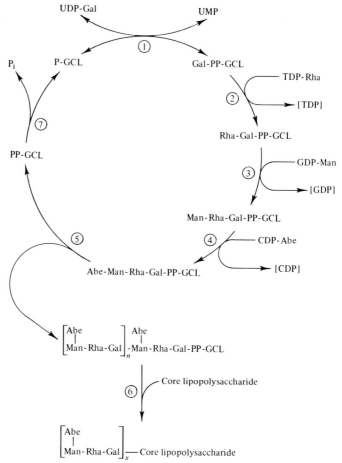

Fig. 6. Pathway of biosynthesis of the O-antigen of *Salmonella typhimurium*.

and Weiner, 1968) yields the completed repeating unit as the oligosaccharide-PP-GCL derivative. It is assumed that these reactions, like the transfer of GlcNAc to MurNAc-(pentapeptide)-PP-GCL (see Section II,2,a,i), are conventional glycosyltransferase reactions which yield the nucleoside diphosphate as product and are not readily reversible, but the reactions have not been characterized in detail. Polymerization of both the tri- and tetrasaccharide intermediates occurs very rapidly above 20°C, but formation and accumulation of the monomeric oligosaccharide–lipid intermediates could be demonstrated by carrying out the incubations at 10°C (Weiner *et al.*, 1965; Osborn and Weiner, 1968). The three transferases responsible for synthesis of trisaccharide-PP-GCL in S. *typhi-*

murium have been obtained in a soluble form dependent on added P-GCL by extraction of the cell envelope with nonionic detergents (Gilbert *et al.,* 1968). The pyrophosphate linkage between the glycosyl unit and the lipid was established by Wright *et al.* (1967). Brief hydrolysis of purified Rha-Gal-PP-GCL in aqueous 0.01 N HCl gave a quantitative yield of free disaccharide and inorganic pyrophosphate, identified enzymatically by treatment with inorganic pyrophosphatase.

ii. **Mechanism of Polymerization of O-Antigen Chains.** As discussed in the preceding section, synthesis of polymeric peptidoglycan is thought to occur by stepwise transfer of repeating units directly from the disaccharide–lipid intermediate to end groups of primer glycan strands. In contrast, synthesis of O-antigen involves initial formation of GCL-linked polymeric intermediates [reaction (5), Fig. 6] from which completed O-specific polysaccharide chains are ultimately transferred to lipopolysaccharide. The occurrence of lipid-linked polymer chains was initially inferred from analysis of the products of O-antigen synthesis in the cell envelope fraction of a mutant which produced an incomplete lipopolysaccharide core structure lacking O-antigen attachment sites (Weiner *et al.,* 1965). The enzymatic product was identified as a polysaccharide with the structure of O-antigen but was not attached to lipopolysaccharide. Attachment of this polymeric product to GCL was suggested by the finding of galactose 1-phosphate reducing ends following extraction of the product from the cell envelope by procedures known to cleave the labile pyrophosphate linkage to the lipid. It is of interest that a similar nonlipopolysaccharide O-specific polymer is accumulated *in vivo* in a variety of mutants blocked in biosynthesis of the lipopolysaccharide core (Beckmann *et al.,* 1964; Osborn, 1968). This so-called haptenic O-antigen material appears to be identical to the enzymatic product; it is firmly bound to cell envelope, and after extraction it also contains galactose 1-phosphate at the reducing termini (Kent and Osborn, 1968a). Direct evidence that the initial product of polymerization is still attached to GCL was provided by isolation of a chloroform–methanol soluble intermediate which contained a dimer of the repeating unit (Osborn and Weiner, 1968). Long-chain polysaccharide–lipid derivatives (which are not soluble in organic solvents) have not yet been isolated in intact form due to degradation of the labile galactose-1-pyrophosphoryl-GCL linkage during conventional extraction procedures.

Conclusive evidence for the postulated mechanism of polymerization emerged from the elegant studies of Bray and Robbins (1967a) on the direction of chain growth in *S. newington.* Pulse-chase experiments *in vivo* and *in vitro* clearly demonstrated that new repeating units are added at the reducing end of the growing chain, as shown in Fig. 7, rather than at the nonreducing end as in the conventional mechanism of polysaccharide synthesis. The mechanism of chain elongation in O-antigen synthesis is therefore analogous to that in fatty acid and polypeptide synthesis and requires that the reducing terminus of the nascent polymer chain remain in an activated form throughout the entire sequence of glycosyl transfer reactions (cf. Fig. 7). This mechanism of polymerization also makes excellent sense in terms of the molecular architecture of the membrane (Robbins *et al.*, 1967); the lipid-linked reducing end of the growing polysaccharide chain can be held in close apposition to the polymerase enzyme and the incoming monomer unit within the organized lipid-rich membrane structure, while the nonreducing end of the polymer is free to extend into the more hydrophilic environment external to the membrane surface and at a distance from the active site of the polymerase.

Experiments of Kanegasaki and Wright (1970) shed additional light on the architectural relationship of the polymerase to other components of the system. These investigators considered two alternative models of O-antigen synthesis. The first postulates a highly organized multienzyme complex containing tightly bound P-GCL. According to this model, carrier lipid molecules would be fixed to the complex and immobile; each step in chain elongation would then require *de novo* synthesis of the incoming repeating unit directly at the site of its incorporation into polymer, and formation of long-chain polymer would be strictly dependent on continuous generation of new repeating units at each site. The alternative model proposes that P-GCL and its derivatives are freely mobile in the membrane and that the sites of synthesis and utilization of oligosaccharide–lipid intermediates may be physically separate. This model would

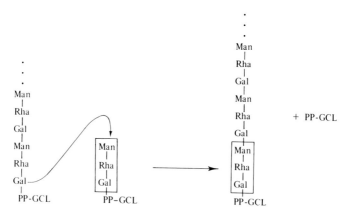

Fig. 7. Mechanism of polymerization of O-antigen.

be consistent with synthesis of repeating units at the inner cytoplasmic surface of the membrane followed by translocation to the external surface for polymerization and attachment of O-antigen chains to lipopolysaccharide. Experiments on utilization of preformed intermediates for polymer synthesis showed that these were efficiently converted to long-chain products, and continuous generation of new repeating units was not required. The results strongly supported the hypothesis that lipid-linked intermediates are mobile in the membrane and freely accessible to the polymerase enzyme. This interpretation was also consistent with the observation that exogenously added intermediates could be efficiently utilized as substrates by the cell envelope in the absence of detergent, provided that the mixture of envelope and lipid substrate was first subjected to repeated cycles of freezing and thawing. This procedure apparently serves to "shock" the added lipid into the membrane matrix, and it may be a technique of general applicability.

iii. **Transfer of O-Antigen Chains to Lipopolysaccharide.** The mechanism of polymerization of O-antigen implies that the lipid-linked polymer is the immediate and obligatory precursor of the O-antigen chains of lipopolysaccharide [reaction (6), Fig. 6]. Further evidence for this hypothesis has been obtained from studies on the transfer of O-specific polysaccharide from GCL to lipopolysaccharide *in vivo* and *in vitro*. Pulse-chase experiments were carried out *in vivo* in a mutant of S. *typhimurium* lacking phosphomannose isomerase, in which exogenous [^{14}C]mannose is incorporated specifically into O-antigen (Kent and Osborn, 1968b). The results showed a rapid incorporation of mannose into lipid-linked polymer during the pulse and a rapid shift of this radioactivity into lipopolysaccharide during the chase period. Transfer of preformed polymer from GCL to an acceptor lipopolysaccharide has also been demonstrated *in vitro* (Cynkin and Osborn, 1968a). The cell envelope fraction of a mutant blocked in synthesis of the core region of lipopolysaccharide was used both for synthesis of O-antigen chains and as a source of the final transferase enzyme. Under these conditions enzymatically synthesized O-antigen accumulated as the polymer-PP-GCL derivative, and transfer of preformed polymer to lipopolysaccharide could be initiated in the absence of nucleotide sugar substrates by addition of an exogenous acceptor lipopolysaccharide containing the complete core structure.

Genetic evidence suggests that the transfer of O-antigen to lipopolysaccharide may be complex. Two classes of mutants have been isolated in S. *typhimurium* which appear to lack this activity (Subbaiah and Stocker, 1964; Wilkinson *et al.*, 1971); one class maps in the *rfa* cluster, which in general determines biosynthesis of the core, and the other maps in the *rfb* region which contains the determinants of O-antigen synthesis. Both classes are able to synthesize the lipid-linked O-specific polymer *in vivo* and *in vitro* but fail to attach O-antigen chains to lipopolysaccharide, even though the isolated lipopolysaccharides are fully active as acceptor of O-antigen chains *in vitro* (Cynkin and Osborn, 1968b). The nature of the enzymatic lesions in the two types of mutants is not yet known.

b. Phage Conversion of O-Antigen. The ability of certain temperate bacteriophages to cause specific modifications in the structure and immunological specificity of the O-antigen of the infected host is a well-known phenomenon, and it is fairly common among *Salmonella* phages (Lüderitz *et al.*, 1966). Extensive genetic and biochemical analysis of the conversion of *S. anatum* by phages ϵ^{15} and ϵ^{34} by Robbins, Wright, and their colleagues has permitted a detailed description of the mechanism of O-antigen conversion in terms of phage-induced modifications of the biosynthetic pathway. Alterations in O-antigen structure associated with conversion by ϵ^{15} and ϵ^{34} (Robbins and Uchida, 1962) are summarized in Fig. 8. In *S. anatum*, the galactosyl residues are O-acetylated, and the linkage between repeating units (i.e., the galactosyl–mannose linkage) is in the α-configuration. Infection with ϵ^{15} results in both disappearance of the O-transacetylase activity and a change in configuration of the galactosyl–mannosyl linkages from α to β. Analysis of mutants of ϵ^{15} resolved three phage-specific functions responsible for these alterations: (a) repression of the host enzyme which catalyzes acetylation of galactosyl residues (Robbins and Uchida, 1965; Robbins *et al.*, 1965), (b) inhibition of the host polymerase (α-polymerase) (Bray and Robbins, 1967b; Losick, 1969), and (c) induction of a phage-specific polymerase (Bray and Robbins, 1967b; Losick and Robbins, 1967). The role of the ϵ^{15} genome in the alteration of the polymerase specificity was clarified

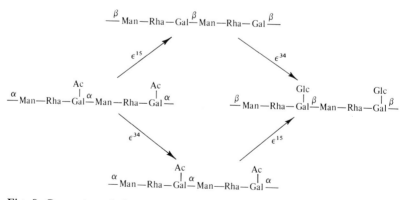

Fig. 8. Conversion of the O-antigen of *Salmonella anatum* by phages ϵ^{15} and ϵ^{34} (Ac represents O-acetyl).

by use of temperature-sensitive mutants of both host (Losick and Robbins, 1967) and phage (Bray and Robbins, 1967b). Evidence that the β-polymerase activity is determined by the phage was obtained by infection of a mutant of S. *anatum* temperature sensitive in α-polymerase activity with wild-type ϵ^{15} (Losick and Robbins, 1967); polymerase activity was restored under the otherwise nonpermissive conditions, and the resulting O-antigen polymer contained only β-galactosyl linkages. Isolation of phage mutants which produced a β-polymerase that was temperature sensitive both *in vivo* and *in vitro* confirmed that the enzyme structure is specified by a phage gene (Bray and Robbins, 1967b). Appearance of the phage-specific β-polymerase activity can be detected within 5 minutes after infection, and by 15–20 minutes, essentially only β-linked polymer is formed *in vivo* or *in vitro* (Bray and Robbins, 1967b). The observed rapid disappearance of host polymerase activity is explained by synthesis of a specific inhibitor of the α-polymerase (Bray and Robbins, 1967b; Losick, 1969) which appears soon after infection and persists in the lysogenic state. Synthesis of the inhibitor appears to be under control of a phage gene separate from the β-polymerase, since loss of host α-polymerase also occurs after infection by phage polymerase mutants. The inhibitor has been purified from lysogenic cells by Losick (1969) and appears to be a low molecular weight protein. The continued presence of the inhibitor in the lysogen adequately accounts for the inability of the cell to synthesize α-linked polymer, but the possibility that the α-polymerase is also subject to phage repression has not been eliminated.

Infection of S. *anatum* lysogenic for ϵ^{15} with a second phage, ϵ^{34}, results in an additional conversion of the O-antigen; namely, glucosylation of the β-linked galactosyl residues of the polymer (Fig. 8). Uchida *et al.* (1965) first demonstrated the appearance of a new glycosyltransferase activity in the cell envelope of infected cells which catalyzed addition of these glucosyl branch units from UDP-glucose to endogenous acceptor O-antigen. The mechanism of glucosylation has recently been elucidated by Wright (1970), who found that a lipid-linked derivative of glucose, glucose-1-P-GCL, participates as an intermediate in the reaction. The evidence strongly supports the following reaction sequence, where (Man-Rha-β-Gal)$_n \cdots$ represents acceptor O-antigen:

$$\text{UDP-Glc} + \text{P-GCL} \overset{\text{Mg}^{++}}{\rightleftharpoons} \text{Glc-1-P-GCL} + \text{UDP} \qquad (4)$$

$$\text{Glc-1-P-GCL} + (\text{Man-Rha-}\beta\text{-Gal})_n \cdots \rightarrow \overset{\displaystyle \text{Glc}}{\overset{\displaystyle |}{(\text{Man-Rha-}\beta\text{-Gal})_n}} \cdots \qquad (5)$$

The mass spectrum of the lipid derived from the purified glucosyl intermediate was indistinguishable from that of the O-antigen carrier lipid, and it is most probable that the same coenzyme participates both in synthesis of the O-antigen repeating unit and its subsequent modification. Interestingly, however, the structure of the intermediates differs in that the glucosyl residue is linked to the GCL via a phosphodiester bond rather than a pyrophosphoryl group, and formation of the intermediate involves reversible transfer of glucose rather than glucose 1-phosphate from UDP-Glc to the lipid. The reaction is therefore similar to that first described by Scher *et al.* (1968) for synthesis of the mannosyl-P-GCL intermediate in biosynthesis of the membrane-bound mannan of *M. lysodeikticus* (see below, Section 5). The exact nature of the O-antigen acceptor of glucosyl residues has not been definitively established, but the specificity of the glucosylation reaction [reaction (5)] indicated that addition of the glucose branch units must occur at the polymer level. Only β-linked galactosyl residues, formed by action of the ϵ^{15}-specific β-polymerase, were glucosylated *in vivo* or *in vitro*. The particulate enzyme system from S. *anatum* singly lysogenic for ϵ^{34} catalyzed synthesis of Glc-P-GCL but was unable to utilize this intermediate for glucosylation of O-antigen, presumably because the α-linked polymer of this organism was unable to act as acceptor. The available evidence that glucosylation occurs on the nascent polymer-PP-GCl chain, perhaps in conjunction with chain elongation. Evidence that both steps in ϵ^{34}-directed glucosylation are phage specific has been obtained by isolation of two classes of phage mutants (Wright, 1970b), which appear to be blocked in reactions (4) and (5), respectively.

Glucosyl-P-GCL has also been shown by Nikaido and Nikaido (1970) to participate as intermediate in biosynthesis of antigen 12_2 in S. *typhimurium*. This is a modification of O-antigen by glucosylation which is not associated with the presence of any known prophage and is present in some strains of S. *typhimurium* and related species. As in the ϵ^{34} conversion, the glucosyl residues are transferred from the lipid intermediate to the 4-position of the galactose residues of the O-antigen (which are, however, there in α-linkage). Conversion of S. *typhimurium* by phage P22 also results in glucosylation of the galactose residues (in this case, at the 6-position) (Lüderitz *et al.*, 1966); the biochemical basis of P22-specific glucosylation has not been investigated, but it presumably involves a similar mechanism.

4. Biosynthesis of Capsular Polysaccharide in Aerobacter aerogenes

The ability to synthesize extracellular capsular polysaccharides is widely distributed in bacteria, and the pioneering studies of Smith *et al.* (1961) on enzymatic synthesis of pneumococcal capsular polysaccharides laid the groundwork for the flood of later investigations on the mechanism of biosynthesis of bacterial wall and envelope polysaccharides. Although little is known about intermediate steps or the possible role of GCL in biosynthesis of the capsular polysaccharides of *Pneumococcus* or other gram-positive bacteria, recent investigations of Frerman *et al.* (1971) have provided a detailed understanding of the mechanism of capsule biosynthesis in the gram-negative rod, *Aerobacter aerogenes.* The pathway of biosynthesis described by these workers is summarized in Fig. 9 and is precisely analogous to that of O-antigen. The capsular polysaccharide of the strain employed is a polymer of the branched tetrasaccharide repeating unit

$$
\begin{array}{c}
\text{GlcUA} \\
| \\
\text{Gal-Man-Gal}
\end{array}
$$

Synthesis of the oligosaccharide repeating unit is again initiated by reversible transfer of galactose 1-phosphate from UDP-galactose to P-GCL [reaction (1), Fig. 9], followed by sequential addition of man-

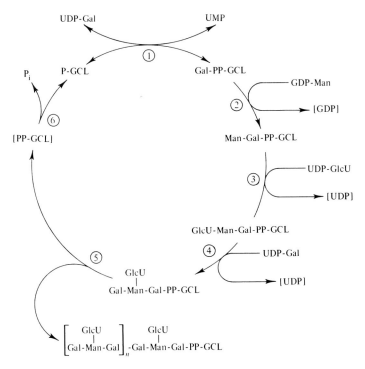

Fig. 9. Pathway of biosynthesis of the capsular polysaccharide of *Aerobacter aerogenes.*

nose, glucuronic acid, and the second galactose residue. The role of P-GCL in this system was established by use of lipid-deficient enzyme preparations. An enzyme system dependent on added lipid for synthesis of the intermediates was obtained by extraction of the cell envelope with acetone, and the active lipid was purified from crude lipid extracts by silicic acid chromatography following alkaline degradation of the bulk phospholipid. Mass spectrometry of the dephosphorylated lipid confirms its identification as the C_{55}-polyisoprenol; a small amount of the C_{60} homolog, was also present. Utilization of exogenous carrier lipid was dependent on addition of the nonionic detergent, Triton-X100; interestingly, the detergent could be replaced by phosphatidylglycerol.

The mechanism of chain elongation of the capsular polysaccharide was not examined in detail, but isolation of a lipid-linked dimer of the repeating unit suggested that the mechanism of polymerization is also similar to that of O-antigen. An unsolved problem of interest relates to the nature of the final acceptor of the finished polymer. The capsular polysaccharide adheres firmly to the cell envelope, suggesting that chains may remain attached to some component of the envelope structure; however, analyses of purified polysaccharide failed to reveal any constituents other than the three known sugars. Although the possibility has not been excluded that the final product of capsule synthesis consists of free polysaccharide chains released by enzymatic hydrolysis of the polymer–lipid linkage, Frerman et al. have suggested the intriguing alternative explanation that P-GCl itself may serve to anchor the polysaccharide to the envelope. According to this hypothesis, long-chain polymer-PP-GCl would represent the end product of capsule biosynthesis, and the lipid would play a structural role in the membrane in addition to its function as biosynthetic coenzyme.

5. Biosynthesis of Membrane-Bound Mannan in Micrococcus lysodeikticus

In the course of studies on the biosynthesis of the mannosyldiglycerides of M. lysodeikticus, Lennarz and Talamo (1966) observed incorporation of mannose from GDP-mannose into an acid-labile lipid product which was clearly unrelated to the diglycerides but could not at that time be identified. Subsequent investigations (Scher et al., 1968) revealed that the new mannolipid was in fact mannosyl-1-phosphorylpolyisoprenol, and functioned as an intermediate in biosynthesis of the membrane-bound mannan of this organism. This was the first example of a GCL-linked intermediate involving attachment of the glycosyl unit to the lipid by a phosphodiester rather than a pyrophosphoryl group; conclusive evidence for the phosphodiester structure was provided by chemical analysis and degradation of purified intermediate as well as identification of GDP as the nucleotide product of its biosynthesis. Enzymatic synthesis of mannosyl-P-GCL was investigated in membrane fractions depleted of lipid by acetone extraction. Synthesis of the intermediate was dependent on added crude lipid extract or purified P-GCL (Lahav *et al.*, 1969) and proceeded according to the following reaction:

$$\text{GDP-mannose} + \text{P-GCL} \overset{\text{Mg}^{++}}{\rightleftharpoons} \text{mannosyl-1-P-GCL} + \text{GDP} \tag{6}$$

With purified P-GCL as substrate the reaction also showed an absolute requirement for a surface-active agent. Phosphatidylglycerol and phosphatidylethanolamine were most effective, but good activity was also obtained with a number of nonionic detergents. It is not yet clear whether these agents simply promote access of the exogenous lipid substrate to the enzyme or interact in some more direct manner with the membrane-bound enzyme protein.

The kinetics of incorporation of mannose from GDP-mannose suggested a precursor–products relationship between mannosyl-P-GCL and polymeric mannan (Scher *et al.*, 1968); this was established by experiments in which purified mannosyl-P-GCL was employed as substrate for mannan synthesis. Differential inhibition of mannosyl-P-GCL synthesis and utilization (by EDTA and Triton-X100, respectively) showed definitively that the lipid intermediate acts as direct donor of mannosyl residues to endogenous acceptor mannan (Scher and Lennarz, 1969). However, analysis of the enzymatic product by methylation and chemical and enzymatic degradation showed that mannose from the lipid was incorporated almost exclusively into nonreducing terminal positions of the acceptor, probably as single branch units. The reaction observed *in vitro* therefore does not represent *de novo* synthesis of mannan and appears rather to be analogous to glucosylation of O-antigen in *Salmonella*. It is also noteworthy that the lipid intermediate in both cases has the phosphodiester structure. No evidence for formation of lipid-linked mannose oligosaccharides could be obtained, and the mechanism of biosynthesis of the backbone chain of the mannan remains to be established.

Enzymatic synthesis of mannosyl-P-GCL by a particulate enzyme preparation from *Mycobacterium tuberculosis* H37Ra has recently been described by Takayama and Goldman (1970). Interestingly, mass spectroscopy of the lipid indicated a preponderance of C_{50}-polyisoprenol, with only a small amount of the C_{55} homolog. The role of mannosyl-P-GCL in *M. tuberculosis* is not yet known; the organism is rich in phosphatidylinositol mannosides and also contains a mannan.

6. Biosynthesis of Teichoic Acids

The cell walls of many gram-positive bacteria contain a teichoic acid or related polymer covalently linked to peptidoglycan (Archibald *et al.*, 1968a; Ghuysen, 1968). The first teichoic acids to be described were polymers of ribitol phosphate or glycerophosphate, in which the polyol residues were linked by phosphodiester bonds and were substituted to varying degrees by glucosyl or N-acetylglucosaminyl residues and by ester-linked D-alanine. The pathways of biosynthesis of ribitol and glycerol teichoic acids have been extensively investigated by Burger and Glaser in membrane–wall fractions of *Lactobacillus plantarum* and *Bacillus* species and by Strominger and associates in *Staphylococcus aureus*. The polyol–phosphate chains are formed by successive transfer of polyol–phosphate units from CDP-ribitol or CDP-glycerol to endogenous acceptor (Burger and Glaser, 1962, 1964; Glaser, 1963, 1964; Chin *et al.*, 1966; Ishimoto and Strominger, 1966), according to the following reaction:

$$n\text{CDP-polyol} + \text{acceptor} \rightarrow (\text{polyol-P})_n\text{-acceptor} + n\text{CMP} \qquad (7)$$

The nature of the endogenous acceptor or primer in these systems was not entirely clear but presumably consisted of end groups of preexisting teichoic acid chains. Chain elongation was shown (Kennedy and Shaw, 1968) to occur by addition of polyol–phosphate units to the polyol–terminal end of the chain:

$$\text{CDP}^*\text{-polyol}^* + \text{polyol-P-(polyol-P)}_n\text{-polyol-P} \rightarrow$$
$$\text{polyol}^*\text{-P}^*\text{-polyol-P-(polyol-P)}_n\text{-polyol-P} + \text{CMP} \quad (8)$$

Addition of glucose or N-acetylglucosamine branch units to polyol residues of the polymer was shown (Nathenson and Strominger, 1962, 1963; Nathenson *et al.*, 1966; Burger, 1963; Glaser and Burger, 1964) to occur by the following reaction where UDP-X represents UDP-glucose or UDP-GlcNAc:

$$\overset{\text{X}}{\underset{|}{n\text{UDP-X}}} + (\text{polyol-P})_n \rightarrow (\text{polyol-P})_n + n\text{UDP} \qquad (9)$$

Although exogenous glucose- or GlcNAc-deficient teichoic acids were effective acceptors for these glucosyltransferase reactions, experiments of Ishimoto and Strominger (1966) strongly suggested that glycosylation normally proceeds concurrently with polymerization. Both the rate and extent of GlcNAc incorporation into the ribitol teichoic acid of *S. aureus* were markedly enchanced when endogenous acceptor was generated *in situ* by addition of CDP-ribitol to the enzyme system.

There is at present no evidence for participation of P-GCL in biosynthesis of the polyribitol phosphate and polyglycerophosphate types of teichoic acid. However, Baddiley and co-workers have shown that lipid-linked intermediates do occur in synthesis of the atypical teichoic acids found in the cell walls of *Staphylococcus lactis*. The teichoic acid of *S. lactis* I3 is composed of alternating residues of glycerol-1-phosphate and N-acetylglucosaminyl-1-phosphate linked by phosphodiester bonds (Archibald *et al.*, 1965):

$$[\text{Glycerol-1-P-GlcNAc-1-P}]_n\text{-glycerol-1-P} \cdots$$

Evidence was obtained (Baddiley *et al.*, 1968; Douglas and Baddiley, 1968, Baddiley, 1970) for the following pathway of biosynthesis:

$$\text{UDP-GlcNAc} + \text{P-lipid} \leftrightarrows \text{GlcNAc-1-PP-lipid} + \text{UMP} \tag{10}$$

$$\text{CDP-glycerol} + \text{GlcNAc-1-PP-lipid} \rightarrow \text{glycerol-1-P-GlcNAc-1-PP-lipid} + \text{CMP} \tag{11}$$

$$\text{Glycerol-1-P-GlcNAc-1-PP-lipid} + \text{acceptor} \rightarrow$$
$$\text{glycerol-1-P-GlcNAc-1-P-acceptor} (+\text{P-lipid}) \tag{12}$$

Incubation of membrane fractions with $[^{32}\text{P}]\text{UDP-}[^{14}\text{C}]\text{GlcNAc}$ in the absence of CDP-glycerol resulted in incorporation of both ^{14}C and ^{32}P into a product identified as a lipid-linked derivative of GlcNAc-1-phosphate. In the presence of CDP-glycerol a small amount of a second lipid product was obtained which yielded glycerol-1-P-GlcNAc on mild acid hydrolysis. Utilization of the first intermediate for polymer synthesis was demonstrated by a two-step incubation in which membranes were first exposed to UDP-$[^{14}\text{C}]$GlcNAc alone, then washed and incubated with CDP-glycerol as sole added substrate. A similar lipid-linked derivative of GlcNAc-1-phosphate has also been found to occur as an intermediate in biosynthesis of the teichoic acid of another strain of *S. lactis*, NCTC 2102 (Brooks and Baddiley, 1969a,b). This teichoic acid is a polymer of phosphodiester-linked GlcNAc-1-phosphate units (Archibald *et al.*, 1968b):

$$\text{GlcNAc-}\alpha\text{-1-P-6-[GlcNAc-}\alpha\text{-1-P-]}_n$$

Synthesis of the intermediate was shown to proceed according to reaction (10), and the pyrophosphate linkage between GlcNAc and the lipid portion was established by isolation of GlcNAc-1-pyrophosphate and inorganic pyrophosphate after brief acid hydrolysis. The nature of the lipid was not established, but the properties of the purified intermediate were consistent with those of other GCL-linked intermediates. The role of GlcNAc-1-PP-lipid as intermediate in polymer synthesis was established by pulse-chase experiments. The direction of chain growth in synthesis of both *S. lactis* polymers has also been investigated by pulse-chase experiments (Hussey *et al.*, 1969). The results indicated that chain elongation occurs by successive addition of single repeating units to the nonreducing terminus of primer chains.

The enzymatic mechanism involved in covalent attachment of teichoic acids and related polysaccharides to peptidoglycan remains an unsolved problem of considerable interest. By analogy with the mechanism of attachment of O-antigen chains to lipopolysaccharide, it is possible that teichoic acid is polymerized independently of peptidoglycan and that the completed polymer is secondarily attached to peptidoglycan strands. Such a mechanism would require activation of the completed polymer chain for the final joining reaction, either by synthesis in the form of a polymer-PP-GCl derivative or by a mechanism analogous to that employed in the DNA ligase reaction. Alternatively, and perhaps more likely, teichoic acid synthesis may be initiated by transfer of the first repeating unit directly to the peptidoglycan acceptor. The observed direction of chain growth is consistent with this mechanism, but it does not exclude the first hypothesis.

7. Structural and Functional Interrelations among Bacterial Glycosyl Carrier Lipids

With the growing number of bacterial systems which are known to utilize polyisoprenol phosphate as coenzyme, the question arises as to whether "glycosyl carrier lipid" represents a family of structurally distinct and functionally specific coenzymes, or whether the same coenzyme is used by all species and for multiple functions in the same cell. The available evidence, though incomplete, suggests that the latter is more nearly correct. Mass spectra characteristics of a C_{55} fully unsaturated polyisoprenol have been obtained from the lipid portion of intermediates of peptidoglycan (Higashi et al., 1967) and mannan biosynthesis (Scher et al., 1968) in M. lysodeikticus; from intermediates of peptidoglycan biosynthesis, as well as total cellular polyisoprenol, in S. aureus (Higashi et al., 1970b); and from intermediates of O-antigen synthesis in Salmonella (Wright et al., 1967) and of capsular polysaccharide synthesis in Aerobacter (Frerman et al., 1971). In addition to the major C_{55} component, traces of lower homologs (C_{50} and C_{45}) were detected in peptidoglycan intermediates from M. lysodeikticus (Higashi et al., 1967); small amounts of C_{50}- and C_{60}-isoprenols were also found in the micrococcal mannosyl-P-GCL (Scher et al., 1968), and a trace of C_{60} was found in the intermediates from A. aerogenes (Frerman et al., 1971). There is at present no evidence to suggest that the lower or higher homologs in these organisms represent a significant fraction of the total polyisoprenol phosphate or that they function in different reactions than the major C_{55} species. It should be noted, however, that the mannosyl-P-GCL of Mycobacterium tuberculosis contains primarily the C_{50}-polyprenol (Takayama and Goldman, 1970). Further evidence for structural similarity between the micrococcal and staphylococcal coenzymes was provided by NMR spectroscopy (Scher *et al.*, 1968; Higashi *et al.*, 1970b) which showed that both lipids contain two internal *trans*-double bonds; NMR data are not yet available for GCL from gram-negative bacteria.

There is also considerable evidence that carrier lipids isolated from different sources are functionally equivalent. Thus, the lipid involved in mannan biosynthesis in *M. lysodeikticus* was fully active in synthesis of capsular polysaccharide in *A. aerogenes* and vice versa (Frerman, Heath, Lahav, Chiu, and Lennarz, quoted in Lahav *et al.*, 1969), and the P-GCL species isolated from *S. aureus* and *M. lysodeikticus* were as effective as the homologous coenzyme in synthesis of O-antigen intermediates in *S. typhimurium* (Osborn, 1967). These results cannot, however, be taken as evidence for structural identity of the coenzymes, since the phospho-monoester of ficaprenol-11, which differs from the bacterial coenzyme in having three internal *trans*-double bonds rather than two, was able to replace the natural coenzyme in synthesis of both mannosyl-P-GCL in *M. lysodeikticus* (Lahav *et al.*, 1969) and intermediates of peptidoglycan synthesis in *S. aureus* (Higashi *et al.*, 1970a).

8. Biosynthesis and Metabolism of Glycosyl Carrier Lipid

a. Biosynthesis of Long-Chain Polyisoprenols. Enzyme systems capable of synthesizing long-chain polyisoprenes from farnesyl pyrophosphate (C_{15}) plus Δ^3-isopentenyl pyrophosphate have been isolated both from *Micrococcus lysodeikticus* (Allen *et al.*, 1967) and *Salmonella newington* (Christensen *et al.*, 1969). Two soluble activities were separated in *M. lysodeikticus*, one catalyzing formation of a C_{20} product, the other leading to products of chain length C_{35}–C_{45} with traces of C_{50}. It is not clear whether these represent precursors of the C_{55} carrier lipid or are instead intermediates in biosynthesis of quinone coenzymes. A soluble enzyme activity yielding relatively short products was also observed in *S. newington*. However, a second activity, associated with the particulate cell envelope fraction, catalyzed formation of a polyisoprenol product which was chromatographically indistinguishable from glycosyl carrier lipid and which served as substrate for synthesis of the O-antigen intermediate, Rha-Gal-PP-GCL. The postulated pathway of biosynthesis of P-GCL is

Farnesyl pyrophosphate + 8 Δ^3-isopentenyl pyrophosphate →
$$\text{glycosyl carrier lipid pyrophosphate (PP-GCL)} + 8PP_i \quad (13)$$

It is assumed that the initial product of biosynthesis is the pyrophosphate derivative and that the coenzymatically active monoester form is generated by subsequent hydrolysis of the terminal phosphate [reaction (15)].

b. Glycosyl Carrier Lipid Pyrophosphate Phosphatase. Pyrophosphoryl-GCL also occurs as a product in polymerization reactions utilizing oligosaccharide-PP-GCL intermediates, as in synthesis of peptidoglycan (Fig. 3), O-antigen (Fig. 6), and capsular polysaccharide in *Aerobacter* (Fig. 9). In these chain elongation reactions only the glycosyl portion of the intermediates is incorporated into the polymer chains, while the phosphate group derived from the initial nucleotide sugar substrate ultimately appears in the medium as P_i (Anderson *et al.*, 1965; Osborn and Weiner, 1967):

$$n(\text{oligosaccharide-PP-GCL}) \rightarrow \text{polymer} + n\text{PP-GCL} \quad (14)$$

$$n\text{PP-GCL} \xrightarrow{\text{H}_2\text{O}} n\text{P-GCL} + nP_i \quad (15)$$

Direct evidence for formation of pyrophosphoryl-GCL as the immediate product of the chain elongation sequence, and for the existence of a membrane-bound phosphatase catalyzing its hydrolysis, emerged from studies of Siewert and Strominger (1967) on the effect of bacitracin on peptidoglycan synthesis. This antibiotic had no effect on synthesis or utilization of lipid-linked intermediates *per se*, but it specifically inhibited release of P_i and regeneration of active P-GCL by reaction (15). As a result, carrier lipid was trapped in the pyrophosphate form and prevented from reentering the reaction cycle. The presence of bacitracin-sensitive PP-GCL phosphatase activity in the membrane was confirmed using isolated PP-GCL as substrate, but the enzyme has not yet been characterized in detail. As would be expected, bacitracin also inhibits *in vitro* synthesis of O-antigen (see Robbins and Wright, 1971; Osborn, 1967) and the *Aerobacter* capsular polysaccharide (Frerman *et al.*, 1971), but it has no effect in systems involving the phosphodiester-linked intermediates, mannosyl- or glycosyl-P-GCL (Scher and Lennarz, 1969; Wright, 1970).

c. Isoprenoid Alcohol Phosphokinase. *Staphylococcus aureus* differs from other bacteria so far investigated in that a large fraction of the total carrier lipid is present as the nonphosphorylated free alcohol (Higashi *et al.*, 1970b). Higashi *et al.* (1970a) have described an ATP-dependent isoprenoid alcohol phosphokinase, which is present in the membrane fraction of *S. aureus* and catalyzes conversion of the free alcohol (HO-GCL) to the coenzymatically active phosphomonoester form:

$$\text{HO-GCL} + \text{ATP} \rightarrow \text{P-GCL} (+\text{ADP}) \quad (16)$$

The enzyme has the very interesting and unusual property of being soluble in butanol and similar solvents. The phosphokinase activity, together with about 3% of the total membrane protein and most of the lipid, was solubilized by extraction of the membrane with acidic butanol. The phosphokinase was soluble in butanol and lower alcohols and was heat stable under these conditions. In aqueous solvents the enzyme was both insoluble and relatively unstable. Fractionation of the butanol extract on DEAE-cellulose in butanol–ammonium acetate resulted in separation into a protein and a phospholipid fraction, both of which were required for phosphokinase activity (Higashi and Strominger, 1970). The phospholipid requirement was specific for phosphatidylglycerol and cardiolipin; phosphatidylethanolamine and other phosphatides tested were essentially inactive. The phosphokinase appears to have a rather broad specificity for its isoprenoid alcohol substrate. In addition to HO-GCL isolated from S. aureus, a mixture of ficaprenols of chain length C_{45}–C_{60} and betulaprenols of C_{30}–C_{45} showed high activity as substrates; the all-trans-C_{45}-alcohol, solanesol, was inactive, as were very long chain polyisoprenols (C_{80}–C_{105}) and the C_{15} compound farnesol.

A variety of polyprenols has been isolated from plant and animal sources. The ficaprenols from leaves of Ficus elasticus (Stone et al., 1967) differ from bacterial GCL in having three internal trans double bonds rather than two and range in chain length from C_{45} to C_{60}; the betulaprenols, obtained from birchwood (Lindgren, 1965; Wellburn and Hemming, 1966) contain predominantly C_{35}- and C_{40}-alcohols; solanesol (Rowland et al., 1956) is an all-trans-C_{45}-alcohol; dolichols, isolated from animal tissues (Burgos et al., 1963; Stone et al., 1967), are very long chain polyprenols (C_{80}–C_{105}) in which the first C_5 unit is saturated; farnesol is the trans-trans-C_{15}-alcohol.

Thus far neither polyisoprenoid alcohol phosphokinase nor the free alcohol form of GCL has been detected in bacteria other than S. aureus, and the function of the phosphokinase in this organism is not known. Higashi et al. (1970a) have put forward the attractive idea that the rate of the cell wall synthesis in S. aureus may be regulated by controlling the level of functional P-GCL through a dephosphorylation–rephosphorylation cycle. A membrane-bound enzyme catalyzing dephosphorylation of C_{55}-ficaprenol phosphate has indeed been detected (see Higashi et al., 1970a). The phosphokinase is discussed further in Chapter 6.

9. Lipid-Linked Intermediates in Fungi and Higher Plants

Yeast cell walls contain two major polysaccharide components, a glucan and a mannan. Behrens and Cabib (1968) have described a particulate enzyme fraction, obtained by lysis of protoplasts of Saccharomyces carlsbergensis, which catalyzes synthesis of mannan from GDP-mannose. Acetolysis of the radioactive enzyme products indicated that mannose was incorporated into internal linkages as well as nonreducing termini and suggested that the enzyme preparation was capable of synthesizing most, if not all, of the polymer chain. Formation of lipid-linked intermediates could not be detected in the enzyme preparation from S. *carlsbergensis*, but convincing evidence for a mannosyl–lipid intermediate has been obtained by Tanner (1969) in S. *cerevisiae*. In this case, the particulate enzyme was isolated after disruption of cells by grinding with alumina. The evidence supported the following reaction sequence:

$$\text{GDP-mannose} + \text{P-lipid} \overset{\text{Mg}^{++}}{\rightleftharpoons} \text{mannosyl-P-lipid} + \text{GDP} \qquad (17)$$

$$\text{Mannosyl-P-lipid} (+\text{acceptor?}) \overset{\text{Mn}^{++}}{\longrightarrow} \text{mannan} \qquad (18)$$

Reversal of reaction (17) by GDP but not GMP and exchange of [^{14}C]GDP (but not [^{14}C]GMP) into GDP-mannose suggested that synthesis of the intermediate was analogous to that of mannosyl-P-GCL in *Micrococcus lysodeikticus* (Scher et al., 1968). The kinetics of incorporation of mannose into lipid and mannan were consistent with the role of the lipid as an intermediate, and this was confirmed by taking advantage of the different metal ion requirements of the two reactions. Thus, formation of the intermediate was supported by Mg^{++}, while its utilization for polymer synthesis specifically required Mn^{++}. Analysis of the enzymatically synthesized mannan showed that [^{14}C]mannose was incorporated into internal linkages, suggesting that the lipid-linked intermediate may participate in synthesis of the backbone of the mannan. The properties of the intermediate were similar to those of bacterial mannosyl-P-GCL, but the lipid has not yet been characterized in detail.

Synthesis and utilization of a mannosyl–lipid intermediate in mung bean (*Phaseolus aureus*) by reactions similar to (17) and (18) has been described by Kauss (1969). Although the lipid was not definitively identified, evidence for a polyisoprenol structure was obtained by purification of ^{3}H-labeled mannosyl–lipid following growth of bean shoots with [^{3}H]mevalonic acid. Villemez and Clark (1969) have also reported incorporation of mannose from GDP-mannose into a lipophilic product in a particulate enzyme system from P. *aureus*. However, the solubility properties of this product differed from those of known mannosyl-P-lipid intermediates in that the product was not extracted from the particles by butanol or chloroform–methanol. The evidence suggested that the material represented oligosaccharide derivatives.

C. Biosynthetic Pathways Not Involving Glycosyl Carrier Lipid: Biosynthesis of the Core Region of Lipopolysaccharide

Although it is now clear that polyisoprenol phosphate coenzymes provide a mechanism of general significance for synthesis of polysaccharides by membrane-bound enzyme systems, it is also clear that not all such processes are mediated by lipid coenzymes. One biosynthetic pathway in which participation of P-GCL has been rigorously excluded is the synthesis of the outer core region of the lipopolysaccharide of *Salmonella typhimurium*. This region consists of a branched pentasaccharide sequence containing glucose, galactose, and N-acetylglucosamine (Fig. 5). Initial characterization of the pathway of biosynthesis depended heavily on the use of mutants blocked in biosynthesis of a required nucleotide sugar precursor, UDP-glucose or UDP-galactose; such mutants form incomplete lipopolysaccharides which contain only that portion of the molecule proximal to the point of the biosynthetic lesion. These studies (for review, see Osborn and Rothfield, 1971) resulted in identification of a series of glycosyltransferases which catalyzed sequential transfer of single sugar residues from the appropriate nucleotide sugar to the incomplete mutant lipopolysaccharide, as indicated in Fig. 10. Two glucosyltransferase activities, I and II, catalyzing reactions (1) and (4) of Fig. 10, respectively, galactosyltransferase I [α,3-galactosyltransferase, reaction (2), Fig. 10], and the N-acetylglucosaminyltransferase [reaction (5), Fig. 10] were demonstrated in the cell envelope fraction of mutants of both *S. typhimurium* (Nikaido, 1962; Osborn *et al.*, 1962; Rosen *et al.*, 1964; Rothfield *et al.*, 1964; Osborn and D'Ari, 1964) and *E. coli* 0111 (Heath *et al.*, 1966; Edstrom and Heath, 1967). Galactosyltransferase II, responsible for addition of the branch α,6-galactosyl residue in *S. typhimurium* [reaction (3), Fig. 10], has not been

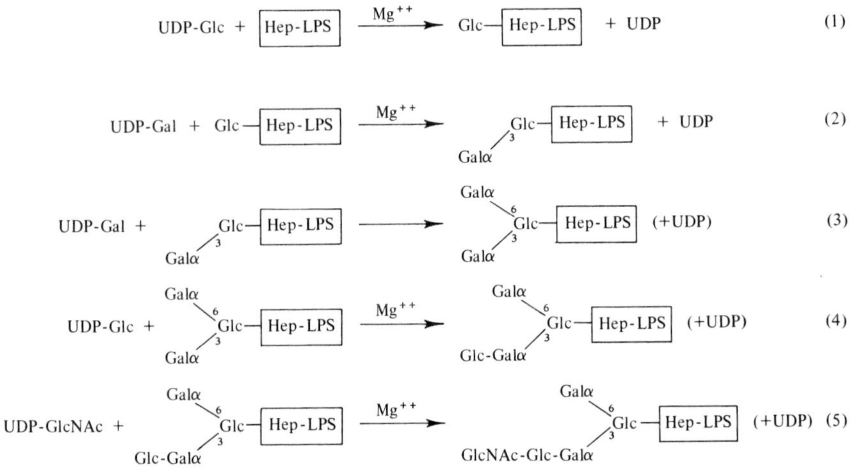

Fig. 10. Pathway of biosynthesis of the outer core region of the lipopolysaccharide of *Salmonella typhimurium*; Hep-Lps represents the backbone–Lipid A portion of the lipopolysaccharide (see Fig. 5).

demonstrated directly; evidence for the existence of this enzyme is based on analysis of mutants deficient in α,3-transferase activity (Wilkinson and Stocker, 1968; Osborn, 1968). The branch galactose residue is not present in *E. coli* (Edstrom and Heath, 1967).

Although the core glycosyltransferase enzymes were primarily associated with the cell envelope fraction, which also contained the endogenous acceptor lipopolysaccharide, certain of the activities (primarily glycosyltransferase I and galactosyltransferase I) were also found in the soluble fraction following disruption of the cells by prolonged sonic oscillation (Rothfield *et al.*, 1964). The soluble galactosyltransferase I (UDP-galactose:lipopolysaccharide α,3-galactosyltransferase) (Endo and Rothfield, 1969a) and glucosyltransferase I (UDP-glucose:lipopolysaccharide glucosyltransferase) (Müller and Rothfield, 1970) have been purified to homogeneity. Both enzymes are dependent on added acceptor lipopolysaccharide and require in addition a phospholipid such as phosphatidylethanolamine. All of the evidence indicates that the sugars are transferred directly from the nucleotide sugar donor to the acceptor lipopolysaccharide. Participation of P-GCL in the purified galactosyl system was rigorously excluded by reconstitution of full activity in a system consisting of highly purified solvent-extracted enzyme, purified solvent-extracted lipopolysaccharide, and synthetic dioleyl phosphatidylethanolamine (Endo and Rothfield, 1969b).

The role of phospholipid in the core transferase reactions is discussed in detail in Chapter 6. Sucrose gradient centrifugation (Weiser and Rothfield, 1968) and monolayer experiments (Romeo *et al.*, 1970a,b) with galactosyltransferase I have provided definitive evidence for initial formation of a lipopolysaccharide–phospholipid complex which is required for effective binding of the transferase enzyme to its lipopolysaccharide substrate. The results have established the following obligatory reaction sequence (where LPS is galactose-deficient lipopolysaccharide, PE is phosphatidylethanolamine, and Enz is α,3-galactosyltransferase):

$$\text{LPS} + \text{PE} \longrightarrow \text{LPS} \cdot \text{PE} \qquad (19)$$

$$\text{Enz} + \text{LPS} \cdot \text{PE} \xrightarrow{\text{Mg}^{++}} \text{Enz} \cdot \text{LPS} \cdot \text{PE} \qquad (20)$$

$$\text{UDP-Gal} + \text{Enz} \cdot \text{LPS} \cdot \text{PE} \xrightarrow{\text{Mg}^{++}} \text{galactosyl-LPS} \cdot \text{PE} + \text{UDP} \ (+\text{Enz}) \qquad (21)$$

Reconstitution of the functional enzyme system in monolayer by stepwise formation of the binary LPS·PE complex and ternary Enz·LPS·PE from an initial monomolecular film of PE (Romeo *et al.*, 1970a,b) strongly suggests that the arrangement of molecules in the monolayer mimics that in the native cell envelope. On the basis of these results and electron microscopic observations of Rothfield and Horne (1967) it seems most reasonable to suppose that the lipid portion of nascent lipopolysaccharide molecules (Lipid A, Fig. 5) is inserted into the nonpolar interior of the membrane phospholipid bilayer with the growing core polysaccharide chain at the polar surface. The mode of binding of the core glycosyltransferases to the membrane is not clearly understood, but it must involve interaction with the lipid bilayer as well as with the acceptor site of the nascent polysaccharide chain. In attempting to rationalize the differences in the mechanisms of biosynthesis of the core portion of lipopolysaccharide and the O-antigen chains, it is tempting to speculate that lipid A may serve to anchor the growing core polysaccharide in the membrane during biosynthesis, as glycosyl carrier lipid is believed to anchor intermediates in O-antigen synthesis. Similarly, if it is assumed that glycosyl transfer from nucleotide sugar donors must take place at the cytoplasmic surface of the membrane, Lipid A may also facilitate transport of the completed core to the external surface, a function analogous to the transport role postulated for GCL. Thus, formation of GCL-linked intermediates would appear to be unnecessary if assembly of the relatively short oligosaccharide chains of the core occurs by stepwise addition of sugars to Lipid A at the internal surface of the membrane where nucleotide sugars are presumably directly available to the transferase enzymes.

III. BIOSYNTHESIS OF PROTEINS AND GLYCOPROTEINS

The mechanisms of biosynthesis of the complex carbohydrate units of the glycoproteins of animal tissues are less clearly understood than the synthesis of bacterial polysaccharides, primarily because of the inherently greater complexity of the animal systems and the difficulties in applying the types of genetic and physiological manipulation which have proved invaluable in bacteria. Nevertheless, progress in defining pathways of biosynthesis has been rapid during the past five years, and a breakthrough in defining intermediate steps and details of molecular mechanisms now appears imminent.

A. Role of the Endoplasmic Reticulum in Glycoprotein Synthesis

There is general agreement (Ginsburg and Neufeld, 1969; Spiro, 1970) that synthesis of many secretory glycoproteins occurs by a multistage process mediated by the endoplasmic reticulum. The polypeptide chains are synthesized on polyribosomes of the rough endoplasmic reticulum, from which completed polypeptides enter the membranes of the endoplasmic reticulum. The proteins then pass progressively through the channels of the rough and smooth endoplasmic reticulum to the Golgi apparatus where final packaging for secretion takes place. Most, if not all, of the sugar residues are added after completion of the polypeptide, either in the course of passage through the endoplasmic reticulum or in the Golgi apparatus, or both. The evidence for this overall view of temporal and spatial relationships was derived primarily from the kinetics of incorporation of radioactive amino acids and sugars in a variety of intact tissues, as determined by radioautography at the electron microscope level (Warshawsky *et al.*, 1963; Caro and Palade, 1964; Nadler *et al.*, 1964; Redman *et al.*, 1966; Zagury *et al.*, 1970) and by subcellular fractionation (Sarcione, 1964; Sarcione *et al.*, 1964; Molnar *et al.*, 1965b; Lawford and Schachter, 1966; Spiro and Spiro, 1966; Simkin and Jamieson, 1967; Melchers and Knopf, 1967; Moroz and Uhr, 1967). Studies on the effects of puromycin on synthesis of membrane glycoproteins in Ehrlich ascites cells (Molnar *et al.*, 1965a; Cook *et al.*, 1965), thyroglobulin (Spiro and Spiro, 1966), liver glycoproteins (Molnar and Sy, 1967), and immunoglobulin (Schenkein and Uhr, 1970) have established that addition of carbohydrate takes place independently of polypeptide synthesis and may require an appreciable period of time for completion. In all cases, incorporation of labeled sugar precursors into glycoprotein products continued for considerable periods of time under conditions in which amino acid incorporation was completely abolished. In general, incorporation of sugars into internal positions of the carbohydrate chains, adjacent to the peptide, was more severely affected by puromycin than was incorporation into the outer portions of the chains. The results were consistent with sequential completion of previously synthesized, partially glycosylated precursor proteins. Such particle-bound precursors have been described for several soluble glycoproteins, including thyroglobulin (Spiro and Spiro, 1966) and plasma glycoproteins (Sarcione *et al.*, 1964; Simkin and Jamieson, 1967; Li *et al.*, 1968) and, indeed, a significant level of endogenous acceptor activity has been observed in many crude particulate glycosyltransferase systems (see below).

Some question still exists as to the level at which the initial sugar is added to the peptide chain. Two general classes of glycopeptide bond are known (Spiro, 1970): N-glycosidic linkage between N-acetylglucosamine and the amide nitrogen of asparagine and O-glycosidic linkage between any of several sugars and a hydroxyamino acid, (serine, threonine, or hydroxylysine). Studies on enzymatic synthesis of carbohydrate units of the latter type have clearly established that glycosylation is initiated after completion of the polypeptide and release from the ribo-

some (see Section III,B, below). However, the site of formation of the
N-acetylglucosaminyl–asparagine linkage is unclear, and there is some
indication that GlcNAc may be attached to ribosome-bound nascent polypeptide chains. *In vivo* labeling studies in liver (Molnar *et al.*, 1965b;
Lawford and Schachter, 1966; Molnar and Sy, 1967; Molnar and Dalisay,
1967) and mouse myeloma tumor cells (Moroz and Uhr, 1967) have
shown incorporation of small amounts of radioactive glucosamine into a
polyribosome fraction, from which the label could be released by incubation with puromycin. Other evidence is, however, in conflict with this
hypothesis (Sarcione *et al.*, 1964; Cook *et al.*, 1965; Sinohara and Sky-
Peck, 1965; Bouchilloux and Cheftel, 1966). Unfortunately, enzymatic
synthesis of the GlcNAc-asparagine bond in cell-free preparations has
not yet been achieved.

B. Mechanism of Biosynthesis of the Carbohydrate Units of Glycoproteins

1. Isolated Glycosyltransferase Systems

Complete or partial synthesis of the carbohydrate units of a number
of glycoproteins has now been accomplished using isolated glycosyltransferase systems and well-defined acceptor proteins. In every case assembly
of the saccharide chain has been shown to occur by sequential addition
of single sugar residues. Synthesis of submaxillary mucins has been extensively studied by Roseman and co-workers and Hagopian and Eylar.
Ovine and bovine submaxillary mucins (OSM and BSM) contain disaccharide units, N-acetylneuraminic acid–N-acetylgalactosamine (NAN-
GalNAc) linked to hydroxyl groups of serine and threonine. Two glycosyltransferase enzymes have been demonstrated in membrane fractions
of sheep submaxillary gland (Carlson *et al.*, 1964; McGuire and Roseman,
1967) which catalyze addition of disaccharide units to serine and threonine residues of exogenous OSM depleted of carbohydrate by treatment
with neuraminidase and acetylhexosaminidase, according to the sequence

$$\text{UDP-GalNAc} + \text{HO-OSM} \rightarrow \text{GalNAc-OSM} + \text{UDP} \tag{22}$$

$$\text{CMP-NAN} + \text{GalNAc-OSM} \rightarrow \text{NAN-GalNAc-OSM} + \text{CMP} \tag{23}$$

The N-acetylgalactosaminyltransferase has also been studied by Hagopian and Eylar (1968, 1969a,b), who have solubilized and partially purified the enzyme from smooth membranes of bovine submaxillary gland.
The oligosaccharide units of porcine submaxillary mucin are more complex, consisting of a branched pentasaccharide containing L-fucose (Fuc)
and galactose in addition to GalNAc and N-glycolylneuraminic acid
(NGN):

$$\overset{\displaystyle \text{Fuc} \quad \text{NGN}}{\underset{\displaystyle \text{GalNAc-Gal-GalNAc-Ser(Thr)}}{\mid \qquad \mid}}$$

A series of five glycosyltransferases responsible for sequential synthesis of the pentasaccharide has recently been found in a particulate fraction from porcine submaxillary gland (Schachter and McGuire, 1968; McGuire and Roseman, 1969). Porcine submaxillary mucin from which part of the carbohydrate unit had been removed by sequential treatment with the appropriate glycosidases was employed as acceptor.

Collagen and basement membranes are glycoproteins containing the disaccharide unit, glucosyl–galactose, linked to hydroxylysine residues of the peptide. Since hydroxylysine is formed by hydroxylation of lysine residues in the completed polypeptide (Prockop and Kivirikko, 1967), glycosylation of these proteins must necessarily occur at a postribosomal stage. Enzymatic synthesis of the disaccharide units of basement membrane and collagen has been shown (Spiro and Spiro, 1968c; Blumenkrantz *et al.*, 1968; Bosmann and Eyler, 1968) to occur by the two-step reaction

$$\text{UDP-Gal} + \text{HO-protein} \rightarrow \text{gal-protein} \ (+\text{UDP}) \qquad (24)$$

$$\text{UDP-Glc} + \text{Gal-protein} \rightarrow \text{Glc-Gal-protein} \ (+\text{UDP}) \qquad (25)$$

The galactosyl- and glucosyltransferase activities were found in both particulate and soluble fractions of kidney cortex and embryonic skin and cartilage, and they utilized as acceptor either basement membrane protein, stripped of carbohydrate by Smith degradation, or native collagen (which normally contains a large fraction of unsubstituted hydroxylysine residues).

Perhaps the most complex units for which the complete pathway of biosynthesis is known are the mucopolysaccharide chains of connective tissue chondromucoproteins (also called proteoglycans or protein–polysaccharides) such as the chondroitin sulfates. The polysaccharide chains consist of alternating residues of hexosamine and uronic acid (N-acetylgalactosamine sulfate and glucuronic acid in the case of chondroitin sulfate). These are attached to serine residues of a polypeptide backbone via a specialized trisaccharide "linkage region" with the structure galactosyl–galactosyl–xylosyl–serine (Ginsburg and Neufeld, 1969):

$$\overset{\displaystyle \text{SO}_4}{\underset{\displaystyle (\text{GlcU-GalNAc})_n\text{-GlcU-Gal-Gal-Xyl-Ser}}{\mid}}$$

Particulate xylosyl- and galactosyltransferases which catalyze stepwise transfer of xylose and galactose from the UDP-sugars to serine residues of endogenous acceptors have been described in mouse mastocytoma

(Grebner *et al.*, 1966a), hen oviduct (Grebner *et al.*, 1966b), and embryonic chick cartilage (Robinson *et al.*, 1966). Recent evidence (Helting and Rodén, 1969a) indicates that separate galactosyltransferases are responsible for addition of the two galactosyl residues; the two activities could be distinguished by their specificities for defined exogenous acceptors. The fact that the main polysaccharide chains consist of repeating disaccharide units raised the attractive possibility that these might be formed by polymerization of a disaccharide intermediate, perhaps lipid linked. However, this possibility appears to have been excluded by experiments of Telser *et al.* (1966), who demonstrated alternate addition of single residues of uronic acid and N-acetylhexosamine from their respective nucleotide sugars to oligosaccharide acceptors of known structure by a particulate enzyme fraction from embryonic chick cartilage. Interestingly, addition of the first glucuronic acid residue to the linkage region appears to be catalyzed by different enzyme than that responsible for incorporation of GlcU into the main chain (Helting and Rodén, 1969b). Sulfation of hexosamine residues occurs at the polysaccharide level, apparently more or less concomitantly with chain elongation (Silbert, 1967; DeLuca and Silbert, 1968; Meezan and Davidson, 1968).

The pathway of biosynthesis of the family of complex asparagine linked carbohydrate units found in many plasma glycoproteins is as yet incompletely understood. These commonly have the terminal sequence, NAN(or Fuc)-Gal-GlcNAc, linked to a core containing mannose and GlcNAc (Spiro, 1970). The core is attached to peptide by linkage between GlcNAc and asparagine. Enzymatic synthesis of the terminal trisaccharide sequence has been investigated by Roseman and co-workers and Spiro and Spiro. The former group first isolated a series of soluble glycosyltransferases from colostrum which catalyze sequential transfer of N-acetylglucosamine (Johnston *et al.*, 1966), galactose (McGuire *et al.*, 1965), and sialic acid (Bartholomew *et al.*, 1964; Roseman *et al.*, 1966) to glycoprotein acceptors stripped of these residues by treatment with glycosidases:

$$\text{UDP-GlcNAc} + (\text{Man,GlcNAc})\text{-protein} \rightarrow$$
$$\text{GlcNAc-(Man,GlcNAc)-protein} \ (+\text{UDP}) \quad (26)$$

$$\text{UDP-Gal} + \text{GlcNAc-(Man,GlcNAc)-protein} \rightarrow$$
$$\text{Gal-GlcNAc-(Man,GlcNAc)-protein} \ (+\text{UDP}) \quad (27)$$

$$\text{CMP-NAN} + \text{Gal-GlcNAc-(Man,GlcNAc)-protein} \rightarrow$$
$$\text{NAN-Gal-GlcNAc-(Man,GlcNAc)-protein} \ +\text{CMP} \quad (28)$$

(Man,GlcNAc)-protein represents the internal core which acts as acceptor. More recently, particulate activities catalyzing reactions (26)–(28) have been characterized in the Golgi-rich smooth membrane fraction of liver (Schachter *et al.,* 1970) and the synaptosome fraction of embryonic chicken brain (Den *et al.,* 1970). The former are believed to participate in synthesis of plasma glycoproteins, while the latter may be involved in synthesis of glycoproteins in nerve endings. A similar series of reactions leading to assembly of the terminal trisaccharide sequence of thyroglobulin has been studied by Spiro and Spiro (1967, 1968a,b). The glycosyltransferases were found in the particulate fraction of thyroid and were solubilized by extraction with detergent. The mechanism of biosynthesis of the mannose-containing core region is not yet known. It is tempting to speculate that the lipid-linked mannose derivative recently described by Caccam *et al.* (1969) may be an intermediate in this process (see below, Section B,2).

Biosynthesis of saccharides with blood group activity has been of particular interest since the oligosaccharide determinants of the ABH and Lewis systems are found as glycolipids in the red cell membrane, as glycoproteins in mucous secretions, and as free oligosaccharides in milk. A number of glocosyltransferases involved in synthesis of ABH and Lewis specificities have been identified in milk (Kobata *et al.,* 1968a,b; Shen *et al.,* 1968; Grollman *et al.,* 1969) and in particulate fractions of gastric mucosa (Grollman and Marcus, 1966; Ziderman *et al.,* 1967; Race *et al.,* 1968; Hearn *et al.,* 1968). The genetic and biochemical evidence indicates that a single gene determines the glycosyltransferase activity responsible for synthesis of a given linkage, irrespective of the tissue localization and whether the final product is glycolipid, glycoprotein, or free oligosaccharide. The same enzyme can therefore exist either as a soluble protein (e.g., in milk) or firmly membrane bound (e.g., gastric mucosa). The factors responsible for these tissue-specific differences remain to be elucidated. This topic is more extensively discussed in Chapter 10.

2. Participation of Polyisoprenol Coenzymes

Recent preliminary reports from several laboratories suggest that polyisoprenol coenzymes analogous to bacterial glycosyl carrier lipids also participate in certain glycosylation reactions related to glycoprotein synthesis. Caccam *et al.* (1969) first observed incorporation of mannose from GDP-mannose into both lipid and protein products in microsomal fractions from a variety of tissues active in synthesis and secretion of mannose-containing glycoproteins. Incorporation of mannose into the lipid product was both more rapid and more extensive than into protein, and preliminary experiments were consistent with a precursor–product relationship. The acceptor lipid has not yet been identified, but the chromatographic and chemical properties of the purified mannolipid were very similar to those of the mannosyl-1-P-GCl intermediate from *M.*

lysodeikticus. More recently Hemming and co-workers (Alam *et al.*, 1970) have surveyed the activity of a variety of known polyisoprenols (see p. 373) as acceptor of mannose in pig liver microsomal preparations. Incorporation of mannose from GDP-mannose into lipid was stimulated two- to fourfold by the monophosphates of pig liver dolichols (C_{85}–C_{105}), betulaprenols (C_{30}–C_{45}), solanesol, and ficaprenols (C_{50}–C_{65}). Farnesol phosphate and cetyl alcohol phosphate were ineffective. The evidence also suggested that the lipid products were able to act as donors of mannosyl residues to endogenous protein acceptors.

Behrens and Leloir (1970) have presented evidence for synthesis and utilization of a glucosyl–lipid intermediate in rat liver microsomes, according to the reaction sequence

$$\text{UDP-Glc} + \text{P-lipid} \rightleftharpoons \text{Glc-P-lipid} + \text{UDP} \tag{29}$$

$$\text{Glc-P-lipid} + \text{protein} \rightarrow \text{Glc-protein} + \text{P-lipid} \tag{30}$$

The acceptor lipid was purified from pig liver and appeared to be identical to synthetic dolichol phosphate. Dolichol is a family of long-chain (C_{80}–C_{105}) polyisoprenols found in animal tissues in which the first C_5 unit is saturated (Burgos *et al.*, 1963). Dolichol phosphate effectively replaced the natural acceptor in reaction (29), and the intermediate thus formed was fully active as glucosyl donor in reaction (30). The evidence indicated that glucose was linked to the lipid through a phosphodiester rather than a pyrophosphoryl bridge. The endogenous acceptor of glucosyl residues in reaction (30) appeared to be a membrane-bound protein, but it has not been definitively identified. Except for collagen and basement membrane, glucose is rarely found in secretory or other extracellular glycoproteins; preliminary evidence suggested that the glucose was not being incorporated into collagen-type units.

Although further work is required to clarify the nature of the acceptor proteins in these lipid-mediated glycosyl transfer reactions and the exact structure(s) of the natural lipid coenzyme or coenzymes, the results appear to open the way to major advances in understanding the mechanism of glycoprotein synthesis and the specific role of the membrane in this process.

C. Role of Membrane-Bound Ribosomes in Protein Synthesis

The role of the membrane in polypeptide synthesis remains an unsolved problem of major importance (for discussion, see Hendler, 1968). While it is clear that polyribosomes of the rough endoplasmic reticulum are responsible for synthesis of specific classes of proteins (Campbell, 1970) including secretory glycoproteins (see above, Section III,A), it is not at all clear what factors might be involved in recognition of specific classes of messenger RNA by membrane-bound ribosomes, or what role might be played by the membrane in this process. It is possible that membrane-bound ribosomes represent a distinct class of ribosomes which contain both specific recognition sites for certain mRNA's and specific binding sites for membrane. In this case, the membrane need only supply ribosome attachment sites, and its role would be essentially a passive one. Alternatively, one might imagine that all ribosomes are identical and that the specificity lies primarily in the membrane. The membrane might then contain specific proteins for recognizing and binding the appropriate mRNA molecules and would in addition be capable of capturing ribosomes from the cytoplasm for their transcription. Data to distinguish these possibilities are largely lacking, although structural analysis of the 30 S subunits of bacterial ribosomes has shown that certain proteins are present in less than stoichiometric amounts (Kurland et al., 1969; Nomura, 1970). This finding suggests the existence of different structural classes, which might then be correlated with differences in function. Membrane-bound ribosomes are known to occur in bacteria and have been shown to account for at least 30% of the total polyribosomes in E. coli (Rouvière et al., 1969) and B. megaterium (Cundliffe, 1970). The function of membrane-bound polyribosomes in bacteria is still entirely unknown, although it is tempting to speculate that they represent the site of synthesis of membrane proteins.

IV. ROLE OF THE MEMBRANE IN DNA SYNTHESIS

In 1963, Jacob et al. proposed the replicon model to account for orderly separation and segregation of sister chromosomes in bacteria in the absence of a specialized mitotic apparatus. The hypothesis states that the chromosome is attached to a specific membrane site, and separation of sister chromosomes after a round of replication is brought about by synthesis of new membrane between the points of attachment. The evidence accumulated over the ensuing years is highly suggestive of an obligatory association of DNA replication with the membrane, although the nature of the association and the biochemical mechanism of replication and segregation remain obscure.

A. Attachment of DNA to Membrane

1. Morphological Evidence

Areas of apparent contact between the bacterial nucleus and membrane have been observed by many investigators in electron microscopic examination of sectioned material (Robinow, 1963). The systematic investigations of Ryter and her colleagues (see Ryter, 1968, for review) in *Bacillus subtilis* and *E. coli* have yielded strong evidence that this contact is a normal and constant feature of the cellular organization. In *B. subtilis* and other gram-positive bacteria which contain mesosomes, the contact is with this structure. Mesosomes are intracytoplasmic membranous structures arising from invaginations of cytoplasmic membrane. Reconstruction of growing cells of *B. subtilis* from serial sections led to the conclusion that contiguities between nucleus and mesosome exist in all cells of the culture and persist throughout the division cycle (Ryter and Jacob, 1964). The data suggested that each chromosome is initially connected to a single mesosome which splits into two during replication; the two mesosomes then move progressively apart, each carrying one of the sister nuclei. A similar picture emerged from reconstruction of electron micrographs of germinating spores of *B. subtilis* (Ryter, 1967). Where mesosomes are absent, as in *E. coli* or protoplasts of *B. subtilis*, contact is made directly with the cytoplasmic membrane (Ryter and Jacob, 1966). In *E. coli* the association was most readily observed in spheroplasts; the number of points of contact approximated one per cell.

2. Isolation of DNA–Membrane Complexes

Association of newly replicated DNA with fragments of bacterial membrane has now been described by many investigators. The early experiments of Goldstein and Brown (1961) with spheroplasts of *E. coli* showed DNA synthesis and enrichment for nascent DNA in a heavy particle fraction probably corresponding to cell envelope. Similar findings were made by Ganesan and Lederberg (1965) in *B. subtilis*. DNA labeled *in vivo* during a short pulse of [^{3}H]thymidine was preferentially recovered in the rapidly sedimenting membrane fraction following lysis of the cells with lysozyme and sodium dodecylsulfate. This fraction also contained about 30% of the total DNA and 25% of the total DNA polymerase activity. That the pulse-labeled material represented nascent DNA was confirmed by pulse-chase experiments which showed a rapid shift of radioactivity into bulk DNA during the case period. The radioactivity was also released by brief treatment of the isolated complex with Pronase. Smith and Hanawalt (1967) obtained a four- to fivefold enrichment of pulse-labeled DNA in the rapidly sedimenting fraction from *E. coli*. Lysozyme lysates were subjected to controlled shear (with or without subsequent pronase digestion) and layered on a 5–20% sucrose gradient over a 62% sucrose shelf. After centrifugation, the pulse-labeled DNA was recovered at the shelf, while the bulk of the DNA was distributed in the upper half of the gradient. The pulse-labeled DNA at the shelf was again characterized as newly replicated material by pulse-chase experiments and isopycnic centrifugation after density labeling with [^{3}H]bromodeoxyuridine. The rapidly sedimenting component to which the DNA was bound was not fully characterized in these experiments, but it was presumed to be membranous since the DNA was released by treatment with deoxycholate. More recently Fuchs and Hanawalt (1970) have reported isolation of a DNA replication complex from *E. coli* in which pulse-labeled nascent DNA was enriched some 100-fold. The complex was obtained by a modified lysis procedure employing the nonionic detergent Brij-58 (Godson and Sinsheimer, 1967), followed by controlled sonication and sucrose gradient centrifugation. The pulse-labeled DNA was recovered in a fraction sedimenting at 100–150 S and containing 1–3% of the total DNA, 7% of the RNA, 9% of the protein, and less than 5% of the DNA polymerase, RNA polymerase, DNase, and RNase of the lysate. Little evidence was obtained for the presence of significant amounts of phospholipid. The radioactive DNA was released into a slowly sedimenting form by treatment with DNase or Pronase, but phospholipase A was without effect.

A somewhat different approach to isolation of DNA–membrane complexes has been developed by Tremblay *et al.* (1969). These workers employed sodium lauryl sarcosinate (Sarkosyl) for lysis of spheroplasts. In the presence of Mg^{++} the detergent crystallizes and in sucrose gradient centrifugation could be recovered as a discrete band (M-band) which also contained 10–30% of the cell membrane, essentially all of the DNA, and a large fraction of the total mRNA and ribosomes. Formation of the M-band complex appeared to depend on the presence of membrane since neither purified nucleic acids nor ribosomes were adsorbed to the detergent crystals directly, while partially purified membranes and phospholipids were. The technique was applied successfully to isolation of complexes from both gram-positive and gram-negative bacteria. The ribosome components of the M-band have been further characterized by Cundliffe (1970). Approximately 65% of the total ribosomes were found in the complex, of which approximately half were identified as polyribosomes attached to DNA by means of nascent mRNA. The remainder appeared to be directly membrane bound, since this fraction was released by Triton-X100, but not by mild treatment with RNase.

Evidence that replication of bacteriophage DNA also involves the membrane has been obtained by techniques similar to those described above. Thus, early intermediates in replication of DNA of phages P22 (Botstein and Levine, 1968), lambda, and ØX174 (Salivar and Sinsheimer, 1969) were found to be associated with rapidly sedimenting structures, presumably membranous in nature, while replicative complexes of T4 (Earhart *et al.*, 1968) and ØII (Linial and Malamy, 1970) were isolated by the Mg^{++} sarcosinate technique. In all cases, the isolated complexes contained parental phage DNA as well as newly synthesized material which could be chased into mature phage DNA or progeny virions.

Although the mode of attachment of the DNA in the isolated complexes is still obscure, and the associated membrane fragments have not yet been characterized in detail, the above results provide strong presumptive evidence that newly replicated DNA, i.e., the region of the replicative fork, is specifically bound to membrane. Evidence that the chromosome may in addition be stably attached to membrane at the replication origin has also been obtained. Sueoka and Quinn (1968) first reported that transforming DNA isolated from the rapidly sedimenting DNA–membrane fraction of *B. subtilis* was enriched approximately twofold for a marker close to the chromosome origin (Ade 16) relative to markers near the center of the chromosome. Some enrichment for a marker (Met) near the terminus was also observed. The slowly sedimenting bulk DNA showed no enrichment for markers near the ends of the chromosome. These observations have been confirmed and extended by Snyder and Young (1970), who isolated DNA from well-washed membrane fractions and from the soluble fraction. Selective enrichment in the membrane fraction for markers at the origin (Ade 16) and terminus (Met) was at least twofold in all cases, and values as high as tenfold were obtained for Ade 16. Experiments of Sueoka and Quinn (1968) in germinating spores of *B. subtilis* also suggested that the origin of the chromosome might be permanently attached to membrane. It is known that during germination the chromosome replicates synchronously from the origin. It was found by pulse labeling with labeled thymidine that the first DNA synthesized in germinating spores was associated, as expected, with the rapidly sedimenting component; however, unlike pulse-labeled DNA from vegetative cells, this fraction was metabolically stable and was not chased into the slowly sedimenting free DNA fraction by subsequent growth in nonradioactive thymidine. More recently Fielding and Fox (1970) have presented evidence for stable membrane attachment of DNA at the replicative origin in *E. coli*. The experiments were based on the assumption that initiation of a new round of replication begins at the chromosome origin at approximately the time of cell division. Synchronous cultures were exposed to [³H]thymidine at various times during the cell cycle and chased with nonradioactive thymidine, and the membrane fraction was isolated following disruption of the cells by sonication. Incorporation of radioactivity into DNA which remained with the membrane fraction under these conditions showed a cyclical pattern, with maximal labeling occurring at approximately the time of cell division. The radioactivity so incorporated remained membrane bound for as long as three generations of subsequent growth in nonradioactive medium.

A model for replication of the bacterial chromosome based on attachment of DNA to membrane at both the chromosome origin and replicative fork has been proposed by Sueoka and Quinn (1968) and is illustrated in Fig. 11. According to this model, the entire replicative apparatus is membrane associated. The origin–terminus locus of the circular chromosome is permanently fixed at this site, and the remainder of the DNA filament is progressively fed through the apparatus as replication proceeds. Attachment of the replicative origin (and terminus) of the chromosome would therefore be metabolically stable, while the newly synthesized DNA at the replicative fork would be transiently associated with the membrane.

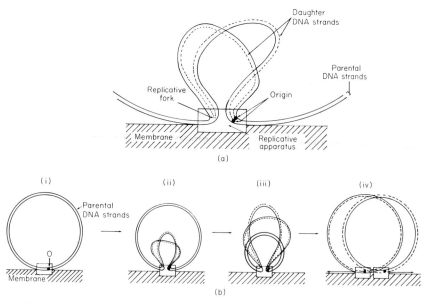

Fig. 11. Postulated model for replication of the bacterial chromosome. (a) Schematic representation of the region of chromosome attachment to the replicative apparatus at the membrane. (b) Stages in chromosome replication: (i) before initiation (O represents chromosome origin); (ii) and (iii) early and late stages in growth of sister chromosomes; (iv) completed sister chromosomes beginning to move apart. (Adapted from Sueoka and Quinn, 1968.)

B. Membrane Alterations in Mutants Temperature Sensitive in DNA Synthesis

Further evidence for a functional association between DNA replication and the membrane has emerged from studies on certain mutants of *E. coli* temperature sensitive in DNA synthesis. Two general classes of such mutants have been described (Kohiyama *et al.*, 1966; Hirota *et al.*, 1968). In one class, DNA synthesis ceases immediately upon shift to high temperature, indicating a block in chain elongation. In the other, the rate of synthesis progressively decreases with time and comes to a halt after an increase in DNA content of 30–40% (Hirota *et al.*, 1968, 1970). The latter class, called *DnaA*, has been identified as defective in initiation of DNA synthesis (Hirota *et al.*, 1970). The defect in initiation of chromosome replication at the restrictive temperature is accompanied by alterations in membrane properties, including abnormal sensitivity to lysis by deoxycholate and greatly increased enhancement of fluorescence with the membrane probe, 1-anilino-8-naphthalene sulfonate (Hirota *et al.*, 1970). Analysis of the membrane proteins in isogenic *DnaA⁺* and *DnaA⁻* strains by disc gel electrophoresis in sodium dodecylsulfate (SDS) has provided additional evidence for a specific alteration in membrane structure (Shapiro *et al.*, 1970). Two independent *DnaA* mutants showed a reproducible decrease in a component of molecular weight ∼32,000 after incubation at 41°C, compared to the mutants at 30°C or the wild type at either temperature. Unfortunately the available evidence does not permit establishment of a causative relationship between the observed membrane changes in *DnaA* mutants and the failure of initiation. The defect in initiation of DNA synthesis was observed immediately after shift to high temperature, while detectable alterations in membrane properties were seen only after an appreciable delay (Hirota *et al.*, 1970). The lag in overt alteration of the gross properties of the membrane can, of course, be rationalized in a number of ways, and the hypothesis that the *DnaA* locus specifies a membrane protein the function of which is required for initiation remains an attractive one.

Another type of temperature-sensitive mutation affecting both DNA synthesis and membrane composition in *E. coli* has been described by Inoue and Guthrie (1969). This mutant stops DNA synthesis immediately at high temperature (Inoue, 1969), and disc gel electrophoresis of the membrane protein by the SDS technique revealed a marked decrease in a major protein band corresponding to molecular weight ∼34,000. Further investigations, however, have suggested that this change in membrane protein probably arises as a secondary effect of the inhibition of DNA synthesis (Inoue and Pardee, 1970). Thus, similar decreases in the 34,000 molecular weight component were observed when DNA synthesis was inhibited in wild-type *E. coli* by any of several means, including thymine starvation, nalidixic acid, 5-diazouracil, and irradiation with ultraviolet light. In addition, a substantial increase in a second protein band of molecular weight ~39,000 was found which appeared to be correlated with inhibition of cell division. Again the alterations in membrane composition appeared gradually, although arrest of DNA synthesis was immediate. It was concluded that DNA synthesis was required for continued production of the 34,000 molecular weight protein, rather than the reverse. The physiological function of this protein is at present entirely unknown.

C. Membrane-Associated DNA Polymerase

The role of DNA polymerase in replication of the bacterial chromosome has been subject to increasing question over the past several years. Purified polymerase is able to carry out accurate repair of single-stranded regions within a duplex template, but efforts to obtain net semiconservative synthesis from a double-stranded template have been unsuccessful. In addition, the rate of DNA synthesis achieved with the polymerase *in vitro* is only 1% of the rate *in vivo* (for review, see Kornberg, 1969; Richardson, 1969). The hypothesis that the replicative apparatus is membrane bound suggested that the required properties might be conferred on the polymerase enzyme by association with the membrane. Indeed, a polymerase complex capable of synthesis of a small amount of biologically active transforming DNA has been isolated from the membrane fraction of *B. subtilis* by Ganesan (1968). However the suspicion that the primary function of the classic DNA polymerase is one of repair rather than replication has been reinforced by recent studies on a mutant of *E. coli* lacking this enzyme. This mutant, called *PolA1⁻*, was isolated by DeLucia and Cairns (1969) and grows quite normally even though DNA polymerase activity is essentially undetectable *in vitro*. Evidence that the *PolA1⁻* mutant is capable of normal semiconservative synthesis of DNA but not repair synthesis has been obtained in whole cells made permeable to exogenous deoxynucleoside triphosphates by treatment with toluene (Moses and Richardson, 1970). Semiconservative replication, as judged by pycnographic analysis of newly synthesized density-labeled DNA, was observed in both *PolA⁺* and *PolA1⁻* strains. Incorporation of nucleotides into this type of product required ATP and was inhibited by N-ethylmaleimide. In contrast repair synthesis, triggered by endogenous or exogenous nuclease activity, required the *PolA*⁺ genotype, was not stimulated by ATP, and was resistant to sulfhydryl reagents. Further evidence that the synthesis observed in toluenized *PolA1*⁻ cells represents replication has come from double mutants containing both the *PolA1*⁻ character and a temperature-sensitive mutation in either DNA initiation (*DnaA*⁻) or elongation (*DnaB*⁻). Incorporation of deoxynucleotide substrates into DNA by toluenized cells fully retained the temperature-sensitivity characteristic of the *DnaA*⁻ or *DnaB*⁻ phenotype (Mordoh *et al.*, 1970).

The properties of the *PolA1*⁻ mutant have given strong impetus to the search for a second "DNA polymerase," responsible for replicative synthesis, and progress toward this end has been very rapid. *In vitro* evidence for the existence and membrane localization of a new polymerase activity, DNA polymerase II, was first presented by Smith *et al.* (1970) and Knippers and Stratling (1970). Smith *et al.* employed gentle lysis of cells embedded in agar. After extensive washing the agar still contained DNA and membranes, which were capable of enzymatic synthesis of DNA. The product had the properties expected for semiconservative replication as determined by pycnographic sedimentation analysis of density-labeled products. The initial rate of synthesis approximated that *in vivo* (although this rate was maintained for only 1–2 minutes). A similar activity was found by Knippers and Stratling (1970) in membrane fractions of the *PolA1*⁻ mutant prepared by gentle lysis of spheroplasts with the nonionic detergent Brij-58. This polymerase activity has been solubilized and partially purified by Knippers (1970). A similar or identical enzyme has also been purified by Kornberg and Gefter (1970) from the soluble fraction of *PolA1*⁻ cells disrupted in a pressure cell. The properties of the partially purified enzyme obtained in the two laboratories were similar but differed from those of the DNA polymerase I of Kornberg in several significant respects. Polymerase II was inhibited by sulfhydryl reagents and high ionic strength, while polymerase I was resistant to both. Polymerase II was not inhibited by antibody against pure *E. coli* DNA polymerase I under conditions giving 90% inhibition of the latter enzyme, nor did added polymerase II act as a hapten inhibitor of the polymerase I–antipolymerase I reaction. The primer requirements of polymerase II were also distinct from those of polymerase I (Knippers, 1970). Characterization of DNA polymerase II is still in an early stage, and the possibility that it represents a mutationally altered form of polymerase I has not yet been rigorously excluded. The activity appears, however, to be a most promising candidate for the long-sought replication enzyme, and its apparent relation to the membrane may provide a major clue to the role of the membrane in replication.

D. Role of Membrane in DNA Replication in Eucaryotic Cells

1. Nuclear Membrane and DNA Synthesis

Electron microscopic evidence for attachment of interphase chromatin fibers to the nuclear membrane has been obtained by several investigators (for review, see Dupraw, 1968), and attachment of chromosomes to the membrane in an ordered fashion has been invoked to explain the reproducible arrangement of chromosomes in metaphase spreads and the close juxtaposition of homologous chromosomes inferred from both genetic and cytological observations (Comings, 1968). Evidence that DNA synthesis also begins at the nuclear membrane has been obtained by Comings and Kakefuda (1968). Autoradiographic studies with synchronous cultures of human amnion cells showed that [³H]thymidine incorporated into DNA during the first 10 minutes of the S period was restricted to the periphery of the nucleus (and to the nucleolus). Further evidence suggesting association of newly synthesized DNA with some cellular component, possibly membranous in nature, was derived from pulse-chase experiments of Friedman and Mueller (1969): DNA pulse-labeled with [³H]thymidine was separated from the bulk DNA both in phenol–water extraction and in sedimentation studies; the properties of the nascent DNA fraction were reminiscent of those of pulse-labeled bacterial DNA. Although the evidence for involvement of the nuclear membrane in DNA synthesis is as yet incomplete and circumstantial, it appears most reasonable that the membrane will prove to play an essential role as mediator of the orderly replication of chromosomes in eucaryotes as well as procaryotes.

2. Role of the Cell Surface in Control of Replication

A more subtle role of the cellular membrane in control of replication has emerged from the elegant studies of Burger on the relation of surface structure to contact inhibition of growth in normal and virally transformed fibroblasts in culture. It is well known that multiplication of normal fibroblasts in tissue culture ceases when the cells have reached a confluent monolayer, while growth of cells transformed by tumor viruses is relatively independent of the inhibitory control mediated by cell–cell contact. The loss of contact inhibition of growth in transformed cells is correlated with an alteration in the surface membrane, which was first detected (Burger and Goldberg, 1967) as an appearance of receptor sites for wheat germ agglutinin on the surface of the transformed cell. This agglutinin was shown to interact specifically with carbohydrates containing N-acetylglucosamine end groups. Later studies (Inbar and Sachs, 1969a; Burger, 1970a; Sela et al., 1970) have shown a similar appearance of a number of other carbohydrate-containing sites recognizable by specific interaction with different agglutinins. A second finding of major importance was that although these agglutinin receptor sites were not exposed at the surface of normal cells, receptor activity equivalent to that of transformed cells could be unmasked by very brief treatment of intact cells with proteolytic enzymes (Burger, 1969; Inbar and Sachs, 1969a,b; Sela *et al.*, 1970). A direct relationship between effective contact inhibition of growth and the presence of the surface layer responsible for masking agglutinin receptor sites has been established in two ways. Transient escape of normal cells from contact inhibition of growth upon removal of the cover layer was demonstrated by brief treatment of non-growing confluent cells in monolayer with low concentrations of proteases (Burger, 1970b). An increase in mitotic index was apparent within 12 hours, and it continued until the cell population had increased approximately twofold, by which time the cover layer had been restored. The reverse phenomenon, phenotypic reversion of transformed cells to a state of normal contact inhibition upon application of an artificial cover layer, has also been demonstrated in an ingenious manner by Burger and Noonan (1970). The experiments were based on the use of monovalent fragments of concanavalin A, which retained the ability to interact with exposed receptor sites but were no longer able to agglutinate cells. Addition of the monovalent agglutinin to the culture medium had no inhibitory effect on the rate of cell growth prior to confluency, but it resulted in complete restoration of the normal contact-dependent inhibition of growth in the confluent state. The effect required adsorption of the agglutinin to the receptor sites, since addition of hapten inhibitors of concanavalin A prior to confluency permitted growth to control levels. The inhibition of growth was also fully reversed by addition of hapten inhibitor to confluent cultures which had already ceased to divide.

The nature of the control exerted by these components of the surface membrane on DNA synthesis and cell division is entirely unknown. Possibilities suggested by Burger and Noonan (1970) include alteration of physicochemical properties of the membrane related to cell mobility and the division process, such as adhesiveness or flexibility; alteration of permeability to nutrients or specific growth factors; and alteration in the transmission from cell to cell of messages related to control of cell division.

References

Alam, A. S., Barr, R. M., Richards, J. B., and Hemming, F. W. (1970). "The Biochemical Society (London) Agenda Papers, October," p. 5.

Allen, C. M., Alworth, W., Macrae, A., and Bloch, K. (1967). *J. Biol. Chem.* **242**, 1895.
Anderson, J. S., and Strominger, J. L. (1966). *Biochem. Biophys. Res. Commun.* **21**, 516.
Anderson, J. S., Matsuhashi, M., Haskin, M. A., and Strominger, J. L. (1965). *Proc. Nat. Acad. Sci. U. S.* **53**, 881.
Anderson, J. S., Meadow, P. M., Haskin, M. A., and Strominger, J. L. (1966). *Arch. Biochem. Biophys.* **116**, 487.
Anderson, J. S., Matsuhashi, M., Haskin, M. A., and Strominger, J. L. (1967). *J. Biol. Chem.* **242**, 3180.
Araki, Y., Shimada, A., and Ito, E. (1966a). *Biochem. Biophys. Res. Commun.* **23**, 518.
Araki, Y., Shimai, R., Shimada, A., and Ishimoto, N. (1966b). *Biochem. Biophys. Res. Commun.* **23**, 466.
Archibald, A. R., Baddiley, J., and Button, D. (1965). *Biochem. J.* **95**, 8C.
Archibald, A. R., Baddiley, J., and Blumson, N. L. (1968a). *Advan. Enzymol.* **30**, 223.
Archibald, A. R., Baddiley, J., Button, D., Heptinsall, S., and Stafford, G. H. (1968b). *Nature (London)* **219**, 5156.
Baddiley, J. (1970). Personal communication.
Baddiley, J., Blumson, N. L., and Douglas, L. J. (1968). *Biochem. J.* **110**, 565.
Bartholomew, B., Jourdian, G. W., and Roseman, S. (1964). *Abstr. Int. Congr. Biochem., 6th* **6**, 503.
Beckmann, I., Subbaiah, T. V., and Stocker, B. A. D. (1964). *Nature (London)* **201**, 1299.
Behrens, N. H., and Cabib, E. (1968). *J. Biol. Chem.* **243**, 502.
Behrens, N. H., and Leloir, L. F. (1970). *Proc. Nat. Acad. Sci. U. S.* **66**, 153.
Best, G. K., and Durham, N. N. (1965). *Arch. Biochem. Biophys.* **111**, 685.
Best, G. K., Grastie, M. K., and McConnell, R. D. (1970). *J. Bacteriol.* **102**, 476.
Blumenkrantz, N., Prockop, D. J., and Rosenbloom, J. (1968). *Abstr. Amer. Chem. Soc., 150th Meeting*, 286.
Bosmann, H. B., and Eylar, E. H. (1968). *Biochem. Biophys. Res. Commun.* **30**, 89.
Botstein, D., and Levine, M. (1968). *Cold Spring Harbor Symp. Quant. Biol.* **33**, 659.
Bouchilloux, S., and Cheftel, S. (1966). *Biochem. Biophys. Res. Commun.* **23**, 305.
Bray, D., and Robbins, P. W. (1967a). *Biochem. Biophys. Res. Commun.* **28**, 334.
Bray, D., and Robbins, P. W. (1967b). *J. Mol. Biol.* **30**, 457.
Brooks, D., and Baddiley, J. (1969a). *Biochem. J.* **113**, 635.
Brooks, D., and Baddiley, J. (1969b). *Biochem. J.* **115**, 307.
Bumsted, R. M., Dahl, J. L., Söll, D., and Strominger, J. L. (1968). *J. Biol. Chem.* **243**, 779.
Burger, M. M. (1963). *Biochim. Biophys. Acta* **71**, 495.
Burger, M. M. (1969). *Proc. Nat. Acad. Sci. U. S.* **62**, 994.
Burger, M. M. (1970a). *In* "Proceedings of the International Congress on Biological Membranes" (A. Bolis, ed.), Chap. II, p. 5. North-Holland Publ., Amsterdam.
Burger, M. M. (1970b). *Nature (London)* **227**, 170.
Burger, M. M., and Glaser, L. (1962). *Biochim. Biophys. Acta* **64**, 575.
Burger, M. M., and Glaser, L. (1964). *J. Biol. Chem.* **239**, 3168.
Burger, M. M., and Goldberg, A. R. (1967). *Proc. Nat. Acad. Sci. U. S.* **57**, 350.
Burger, M. M., and Noonan, K. D. (1970). *Nature (London)* **228**, 512.
Burgos, J., Hemming, F. W., Pennock, J. F., and Norton, R. A. (1963). *Biochem. J.* **88**, 4680.

Caccam, J. F., Jackson, J. J., and Eylar, E. II. (1969). *Biochem. Biophys. Res. Commun.* 35, 505.
Campbell, P. N. (1970). *FEBS Lett.* 7, 1.
Carlson, D. M., McGuire, E. J., Jourdian, G. W., and Roseman, S. (1964). *Fed. Proc. Fed. Amer. Soc. Exp. Biol.* 23, 380.
Caro, L. G., and Palade, G. E. (1964). *J. Cell Biol.* 20, 473.
Chatterjee, A. W., and Park, J. T. (1964). *Proc. Nat. Acad. Sci. U. S.* 51, 9.
Chin, T., Burger, M. M., and Glaser, L. (1966). *Arch. Biochem. Biophys.* 116, 358.
Christensen, J. G., Gross, S. K., and Robbins, P. W. (1969). *J. Biol. Chem.* 244, 5436.
Comings, D. E. (1968). *Amer. J. Hum. Genet.* 20, 440.
Comings, D. E., and Kakefuda, T. (1968). *J. Mol. Biol.* 33, 225.
Cook, G. M. W., Laico, M. T., and Eylar, E. H. (1965). *Proc. Nat. Acad. Sci. U. S.* 54, 247.
Cundliffe, E. (1970). *J. Mol. Biol.* 52, 467.
Cynkin, M. A., and Osborn, M. J. (1968a). *Fed. Proc. Fed. Amer. Soc. Exp. Biol.* 27, 293.
Cynkin, M. A., and Osborn, M. J. (1968b). Unpublished observations.
Dankert, M., Wright, A., Kelley, W. S., and Robbins, P. W. (1966). *Arch. Biochem. Biophys.* 116, 425.
DeLuca, S., and Silbert, J. E. (1968). *J. Biol. Chem.* 243, 2725.
DeLucia, R., and Cairns, J. (1969). *Nature (London)* 224, 1164.
Den, H., Kaufman, B., and Roseman, S. (1970). *J. Biol. Chem.* 245, 6607.
Dietrich, C. P., Colucci, A. V., and Strominger, J. L. (1967). *J. Biol. Chem.* 242, 3218.
Douglas, L. J., and Baddiley, J. (1968). *FEBS Lett.* 1, 114.
Dupraw, E. J. (1968). "Cell and Molecular Biology." Academic Press, New York.
Earhart, C. F., Tremblay, G. Y., Daniels, M. J., and Schaechter, M. (1968). *Cold Spring Harbor Symp. Quant. Biol.* 33, 707.
Edstrom, R. D., and Heath, E. C. (1967). *J. Biol. Chem.* 242, 3581.
Endo, A., and Rothfield, L. (1969a). *Biochemistry* 12, 3500.
Endo, A., and Rothfield, L. (1969b). *Biochemistry* 12, 3508.
Fielding, P., and Fox, C. F. (1970). *Biochem. Biophys. Res. Commun.* 41, 157.
Frerman, F. E., Troy, F. A., and Heath, E. C. (1971). *J. Biol. Chem.* 246, 118.
Friedman, D. L., and Mueller, G. C. (1969). *Biochim. Biophys. Acta* 174, 253.
Fuchs, E., and Hanawalt, P. C. (1970). *J. Mol. Biol.* 52, 301.
Ganesan, A. T. (1968). *Cold Spring Harbor Symp. Quant. Biol.* 33, 45.
Ganesan, A. T., and Lederberg, J. (1965). *Biochem. Biophys. Res. Commun.* 18, 824.
Ghuysen, J. M. (1968). *Bacteriol. Rev.* 32, 425.
Ghuysen, J. M., Strominger, J. L., and Tipper, D. J. (1968). *In* "Comprehensive Biochemistry" (M. Florkin and E. H. Stotz, eds.), Vol. 26A, pp. 53–104. Elsevier, New York.
Gilbert, J., Bevill, R. D., and Osborn, M. J. (1968). Unpublished experiments.
Ginsburg, V., and Neufeld, E. F. (1969). *Annu. Rev. Biochem.* 38, 371.
Glaser, L. (1963). *Biochim. Biophys. Acta* 71, 237.
Glaser, L. (1964). *J. Biol. Chem.* 239, 3178.
Glaser, L., and Burger, M. M. (1964). *J. Biol. Chem.* 239, 3187.
Godson, G. N., and Sinsheimer, R. L. (1967). *Biochim. Biophys. Acta* 149, 476.
Goldstein, A., and Brown, B. J. (1961). *Biochim. Biophys. Acta* 53, 19.
Grebner, E. E., Hall, C. W., and Neufeld, E. F. (1966a). *Biochem. Biophys. Res. Commun.* 22, 672.

Grebner, E. E., Hall, C. W., and Neufeld, E. F. (1966b). *Arch. Biochem. Biophys.* **116**, 391.

Grollman, A. P., and Marcus, D. M. (1966). *Biochem. Biophys. Res. Commun.* **25**, 542.

Grollman, E. F., Kobata, A., and Ginsburg, V. (1969). *J. Clin. Invest.* **48**, 1489.

Hagopian, A., and Eylar, E. H. (1968). *Arch. Biochem. Biophys.* **128**, 422.

Hagopian, A., and Eylar, E. H. (1969a). *Arch. Biochem. Biophys.* **129**, 447.

Hagopian, A., and Eylar, E. H. (1969b). *Arch. Biochem. Biophys.* **129**, 515.

Hearn, V. M., Smith, Z. G., and Watkins, W. M. (1968). *Biochem. J.* **109**, 315.

Heath, E. C., Mayer, R. M., Edstrom, R. D., and Beaudreau, C. A. (1966). *Ann. N. Y. Acad. Sci.* **133**, 315.

Helting, T., and Rodén, L. (1969a). *J. Biol. Chem.* **244**, 2790.

Helting, T., and Rodén, L. (1969b). *J. Biol. Chem.* **244**, 2799.

Hendler, R. W. (1968). "Protein Biosynthesis and Membrane Biochemistry." Wiley, New York.

Heydanek, M. G., Jr., and Neuhaus, F. C. (1969). *Biochemistry* **8**, 1474.

Heydanek, M. G., Jr., Struve, W. G., and Neuhaus, F. C. (1969). *Biochemistry* **8**, 1214.

Higashi, Y., and Strominger, J. L. (1970). *J. Biol. Chem.* **245**, 3691.

Higashi, Y., Strominger, J. L., and Sweeley (1967). *Proc. Nat. Acad. Sci. U. S.* **57**, 1878.

Higashi, Y., Siewert, G., and Strominger, J. L. (1970a). *J. Biol. Chem.* **245**, 3683.

Higashi, Y., Strominger, J. L., and Sweeley, C. C. (1970b). *J. Biol. Chem.* **245**, 3697.

Hirota, Y., Ryter, A., and Jacob, F. (1968). *Cold Spring Harbor Symp. Quant. Biol.* **33**, 677.

Hirota, Y., Mordoh, J., and Jacob, F. (1970). *J. Mol. Biol.* **53**, 369.

Hussey, H., Brooks, D., and Baddiley, J. (1969). *Nature (London)* **221**, 665.

Inbar, M., and Sachs, L. (1969a). *Proc. Nat. Acad. Sci. U. S.* **63**, 1418.

Inbar, M., and Sachs, L. (1969b). *Nature (London)* **223**, 710.

Inoue, M. (1969). *J. Bacteriol.* **99**, 842.

Inoue, M., and Guthrie, J. P. (1969). *Proc. Nat. Acad. Sci. U. S.* **64**, 957.

Inoue, M., and Pardee, A. (1970). *J. Biol. Chem.* **245**, 5813.

Ishimoto, N., and Strominger, J. L. (1966). *J. Biol. Chem.* **241**, 639.

Izaki, K., and Strominger, J. L. (1968). *J. Biol. Chem.* **243**, 3193.

Izaki, K., Matsuhashi, M., and Strominger, J. L. (1966). *Proc. Nat. Acad. Sci. U. S.* **55**, 656.

Jacob, F., Brenner, S., and Cuzin, F. (1963). *Cold Spring Harbor Symp. Quant. Biol.* **28**, 329.

Johnston, I. R., McGuire, E. J., Jourdian, G. W., and Roseman, S. (1966). *J. Biol. Chem.* **241**, 5735.

Kamiryo, T., and Matsuhashi, M. (1969). *Biochem. Biophys. Res. Commun.* **36**, 215.

Kanegasaki, S., and Wright, A. (1970). *Proc. Nat. Acad. Sci. U. S.* **67**, 951.

Katz, W., Matsuhashi, M., Dietrich, C. P., and Strominger, J. L. (1967). *J. Biol. Chem.* **242**, 3207.

Kauss, H. (1969). *FEBS Lett.* **5**, 81.

Kennedy, L. D., and Shaw, D. R. D. (1968). *Biochem. Biophys. Res. Commun.* **32**, 861.

Kent, J. L., and Osborn, M. J. (1968a). *Biochemistry* **7**, 4396.

Kent, J. L., and Osborn, M. J. (1968b). *Biochemistry* **7**, 4419.

Knippers, R. (1970). *Nature (London)* **228**, 1050.

Knippers, R., and Stratling, W. (1970). *Nature* (*London*) **226**, 713.

Kobata, A., Grollman, E. F., and Ginsburg, V. (1968a). *Biochem. Biophys. Res. Commun.* **32**, 272.

Kobata, A., Grollman, E. F., and Ginsburg, V. (1968b). *Arch. Biochem. Biophys.* **124**, 609.

Kohiyama, M., Cousin, D., Ryter, A., and Jacob, F. (1966). *Ann. Inst. Pasteur, Paris* **110**, 465.

Kornberg, A. (1969). *Science* **163**, 1410.

Kornberg, T., and Gefter, M. L. (1970). *Biochem. Biophys. Res. Commun.* **40**, 1348.

Kurland, C. G., Voynow, P., Hardy, S. J. S., Randall, L., and Lutter, L. (1969). *Cold Spring Harbor Symp. Quant. Biol.* **34**, 17.

Lahav, M., Chiu, T. H., and Lennarz, W. J. (1969). *J. Biol. Chem.* **244**, 5890.

Lawford, G. P., and Schachter, H. (1966). *J. Biol. Chem.* **241**, 5408.

Lawrence, P. I., and Strominger, J. L. (1970a). *J. Biol. Chem.* **245**, 3660.

Lawrence, P. J., and Strominger, J. L. (1970b). *J. Biol. Chem.* **245**, 3653.

Lennarz, W. J., and Talamo, B. (1966). *J. Biol. Chem.* **241**, 2707.

Li, Y. T., Li, S. C., and Shetlar, M. R. (1968). *J. Biol. Chem.* **243**, 656.

Lindgren, B. O. (1965). *Acta Chem. Scand.* **19**, 1317.

Linial, M., and Malamy, M. (1970). *J. Virol.* **6**, 72.

Losick, R. (1969). *J. Mol. Biol.* **42**, 237.

Losick, R., and Robbins, P. W. (1967). *J. Mol. Biol.* **30**, 445.

Lüderitz, O., Staub, A. M., and Westphal, O. (1966). *Bacteriol. Rev.* **30**, 193.

McGuire, E. J., and Roseman, S. (1967). *J. Biol. Chem.* **242**, 3745.

McGuire, E. J., and Roseman, S. (1969). *Abstr. Amer. Chem. Soc., 158th Meeting,* 52.

McGuire, E. J., Jourdian, G. W., Carlson, D. M., and Roseman, S. (1965). *J. Biol. Chem.* **240**, PC4112.

Martin, H. H. (1966). *Annu. Rev. Biochem.* **35**, 457.

Matsuhashi, M., Dietrich, C. P., and Strominger, J. L. (1967). *J. Biol. Chem.* **242**, 3191.

Meadow, P. M., Anderson, J. S., and Strominger, J. L. (1964). *Biochem. Biophys. Res. Commun.* **14**, 382.

Meezan, E., and Davidson, E. A. (1968). *J. Biol. Chem.* **242**, 4956.

Melchers, F., and Knopf, P. M. (1967). *Cold Spring Harbor Symp. Quant. Biol.* **32**, 255.

Mirelman, D., and Sharon, N. (1966). *Biochem. Biophys. Res. Commun.* **24**, 237.

Mirelman, D., and Sharon, N. (1967). *J. Biol. Chem.* **242**, 3414.

Molnar, J., and Dalisay, S. (1967). *Biochemistry* **6**, 1941.

Molnar, J., and Sy, D. (1967). *Biochemistry* **6**, 1941.

Molnar, J., Lutes, R. A., and Winzler, R. J. (1965a). *Cancer Res.* **25**, 1438.

Molnar, J., Robinson, G. B., and Winzler, R. J. (1965b). *J. Biol. Chem.* **240**, 1882.

Mordoh, J., Hirota, Y., and Jacob, F. (1970). *Proc. Nat. Acad. Sci. U. S.* **67**, 773.

Moroz, C., and Uhr, J. W. (1967). *Cold Spring Harbor Symp. Quant. Biol.* **32**, 263.

Moses, R. E., and Richardson, C. C. (1970). *Proc. Nat. Acad. Sci.* **67**, 674.

Müller, E., and Rothfield, L. (1970). Personal communication.

Muñoz, E., Ghuysen, J. M., Leyh-Bouille, M., Petit, J. F., Heymann, H., Bricas, E., and Lefrancier, P. (1966). *Biochemistry* **5**, 3748.

Nadler, N. J., Young, B. A., Leblond, C. P., and Mitmaker, B. (1964). *Endocrinology* **74**, 333.

Nathenson, S. G., and Strominger, J. L. (1962). *J. Biol. Chem.* **237**, PC3839.

Nathenson, S. G., and Strominger, J. L. (1963). *J. Biol. Chem.* **238**, 3161.
Nathenson, S. G., Ishimoto, N., Anderson, J. S., and Strominger, J. L. (1966). *J. Boil. Chem.* **241**, 651.
Nikaido, H. (1962). *Proc. Nat. Acad. Sci. U. S.* **48**, 1337, 1542.
Nikaido, K., and Nikaido, H. (1970). Personal communication.
Nomura, M. (1970). *Bacteriol. Rev.* **34**, 228.
Osborn, M. J. (1967). Unpublished experiments.
Osborn, M. J. (1968). *Nature (London)* **217**, 957.
Osborn, M. J. (1969). *Annu. Rev. Biochem.* **38**, 501.
Osborn, M. J., and D'Ari, L. (1964). *Biochem. Biophys. Res. Commun.* **16**, 568.
Osborn, M. J., and Rothfield, L. (1971). In "Microbiol Toxins" (S. J. Ajl, ed.), Vol. IV, p. 331. Academic Press, New York.
Osborn, M. J., and Weiner, I. M. (1967). *Fed. Proc. Fed. Amer. Soc. Exp. Biol.* **26**, 70.
Osborn, M. J., and Weiner, I. M. (1968). *J. Biol. Chem.* **243**, 2631.
Osborn, M. J., and Yuan Tze-Yuen, R. (1968). *J. Biol. Chem.* **243**, 5145.
Osborn, M. J., Rosen, S. M., Rothfield, L., and Horecker, B. L. (1962). *Proc. Nat. Acad. Sci. U. S.* **48**, 1831.
Petit, J. F., Strominger, J. L., and Söll, D. (1968). *J. Biol. Chem.* **243**, 757.
Plapp, R., and Strominger, J. L. (1970a). *J. Biol. Chem.* **245**, 3667.
Plapp, R., and Strominger, J. L. (1970b). *J. Biol. Chem.* **245**, 3675.
Prockop, D. J., and Kivirikko, K. E. (1967). *Ann. Intern. Med.* **66**, 1243.
Race, C., Ziderman, D., and Watkins, W. M. (1968). *Biochem. J.* **107**, 733.
Redman, C. M., Siekevitz, P., and Palade, G. E. (1966). *J. Biol. Chem.* **241**, 1150.
Richardson, C. C. (1969). *Annu. Rev. Biochem.* **38**, 795.
Robbins, P. W., and Uchida, T. (1962). *Biochemistry* **1**, 323.
Robbins, P. W., and Uchida, T. (1965). *J. Biol. Chem.* **240**, 375.
Robbins, P. W., and Wright, A. (1971). In "Microbiol Toxins" (S. J. Ajl, ed.), Vol. IV, p. 351. Academic Press, New York.
Robbins, P. W., Keller, J. M., Wright, A., and Bernstein, R. L. (1965). *J. Biol. Chem.* **240**, 384.
Robbins, P. W., Bray, D., Dankert, M., and Wright, A. (1967). *Science* **158**, 1536.
Roberts, W. S. L., Petit, J. F., and Strominger, J. L. (1968a). *J. Biol. Chem.* **243**, 768.
Roberts, W. S. L., Strominger, J. L., and Söll, D. (1968b). *J. Biol. Chem.* **243**, 749.
Robinow, C. F. (1963). In "The Cell" (J. Bracket and A. E. Mirsky, eds.), Vol. 4, p. 45. Academic Press, New York.
Robinson, H. C., Telser, A., and Dorfman, A. (1966). *Proc. Nat. Acad. Sci. U. S.* **56**, 1859.
Romeo, D., Girard, A., and Rothfield, L. (1970a). *J. Mol. Biol.* **53**, 475.
Romeo, D., Hinckley, A., and Rothfield, L. (1970b). *J. Mol. Biol.* **53**, 491.
Roseman, S., Carlson, D. M., Jourdian, G. W., McGuire, E. J., Kaufman, B., Basu, S., and Bartholomew, B. (1966). *Methods Enzymol.* **8**, 354.
Rosen, S. M., Osborn, M. J., and Horecker, B. L. (1964). *J. Biol. Chem.* **239**, 3196.
Rothfield, L., and Horne, R. W. (1967). *J. Bacteriol.* **93**, 1705.
Rothfield, L., Osborn, M. J., and Horecker, B. L. (1964). *J. Biol. Chem.* **239**, 2788.
Rouvière, J., Lederberg, S., Granboulan, P., and Gros, F. (1969). *J. Mol. Biol.* **46**, 413.
Rowland, R. L., Latimer, P. H., and Giles, J. A. (1956). *J. Amer. Chem. Soc.* **78**, 4680.

Ryter, A. (1967). *Folia Microbiol. Prague* 12, 283.

Ryter, A. (1968). *Bacteriol. Rev.* 32, 39.

Ryter, A., and Jacob, F. (1964). *Ann. Inst. Pasteur, Paris* 107, 384.

Ryter, A., and Jacob, F. (1966). *Ann. Inst. Pasteur, Paris* 110, 801.

Salivar, W. O., and Sinsheimer, R. L. (1969). *J. Mol. Biol.* 41, 39.

Sarcione, E. J. (1964). *J. Biol. Chem.* 239, 1686.

Sarcione, E. J., Bohne, M., and Leahy, M. (1964). *Biochemistry* 3, 1973.

Schachter, H., and McGuire, E. J. (1968). *Fed. Proc. Fed. Amer. Soc. Exp. Biol.* 27, 345.

Schachter, H., Inderjit, J., Hudgin, R. L., Pinteric, L., McGuire, E. J., and Roseman, S. (1970). *J. Biol. Chem.* 245, 1090.

Schenkein, I., and Uhr, J. W. (1970). *J. Cell Biol.* 46, 42.

Scher, M., and Lennarz, W. J. (1969). *J. Biol. Chem.* 244, 2777.

Scher, M., Lennarz, W. J., and Sweeley, C. C. (1968). *Proc. Nat. Acad. Sci. U. S.* 59, 1313.

Sela, B. A., Lis, H., Sharon, N., and Sachs, L. (1970). *Membrane Biol.* 3, 267.

Shapiro, B., Siccardi, A., Hirota, Y., and Jacob, F. (1970). *J. Mol. Biol.* 52, 75.

Shen, L., Grollman, E. F., and Ginsburg, V. (1968). *Proc. Nat. Acad. Sci. U. S.* 59, 224.

Siewert, G., and Strominger, J. L. (1967). *Proc. Nat. Acad. Sci. U. S.* 57, 767.

Siewert, G., and Strominger, J. L. (1968). *J. Biol. Chem.* 243, 783.

Silbert, J. E. (1967). *J. Biol. Chem.* 242, 5146, 5153.

Simkin, J. L., and Jamieson, J. C. (1967). *Biochem. J.* 103, 153.

Sinha, R. K., and Neuhaus, F. C. (1968). *J. Bacteriol.* 96, 374.

Sinohara, H., and Sky-Peck, H. H. (1965). *Biochim. Biophys. Acta* 101, 90.

Smith, D. W., and Hanawalt, P. C. (1967). *Biochim. Biophys. Acta* 149, 519.

Smith, D. W., Schaller, H., and Bonhoeffer, F. J. (1970). *Nature (London)* 226, 711.

Smith, E. E. B., Mills, G. T., and Bernheimer, H. P. (1961). *J. Biol. Chem.* 236, 2179.

Snyder, R. W., and Young, F. E. (1969). *Biochem. Biophys. Res. Commun.* 35, 354.

Spiro, M. J., and Spiro, R. G. (1967). *Abstr. Int. Congr. Biochem., 7th* 4, 710.

Spiro, M. J., and Spiro, R. G. (1968a). *J. Biol. Chem.* 243, 6520.

Spiro, M. J., and Spiro, R. G. (1968b). *J. Biol. Chem.* 243, 6529.

Spiro, M. J., and Spiro, R. G. (1968c). *Fed. Proc. Fed. Amer. Soc. Exp. Biol.* 27, 345.

Spiro, R. G. (1970). *Annu. Rev. Biochem.* 39, 599.

Spiro, R. G., and Spiro, M. J. (1966). *J. Biol. Chem.* 241, 1271.

Stone, K. J., Butterworth, P. H. W., and Hemming, F. W. (1967a). *Biochem. J.* 102, 443.

Stone, K. J., Wellburn, A. R., Hemming, F. W., and Pennock, J. F. (1967b). *Biochem. J.* 102, 325.

Struve, W. G., and Neuhaus, F. C. (1965). *Biochem. Biophys. Res. Commun.* 18, 6.

Struve, W. G., Sinha, R. K., and Neuhaus, F. C. (1966). *Biochemistry* 5, 82.

Subbaiah, T. V., and Stocker, B. A. D. (1964). *Nature (London)* 201, 1298.

Sueoka, N., and Quinn, W. G. (1968). *Cold Spring Harbor Symp. Quant. Biol.* 33, 695.

Takayama, K., and Goldman, D. S. (1970). *J. Biol. Chem.* 245, 6251.

Tanner, W. (1969). *Biochem. Biophys. Res. Commun.* 35, 144.

Telser, A., Robinson, H. C., and Dorfman, A. (1966). *Arch. Biochem. Biophys.* 116, 458.

Thorndike, J., and Park, J. T. (1969). *Biochem. Biophys. Res. Commun.* 35, 642.

Tipper, D. J., and Strominger, J. L. (1965). *Proc. Nat. Acad. Sci. U. S.* **54**, 1133.
Tipper, D. J., and Strominger, J. L. (1968). *J. Biol. Chem.* **234**, 3169.
Tipper, D. J., Strominger, J. L., and Ghuysen, J. M. (1967). *Biochemistry* **6**, 921.
Tremblay, G. Y., Daniels, M. J., and Schaechter, M. (1969). *J. Mol. Biol.* **40**, 65.
Uchida, T., Makino, T., Kurahashi, K., and Uetake, O. (1965). *Biochem. Biophys. Res. Commun.* **21**, 354.
Villemez, C. L., and Clark, A. F. (1969). *Biochem. Biophys. Res. Commun.* **36**, 57.
Warshawsky, H., Leblond, C. P., and Droz, B. (1963). *J. Cell Biol.* **16**, 1.
Weidel, W., and Pelzer, H. (1964). *Advan. Enzymol.* **26**, 193.
Weiner, I. M., Higuchi, T., Rothfield, L., Saltmarsh-Andrew, M., Osborn, M. J., and Horecker, B. L. (1965). *Proc. Nat. Acad. Sci. U. S.* **54**, 228.
Weiser, M. M., and Rothfield, L. (1968). *J. Biol. Chem.* **243**, 1320.
Wellburn, A. R., and Hemming, F. W. (1966). *Nature (London)* **212**, 1364.
Wilkinson, R. G., and Stocker, B. A. D. (1968). *Nature (London)* **218**, 954.
Wilkinson, R. G., Gemski, P., Jr., and Stocker, B. A. D. (1971). *J. Bacteriol.* In press.
Wise, E. M., and Park, J. T. (1965). *Proc. Nat. Acad. Sci. U. S.* **54**, 75.
Wright, A. (1970). Personal communication.
Wright, A. (1971). *J. Bacteriol.*, In press.
Wright, A., Dankert, M., and Robbins, P. W. (1965). *Proc. Nat. Acad. Sci. U. S.* **54**, 235.
Wright, A., Dankert, M., Fennessey, P., Robbins, P. W. (1967). *Proc. Nat. Acad. Sci. U. S.* **57**, 1798.
Zagury, D., Uhr, J. W., Jamieson, J. C., and Palade, G. E. (1970). *J. Cell Biol.* **46**, 52.
Ziderman, D., Gompertz, S., Smith, Z. G., and Watkins, W. M. (1967). *Biochem. Biophys. Res. Commun.* **29**, 56.

9

STRUCTURAL AND FUNCTIONAL ORGANIZATION OF MITOCHONDRIAL MEMBRANES

J. M. FESSENDEN-RADEN and E. RACKER

I. INTRODUCTION

Before discussing the mitochondrial membranes which are the subject of this review, it may be worthwhile to reflect on the meaning of the term membrane. Like other intracellular structures, membranes have undergone changes in the course of evolution and it is not easy to differentiate between components of what might be called a "primary" membrane and components that have been acquired later. A strong case for a primary membrane can be made in yeast promitochondria (cf. Schatz, 1970) which have been depleted of respiratory pigments during exposure of yeast cells to anaerobic conditions in the absence of unsaturated fatty acids but have retained an oligomycin-sensitive ATPase. Similarly, chemical resolution of bovine heart mitochondria has yielded a vesicular membrane preparation with an oligomycin-sensitive ATPase, resembling the inner mitochondrial membrane though lacking respiratory pigments (Kagawa and Racker, 1966b). The basic properties of such "primary" membranes which contain hydrophobic proteins and phospholipids are still ill understood and many questions remain to be answered. How are the hydrophobic proteins organized? How is their function modified by interaction with phospholipids? What is the role of phospholipids in the organization of structure? Does the chemical composition of phospholipids control their function directly or, indirectly, by conditioning their physical state? Investigations addressing themselves to some of these questions have generally yielded ambiguous answers. For example, it was proposed (Fleischer et al., 1967; Napolitano et al., 1967) that virtually all of the phospholipids of mitochondrial membranes can be removed with acetone without alteration of the basic membrane structure as viewed in an electron microscope. Although it is probable that proteins could assemble a rigid membrane structure, evidence for this must be obtained without the use of agents (such as acetone) which are known to denature some proteins of the mitochondrial membrane. Thus, on examining a hard-boiled egg one might challenge the role of the shell in maintaining the structure of the egg. Experiments on the reconstitution of the inner membrane of mitochondria (Kawaga and Racker, 1966c) strongly favor the notion that the phospholipids play an important role in the structural organization.

Controversies over the molecular organization of membranes have been partly real and partly semantic. The well-known model of Danielli and Davson (1935) with a central phospholipid bilayer core covered by proteins on each side has been the center of many of these discussions. The proponents of the Danielli-Davson model have been fully aware of the existence of several transport systems which necessitate a mosaic structure of the membrane permitting passage of hydrophilic substances. The existence of "protein channels" across the membrane is entirely consistent with the model. The fundamental issue at stake is continuity. Do phospholipid bilayers provide such continuity in the two-dimensional organizations of membranes or is the membrane an assembly of subunits consisting of proteins (Frey-Wyssling, 1955) or lipoproteins (Lucy, 1964; Green and Perdue, 1966) without the presence of phospholipid bilayers? This question is important not only for conceptual considerations, but also for the design of experiments with model membranes and on reconstitution of natural membranes from isolated components. In our opinion the available evidence seems to favor the existence of phospholipid bilayers. This evidence has been persuasively presented in several reviews (Mueller and Rudin, 1969; Thompson and Henn, 1970; Stoeckenius, 1970) and need not be repeated here; but it should be emphasized that not enough consideration has been given in many discussions to the fact that many constituents of isolated membranes are secondary and can be removed without destroying the structural continuity or the physiological activity of the membrane. For example, phospholipases which are too large to penetrate the mitochondrial membrane can hydrolyze surface phospholipid without destruction of the potential capacity to catalyze oxidative phosphorylation (Burstein et al., 1970).

An understanding of the molecular architecture of membranes may have to wait until a successful reconstitution of fully functional membranes has been achieved. In the meantime, emphasis should be placed on the great individuality of different membranes and on the difficulties in arriving at general conclusions applicable to all membranes. This point will emerge clearly from the discussion of the inner and outer membrane of mitochondria. These membranes are structurally and functionally closely associated and yet each has properties that clearly distinguish it from the others.

II. THE INTACT MITOCHONDRION

A. Structural Organization

Nearly two decades ago, Palade (1952) published the first high-resolution electron micrograph of mitochondria from a variety of species showing a convoluted membrane filled with structureless material (matrix). Use of ultrathin sections for electron microscopy by Sjöstrand (1953) revealed the presence of a double membrane. Palade (1953) confirmed the presence of a double membrane and interpreted the convolutions (cristae) to be folds of the inner membrane. The mitochondrial structure was considered by Palade (1953, 1956) to be two chambers bounded by two separate membranes. The outer chamber (intermembrane space) is delineated by the outer and the inner membrane, while the inner chamber or matrix is completely enclosed by the inner membrane whose cristae protrude towards the interior (Fig. 1).

A striking morphological difference between the two membranes was

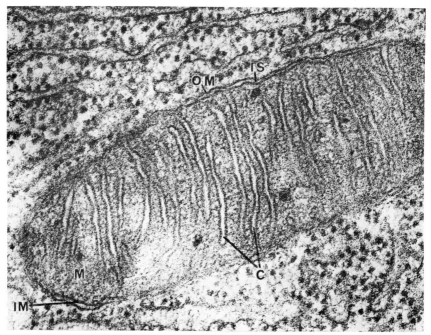

Fig. 1. Mitochondrion of an exocrine cell of the pancreas. The material was pre-pared by fixing in osmium tetroxide followed by treatment with uranyl acetate, gly-cerinating and transferring via ethanol to Epon; it was then stained with lead hy-droxide. The method was developed by Mr. R. Hebert. Its chief advantage is that it gives good visualization of the triple-layer structure of membranes. Abbreviations: OM, outer membrane; IS, intermembrane space; IM, inner membrane; C, cristae; and M, matrix ($\times$120,000). Micrograph was kindly provided by Dr. Stanley Bullivant of the University of Aukland, Aukland, New Zealand.

discovered by Fernández-Morán (1962). He observed that the negatively stained inner mitochondrial membrane exhibited small mushroomlike subunits projecting at regular intervals into the matrix space. These inner membrane spheres were 85 Å in diameter and have been identified with the mitochondrial ATPase (Racker et al., 1965). These distinctive struc-tures were not found on the outer membrane; however, smaller and less regular projecting subunits (60 Å) have been observed on the outer membrane by Parsons (1965, Parsons et al., 1967).

The formulation of mitochondrial structure proposed by Palade (1956) has been widely accepted, although several alternatives have been sug-gested by Green and co-workers. According to their most recent formula-tion (Penniston et al., 1968) there is a double outer membrane made from the outer boundary membrane and inner boundary membrane and a system of inner membranes consisting of only the cristae which are attached to the "outer membrane system" about an orifice. Since in our opinion this formulation represents more a departure in nomenclature than in concept, we prefer to accept Palade's formulation.

B. Reversible Structural Alterations

Mitochondria from a variety of sources have been shown to undergo shape and volume changes in response to various conditions. Hackenbrock (1966) reported that in rat liver mitochondria, the ultrastructure of the inner mitochondrial membrane is influenced by the metabolic state. Two distinct mitochondrial forms were visible by electron microscopy, a "high-energy" orthodox state and a "low-energy" condensed state (Hackenbrock, 1966, 1968a,b). In the "condensed" state the matrix is squeezed together and therefore stains more heavily. When these mitochondria are allowed to respire in the presence of substrate and P_i, the inner membrane expands to the "orthodox" state. The mitochondria return to the condensed state on addition of ADP (Fig. 2).

Green et al. (1968a) have described three conformational states in bovine heart mitochondria: a nonenergized (condensed), an energized (orthodox), and an energized-twisted state obtained in the presence of inorganic phosphate. It was suggested that these structural changes of the inner mitochondrial membrane represent the conversion of the energy of electron transport into conformational energy which could be used directly for ATP synthesis. These changes could be prevented either by uncouplers or inhibitors. The energized state is proposed to be the functional equivalent of the high-energy intermediates of Slater (1953). More recently, structural changes have been reported (Blondin and Green, 1969) which are not energy linked and which are proposed to be in a "pseudoenergized" state caused by passive or facilitated diffusion of alkali metal ions. Although morphologically indistinguishable from the energized state, the pseudoenergized state is reported to be insensitive to uncouplers and inhibitors.

Packer and associates (Deamer et al., 1967; Packer et al., 1968) earlier reported reversible ultrastructural changes in mitochondria which were attributed to osmotic changes due to "energized ion movements." Similar osmotic changes during energy-driven Ca^{++} accumulation were reported by Hackenbrock and Caplan (1969) and were claimed to represent a second type of ultrastructural change.

Mintz et al. (1967) failed to see ultrastructural changes in tumor or fetal mitochondria which lacked respiratory control and ion transport. With the tumor mitochondria, however, some structural changes could be observed if Mg^{++} and serum albumin were added to partially restore

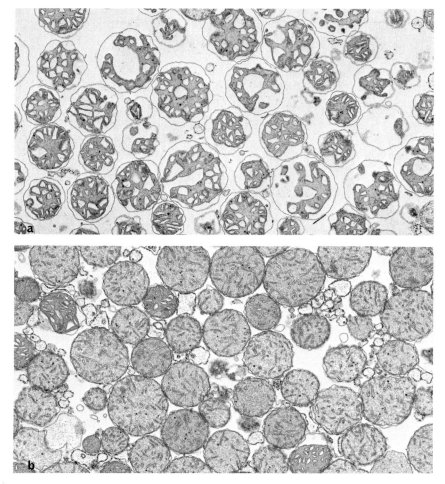

Fig. 2. (a) Condensed conformation of rat liver mitochondria (×27,000). (b) Orthodox conformation of rat liver mitochondria (×27,000) (Hackenbrock 1968a). Micrograph was kindly provided by Dr. Charles R. Hackenbrock of Johns Hopkins University School of Medicine, Baltimore.

respiratory control and ion transport. Weinbach *et al.* (1967) reported that although oxidative phosphorylation could be restored to uncoupled rat liver mitochondria by serum albumin, there was no restoration of normal structure. These workers concluded that "ATP or other high energy intermediates generated during the recoupled phosphorylation had little discernable effect on the alterations of mitochondrial morphology."

Stoner and Sirak (1969) exposed bovine heart mitochondria to varying osmotic pressures and observed by electron microscopy reversible changes similar to the "nonenergized" and "energized" forms. The "energy-twisted" state could then be induced if nonenergized osmotically swollen mitochondria were exposed to 0.15% glutaraldehyde during osmotic contraction. Pfaff *et al.* (1968) have also reported similar osmotically directed structural changes in liver and kidney mitochondria. Working with bovine heart mitochondria in various metabolic states, Weber and Blair (1969) have found that addition of ADP alone was sufficient to convert all mitochondria to a condensed ("nonenergized") form which was not altered on addition of substrate and P_i. Thus, ATP was being formed while the ultrastructure of the inner membrane, as viewed by electron microscopy, remained stable. They suggested that in steady-state conditions any ultrastructural alterations result from osmotic changes induced by the metabolic processes within the mitochondria.

Wrigglesworth and Packer (1969) have shown that reversible conformational alterations can also be induced in submitochondrial particles by changes in pH. Since proton movements occur during respiration (Mitchell, 1966a), these conformational changes may be secondary to electron transport and not directly related to energy conservation. Packer and Utsumi (1969) have further reported that rat liver mitochondria treated with 0.8% glutaraldehyde were in a fixed state in relation to osmotic changes in the membrane but could still oxidize substrates and utilize this energy for Ca^{++} accumulation.

It is apparent from these reports that the striking morphological variations seen in the structure of the mitochondrial inner membrane are real but cannot be unambiguously related to metabolic states. In fact, the majority of investigators seem to favor the view that the observed ultrastructural changes are osmotic in origin, secondary to ion movements. Obviously the latter are related to the metabolic state of mitochondria so that it is not surprising that under controlled conditions correlation between the metabolic state and the conformation of the inner mitochondrial membrane can be discerned. The major point we wish to make is that we feel that it is not possible to determine from electron micrographs whether or not structures are "energized." Therefore, the implication arising from the use of such functional terms in electron microscopy is unfortunate. The possibility still remains, however, that ultrastructural changes at a molecular level could be occurring during oxidative phosphorylation (Boyer, 1965). It is unlikely, however, that these could be discernable in the electron microscope with the present limits of resolution.

C. Irreversible Structural Alterations

Prior to the discovery of the readily reversible "low-amplitude" volume changes described above, numerous reports were published (cf. Lehninger, 1962) on "large-amplitude" changes usually referred to as swelling and contraction of mitochondria. Some of these changes are also reversible; others are not and may result in the fracture of the outer membrane. In fact, swelling and contraction followed by density gradient centrifugation has been developed as an effective tool for the separation of outer and inner membrane (Parsons *et al.*, 1966; Sottacasa *et al.*, 1967a).

In early studies correlations were made between volume changes and metabolic changes induced by substrates, ATP, and uncouplers (i.e., fatty acids, thyroxine, etc.). In recent years more emphasis has been placed on correlations between ion movements and volume changes. Chappell and his collaborators (cf. Chappell, 1968) have successfully used volume changes to analyze anion movements and discovered several anion-specific channels within the inner mitochondrial membrane.

More drastic procedures result in fracture of the mitochondrial membranes and formation of inner membrane submitochondrial particles. It was shown that rat liver mitochondria disrupted with digitonin retained phosphorylating activity (Raw, 1955; Cooper and Lehninger, 1956). Submitochondrial particles prepared by this procedure are believed to retain the structural relationship of mitochondria, i.e., the inner membrane spheres face the interior of the vesicles (cf. Mitchell, 1966a), whereas submitochondrial particles made by sonic oscillation are "inside out," i.e., the inner membrane spheres face the medium (Löw and Vallin, 1963; Lee and Ernster, 1966). Morphological evidence for this difference between digitonin particles and sonicated particles has been presented (Malviya *et al.*, 1968). On the other hand, submitochondrial particles have been prepared by disruption of bovine heart mitochondria with digitonin and were shown to be inside out by immunological and morphological analysis (Loyter *et al.*, 1969, Christiansen *et al.*, 1969). Thus, conditions of fracture, source of mitochondria, and procedure of isolation may influence the structural organization of the isolated inner mitochondrial membrane. It should also be stressed that preparations of submitochondrial particles may be quite heavily contaminated with vesicles derived from the outer membrane. Particularly in bovine heart mitochondria, which are widely employed because of their superior stability, a complete separation of inner and outer membranes is very difficult.

An interesting morphological alteration of the inner membrane has been induced in bovine heart mitochondria by treating the intact mitochondria with phospholipase C (Racker *et al.*, 1970). The resulting structures seem to represent an intermediate state between mitochondria and submitochondrial particles. While they behaved like mitochondria upon negative staining and in their sedimentation pattern, they responded like submitochondrial particles in functional tests, such as in the sensitivity of phosphorylation to valinomycin, nigericin, or Ca^{++}. Serial sections revealed that these structures, indeed, consisted of inner membrane submitochondrial particles packaged within an outer mitochondrial membrane. Similar morphological alterations of mitochondria following phospholipase A treatment have been previously reported by Bachmann *et al.* (1967) but not correlated with functional activities.

More drastic alterations of mitochondrial structure resulting in the resolution of components of oxidative phosphorylation from the inner membrane will be discussed later. It is, however, worthwhile to emphasize that major alterations can be induced in the membrane without loss of functional activity, i.e., catalysis of oxidative phosphorylation. Thus, it is felt that continuous comparison of changes in structure and function during manipulation and treatments of the mitochondrial membranes is a fruitful approach to the exploration of the molecular biology of membranes.

III. STRUCTURE OF THE OUTER MITOCHONDRIAL MEMBRANE

A. Separation of the Inner and Outer Membrane

Several methods for separation of the outer and inner membrane of mitochondria have been reported. Parsons and associates (cf. Parsons and Williams, 1967) observed that exposure of liver mitochondria to a hypotonic phosphate medium caused swelling and concomitant rupture of the outer membrane. The two membrane fractions could then be separated from each other as well as from unbroken mitochondria by centrifugation through a sucrose density gradient. Several laboratories have modified this procedure in attempts to prepare pure membrane fractions. Other methods employed have included a swelling–contraction cycle followed by gentle sonication (cf. Sottocasa *et al.*, 1967b), mild digitonin treatment (Lévy *et al.*, 1967), osmotic lysis (Schnaitman *et al.*, 1967), or freezing the swollen mitochondria at liquid-air temperatures (Brdiczka *et al.*, 1968). Separation of the membranes is usually accomplished by density gradient centrifugation. More recently, Heidrich *et al.* (1970) have used the swelling–contraction technique for the disruption of mitochondria followed by free-flow electrophoresis to separate the membranes. Other procedures such as treatment with phospholipase A (cf. Allmann and Bachmann, 1967), diethylstilbesterol (Byington *et al.*, 1968), or oleate (Allmann *et al.*, 1968) have been used for the disruption of mitochondria followed by centrifugal separation of membrane fractions.

B. Protein Constituents

Controversies regarding the localizations of various membrane constituents have somewhat clouded the field. However, the majority of workers is in agreement with the enzyme distribution indicated in Table I (for an in-depth discussion and references, see the review by Ernster and Kuylenstierna, 1970). Here we should only like to point out a few of the major controversies concerning the proteins of the outer membrane.

While it is widely accepted that the enzymes of the citric acid cycle (except succinate dehydrogenase), as well as those of fatty acid synthesis, are in the matrix, Green *et al.* (1966) place the citric acid cycle enzymes and all of the fatty acid oxidation enzymes in the outer membrane. Indeed, it has been reported by Green and co-workers (Allmann *et al.*, 1968) that the outer membrane accounts for over 30% of the total mitochondrial protein and there is virtually no protein within the matrix space. However, a location of the citric acid cycle enzymes in the outer membrane is very difficult to accept in view of evidence from intact mitochondria. (a) Intact mitochondria do not oxidize the citric acid cycle substrates as readily as damaged mitochondria do (Ernster, 1959). (b) Independent permeability studies have shown the outer membrane to be freely permeable to small molecular weight substances (e.g. nucleotides) while the inner membrane is not (cf. Klingenberg and Pfaff, 1966). (c) It has been reported that the inner membrane has several specific transport systems, e.g., for ATP and citrate (cf. Chappell, 1968). (d) Pyridine nucleotides which are required in the citric acid cycle appear to be located in the matrix space (cf. Ernster and Kuylenstierna, 1970). These observations explain the "latency" of the citric acid cycle enzymes as well as the need for substrate-specific transport systems within the inner membrane. If these enzymes were in the outer membrane as proposed by Green, the latency phenomena would be difficult to explain.

The major reason for localization of the citric acid cycle enzymes in the outer membrane is based on the isolation procedure. Upon reviewing the methods of separation used by Green and co-workers, such findings are not surprising since treatment with cholate or phospholipase A would be expected to damage both the outer and inner membrane and thus

TABLE I[a]

LOCALIZATION OF ENZYMES IN LIVER MITOCHONDRIA

Outer membrane	Intermembrane space	Inner membrane	Matrix
"Rotenone-insensitive" DPNH-cytochrome c reductase (DPNH-cytochrome b_5 reductase, cytochrome b_5)	Adenylate kinase	Respiratory chain (cytochromes b, c_1, c, a, a_3; succinate dehydrogenase; succinate–cytochrome c reductase, succinate oxidase, "rotenone-sensitive" DPNH-cytochrome c reductase; DPNH oxidase; choline–cytochrome c reductase; cytochrome c oxidase; respiratory chain-linked phosphorylation)	Malate dehydrogenase
Monoamine oxidase	Nucleoside diphosphokinase		Isocitrate dehydrogenase (TPN specific)
Kynurenine hydrolylase	Nucleoside monophosphokinase		Isocitrate dehydrogenase (DPN specific)
ATP-Dependent fatty acyl-CoA synthetase	Xylitol dehydrogenase (DPN specific)?		Glutamate dehydrogenase
Glycerol phosphate acyltransferase			α-Ketoglutarate dehydrogenase
Lysophosphatidate acyl transferase			Citrate synthetase
Lysolecithin acyltransferase			Aconitase
Choline phosphotransferase		β-Hydroxybutyrate dehydrogenase	Fumarase
Phosphatidate phosphatase		Ferrochelatase	Pyruvate carboxylase
Phospholipase A_{11}		α-Aminolevulinic acid synthetase?	Phosphopyruvate carboxylase
Nucleoside diphosphokinase		Carnitine palmityltransferase	Aspartate aminotransferase
Fatty acid elongation system		Fatty acid oxidation system?	Ornithine carbamoyltransferase
Xylitol dehydrogenase (DPN specific)?		Fatty acid elongation system	Fatty acyl-CoA synthetase(s)
		Xylitol dehydrogenase (TPN specific)?	Fatty acid oxidation systems?
			Xylitol dehydrogenase (TPN specific)?

[a] Taken from Ernster and Kuylenstierna (1970).

allow the enclosed matrix material to escape and become associated with membrane fragments. However, some of the data cannot be entirely explained by differences in techniques or definitions. For example, the discrepancy with regard to monoamine oxidase is particularly striking and deserves discussion, since it points to the necessity of using additional criteria for the association between enzymes and membranes. While many investigators in the field concur that monoamine oxidase (as well as the rotenone-insensitive DPNH-cytochrome c reductase) is located in the outer membrane, Green *et al.* (1968b) insist that this conclusion is erroneous. Ernster and Kuylenstierna (1970) have recently summarized the overwhelming evidence for the localization of these enzymes in the outer membrane. In experiments on the reassociation of purified monoamine oxidase with beef kidney cortex mitochondrial membranes (Racker and Procter, 1970), a specific reconstitution of monoamine oxidase with the outer membrane was observed. Moreover, inspection of the specific activity of the isolated membrane fractions as well as reconstituted membranes left little room for doubt that monoamine oxidase is specifically associated with the outer membrane. Parallel experiments with bovine heart mitochondrial ATPase (F_1) yielded equally unambiguous data on the specific binding of F_1 to the inner membrane. Since there is general agreement that the ATPase is a constituent of the inner membrane, it is apparent that monoamine oxidase is associated with a different membrane which, by generally accepted criteria and morphological appearances, is the outer membrane. Reconstitution experiments of this kind are particularly useful when an enzyme used as a marker for a membrane dissociates during various manipulative processes, thus lowering the final specific activities of the membrane-associated enzyme. It should be pointed out, however, that demonstration of the preferential association of monoamine oxidase with the outer membrane does not preclude the presence of a small fraction of monoamine oxidase in the inner membrane. Since there is increasing evidence for the existence of multiple forms of monoamine oxidase, first reported by Gorkin (1963), a minor component may, indeed, be located in the inner membrane.

C. Lipid Constituents

According to Parsons *et al.* (1967) the outer mitochondrial membrane contains 0.83 mg phospholipid/mg protein. This is about three times higher than the phospholipid content of inner membrane (0.28 mg/mg protein), which is similar to the phospholipid content of microsomes (0.39 mg/mg protein). An analysis of the phospholipid composition of mitochondria (Parsons *et al.*, 1967; Stoffel and Schiefer, 1968) has revealed that there is at least seven times as much cardiolipin in the inner membrane as in the outer membrane, whereas phosphatidylinositol and phosphatidylserine are higher in the outer membrane. Phosphatidylcholine and phosphatidylethanolamine seem to be relatively evenly distributed between the two membranes. The composition of the outer membrane phospholipids resembles that of microsomes. The cholesterol content of the outer membrane (30 μg/mg protein) also is similar to that of microsomes and is about six times greater than that of the inner mitochondrial membrane.

IV. POSSIBLE FUNCTIONS OF THE OUTER MITOCHONDRIAL MEMBRANE

The striking similarities between the outer mitochondrial membrane and the endoplasmic reticulum have led to the postulation that the outer membrane may have originally derived from endoplasmic reticulum (cf. Parsons *et al.*, 1967). Not only are there similarities in lipid composition (cf. above) but also in enzyme composition, e.g., both membrane types contain a cytochrome b_5 and a rotenone-insensitive DPNH-cytochrome c reductase. However, there are several enzymes which are found only in one of the membranes; monoamine oxidase is in the outer mitochondrial membrane while the endoplasmic reticulum contains a TPNH-cytochrome c reductase, a cytochrome P_{450}, and glucose-6-phosphatase as well as a mixed-function oxidase.

Recent studies have indicated a phospholipid exchange between microsomes and mitochondria (Wirtz and Zilversmit, 1968). McMurray and Dawson (1969) observed that the specific radioactivity of phosphatidylcholine and phosphatidylethanolamine isolated from the outer membrane was higher than that of either the inner membrane or the whole mitochondria, suggesting a possible preferential exchange between microsomes and the outer membrane.

Electrophoretic comparison of inner and outer membranes, intact mitochondria, and microsomes by Heidrich *et al.* (1970) showed that the outer membrane and intact mitochondria had the same electrophoretic mobility which was distinct from that of the inner membrane, thereby allowing clear separation of inner and outer membranes. Microsomes moved only slightly more towards the anode than the outer membrane did and could not be separated from it by this method.

Robertson (1960) reported connections between the endoplasmic reticulum and the outer membrane of mitochondria. Although Parsons *et al.* (1967) observed that sheets of rough endoplasmic reticulum often envelop mitochondria, they could not establish continuity between the endoplasmic reticulum and the outer mitochondrial membrane.

A. Function as a Protective Envelope

The similarity of the outer mitochondrial membrane and the endoplasmic reticulum would be in line with the suggestion (cf. Lehninger,

1965) that mitochondria may have evolved from invading microorganisms. The size and shape of mitochondria are similar to that of bacteria. There are several similarities between bacterial membranes and the inner mitochondrial membrane. Both membranes are capable of invaginations, contain relatively large amounts of cardiolipin, and house the oxidative phosphorylation apparatus of the cell. Like bacteria, mitochondria contain DNA and can synthesize nucleic acids and proteins. The similarities of mitochondrial DNA with bacterial DNA are as striking as its differences from nuclear DNA. Both mitochondrial and bacterial DNA have been shown to be circular, whereas nuclear DNA is linear (cf. Nass, 1969). Mitochondrial protein synthesis is sensitive to inhibitors (e.g., chloramphenicol) of bacterial, but not of mammalian, cytoplasmic protein synthesis (e.g., cycloheximide). The relative size of the mitochondrial ribosomes is often similar to that of bacterial ribosomes (cf. Schatz, 1970), while mammalian cytoplasmic ribosomes are generally somewhat larger. It has been demonstrated by Luck (1965) that in exponentially growing *Neurospora crassa* cells, mitochondria grow and divide during cell multiplication without any indication of *de novo* synthesis. In oxygen-adapting yeast cells, respiring mitochondria are made from promitochondria (Criddle and Schatz, 1969) rather than by *de novo* synthesis.

While the bacterial origin of mitochondria has become a popular hypothesis, several aspects remain to be explained. If the presence of an oxidative phosphorylation apparatus in each is an argument in favor of a common evolutionary origin of mitochondria and bacteria, why is the cytochrome *c* of the inner membrane (Sherman *et al.*, 1966), as well as the inner membrane spheres (F_1) (Schatz, 1968), coded for by nuclear DNA rather than by mitochondrial DNA? Indeed, present evidence indicates that the majority of the mitochondrial proteins are coded for by nuclear genes and synthesized on cytoplasmic ribosomes (Clark-Walker and Linnane, 1967). One can claim that during evolution much of the initial genetic information of the mitochondrion was transferred to the nucleus, and Wilkie (1964) has postulated an interaction between nuclear and mitochondrial DNA. Borst (1971) has pointed out that as one moves up the phylogenetic ladder from yeast to mammals, both the size and the informational content of the mitochondrial DNA become progressively smaller.

If mitochondria have a parasitic origin, then the outer membrane may well represent the cellular contribution to the symbiotic relationship, serving to isolate and contain the inner membrane. This would explain some of the similarities mentioned above between the endoplasmic reticulum and the outer membrane. The presence of new components in the outer membrane, e.g., monoamine oxidase and in some cases hexokinase, would represent acquisition of secondary properties during evolution.

B. Function as a Permeability Barrier

In contrast to the inner membrane, the outer membrane has been shown to be freely permeable to small molecules such as nucleotides, sucrose, and salts (Werkheiser and Bartley, 1957; Bücher and Klingenberg, 1958). This relatively high permeability may be related to the phospholipid composition of the outer membrane. Chappell *et al.* (1968) have reported that addition of cardiolipin to phospholipid vesicles reduces their permeability to anions and suggested that the lack of cardiolipin in the outer membrane may be a factor in its permeability to anions.

Larger molecules, however, such as polyglucose and albumin (Werkheiser and Bartley, 1957) are believed not to permeate the outer membrane, but perhaps inulin does (Pfaff, 1965; for a discussion of this problem, see Lehninger, 1967). Wojtczak and Zaluska (1969) have recently reported that cytochrome *c* does not penetrate the outer membrane of rat liver mitochondria. Studies with bovine heart mitochondria, on the other hand, have indicated that the outer membrane is permeable to proteins such as cytochrome *c* and γ-globulins (Racker, *et al.* 1970); however, damage of the outer membrane during isolation could have occurred. At this time the role of the outer membrane as a permeability barrier is therefore still unclear.

C. Function as an Anchor for Enzymes

The outer membrane appears to contain a curious conglomerate of enzymes (see Table I) which are unrelated in function and do not appear to interact with one another. One possible reason for their positioning on a membrane is immobilization. Such topographical fixation at a given place could control enzyme activity in several ways. For example, norepinephrine released from storage granules into the synapse is protected against the action of monoamine oxidase which is located in the mitochondria. Alternatively, the fixation of an enzyme may increase its proximity to a substrate. For example, hexokinase of the outer membrane (Rose and Warms, 1967; Kropp and Wilson, 1970) would be in a favored position to compete for ATP generated within mitochondria and would facilitate the return of the product, ADP, to the coupling device of the inner mitochondrial membrane. Still another possible reason for an enzyme being membrane bound would be to change its properties, and several examples for such allotopic properties of enzymes in the inner membrane have been described (Racker, 1967a). This may induce stabilization of the protein, alteration of susceptibility to inhibitors, as well as changes in K_m and substrate specificity.

V. STRUCTURE OF THE INNER MITOCHONDRIAL MEMBRANE

A. Lipid Constituents

1. Ubiquinone (Q)

Ubiquinone is present in relatively large amounts in mitochondria (4–5 nmole/mg protein). On a molar basis about 5–10 times more Q_{10} than cytochrome c is present in mammalian mitochondria. It was pointed out by Klingenberg (1968) that the high amounts of Q_{10} and DPN are in line with their function as hydrogen acceptors in multiple channels. Ubiquinone is required for the oxidation of both succinate and DPNH (Szarkowska, 1966). There are sufficient indications to suspect that the role of Q in these two pathways may not be equivalent. For example, it was shown (Schatz and Racker, 1966) that Q_0 functions as external hydrogen acceptor for DPNH during oxidation coupled to phosphorylation. Yet reduction of Q_0 by succinate is not associated with phosphorylation although the process is antimycin-sensitive (Christiansen and Racker, 1968). It is, therefore, possible that like DPN, Q is specifically associated with certain protein constituents of the membrane which are responsible for specificity. Exploration of such a topographical distribution of Q within the mitochondrial membrane is at present technically difficult.

2. Phospholipids

The chemistry and physical chemistry of phospholipids in membranes has been recently reviewed (Chapman and Leslie, 1970). Therefore only some special features concerning phospholipids of the inner mitochondrial membrane will be discussed here. The inner mitochondrial membrane has a relatively high content of cardiolipin, but it contains negligible amounts of cholesterol. It was shown (Burstein et al., 1970) that by sequential exposure of the two sides of the mitochondrial membrane to phospholipase C from Clostridium welchii virtually all of the phosphatidylcholine and phosphatidylethanolamine can be cleaved. The resulting particles, which have lost two-thirds of their phospholipids, catalyzed oxidative phosphorylation with an efficiency similar to untreated control particles provided that coupling factors were added. It is apparent from these experiments that both sides of the membrane have phospholipids that are expendable and that the role of these phospholipids can be fulfilled by the residual diglyceride which is retained by the particles after phospholipase C digestion.* On the other hand, exposure of submitochondrial particles that have been exhaustively digested by phospholipase C from *C. welchii* to a phospholipase C from *Bacillus cereus* which is capable of cleaving cardiolipin as well as phosphatidylethanolamine and phosphatidylcholine resulted in a very marked loss in phosphorylating capacity. It can therefore be concluded that although a large fraction of phospholipids on both sides of the membrane is not essential for oxidative phosphorylation, a small specific fraction sensitive to the *B. cereus* phospholipase C is required for the phosphorylation process. Thus by the use of phospholipases with different substrate specificity, a systematic analysis of phospholipid components of membranes can be carried out.

In contrast to the apparent specific role of phospholipids in phosphorylation, the effect of phospholipids on the respiratory chain seem to be rather nonspecific (Fleischer *et al.,* 1962). Although a complete dependency on added phospholipids in the reconstituted succinoxidase complex was demonstrated (Yamashita and Racker, 1968), the lack of specificity was also pointed out. Phospholipids from various sources and of different structure could be used for the reconstitution of this antimycin-sensitive respiratory chain segment.

The oligomycin-sensitive ATPase of the inner mitochondrial membrane, which also requires phospholipids, exhibited a similar broad-range response to various phospholipids (Kagawa and Racker, 1966b; Bulos and Racker, 1968b).

In relation to the presence of phospholipids in the inner mitochondrial membrane, the remarkable responses of artificial model membrane consisting of phospholipid bilayers to uncoupling agents are of interest (cf. Mueller and Rudin, 1969; Thompson and Henn, 1970; Pressman, 1970). Dinitrophenol and valinomycin plus potassium decrease the resistance of such membranes at concentrations similar to those required for uncoupling. These findings bring up again the possible participation of phospholipid bilayers in the assembly of the inner mitochondrial membrane discussed earlier.

B. Protein Constituents

The members of the respiratory chain are associated with the inner mitochondrial membrane. Some of these enzymes have been used as markers for the inner membrane. It was, however, pointed out earlier that they must be viewed as secondary acquisitions, since membranes

---

* Burstein *et al.* (1971) have found that digestion of the residual diglycerides with lipase results in a rapid loss of phosphorylating activity.

lacking several or all respiratory components have been isolated (Kagawa and Racker, 1966a,b,c; Criddle and Schatz, 1969). Since the majority, if not all, of these components are synthesized in the microsomes outside the mitochondria, they must be transported and incorporated into the membrane by a rather specific mechanism that allows for the assymetric assembly, which will be discussed later. Virtually nothing is known about how this remarkable task is accomplished, and its elucidation remains one of the challenging problems in biochemistry.

The composition of the inner membrane and the properties of the various components of the respiratory chain have been repeatedly reviewed (cf. Klingenberg, 1968) and only some special features and recent developments will be dealt with in this chapter. Some of the discussions of the individual constituents of the respiratory chain will be included in Section VI.

It has been frequently claimed that the components of the respiratory chain are present in fixed proportions, and speculations have been advanced explaining these relations in terms of the mechanism of oxidative phosphorylation. It is becoming increasingly apparent, however, that the ratios of constituents vary considerably in mitochondria from different sources (Klingenberg, 1968) and these speculations seem premature. Although various cytochromes are present in similar amounts (about 1 nmole/mg mitochondrial protein), estimates for cytochrome c_1 as low as 0.1 nmole/mg have been reported. Accurate estimations of the individual flavoproteins are not available but judging from the flavin content, these constituents are on a molar basis considerably lower than the cytochromes. More important from a functional viewpoint is the fact that the reaction rates of the individual steps are increasingly more rapid as electrons approach oxygen (cf. Chance and Williams, 1956). This is in line with the observation that the last components (cytochrome a and a_3) are the most abundant and the flavoproteins at the beginning of the chain are the least abundant. This kinetic feature is characteristic for many multienzyme systems (e.g., glycolysis) and is responsible for the pattern of the redox state of the individual constituents during active turnover. As in other multienzyme systems, the analysis of the turnover of the individual components is complicated by the presence of regulatory mechanisms that effect the various steps to a different degree.

Chance and Williams (1956) have proposed that some of the reduced carriers are in a controlled state ($C_{red} \sim I$) which is released simultaneously with the generation of ATP:

$$C_{red} \sim I \rightarrow X \sim I \xrightarrow{ADP+P_i} X + I + ATP$$

Although this specific formulation is not favored at present, the concept of a conformational change of a respiratory carrier during catalysis resulting in profound effects on the function as well as the structure of mitochondria is very fashionable. A concrete example of such a conformational alteration during catalysis has been described for glyceraldehyde-3-P dehydrogenase (Krimsky and Racker, 1963), which markedly loses its ability to interact with DPN when acylated. On deacylation (in the presence of P_i and ADP) the high affinity for DPN is restored. Although less clearly defined, respiratory components also undergo changes in properties. It was shown in several studies (Tyler, *et al.* 1965; Fessenden-Raden, 1969) that exposure to DPNH has profound effects on the sensitivity of the respiratory chain to mersalyl even after the substrate has been completely oxidized. A similar "memory" effect after exposure to DPNH has been observed with submitochondrial particles in the response to low concentrations of rotenone (Horstman and Racker, 1968). In an interesting study Lee *et al.* (1969) have shown that when inhibitors of oxidation are used (e.g., KCN, antimycin, etc.) respiratory control can be induced by energy-transfer inhibitors (e.g., oligomycin) and maintained at very low oxidation rates. On the other hand, if respiration is inhibited by lowering the input of reducing equivalents (DPNH), respiratory control induced by energy-transfer inhibitors is lost. Although this phenomenon has not been elucidated, the data suggests a relationship between the redox state of the carriers and control of mechanisms (cf. Chance and Pring, 1968).

In addition to the components of the respiratory chain, the inner membrane contains the components of the coupling device which are responsible for harvesting the energy generated during oxidation. This coupling device appears to be closely associated with the membrane proper and includes several protein components with pronounced hydrophobic properties which resist resolution without loss of activity. These hydrophobic proteins may be indeed the true "structural proteins" of the inner mitochondrial membrane. It was shown (Bulos and Racker, 1968a) that they are very labile in the presence of proteolytic agents and at temperatures above 50°C. It was also found that these insoluble constituents of the membranes interact with the more soluble components of the coupling device, e.g., mitochondrial ATPase. Evidence for this interaction was obtained by studies on the stability of the hydrophobic proteins, which were protected against heat inactivation by addition of coupling factors. The hydrophobic proteins also bind phospholipids (Kagawa and Racker, 1966b). Experiments on the specificity of interaction indicate that the physical state of the phospholipids is of greater importance than the chemical structure.

What is the significance of the "structural protein" (SP) of mitochondria which has been prepared by several investigators at the Enzyme Institute at the University of Wisconsin? It is apparent from a review of this work (Lenaz *et al.*, 1968a,b) and from the earlier work of Haldar *et al.* (1966) that preparations of SP obtained after solubilization with dodecylsulfate (Criddle *et al.*, 1962) or after treatment with acetone (Richardson *et al.*, 1963) are mixtures of proteins. Incisive experiments by Schatz and Saltzgaber (1969) have demonstrated that a major component of SP is denatured mitochondrial ATPase. This was unambiguously established by reconstitution of submitochondrial particles with radioactive but functionally active ATPase and the subsequent isolation of SP by several described procedures. Since it is well known that dodecylsulfate as well as acetone causes denaturation of many proteins it is not surprising to find that preparations of SP are rather insoluble. There is no available evidence that any component of SP participates in the assembly of mitochondrial membranes. Although the concept of a membrane structural protein is both old and appealing (cf. Racker, 1967b), the use of this name in the case of mitochondria should be withheld until direct experimental evidence for such an organizational protein becomes available. Although the hydrophobic proteins present in the resolved preparations of the oligomycin-sensitive ATPase may have such a function, we have refrained from calling them structural proteins in order to avoid compounding the already existing confusion.

Some of the properties of the individual coupling factors will be discussed in Section VI,B, but one feature requires emphasis here in relation to the membrane structure. Thus far, five different soluble proteins (coupling factors) have been extracted from mitochondria. Every one of them can be reconstituted into the membrane just by addition to depleted membranes. Although a systematic study of the order of addition has not been undertaken, it appears that a functionally active assembly is organized no matter what the order of addition is. It seems to hold in general that as the resolution of multiple coupling factors from the membrane becomes more extensive, the time required for reconstitution of oxidative phosphorylation increases from 5 minutes (Penefsky *et al.*, 1960) to 30 minutes (Fessenden-Raden, 1969; Racker *et al.*, 1969).

VI. FUNCTIONS OF THE INNER MITOCHONDRIAL MEMBRANE

A. Oxidation

The respiratory chain of the inner mitochondrial membrane (see Fig. 3) has been analyzed by the following three approaches, which differ from each other in several aspects. (a) Isolation of pure proteins capable of catalyzing oxidoreductions with artificial electron donors and acceptors, (b) isolation of complexes which catalyze segments of the respiratory chain but which can be used to reconstitute the entire respiratory chain oxidizing natural substrates such as DPNH or succinate, and (c) isolation of individual components of the respiratory chain which can be used to reconstitute the entire respiratory chain catalyzing the oxidation of natural substrates, e.g., succinate.

1. Isolation of Members of the Respiratory Chain

The first approach has yielded highly purified proteins which have been characterized. It has, however, become increasingly apparent that the isolated proteins were substantially altered and their physiological function impaired.

Succinate dehydrogenase prepared in the absence of succinate (Bernath and Singer, 1962) readily oxidizes succinate in the presence of artificial electron acceptors, but the enzyme cannot recombine with the inner mitochondrial membrane. Inclusion of succinate during the purification procedure (King, 1963) preserves this important property of the enzyme.

DPNH dehydrogenase has been isolated in multiple forms from bovine heart mitochondria (cf. Singer, 1968). The enzyme contains flavin, non-heme iron, and labile sulfide but the proportions of these components vary with different preparations. Hatefi and Stempel (1969) isolated a DPNH-dehydrogenase which is active in the reconstitution of complex I (Hatefi and Stempel, 1967). A rotenone-sensitive DPNH-cytochrome c reductase was reconstituted by Albracht and Slater (1969) from Q_{10} and preparations of complex I and complex III that had been depleted of Q_{10} by pentane extraction prior to cleavage with cholate. Evidence for both a pure and water-soluble preparation of DPNH-dehydrogenase which is active in reconstitution of a DPNH-oxidase complex is, however, not available.

Cytochrome b preparations isolated by different procedures have dissimilar properties. If solubilized with the aid of proteolytic enzymes (Ohnishi, 1966) the enzyme can be reduced by lactate in the presence of yeast lactate dehydrogenase. If solubilization is achieved with dodecylsulfate (Goldberger *et al.*, 1961) the enzyme can be reduced with dithionite but not by yeast lactate dehydrogenase. Interestingly enough, the latter cytochrome b preparation recombined with succinate dehydrogenase and phospholipids to yield a complex capable of reducing Q_{10} on addition of succinate, although no oxidoreduction of the cytochrome b took place. In the same system the preparation of cytochrome b described by Ohnishi was reconstitutively inactive (Bruni and Racker, 1968). This difference between the two cytochrome b preparations was interpreted as follows. The cytochrome b preparation treated with dodecylsulfate has lost its catalytic activity but has retained a structural role as a membrane component, whereas the cytochrome b exposed to proteolysis has retained its catalytic activity but has lost its structural role (i.e., it cannot react with other membrane components). Neither of these two cytochrome b preparations could be used to reconstitute an active complex of succin-oxidase. To prepare such a complex, addition of a cytochrome b preparation which had not been exposed to either dodecylsulfate or to proteolytic enzymes was required (Yamashita and Racker, 1969). It is of interest to point out that F_1, another component of the inner mitochondrial membrane, also loses its capacity to function as a structural component after exposure to a proteolytic enzyme without losing its ATPase activity (Horstman and Racter, 1970). Certain chemical modifications, on the other hand, destroy the ATPase activity of F_1 without changing its capacity to serve as a structural component of the membrane (Penefsky, 1967).

Cytochrome c_1 has been isolated as a highly purified protein (Orii and Okunuki, 1962) which could be reduced by artificial electron donors but which showed little or no activity in reconstitution. In contrast, a preparation of cytochrome c_1 obtained by a more gentle isolation procedure was active in reconstitution (Yamashita and Racker, 1969).

Cytochrome c isolated by conventional procedures has been shown to be reconstitutively active (Jacobs and Sanadi, 1960).

Cytochrome oxidase has been purified by a large number of variants of a procedure depending on solubilization with cholate or deoxycholate. The preparation oxidizes ascorbate in the presence of cytochrome c, but the amount of cytochrome c required to obtain equivalent oxidation rates is an order of magnitude higher than that needed in respiring mitochondria or submitochondrial particles. Although such a preparation of cytochrome oxidase has been used in the reconstitution of an active succin-oxidase complex (Yamashita and Racker, 1969), it was not established that the enzyme had acquired the properties of membrane-bound cytochrome oxidase. In fact, the high concentration of cytochrome c required to saturate the system suggested that it had not.

Cytochrome oxidase has been one of the most controversial enzymes; it has excited scientific feuds ever since it was discovered by MacMunn (1887) and elevated to scientific respectability by Keilin (1925). It is now almost uniformly accepted that it consists of two components, cytochrome a and a_3, which are present in about equal amounts. Since all the evidence points to a single heme component, the properties distinguishing cytochromes a and a_3, including the reactivity of the latter with carbon monoxide, must be due to differences in the proteins. However, it has not been established whether these differences are located in the primary protein structure or are the result of subunit interactions. No separation of specific species of cytochrome a and a_3 has been accomplished in spite of numerous attempts. Recent investigations (Chan et al., 1970) suggest that a low molecular weight species of cytochrome oxidase, which has properties of cytochrome a_3, can be prepared and that association of these subunits is responsible for the emergence of the divergent property of cytochrome a. These most interesting findings should influence greatly our thinking about the conformation of cytochrome oxidase within the membrane, and more will be said about this in Section VII.

2. Isolation of Complexes of the Respiratory Chain

The second approach to the resolution of the respiratory chain has been directed toward the separation of complexes which catalyze partial reactions. Four such complexes have been described (Hatefi et al., 1962), which can be reconstituted to give rise to a coordinated respiratory chain which catalyzes the oxidation of DPNH or succinate. Complex I contains DPNH-dehydrogenase, which catalyzes the oxidation of DPNH in the presence of a variety of artificial electron acceptors such as ferricyanide or phenazine methosulfate. Complex I also contains a nonheme iron protein and Q_{10}. In recent experiments (Hatefi and Stempel, 1967) the two proteins have been separated from each other and shown to act sequentially in the reduction of Q_{10}. Complex II contains succinate dehydrogenase and cytochrome b. It catalyzes the reduction of Q_{10} by succinate. Complex III contains cytochrome b, cytochrome c_1, and a nonheme iron protein. It catalyzes the reduction of cytochrome c by reduced Q. Complex IV is a preparation of purified cytochrome oxidase. It contains cytochrome a and a_3 and catalyzes the oxidation of reduced cytochrome c.

All four complexes contain phospholipids and under appropriate conditions interact with each other to reconstitute an active respiratory chain. Complexes I, III, and IV, when reconstituted, catalyze DPNH oxidation; complexes II, III, and IV catalyze succinate oxidation. These reconstituted systems have the properties of the membranous respiratory chain with regard to susceptibility to inhibitors such as rotenone and antimycin. The reconstituted complexes, however, do not catalyze oxidative phosphorylation.

3. Isolation of Individual Components Which are Reconstitutively Active

The third approach is to isolate individual components of the respiratory chain which are reconstitutively active. Thus far, reconstitution has been defined in terms of oxidative activity, but it is conceivable and, indeed, likely that an extension of the definition to include oxidative phosphorylation may be required in the future. Five enzymes of the respiratory chain can be separated from each other without loss of reconstitutive activity. They are succinate dehydrogenase, cytochrome b, cytochrome c_1, cytochrome c, and cytochrome oxidase. Of these, only succinate dehydrogenase (Racker et al., 1969), cytochrome c (Jacobs and Sanadi, 1960), and cytochrome oxidase (Arion and Racker, 1970) have been shown to be reconstitutively active in phosphorylation as well as in oxidation. The mixing of these five components in the presence of phospholipids and Q_{10} results in the formation of a complex which catalyzes the oxidation of succinate at a rate comparable to that of phosphorylating submitochondrial particles (Yamashita and Racker, 1969). The oxidation is completely inhibited by antimycin A or by cyanide.

There are several aspects of this system worthy of emphasis. One is that all the components are soluble except cytochrome b. Electron micrographs of cytochrome b (Yamashita and Racker, 1968) show that the preparation consists of vesicular structures which may well contribute to the organization of the other components of the respiratory chain. Attempts to solubilize cytochrome b have thus far invariably resulted in loss of reconstitutive activity. The second point of interest is that the nonheme iron protein present in complex III (Rieske et al., 1964) does not appear to be required for the reconstituted antimycin-sensitive electron transport chain which operates at a rate comparable to that of phosphorylating submitochondrial particles. The close association of this protein with complex III thus suggests that it may have an auxiliary role, perhaps as a link to the phosphorylating coupling device. Such a role has been suggested for nonheme iron because of other observations (Butow and Racker, 1965; Boyer, 1968).

An interesting aspect of the reconstitution of the succinoxidase complex is the role of phospholipids and the length of time required for achieving maximal activity. Although a variety of different phospholipids were suitable in these reconstitution experiments, it was necessary to establish with each the appropriate concentration as well as the optimal time and temperature of incubation with the proteins (Yamashita and Racker, 1969). In some cases, e.g., with crude mixtures of phospholipids from soybean which gave the best results, 3–4 hours of incubation at $37°C$ were needed to attain the maximal rates of succinate oxidation.

Various attempts to induce coupled phosphorylation in the reconstituted succinoxidase by addition of coupling factors or to induce respiratory control by addition of oligomycin have thus far failed. One possible reason for this failure is an improper alignment of the respiratory chain. As mentioned previously the components of the inner mitochondrial membrane are organized in a characteristic topography, with cytochrome c on the inside of submitochondrial particles and mitochondrial ATPase (F_1) on the outside. It is therefore not possible to remove cytochrome c by repeated washing of submitochondrial particles. Moreover, the oxidation of DPNH or succinate by these particles is insensitive to an antibody against cytochrome c (Racker et al., 1970). The reconstituted complex is sensitive to the antibody against cytochrome c and repeated washing of the complex results in loss of succinoxidase activity, which can be restored by addition of cytochrome c (Cunningham and Racker, 1969).

A gentle method for the removal of cytochrome c from submitochondrial particles has been developed by Racker et al. (1970). Submitochondrial particles exposed to appropriate concentrations of cholate or deoxycholate can be depleted of cytochrome c, and they can be reconstituted if excess cytochrome c is added back in the presence of the detergent. After removal of excess detergents by washing, these submitochondrial particles respire and are once more insensitive to the antibody against cytochrome c. Attempts are being made to induce such a topographical orientation of cytochrome c in the reconstituted complex of succinoxidase.

B. Phosphorylation

1. Current Hypotheses on the Function of the Coupling Device

The coupling device which links oxidation to phosphorylation and to the generation of ATP from ADP and inorganic orthophosphate has multiple components (Racker, 1970). A central role has been given to the oligomycin-sensitive ATPase (F_1). According to the chemical hypothesis the ATPase catalyzes the transphosphorylation between ADP and an unknown phosphorylated intermediate $X \sim P$, whereas according to the chemiosmotic hypothesis, it is responsible for the translocation of protons across the membrane in the direction opposite to that catalyzed by the respiratory chain (Mitchell, 1966a). The details of the proposed mechanism of the action of ATPase are complex and very speculative but the principal thought behind it is both simple and ingenious. The reversible reaction catalyzed by ATPase is

$$ATP + H_2O \rightleftharpoons ADP + P_i$$

The equilibrium of this reaction in a water medium is far toward the formation of ADP and P_i. If the water concentration could be lowered from 55 M to below 5 μM, reversal of the reaction would take place quite readily (Mitchell, 1966b). Since formally the translocation of protons across the membrane is equivalent to the expulsion of OH⁻ on one side and of H⁺ on the other side, we are dealing with an effective dehydration mechanism. The physiological utilization of ATP with continuous regeneration of ADP and P_i further aids the reaction in the direction of reversal. The detailed formulation of the translocation mechanism via the ATPase will not be discussed here, because it has been thoroughly reviewed by Greville (1969). Moreover, in our opinion it is the least attractive part of the Mitchell hypothesis. The formulation includes the participation of two unknown ions YO^- and X^- as well as a concerted reaction involving an unknown high-energy intermediate ($X \sim Y$), ADP, and P_i. As pointed out in detail elsewhere (Racker, 1970), the various partial reactions of oxidative phosphorylation, particularly the [^{14}C]ADP–ATP exchange and the exchange of [^{18}O]water with oxygen in ATP, are more easily explained by assuming the formation of a phosphorylated high-energy intermediate.

The key feature common to the chemical and chemiosmotic hypotheses of oxidative phosphorylation is the formation of a high-energy nonphosphorylated intermediate $X \sim Y$. The experimental evidence that has led to the inclusion of this intermediate in both hypotheses has come from studies of energy-dependent reactions that can be performed without the aid of ATP. These include ion translocations, the reversal of electron transport (e.g., the reduction of DPNH by succinate), and the energy-linked transhydrogenation reaction between DPNH and TPN. These processes take place in the presence of energy-transfer inhibitors such as oligomycin which preclude the generation of ATP. Submitochondrial particles that are completely stripped of coupling factors can perform some of these energy-dependent functions without addition of adenine nucleotides or inorganic phosphate. Indeed, addition of oligomycin to such submitochondrial particles greatly facilitates these energy-dependent reactions because the energy-transfer inhibitor prevents the dissipation of $X \sim Y$ by either H_2O or P_i. Yet these inhibitors do not interfere with the utilization of $X \sim Y$ in processes not requiring P_i. Submitochondrial particles which are loosely coupled (i.e., they oxidize substrates dissipating the generated energy as heat) cannot efficiently catalyze oxidative energy-linked reactions without addition of a compound such as oligomycin which inhibits the dissipation of $X \sim Y$ (Racker and Monroy, 1964, Lee et al., 1964). Under controlled conditions (low salt), oligomycin inhibits respiration which is released on addition of an uncoupler (Lee and Ernster, 1966). It has been shown that coupling factors can be substituted for oligomycin in facilitating energy-linked reactions, e.g., the energy-dependent transhydrogenase (Fessenden-Raden, 1969), and in the maintenance of respiratory control (Cockrell and Racker, 1969).

While the observations on the utilization of $X \sim Y$ have made the inclusion of a high-energy intermediate in the chemiosmotic hypothesis imperative and have thus narrowed the gap between the chemiosmotic and chemical hypotheses, the key question has remained unresolved. Are the translocation of protons across the membranes and the formation of a membrane potential the primary events leading to X $\sim$ Y formation? Or is the primary event a consequence of the oxidation process at the molecular level which leads to the formation of a high-energy intermediate of the oxidation chain? Such an intermediate may be formed by interaction of a respiratory catalyst with an unknown compound X, to form A $\sim$ X, or it may simply be an expression of a change in the conformation of the catalyst.

An important distinction between these two major hypotheses is that the chemiosmotic hypothesis requires a separation of charges across a membrane. Thus, oxidative phosphorylation could not take place in open-ended flat pieces of membranes without a delineation of compartments. The chemical hypothesis, on the other hand, does not have such restrictions and could permit oxidative phosphorylation even in a completely soluble system analogous to substrate-level phosphorylation. Even if, for the sake of efficiency, a membrane should be required, the organization of the respiratory carriers could be parallel to the membrane surface, whereas the formulation of Mitchell demands an organization across the membrane. Considerations of this type have led to systematic studies of the topography of the inner membrane discussed earlier which clearly established an organization transversing the membrane (Racker et al., 1970). Unfortunately, such findings do not determine whether one or the other hypothesis is correct.

Just as it is possible to include one or more high-energy intermediates in the chemiosmotic hypothesis without damaging its basic and ingenious concept, one can reconcile the experimental findings of the organization of the respiratory carriers across the membrane in terms of an integral proton pump in parallel with the phosphorylation system. Although it is not immediately apparent what the function of such a proton pump in the inner membrane would be, it should not be too difficult to imagine one. For example, one could postulate the secondary installment of a pump for regulatory purposes, e.g., respiratory control. Indeed, Mitchell has explained the phenomenon of respiratory control in terms of the build-up of a membrane potential (Mitchell, 1966a). Inclusion of this mechanism should not damage the chemical hypothesis any more than inclusion of X $\sim$ Y has detracted from the chemiosmotic hypothesis. As has been pointed out previously (Racker, 1967a) the presence of a parallel proton pump also embraces Mitchell's attractive explanation for the mode of action of uncoupling agents as modifiers of membrane permeability (Mitchell, 1966a).

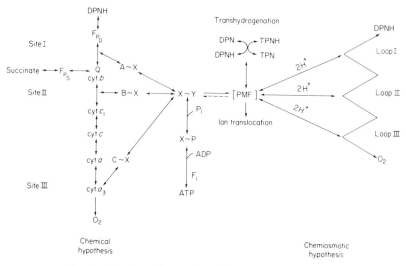

Fig. 3. Two hypotheses of oxidative phosphorylation.

A general scheme for the operation of oxidative phosphorylation according to a chemical and a chemiosmotic hypothesis is shown in Fig. 3. This formulation includes several compromises, such as a variable proton motive force [PMF]. This permits an explanation for the experimental observation that the reversal of the transhydrogenase reaction can drive ion transport (Mitchell and Moyle, 1965), while it cannot drive ATP generation from ADP and P_i. On the other hand, the formulation also includes an intermediate $X \sim P$, which appears more compatible with the data available from exchange reactions than the concerted reaction visualized by Mitchell (1966a).

In view of these expansions of both the chemical and chemiosmotic hypotheses, a decisive experimental approach to differentiate between the two has become increasingly difficult. The kinetic approach, measuring the sequences of electron and ion translocations, has led to controversies of interpretation (Chance et al., 1967; Mitchell, 1967). Applied to mitochondria, which catalyze numerous metabolic processes in addition to electron transport, kinetic evidence is at best mildly persuasive. In our opinion, decisive experiments will therefore be possible only after a thorough resolution and reconstitution of the system has been achieved.

2. Resolution and Possible Function of Coupling Factors

The most effective resolution and reconstitution of the system of oxidative phosphorylation has been achieved by a simple procedure (Racker et al., 1969). Submitochondrial particles exposed to silicotungstate (STA-particles) are depleted with respect to five coupling factors as well as succinate dehydrogenase. In view of the multiple systems of analysis and preparative procedures that have been used in the past, there have been some ambiguities regarding the distinctiveness of the various coupling factors. With the availability of STA-particles, these ambiguities have been eliminated since the requirement for at least five distinct protein factors (F_1, F_2, F_3, F_5, and F_6) can be established under identical experimental conditions. It has also become possible to make comparisons with other coupling factors reported in the literature. In recent collaborative experiments with Drs. Sanadi and Lam (Racker et al., 1971) we have established that F_2, which stimulates oxidative phosphorylation and the $^{32}P_i$–ATP exchange in STA particles, can be effectively replaced by factor B (Lam et al., 1967). Preparations of F_2 can also replace factor B in the rather specific assay for B activity (Lam et al., 1967). Moreover, an antibody against factor B was found to inhibit the stimulation by F_2. However, the presence of yet another factor has not been excluded by these experiments.

What is the function of these coupling factors? According to the chemical hypothesis, coupling factors catalyze the transformations of high-energy intermediates. According to the chemiosmotic hypothesis, the formation of $X \sim Y$ and its utilization for ATP generation are also enzyme-catalyzed reactions which justify the participation of multiple catalysts. Thus far, the only catalytic activity discovered in coupling factors is the ATPase of F_1.

In addition to the catalytic functions of coupling factors, structural functions have been demonstrated (cf. Racker, 1970). If respiratory control is dependent on the formation of a membrane potential (Mitchell, 1966a) the requirement for coupling factors (Cockrell and Racker, 1969) implies that these factors participate specifically in the translocation of protons and the formation of a membrane potential or that they have an indirect structural role by preventing proton leakage. The latter explanation is more likely in view of the effectiveness of oligomycin in inducing respiratory control (Lee and Ernster, 1966). This would not, however, eliminate additional roles of coupling factors as catalysts in the transformation of high-energy intermediates, particularly since it has been established that at least one factor, namely F_1, has a catalytic as well as a structural function.

Several new properties are conferred on F_1 after attachment to the membrane. The most useful of these "allotopic" properties is the sensitivity of the ATPase activity to energy-transfer inhibitors such as oligomycin or dicyclohexylcarbodiimide. Oligomycin sensitivity of F_1 has not as yet been demonstrated in a truly soluble system, but at least two soluble proteins have been shown to be required in addition to an insoluble membrane containing hydrophobic proteins. One of these soluble proteins, described first by Kagawa and Racker (1966a), has been referred to as F_c (Bulos and Racker, 1968a) or OSCP (MacLennan and Tzagoloff, 1968). A second protein, extracted by treatment of particles with thiocyanate, was shown to be required for oligomycin sensitivity and was replaced by highly purified preparations of F_6 (Knowles et al., 1971). In view of the complexity of the oligomycin-sensitive ATPase and the participation of multiple coupling factors, it seems likely that it represents an integral part of the system of oxidative phosphorylation. According to the chemiosmotic hypothesis the ATPase is responsible for proton translocation across the membrane by means of a hydrodehydration reaction, and oligomycin inhibits proton translocation (Mitchell, 1967). According to the chemical hypothesis the oligomycin-sensitive ATPase catalyzes the enzymic transformation of high-energy intermediates into ATP. Further exploration of the properties of the purified oligomycin-sensitive ATPase might therefore yield clues to the mechanism of action of the coupling factors.

Before closing the discussion of the phosphorylation process, some comments on the natural inhibitor of mitochondrial ATPase (Pullman and Monroy, 1963) seem appropriate. It is obvious that the process of harvesting the energy of oxidation would be wasted if the product, ATP, were hydrolyzed by ATPase. Moreover, since mitochondria always contain ATP, respiratory control could not be achieved unless the ATPase activity were kept in check. Such a function in respiratory control and energy conservation has been assigned to a small molecular weight protein which has the remarkable property of inhibiting ATPase activity without inhibiting oxidative phosphorylation. This protein has been obtained in homogeneous form yielding a single band in acrylamide gel electrophoresis (Horstman and Racker, 1970). Both ATP and Mg^{++} were shown to be required for this protein to be an effective inhibitor of ATPase activity. Other triphosphonucleosides which are hydrolyzed by F_1 are also effective in this system. Membrane-bound F_1 is much more susceptible to inhibition than soluble F_1. Since the inhibitor also depresses the $^{32}P_i$–ATP exchange reaction while oxidative phosphorylation remains unimpaired (Pullman and Monroy, 1963), it is likely that it changes the conformation of F_1 on the membrane and lowers its reactivity with H_2O or ATP or both. All attempts to implicate this inhibitor in a direct function in oxidative phosphorylation other than by preventing ATP utilization have thus far failed (Horstman and Racker, 1970).

C. Ion Translocation

It is apparent from the organization of the components catalyzing the Krebs cycle and the oxidation of DPNH that the matrix space enveloped by the inner mitochondrial membrane must be accessible to a variety of ions and substrates. On the other hand, the operation of a proton pump linked either in series or in parallel with $X \sim Y$ implies a certain degree of ion impermeability which precludes the free diffusion of ions. It is therefore not surprising that a multitude of specific ion transport systems are being described and chemicals (e.g., valinomycin) are being discovered which influence the movement of certain ions (e.g., potassium) into the matrix.

The use of cation-sensitive electrodes for cation transport in mitochondria (Moore and Pressman, 1964) and the design of simple methods for the study of anion transport (Chappell, 1968) have opened the door to a number of important investigations of these systems. These studies, which have been extensively reviewed (Pressman, 1970; Chappell, 1968), clearly point to the presence of specific channels for the translocation of cations such as K^+, Mg^{++}, and Ca^{++} and of anions such as adenine nucleotides, phosphate, and intermediates of the Krebs cycle. For a discussion of the properties of these ion transport systems the reader is referred to the above-mentioned reviews.

VII. THE RELATIONSHIP BETWEEN STRUCTURE AND FUNCTION OF THE INNER MITOCHONDRIAL MEMBRANE

According to Mitchell (1966a) the members of the respiratory chain are organized in such a manner that during the oxidation of DPNH six protons are conducted across the membrane. Mitchell therefore suggested that the respiratory chain is folded into three loops. Since the proton translocation in mitochondria takes place from the inside to the outside, the substrates must interact with the respective dehydrogenases on the matrix side of the membrane. The observed pH changes during cytochrome oxidase activity (Mitchell, 1966a) suggests that this enzyme is also located at the matrix side of the inner membrane. Thus the beginning and the end of the oxidative events take place in the matrix. Cytochrome c is located on the side of the membrane which faces the outer membrane. This was deduced from the fact that cytochrome c is readily removed by salt extraction from mitochondria but not from submitochondrial particles. More direct evidence was obtained from studies with an antibody against cytochrome c (Racker et al., 1970). Accordingly, in the third loop of the respiratory chain (Mitchell, 1966a), electrons are conducted across the membrane from cytochrome c to cytochrome a to copper to cytochrome a_3, which reacts with oxygen at the matrix side of the membrane.

In view of the notable lack of success of attempts to induce the reconstituted succinoxidase (Yamashita and Racker, 1969) to catalyze coupled phosphorylation, a systematic analysis of the topography of the respiratory carriers in the inner mitochondrial membrane has been initiated. Macromolecular probes such as antibodies which were shown to react specifically with surface components were used to establish the localization of various membrane components. Although this work is still in progress, it is already apparent both from inhibition studies with antibodies (Racker *et al.*, 1970) and from reconstitution experiments with highly resolved particles (Racker *et al.*, 1969) that the coupling factors as well as succinate dehydrogenase are located at the matrix side of the membrane, whereas cytochrome *c* is located at the other side. Of particular interest is the finding that addition of deoxycholate or cholate under suitable conditions makes cytochrome *c* accessible to the antibody. Since the inhibitory effect of deoxycholate and particularly of cholate on the P:O ratio of submitochondrial particles is fully reversible (Arion and Racker, 1970), it is apparent that in the absence of these detergents even small protein components such as cytochrome *c* must be in a fixed position with very restricted mobility.

In the light of these findings the reactivity of mitochondria as well as of submitochondrial particles with reduced cytochrome *c* points to the presence of cytochrome oxidase on both sides of the membrane. This conclusion is borne out by the inhibition of ferrocytochrome *c* oxidation in both mitochondria and submitochondrial particles by antibody against cytochrome oxidase or by very low amounts of high molecular weight

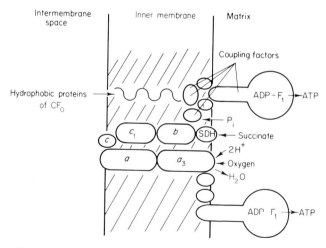

Fig. 4. Topography of the inner mitochondrial membrane.

polylysine, which is a rather specific inhibitor of cytochrome oxidase under these conditions (Minnaert and Smith, 1962).

A scheme of our present view of the topography of the inner mitochondrial membrane is shown in Fig. 4. It should be pointed out that the position of cytochromes a, a_3, b, and c_1 is still ambiguous. All experimental findings would also be consistent with a formulation that permits each of these constituents to reach from one side of the membrane to the other. These questions will be resolved only when the exact dimensions of the native heme proteins and their relationship to the two surfaces is established.

Acknowledgments

We are grateful to Dr. G. Schatz for his stimulating discussions and valuable criticism of the manuscript during its preparation. We also wish to thank Dr. S. Bullivant and Dr. C. R. Hackenbrock for providing prints of their electron micrographs.

This chapter was written during tenure of a Career Development Award (AM12614) from the National Institute of Arthritis and Metabolic Diseases to one of us (J. M. F.-R.). Research quoted from the authors' laboratory was supported by United States Public Health Service Research Grants AM-11715 from the National Institute of Arthritis and Metabolic Diseases and CA-08964 from the National Cancer Institute. Support from the Albert Einstein Award to one of us (E. R.) is also gratefully acknowledged.

References

Albracht, S. P. J., and Slater, E. C. (1969). Biochim. Biophys. Acta 189, 308.
Allmann, D. W., and Bachmann, E. (1967). Methods Enzymol. 10, 438.
Allmann, D. W., Bachmann, E., Orme-Johnson, N., Tan, W. C., and Green, D. E. (1968). Arch. Biochem. Biophys. 125, 981.
Arion, W. J., and Racker, E. (1970). J. Biol. Chem., 245, 5186.
Bachmann, E., Lenaz, G., Perdue, J. F., Orme-Johnson, N. and Green, D. E. (1967). Arch. Biochem. Biophys. 121, 73.
Bernath, P., and Singer, T. P. (1962). Methods Enzymol. 5, 597.
Blondin, G. A., and Green, D. E. (1969). Arch. Biochem. Biophys. 132, 509.
Borst, P. (1971). In "Autonomy and Biogenesis of Mitochondria and Chloroplasts, Canberra, December 1969." (N. K. Boardman, A. W. Linnane, and R. M. Smillie, eds.), p. 260. North-Holland Publ., Amsterdam.
Boyer, P. D. (1965). In "Oxidases and Related Redox Systems" (T. E. King, H. S. Mason, and M. Morrison, eds.), Vol. II, p. 994. Wiley, New York.
Boyer, P. D. (1968). In "Biological Oxidation" (T. P. Singer, ed.), p. 193. Wiley (Interscience), New York.
Brdiczka, D., Pette, D., Brunner, G., and Miller, F. (1968). Eur. J. Biochem. 5, 294.
Bruni, A., and Racker, E. (1968). J. Biol. Chem. 243, 962.
Bücher, T., and Klingenberg, M. (1958). Angew. Chem. 70, 552.
Bulos, B., and Racker, E. (1968a). J. Biol. Chem. 243, 3891.

Bulos, B., and Racker, E. (1968b). *J. Biol. Chem.* **243**, 3901.

Burstein, C., Morris, R. G., and Racker, E. (1970). *Fed. Proc. Fed. Amer. Soc. Exp. Biol.* **29**, 605.

Burstein, C., Kandrach, M. A., and Racker, E. (1971). *J. Biol. Chem.*, in press.

Butow, R. A., and Racker, E. (1965). *J. Gen. Physiol. Part 2* **49**, 149.

Byington, K. H., Smoly, J. M., Morey, A. V., and Green, D. E. (1968). *Arch. Biochem. Biophys.* **128**, 762.

Chan, S., Love, B., and Stotz, E. (1970). *J. Biol. Chem.* **245**, 6669.

Chance, B., and Pring, M. (1968). In "Biochemie des Sauerstoffs" (B. Hess and Hj. Staudinger, eds.), p. 102. Springer, Berlin.

Chance, B., and Williams, G. R. (1956). *Advan. Enzymol.* **17**, 65.

Chance, B., Lee, C-p., and Mela, L. (1967). *Fed. Proc. Fed. Amer. Soc. Exp. Biol.* **26**, 1341.

Chapman, D., and Leslie, R. B. (1970). In "Membranes of Mitochondria and Chloroplasts" (E. Racker, ed.), p. 91. Van Nostrand-Reinhold, New York.

Chappell, J. B. (1968). *Brit. Med. Bull.* **24**, 150.

Chappell, J. B., Henderson, P. J. F., McGivan, J. D., and Robinson, B. H. (1968). In "The Interaction of Drugs and Subcellular Components" (P. N. Campbell, ed.) p. 71. Churchill, London.

Christiansen, R. O., and Racker. (1968). Unpublished observations.

Christiansen, R. O., Loyter, A., Steensland, H., Saltzgaber, J., and Racker, E. (1969). *J. Biol. Chem.* **244**, 4428.

Clark-Walker, C. D., and Linnane, A. W. (1967). *J. Cell Biol.* **34**, 1.

Cockrell, R. S., and Racker, E. (1969). *Biochem. Biophys. Res. Commun.* **35**, 414.

Cooper, C., and Lehninger, A. L. (1956). *J. Biol. Chem.* **219**, 489.

Criddle, R. S., and Schatz, G. (1969). *Biochemistry* **8**, 322.

Criddle, R. S., Bock, R. M., Green, D. E., and Tisdale, H. (1962). *Biochemistry* **1**, 827.

Cunningham, C., and Racker, E. (1969). Unpublished observations.

Danielli, J. F., and Davson, H. (1935). *J. Cell. Comp. Physiol.* **5**, 495.

Deamer, D. W., Utsumi, K., and Packer, L. (1967). *Arch. Biochem. Biophys.* **121**, 641.

Ernster, L. (1959). *Biochem. Soc. Symp.* **16**, 54.

Ernster, L., and Kuylenstierna, B. (1970). In "Membranes of Mitochondria and Chloroplasts" (E. Racker, ed.), p. 172. Van Nostrand-Reinhold, New York.

Fernández-Morán, H. (1962). *Circulation* **26**, 1039.

Fessenden-Raden, J. M. (1969). *J. Biol. Chem.* **244**, 6662.

Fleischer, S., Brierley, G., Klouwen, H., and Slautterback, D. B. (1962). *J. Biol. Chem.* **237**, 3264.

Fleischer, S., Fleischer, B., and Stoeckenius, W. (1967). *J. Cell Biol.* **32**, 193.

Frey-Wyssling, A. (1955). "Die submikroskopische Strukur des Cytoplasmas, Protoplasmatologia II A 2." Springer, Vienna.

Goldberger, R., Smith, A. L., Tisdale, H., and Bomstein, R. (1961). *J. Biol. Chem.* **236**, 2788.

Gorkin, V. Z. (1963). *Nature (London)* **200**, 77.

Green, D. E., and Perdue, J. F. (1966). *Ann. N. Y. Acad. Sci.* **137**, 667.

Green, D. E., Bachmann, E., Allmann, D. W., and Perdue, J. F. (1966). *Arch. Biochem. Biophys.* **115**, 172.

Green, D. E., Asai, J., Harris, R. A., and Penniston, J. T. (1968a). *Arch. Biochem. Biophys.* **125**, 684.

Green, D. E., Allmann, D. W., Harris, R. A., and Tan, W. C. (1968b). *Biochem. Biophys. Res. Commun.* **31**, 368.

Greville, G. D. (1969). *In* "Current Topics in Bioenergetics" (D. R. Sanadi, ed.), Vol. 3, p. 1. Academic Press, New York.

Hackenbrock, C. R. (1966). *J. Cell Biol.* **30**, 269.

Hackenbrock, C. R. (1968a). *J. Cell Biol.* **37**, 345.

Hackenbrock, C. R. (1968b). *Proc. Nat. Acad. Sci. U. S.* **61**, 598.

Hackenbrock, C. R., and Caplan, A. I. (1969). *J. Cell Biol.* **42**, 221.

Haldar, D., Freeman, K., and Work, T. S. (1966). *Nature (London)* **211**, 9.

Hatefi, Y., and Stempel, K. E. (1967). *Biochem. Biophys. Res. Commun.* **26**, 301.

Hatefi, Y., and Stempel, K. E. (1969). *J. Biol. Chem.* **244**, 2350.

Hatefi, Y., Haavik, A. G., Fowler, L. R., and Griffiths, D. E. (1962). *J. Biol. Chem.* **237**, 2661.

Heidrich, H.-G., Stahn, R., and Hanning, K. (1970). *J. Cell Biol.* **46**, 137.

Horstman, L. L., and Racker, E. (1968). Unpublished observations.

Horstman, L. L., and Racker, E. (1970). *J. Biol. Chem.* **245**, 1336.

Jacobs, E. E., and Sanadi, D. R. (1960). *J. Biol. Chem.* **235**, 531.

Kagawa, Y., and Racker, E. (1966a). *J. Biol. Chem.* **241**, 2461.

Kagawa, Y., and Racker, E. (1966b). *J. Biol. Chem.* **241**, 2467.

Kagawa, Y., and Racker, E. (1966c). *J. Biol. Chem.* **241**, 2475.

Keilin, D. (1925). *Proc. Roy. Soc. London* **B98**, 312.

King, T. E. (1963). *J. Biol. Chem.* **238**, 4037.

Klingenberg, M. (1968). *In* "Biological Oxidations" (T. P. Singer, ed.), p. 3. Wiley (Interscience), New York.

Klingenberg, M., and Pfaff, E. (1966). *In* "Regulation of Metabolic Processes in Mitochondria" (J. M. Tager, S. Papa, E. Quagliariello, and E. C. Slater, eds.), *Biochim. Biophys. Acta Library* **7**, 180. Elsevier, Amsterdam.

Knowles, A. F., Guillory, R. J., and Racker, E. (1971). *J. Biol. Chem.* **246**, 2672.

Krimsky, I., and Racker, E. (1963). *Biochemistry* **2**, 512.

Kropp, E. S., and Wilson, J. E. (1970). *Biochem. Biophys. Res. Commun.* **38**, 74.

Lam, K. W., Warshaw, J. B., and Sanadi, D. R. (1967). *Arch. Biochem. Biophys.* **119**, 477.

Lee, C-p., and Ernster, L. (1966). *In* "Regulation of Metabolic Processes in Mitochondria" (J. M. Tager, S. Papa, E. Quagliariello, and E. C. Slater, eds.), *Biochim. Biophys. Acta Library*, **7**, 218. Elsevier, Amsterdam.

Lee, C-p., Azzone, G. F., and Ernster, L. (1964). *Nature (London)* **201**, 152.

Lee, C-p., Ernster, L., and Chance, B. (1969). *Eur. J. Biochem.* **8**, 153.

Lehninger, A. L. (1962). *Physiol. Rev.* **42**, 467.

Lehninger, A. L. (1965). "The Mitochondrion." Benjamin, New York.

Lehninger, A. L. (1967). *In* "Round Table Discussion on Mitochondrial Structure and Compartmentation" (E. Quagliariello, S. Papa, E. C. Slater, and J. M. Tager, eds.), p. 198. Adriatica Editrice, Bari.

Lenaz, G., Haard, N. F., Lauwers, A., Allmann, D. W., and Green, D. E. (1968a). *Arch. Biochem. Biophys.* **126**, 746.

Lenaz, G., Haard, N. F., Silman, H. I., and Green, D. E. (1968b). *Arch. Biochem. Biophys.* **128**, 293.

Lévy, M., Toury, R., and André, J. (1967). *Biochim. Biophys. Acta* **135**, 599.

Löw, H., and Vallin, I. (1963). *Biochim. Biophys. Acta* **69**, 361.

Loyter, A., Christiansen, R. O., Steensland, H., Saltzgaber, J., and Racker, E. (1969). J. Biol. Chem. 244, 4422.

Luck, D. J. L. (1965). J. Cell Biol. 24, 461.

Lucy, J. A. (1964). J. Theor. Biol. 7, 360.

MacLennan, D. H., and Tzagoloff, A. (1968). Biochemistry 7, 1603.

MacMunn, C. A. (1887). J. Physiol. 8, 51.

McMurray, W. C., and Dawson, R. M. C. (1969). Biochem. J. 112, 91.

Malviya, A. N., Parsa, B., Yodaiken, R. E., and Elliott, W. B. (1968). Biochim. Biophys. Acta 162, 195.

Minnaert, K., and Smith, L. (1962). In "Abstracts of the 5th International Congress of Biochemistry, Moscow, 1961" (N. M. Sisakian, ed.), p. 467. Macmillan (Pergamon), London.

Mintz, H. A., Yawn, D. H., Safer, B., Bresnick, E., Liebelt, A. G., Blailock, Z. R., Rabin, E. R., and Schwartz, A. (1967). J. Cell Biol. 34, 513.

Mitchell, P. (1966a). Biol. Rev. Cambridge Phil. Soc. 41, 445.

Mitchell, P. (1966b). In "Regulation of Metabolic Processes in Mitochondria" (J. M. Tager, S. Papa, E. Quagliariello, and E. C. Slater, eds.), Biochim. Biophys. Acta Library 7, 65. Elsevier, Amsterdam.

Mitchell, P. (1967). Fed. Proc. Fed. Amer. Soc. Exp. Biol. 26, 1370.

Mitchell, P., and Moyle, J. (1965). Nature (London) 208, 1205.

Moore, C., and Pressman, B. C. (1964). Biochem. Biophys. Res. Commun. 15, 562.

Mueller, P., and Rudin, D. O. (1969). In "Current Topics in Biochemistry" (D. R. Sanadi, ed.), Vol. 3, p. 157. Academic Press, New York.

Napolitano, L., Lebaron, F., and Scaletti, J. (1967). J. Cell Biol. 34, 817.

Nass, S. (1969). Int. Rev. Cytol. 24, 55.

Ohnishi, K. (1966). J. Biochem. Tokyo 59, 1.

Orii, Y., and Okunuki, K. (1962). J. Jap. Biochem. Soc. 35, 481.

Packer, L., and Utsumi, K. (1969). Arch. Biochem. Biophys. 131, 386.

Packer, L., Wrigglesworth, J. M., Fortes, P. A. G., and Pressman, B. C. (1968). J. Cell Biol. 39, 382.

Palade, G. E. (1952). Anat. Rec. 114, 427.

Palade, G. E. (1953). J. Histochem. Cytochem. 1, 188.

Palade, G. E. (1956). In "Enzymes: Units of Biological Structure and Function" (O. H. Gaebler, ed.), p. 185. Academic Press, New York.

Parsons, D. F. (1965). Int. Rev. Exp. Pathol. 4, 1.

Parsons, D. F., and Williams, G. R. (1967). Methods Enzymol. 10, 443.

Parsons, D. F., Williams, G. R., and Chance, B. (1966). Ann. N. Y. Acad. Sci. 137, 643.

Parsons, D. F., Williams, G. R., Thompson, W., Wilson, D., and Chance, B. (1967). In "Round Table Discussion on Mitochondrial Structure and Compartmentation" (E. Quagliariello, S. Papa, E. C. Slater, and J. M. Tager, eds.), p. 29, Adriatica Editrice. Bari.

Penefsky, H. S. (1967). J. Biol. Chem. 242, 5789.

Penefsky, H. S., Pullman, M. E., Datta, A., and Racker, E. (1960). J. Biol. Chem. 235, 3330.

Penniston, J. T., Harris, R. A., Asai, J., and Green, D. E. (1968). Proc. Nat. Acad. Sci. U. S. 59, 624.

Pfaff, E. (1965). Doctoral Thesis. Universität Marburg, Germany.

Pfaff, E., Klingenberg, M., Ritt, E., and Vogell, W. (1968). Eur. J. Biochem. 5, 222.

Pressman, B. C. (1970). In "Membranes of Mitochondria and Chloroplasts" (E. Racker, ed.), p. 213. Van Nostrand-Reinhold, New York.

Pullman, M. E., and Monroy, G. C. (1963). J. Biol. Chem. 238, 3762.

Racker, E. (1967a). Fed. Proc. Fed. Amer. Soc. Exp. Biol. 26, 1335.

Racker, E. (1967b). In "Organizational Biochemistry" (H. J. Vogel, J. O. Lampen, and V. Bryson, eds.), p. 487. Academic Press, New York.

Racker, E. (1970). In "Membranes of Mitochondria and Chloroplasts" (E. Racker, ed.), p. 127. Van Nostrand-Reinhold, New York.

Racker, E., and Monroy, G. (1964). Abstr. Int. Congr. Biochem., 6th, 1963 32, 760.

Racker, E., and Proctor, H. (1970). Biochem. Biophys. Res. Commun. 39, 1120.

Racker, E., Tyler, D. D., Estabrook, R. W., Conover, T. E., Parsons, D. F., and Chance, B. (1965). In "Oxidases and Related Redox Systems" (T. S. King, H. S. Mason, and M. Morrison, eds.), Vol. II, p. 1077. Wiley, New York.

Racker, E., Horstman, L. L., Kling, D., and Fessenden-Raden, J. M. (1969). J. Biol. Chem. 244, 6668.

Racker, E., Burstein, C., Loyter, A., and Christiansen, R. O. (1970). In "Electron Transport and Energy Conservation" (J. M. Tager, S. Papa, E. Quagliariello, and E. C. Slater, eds.), p. 235. Adriatica Editrice, Bari.

Racker, E., Fessenden-Raden, J. M., Kandrach, M. A., Lann, K. W., and Sanadi, D. B. (1971). Biochem. Biophys. Res. Commun. 41, 1474.

Raw, I. (1955). J. Amer. Chem. Soc. 77, 503.

Richardson, S. H., Hultin, H. O., and Green, D. E. (1963). Proc. Nat. Acad. Sci. U. S. 50, 821.

Rieske, J. S., Hansen, R. E., and Zaugg, W. S. (1964). J. Biol. Chem. 239, 3017.

Robertson, J. D. (1960). J. Physiol. London 153, 58P.

Rose, I. A., and Warms, J. V. B. (1967). J. Biol. Chem. 242, 1635.

Schatz, G. (1968). J. Biol. Chem. 243, 2192.

Schatz, G. (1970). In "Membranes of Mitochondria and Chloroplasts" (E. Racker, ed.), p. 251. Van Nostrand-Reinhold, New York.

Schatz, G., and Racker, E. (1966). J. Biol. Chem. 241, 1429.

Schatz, G., and Saltzgaber, J. (1969). Biochim. Biophys. Acta 180, 186.

Schnaitman, C., Erwin, V. G., and Greenawalt, J. W. (1967). J. Cell Biol. 32, 719.

Sherman, F., Stewart, J. W., Margoliash, E., Parker, J., and Campbell, W. (1966). Proc. Nat. Acad. Sci. U. S. 55, 1498.

Singer, T. P. (1968). In "Biological Oxidations" (T. P. Singer, ed.), p. 339. Wiley (Interscience), New York.

Sjöstrand, F. S. (1953). Nature (London) 171, 30.

Slater, E. C. (1953). Nature (London) 172, 975.

Sottocasa, G. L., Kuylenstierna, B., Ernster, L., and Bergstrand, A. (1967a). J. Cell Biol. 32, 415.

Sottocasa, G. L., Kuylenstierna, B., Ernster, L., and Bergstrand, A. (1967b). Methods Enzymol. 10, 448.

Stoeckenius, W. (1970). In "Membranes of Mitochondria and Chloroplasts" (E. Racker, ed.), p. 53. Van Nostrand-Reinhold, New York.

Stoffel, W., and Schiefer, H.-G. (1968). Hoppe-Seyler's Z. Physiol. Chem. 349, 1017.

Stoner, C. D., and Sirak, H. D. (1969). Biochem. Biophys. Res. Commun. 35, 59.

Szarkowska, L. (1966). Arch. Biochem. Biophys. 113, 519.

Thompson, T. E., and Henn, F. A. (1970). In "Membranes of Mitochondria and Chloroplasts" (E. Racker, ed.), p. 1. Van Nostrand-Reinhold, New York.

Tyler, D. D., Gonze, J., Estabrook, R. W., and Butow, R. A. (1965). *In* "A Symposium on Non-heme Iron Proteins" (A. San Pietro, ed.), p. 447. Antioch Press, Yellow Springs, Ohio.

Weber, N. E., and Blair, P. V. (1969). *Biochem. Biophys. Res. Commun.* **36**, 987.

Weinbach, E. C., Garbus, J., and Sheffield, H. G. (1967). *Exp. Cell Res.* **46**, 129.

Werkheiser, W. C., and Bartley, W. (1957). *Biochem. J.* **66**, 79.

Wilkie, D. (1964). "The Cytoplasm in Heredity." Wiley, New York.

Wirtz, K. W. A., and Zilversmit, D. B. (1968). *J. Biol. Chem.* **243**, 3596.

Wrigglesworth, J. M., and Packer, L. (1969). *Arch. Biochem. Biophys.* **133**, 194.

Wojtczak, L., and Zaluska, H. (1969). *Biochim. Biophys. Acta* **193**, 64.

Yamashita, S., and Racker, E. (1968). *J. Biol. Chem.* **243**, 2446.

Yamashita, S., and Racker, E. (1969). *J. Biol. Chem.* **244**, 1220.

10

STRUCTURE AND FUNCTION OF SURFACE COMPONENTS OF MAMMALIAN CELLS

V. GINSBURG and A. KOBATA

ABBREVIATIONS

Fuc	L-Fucose*
Gal	Galactose
GalNAc	N-Acetylgalactosamine
Glc	Glucose
GlcNAc	N-Acetylglucosamine
GlcUA	Glucuronic acid
Gm	Glucosamine
Man	Mannose
NANA	N-Acetylneuraminic acid
NGNA	N-Glycolylneuraminic acid
Xyl	Xylose

* Sugars are in the D-configuration unless otherwise specified.

I. INTRODUCTION

The surfaces of mammalian cells are in contact with other cells or with their environment, and in addition to carrying on metabolic processes necessary for life they must also exhibit the specificity that distinguishes cells from different tissues as well as the same cell types from different individuals. The "cell surface," as used in this chapter, includes those structures that are not shielded from the environment by a permeability barrier—that is to say, that are external to the lipid membrane common to all cells. Structures are assumed to be in this location if they can be detected by reactions with specific reagents such as antibodies or lectins under conditions in which the permeability of the cell is unchanged or if they can be liberated from intact cells by treatment with various enzymes. These criteria for surface location are somewhat arbitrary because it is not possible to draw a sharp line between surface and environment. For example, one might expect that the surface contains complex carbohydrates destined for excretion by the cell but temporarily associated with it or, conversely, material of extracellular origin that adheres loosely to the surface. Both materials may be structurally or functionally important, and there is no reason to exclude either as normal components of that region of the cell which electron microscopists have referred to as a "fuzzy coat" (Revel and Ito, 1967).

The fuzzy coat is found on all mammalian cells (Rambourg et al., 1966; Pease, 1966) and presumably contains the glycoproteins and glycolipids that carry the carbohydrate structures which are the combining sites (determinants) for many antibodies and lectins. In view of the fact that certain sequences of sugars are common to glycoproteins and glycolipids, it is not surprising that the same determinants sometimes occur in both types of molecule. While acting as specific determinants for various reagents is not a natural function of these structures, it is an indication of their complex nature and suggests their possible participation in biologic phenomena that require cellular specificity.

Glycoproteins and glycolipids found on cell surfaces are rather fancifully depicted in Fig. 1. In general, glycoproteins consist of sugars (represented by the small circles) covalently linked to a polypeptide chain, while glycolipids consist of sugars covalently linked to lipids. The glycoprotein is shown in Fig. 1 as being anchored in the lipid membrane of the cell by a hydrophobic region at one end of the polypeptide chain while the other end, which contains the hydrophilic carbohydrate chains, extends outward into the environment. This type of picture arises mainly from studies on the major glycoproteins of red cell membrane (Mora-

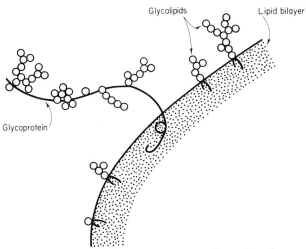

Fig. 1. Diagrammatic representation of a cell surface.

wiecki, 1964; Winzler, 1969). Treatment of these macromolecules with proteases results in two types of fragment: (a) carbohydrate-rich soluble peptides and (b) insoluble peptides, low in carbohydrate and with a high content of hydrophobic amino acids. Glycolipids are depicted as integral parts of the lipid membrane with their carbohydrate moieties extending outward. As far as specificity is concerned, the most important feature of the carbohydrate chain is the nature of the sugars at the distal (or "non-reducing") ends of the chain. The sugars farther in have less effect on specificity, and serologic identity seems independent of whether the chain is linked to a polypeptide or to a lipid (Watkins *et al.*, 1962).

II. GLYCOPROTEINS AND GLYCOLIPIDS

A. Glycoproteins

The structure of glycoproteins has been the subject of several reviews (e.g., Gottschalk, 1966; Spiro, 1969). The isolated cell membrane, at least from red cells, is about 50% protein, of which about 10% is glycoprotein (Winzler, 1969). While some of the protein may be internal to the lipid barrier, as in the case of the presumed structural protein "spectrin" (Marchesi, 1969), it is clear from the amount of glycopeptides released from intact cells by proteinases that most of the glycoproteins are external (Kraemer, 1971).

Glycoproteins comprise a diverse class of compounds ranging in molecular weight from 15,000 to over a million and in carbohydrate content from 1 to 85%. Some glycoproteins contain a single oligosaccharide chain while others may have up to several hundred. Seven sugars account for nearly all of the carbohydrates found in mammalian glycoproteins: L-fucose, mannose, galactose, glucose, N-acetylglucosamine, N-acetylgalactosamine, and sialic acid. Other naturally occurring sugars in mammalian complex carbohydrates, chiefly mucopolysaccharides, include glucuronic acid, galacturonic acid, L-iduronic acid, xylose, and L-arabinose. The occurrence of glucose appears to be restricted to structural collagen-type proteins. In general, fucose and sialic acid, when they occur, are found only at the nonreducing ends of the chains. The oligosaccharides are glycosidically linked to the protein at their reducing end by one of four different linkages. Based on a limited number of known structures, certain generalizations can be made about the nature of the carbohydrate chain attached by each type of linkage.

1. N-Acetylglucosaminylasparagine

Most of the plasma glycoproteins and several hormones contain this linkage, which is a glycosylamine bond between N-acetylglucosamine and the β-amide group of asparagine.

N-Acetylglucosaminylasparagine

The oligosaccharide often has a core structure containing several mannosyl residues to which side chains are attached. Common side chains include NANA → Gal → GlcNAc and Fuc → Gal → GlcNAc. An example is the chain found in human chorionic gonadotropin (Bahl, 1969).

Human chorionic gonadotropin

2. N-Acetylgalactosaminylserine (or threonine)

This bond is a β-glycosidic linkage between N-acetylgalactosamine and the hydroxyl group of either serine or threonine. The major glycoprotein of erythrocyte membranes contains many oligosaccharides with the following structure (Winzler, 1969):

$$\text{NANA-}(\alpha2\to3)\text{-Gal}(\beta1\to3)\text{-GalNAc-}\beta1\to \text{Ser (or Thr)}$$
$$\text{NANA-}(\alpha2\to6)$$

A more complex chain is found in the glycoproteins of mucous secretions that exhibit ABH and Lewis specificities (the "soluble blood group substances"). Its basic structure is as follows (Lloyd and Kabat, 1968):

Gal-($\beta1\to$ 3, 4)-GlcNAc-($\beta1\to$ 6)

Gal-($\beta1\to$4)-GlcNAc-($\beta1\to$ 6)

Gal-($\beta1\to$3)-GlcNAc-($\beta1\to$4)-Gal-($\beta1\to$3)-GlcNAc-($\beta1\to$3)-Gal-($\beta1\to$3)   GalNAc$\to$Ser (or Thr)

Gal-($\beta1\to$3)-GlcNAc-($\beta1\to$3)

N-Acetylglactosaminylserine

Depending on the blood type of the donor, various sugar residues responsible for blood group activity are attached to the above structure (see Fig. 3).

3. Xylosylserine

Mucopolysaccharide–protein complexes ("protein–polysaccharides" or "proteoglycans") contain this linkage, which is a β-glycosidic bond between xylose and serine (Lindahl and Rodén, 1966). For example, the linkage between polysaccharide and protein in chondroitin sulfate A, B, and C, heparin, and heparitin sulfate appears to be as follows:

$$\cdots \text{GlcUA-}(\beta1 \to 3)\text{-Gal-}(\beta1 \to 3)\text{-Gal-}(\beta1 \to 4)\text{-Xyl-}(\beta1 \to)\text{Ser}$$

Attached to the glucuronic acid residues are long polymers, chiefly composed of repeating disaccharide units characteristic for each type of mucopolysaccharide (for the chemistry of these macromolecules see Quintarelli, 1968). L-Arabinose in place of xylose may occur in some mucopolysaccharides (Wardi et al., 1969).

4. Galactosylhydroxylysine (or hydroxyproline)

This linkage is found only in structural proteins such as collagen, basement membrane, and vitreous membrane and consists of a β-glycosidic bond between galactose and the hydroxyl group of the appropriate amino acid. The carbohydrate unit is either the monosaccharide galactose or the disaccharide, Glc-(α1 → 2)-Gal (Spiro, 1967).

Some glycoproteins contain oligosaccharides attached by only one type of linkage while others contain oligosaccharides attached by more than one. For example, immunoglobulin G contains two oligosaccharides, one linked by N-acetylglucosamine to asparagine and the other by N-acetyl-galactosamine to threonine (Smyth and Utsumi, 1967). Myeloma globulin A contains three oligosaccharides, two of the first type and one of the second (Dawson and Clamp, 1968). The oligosaccharides obtained from these proteins are therefore basically different. This type of heterogeneity has been called "central" (Clamp et al., 1968) to distinguish it from the "peripheral" heterogeneity (also called "microheterogeneity"; see Gottschalk, 1969) in which the oligosaccharides have the same basic structure but differ in their content of certain sugars at or near nonreducing ends. The latter situation is quite common and may in large part reflect the stepwise manner in which the heterosaccharide chains are synthesized (see Section II,C).

B. Glycolipids

Glycosphingolipids are responsible for most of the serological specificity exhibited by mammalian cells (cf. Rapport and Graf, 1969). These compounds consist of a carbohydrate composed of one or more monosaccharides glycosidically linked to the terminal hydroxyl of ceramide (N-acylsphingosine), which has the following structure:

$$\underset{\substack{| \\ \text{NH}-\text{C}-\text{R} \\ \underset{\text{O}}{\|}}}{\text{HO}-\text{CH}_2-\text{CH}-\overset{\overset{\text{OH}}{|}}{\text{CH}}-\text{CH}=\text{CH}-(\text{CH}_2)_{12}-\text{CH}_3}$$

N-Acylsphingosine

The N-acyl group can be one of several different fatty acids. Some simple glycosphingolipids are shown in Table I. Most of the more complex ones are derivatives of the last three tetraose ceramides listed. Glycosphingolipids contain all the sugars commonly found in glycoproteins with the exception of mannose.

The distribution of the various glycolipids differs markedly from tissue to tissue and in some cases the glycolipids are responsible for organ-specific antigenicity. For example, galactocerebroside is a major component of brain glycolipids and accounts for the organ specificity of

TABLE 1
SOME REPRESENTATIVE GLYCOSPHINGOLIPIDS[a]

Name	Structure
Glycocerebroside	Glc → ceramide
Galactocerebroside	Gal → ceramide
Cytolipin H (lactosylceramide)	Gal-(β1 → 4)-Glc → ceramide
Hematoside	NANA-(2 → 3)-Gal-(β1 → 4)-Glc → ceramide
Globoside 1	GalNAc-(β1 → 3)-Gal-(β1 → 4)-Gal-(β1 → 4)-Glc → ceramide
Ganglio-N-tetraose ceramide	Gal-(β1 → 3)-GalNAc-(β1 → 4)-Gal-(β1 → 4)-Glc → ceramide
Lacto-N-neotetraose ceramide	Gal-(β1 → 4)-GlcNAc-(β1 → 3)-Gal-(β1 → 4)-Glc → ceramide
Lacto-N-tetraose ceramide	Gal-(β1 → 3)-GlcNAc-(β1 → 3)-Gal-(β1 → 4)-Glc → ceramide

[a] From Sweeley and Dawson (1969).

antibrain sera (Joffe et al., 1963). The immunological activity of galacto-cerebroside, chemically one of the simplest glycosphingolipids, suggested that the other more complex members of the class would also be immunologically active. This is now well documented and antisera prepared against individual glycosphingolipids are being used to investigate the architecture of cell surfaces (Hakomori, 1969). Interestingly, the distribution of glycolipids isolated from kidneys of different strains of mice are also quite different (Adams and Gray, 1968).

C. Biosynthesis

The activated forms of monosaccharides are nucleotide-linked sugars and take part in the synthesis of glycoproteins and glycolipids by functioning as glycosyl donors. They are particularly well suited for this role because the free energy of hydrolysis of their glycosyl residues is quite high (about −8000 cal) and most of the synthetic reactions in which they participate are essentially irreversible. These compounds are also intermediates in the synthesis of most monosaccharides found in nature. Figure 2 summarizes the relevant pathways leading to sugars found in mammalian glycoproteins and glycolipids. Only glucose, mannose, and N-acetylglucosamine arise directly from fructose 6-phosphate; the remaining sugars are formed from these three after their incorporation into the appropriate nucleotide. It appears that three bases, GDP, UDP, and CMP, are involved and a given sugar is always associated with only one base (e.g., L-fucose is always carried by GDP, or sialic acid is always carried by CMP).

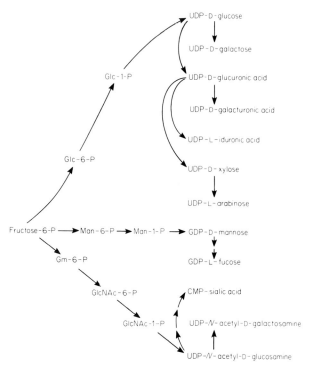

Fig. 2. Biosynthesis of monosaccharides found in mammalian glycoproteins and glycolipids (for references to these reactions see Ginsburg, 1964).

In contrast, plants and bacteria also utilize ADP, dTDP, and CDP to carry sugars (in addition to GDP, UDP, and CMP) and the same sugar is often carried by different bases. Glucose, for instance, is found to be linked to UDP, GDP, ADP, CDP, as well as dTDP, and enzymes specific for each nucleotide have been described. The factors that govern the choice of a particular base for a given reaction are not apparent, but it is clear that the use of different bases sharply separates pathways of biosynthesis and may be advantageous to the organism in that it offers a means for the independent control of the various pathways. For example, ADP-glucose gives rise to $\alpha 1 \rightarrow 4$ glucans in bacteria and plants (glycogen and starch, respectively) and it appears that the synthesis of these polymers is regulated by activation and inhibition of ADP-glucose pyrophosphorylase by various metabolites (Preiss, 1969). Regulation at this step in the biosynthetic sequence implies not only that the pyrophosphorylase reaction is rate limiting, but also that the ADP-glucose which is produced in the reaction serves only for the formation of these products and is neither an intermediate in the synthesis of other sugars or a glycosyl donor for the synthesis of other complex saccharides. Use of ADP as a carrier for the glucose allows the pyrophosphorylase step to be unique to the synthesis of $\alpha 1 \rightarrow 4$ glucans and eminently suited for a regulatory role. In animals, UDP-glucose gives rise to glycogen, and possibly because UDP-glucose has so many other functions, control of glycogen synthesis in animals appears to be at the level of the glycosyltransferase. Several examples in animals and bacteria of control by negative feedback have been described in which nucleotide-linked sugars regulate their own concentration by inhibition of an enzyme catalyzing the irreversible formation of a precursor (cf. Neufeld and Ginsburg, 1965). Again, separation of pathways by the use of different bases creates reactions unique to the synthesis of certain intermediates and therefore effective sites for control.

Another possible advantage of using different bases is suggested by the mechanism of synthesis of complex saccharides. These complicated structures arise by the ordered and stepwise addition of glycosyl residues from nucleotide-linked sugars catalyzed by glycosyltransferases. For example, the synthesis of ganglioside G_{M1}, an important glycolipid of brain cells, is formed by the stepwise addition of sugars to the terminal hydroxyl of ceramide as shown in Fig. 3. Clearly the inherent specificities of five glycosyltransferases determine the final structure that is formed. However, enzyme specificity is a relative concept and glycosyltransferases make mistakes. The glycogen of chickens fed galactose contains galactose

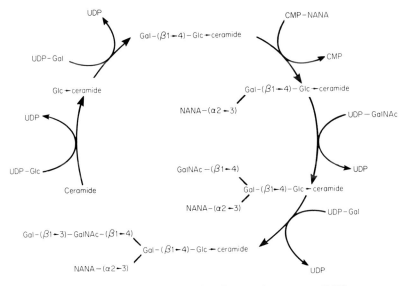

Fig. 3. Biosynthesis of ganglioside G_{M1} (Roseman, 1968).

(Nordin and Hansen, 1963), while the glycogen of rats fed glucosamine contains glucosamine (Maley *et al.*, 1966). Glycogen synthetase, whose normal substrate is UDP-glucose, will use the UDP-galactose and UDP-glucosamine that accumulates in the liver of these animals under the above conditions. Carrying monosaccharides on different bases may insure the accurate synthesis of complex saccharides by decreasing the probability of error; for instance, a transglycosylase responsible for the transfer of L-fucose from GDP-L-fucose is much less likely to make a mistake and transfer galactose from UDP-galactose than if galactose were also carried by GDP.

The question as to which base is involved in a given reaction has been complicated by the finding that polyanions can influence the base specificity of glycosyltransferases (Liu *et al.*, 1969). An enzyme preparation from *Mycobacterium smegmatis* catalyzes the synthesis of the disaccharide, trehalose phosphate, according to the following equation:

$$\text{GDP-glucose} + \text{glucose-6-P} \rightarrow \text{trehalose-P} + \text{GDP}$$

The enzyme is specific for GDP-glucose and will not utilize UDP-glucose. However, upon addition of polyanions to the enzyme preparation, UDP-glucose becomes a substrate.

Another type of change in specificity has been shown with the galactosyltransferase found in milk that is responsible for the synthesis of lactose (Brew *et al.*, 1968). This enzyme catalyzes the reaction UDP-galactose + glucose → lactose + UDP and can be resolved into two proteins which are inactive individually but which regain full activity when complexed. One of the proteins is α-lactalbumin and has no enzymatic activity. The other protein, while inactive in the above reaction, is a galactosyltransferase that uses N-acetylglucosamine as an acceptor in place of glucose and is probably involved in the synthesis of glycoproteins. Thus, α-lactalbumin, which has been termed a "specifier" protein, is able to change the specificity and function of a galactosyltransferase. This type of switch may occur with other glycosyltransferases (Kobata and Ginsburg, 1969).

As mentioned earlier, glycolipids and glycoproteins sometimes have common serologic properties because identical sugar sequences can occur in each. The common sequences in both groups appear to be synthesized by the same glycosyltransferases. For example, the structure GalNAc-(α1 → 3)-Gal ··· occurs in glycolipids of erythrocytes, in glycoproteins of mucous secretions, and in oligosaccharides of urine and milk from individuals with blood type A; it is not found in similar material from individuals with blood type B or O (cf. Kobata and Ginsburg, 1970). Family studies show that the inheritance of blood type A is controlled by one gene (Race and Sanger, 1958), and it follows that the N-acetylgalac-

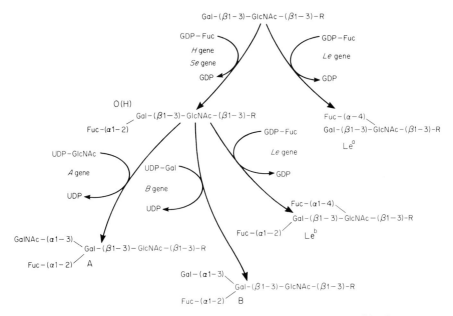

Fig. 4. Biosynthesis of structures responsible for ABO and Lewis blood types in man (after Kobata *et al.*, 1968).

tosaminyltransferase coded for by this gene as shown in Fig. 4 is involved in the synthesis of many different products.

The carbohydrate chains in glycoproteins are probably synthesized in the same way as those in glycolipids. The kinetics of incorporation of labeled sugars and amino acids into various subcellular fractions has led to the view that most of the sugar residues are added after the polypeptide chains are completed and released from ribosomes (Sarcione, 1964). There is still some uncertainty about the sugar residue linked directly to the polypeptide. Some experiments indicate that this sugar is added before the polypeptide is released (Sherr and Uhr, 1969). Also, it is possible that the synthesis of some glycoproteins may involve lipid intermediates (Caccam *et al.*, 1969) similar to those involved in the synthesis of certain complex carbohydrates in bacteria (Osborn, 1969; see Chapter 8).

D. Genetics

From a genetic viewpoint, the carbohydrate structures of cell surfaces can be considered to be secondary gene products in that the primary gene products are enzymes and these enzymes, working in concert, de-

termine which specific structures are formed. This situation is best illustrated by the synthesis of the determinants of ABO and Lewis blood
types in man (Watkins, 1966; Marcus, 1969), a simplified version of
which is shown in Fig. 4. The carbohydrate structures that are depicted
along with their associated serologic activity occur in both glycoproteins
and glycolipids. Most individuals have the enzymes necessary to synthesize the structure shown at the top of Fig. 4 but differ in their complement of the enzymes which are able to modify this structure.

Family studies indicate that inheritance of ABO and Lewis blood types
is controlled by the action of genes at four independent loci, the *Lele*,
the *Hh*, the *Sese*, and the *ABO* loci. As shown in Fig. 4, the *Le* gene
produces a fucosyltransferase that attaches L-fucose to N-acetylglucosamine in $\alpha 1 \rightarrow 4$ linkage. Individuals with the genotype *lele* do not
produce this enzyme and cannot synthesize either the Le^a or the Le^b
determinant, both of which control $\alpha 1 \rightarrow 4$ linked L-fucose. The *H* gene
produces another fucosyltransferase that adds fucose to galactose in
$\alpha 1 \rightarrow 2$ linkage to form the glycolipid of cell surfaces responsible for
blood type O. The genotype *hh*, in which this fucosyltransferase is missing, is very rare and results in the "Bombay" phenotype. For the same
enzyme to take part in the synthesis of glycoproteins with O(H) activity
an additional gene, the *Se* gene, is required. The nature or function of
the *Se* gene is not known but in some way it allows the expression of the
H gene (in this sense, the formation of an $\alpha,2$-fucosyltransferase) in
secretory organs. The *H* gene is not expressed in the secretory organs of
individuals with the genotype *sese*. Curiously, the *Se* gene is also required for the synthesis of the Le^b determinant even though it appears
to be a glycolipid like the ABH determinants (Marcus, 1969). However,
while the ABH glycolipids are integral parts of the cell membranes and
synthesized by the cells on which they occur, the Le^b (and Le^a) glycolipids appear to be loosely adsorbed onto the cells from the serum where
they are associated with lipoproteins. Their site of synthesis is not known.
Finally, two allelic genes, the *A* and *B* genes are responsible for the formation of an N-acetylgalactosaminyltransferase and a galactosyltransferase, respectively, that attach the appropriate hexose galactose in
$\alpha 1 \rightarrow 3$ linkages as shown. Allelic with *A* and *B* is the *O* gene and individuals with the genotype *OO* produce neither enzyme.

As can be seen from Fig. 4, the genetically determined presence or
absence of four glycosltransferases determine the ABO and Lewis blood
types. This mechanism of control provides a biochemical explanation for
antigens produced by "gene interaction"—that is to say, for the appearance of antigens in the hybrid which are absent in both homozygous

parents. The Le^b determinant is one of these. From family studies, Cep-pellini (1954) proposed that the Le(b+) character was an interaction product of two genes which in the scheme of Watkins and Morgan (Watkins, 1966) would be the H gene, whose expression for the synthesis of Le^b determinant requires the Se gene and the Le gene. As nearly everyone possesses an H gene, the Le^b determinant is, in effect, an inter-action product of the Se gene and the Le gene. These genes control the formation of the two specific fucosyltransferases that are necessary for the synthesis of the Le^b determinant (see Fig. 3). Both fucose residues are required for Le^b activity and so if each parent provided a gene for only one of the two enzymes, the determinant would clearly be an inter-action product. In addition to the Le(b+) character, there are several reports of interactions of other genes resulting in new cellular antigens. These include interactions within the Rh system and interactions of the I gene with the A, B, or H genes (Race and Sanger, 1958). The formation of unique carbohydrate structures by the concerted action of glycosyl-transferases provided by both parents may be a factor in the expression of individuality. It should be kept in mind that the glycolipids responsible for ABO and Lewis specificities are only a small part of the total glyco-lipid in the cell membrane (perhaps less than 1%) and variations, both qualitative, also can be expected to occur in the other types of glycolipid.

One reason for the "peripheral" heterogeneity that occurs in many glycoproteins as mentioned earlier (Section I,B) appears to be due to the incomplete synthesis of their carbohydrate chains. An extreme ex-ample of this situation is the glycoprotein from porcine submaxillary gland (Carlson, 1968) which contains different carbohydrate units rang-ing in complexity from one N-acetylgalactosamine residue to the pentasaccharide,

$$\text{GalNAc-}(\alpha1 \rightarrow 3)\text{-Gal-}(\beta1 \rightarrow 3)\text{-GalNAc} \cdots$$
$$\overset{}{\underset{\text{Fuc-}(\alpha1 \rightarrow 2)}{\diagup}} \quad \overset{}{\underset{\text{NGNA-}(\alpha2 \rightarrow 6)}{\diagup}}$$

The smaller units probably represent successive stages in the synthesis of the more complex structure.

Some of the multiple blood group specificities found in single glyco-proteins may also be explained by incompletely synthesized chains. How-ever, another cause of this phenomenon is probably the acceptor specificity of the transglycosylases. For example, the N-acetylgalacto-saminyltransferase that is responsible for the formation of the determi-nant of blood type A (Fig. 4) is not able to add N-acetylgalactosamine to the Le^b-active structures (Kobata and Ginsburg, 1970), presumably because of steric hindrance by the L-fucose residue on the sugar adjacent to the galactose. Because of this, once the Le^b-active structure is formed it will persist since the chain is no longer a substrate for further lengthening.

III. SURFACE STRUCTURES

Surface carbohydrates can be detected by the use of two general types of protein. The first type includes antibodies, either naturally occurring or those that result from deliberate immunization of experimental animals. Immunization can be with preparations of whole cells or cell surface components such as purified glycolipids. The second type of protein includes the "agglutinins" obtained from plants and lower animals. These mimic some antibodies in their ability to bind certain sugars and to agglutinate suspensions of cells that contain these sugars exposed on their surfaces. Some investigations involving both antibodies and agglutinins will be discussed in this section.

An agglutinin from the red kidney bean *Phaseolus vulgaris* agglutinates erythrocytes and leukocytes and in addition induces lymphocytes *in vitro* to undergo mitosis. The receptor site for this protein is liberated from erythrocytes by treatment with trypsin as a glycopeptide with the following probable structure (Kornfeld and Kornfeld, 1969):

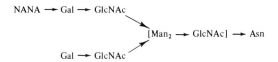

The glycopeptide binds to the agglutinin and inhibits both its agglutinating and its lymphocyte-stimulating properties. Studies with glycosidases indicate that the dominant sugar involved in the binding is the galactose that is covered by the sialic acid in the above structure. The other galactose or the sialic acid can be liberated without loss of inhibitory activity. The glycoprotein fetuin, which contains the sequence NANA → Gal → GlcNAc··· (Spiro, 1962), is also a potent inhibitor of hemagglutination.

Immunization of animals with preparations from human tumors often results in the formation of antibodies directed against the lactosyl moiety of cytolipin H (Rapport *et al.*, 1959). Although this glycolipid accounts for much of the serological difference between normal and tumor tissues, and was originally isolated from tumors, it also occurs in low concentrations in normal tissues as well (Rapport *et al.*, 1962). Interestingly, cytolipin H is also involved in another phenomenon. In the sera of cancer patients as well as pregnant women at terms, there occurs an "agglutination factor" which is a protein associated with the β-globulins. This protein agglutinates suspensions of several different tumor cells but does not agglutinate some suspensions of normal human liver or kidney cells (Tal, 1965). The receptors on the tumor cells appear to be cytolipin H since agglutination is specifically inhibited by lactose and since the factor can be absorbed from serum by cytolipin H but not by other closely related glycolipids. The origin of the serum factor is unknown but it may arise from self-immunization by the cancer or fetal cells.

A related phenomenon is the appearance of an antibody in a large proportion of patients with primary nonmetastatic cancers of the digestive system as well as in pregnant women. This antibody is directed against an antigen (the "carcinoembryonic antigen") found only in embryonic tissues in the first two trimesters of pregnancy and in adenocarcinomas arising from digestive system epithelium (Thomson et al., 1969). The antigen appears to be a glycoprotein but the chemical nature of the structure responsible for antigenicity is not known.

As discussed in Section II,D, differences among individuals in some cell surface determinants such as those responsible for ABO and Lewis blood types reflect genetic differences that can be translated into the presence or absence of specific glycosyltransferases. However, variations in surface carbohydrates also occur among cells with presumably the same genome. For example, variations in surface carbohydrates occur during embryonic development as evidenced by gains or losses of certain blood group antigens (Szulman, 1965). Cultured cells also show a great variability; some carbohydrate antigens are lost while others are maintained (Franks, 1968). There is also variability of a given antigen within a culture. The A-antigen, GalNAc-($\alpha1 \rightarrow 3$)-Gal $\cdots$, in established rabbit cell lines is not expressed in all cells; some cells are A-positive while others are A-negative. Cloning of both positive and negative cells again gives rise to heterogeneous populations (Franks and Dawson, 1966).

An antigen structurally related to the A-antigen that also shows similar variability in cultured cells is the Forssmann antigen (Fogel and Sachs, 1964). The determinant responsible for Forssmann activity, at least in horse kidney and spleen, is a sphingoglycolipid with the following structure (Makita et al., 1966):

$$\text{GalNAc-}(\alpha1 \rightarrow 3)\text{-Gal-}(\beta1 \rightarrow 4)\text{-Gal-}(\beta1 \rightarrow 4)\text{-Glc-ceramide}$$

In other cells, glycoproteins may be responsible for Forsmann activity.

In addition to the foregoing examples, there are changes in the surface carbohydrates of tumor cells as evidenced by loss of blood group antigens (Hakomori et al., 1967).

In considering possible mechanisms for changes in cell surface carbohydrates, it is important to know whether there is an actual quantitative
change in the surface structures or whether their apparent appearance
or disappearance is caused by other material uncovering or covering the
surface structures. For instance, cells treated with proteinases often show
enhanced agglutinability by specific antisera (Springer, 1963). In some
cases the proteinases may act unspecifically by decreasing the strong
negative charge on the cells, thereby rendering them more prone to
agglutination. In other cases, proteinases uncover sites which are normally masked and new specificities emerge. For example, antisera
directed against globoside 1 or hematoside (see Table I) show greater
reactivity with fetal erythrocytes than with adult erythrocytes as measured by agglutination or by absorption capacity (Hakomori, 1969).
Treatment of adult erythrocytes with trypsin increases their reactivity to
the same level as that of fetal erythrocytes, while the reactivity of fetal
erythrocytes is not affected by the enzyme treatment. These results suggest that the globoside and hematoside of fetal cells are directly exposed,
while those of the adult erythrocyte are masked by a trypsin-sensitive
covering. For this reason, analysis of the total glycolipids of a cell may
actually reveal little about functionally exposed surface structures. A case
in point is the level of hematoside in normal cells and in cells transformed by virus (Hakomori et al., 1968). The cells transformed by virus
have significantly lower levels of hematoside than the original normal
cells and yet the reactivity of antihematoside sera is much greater against
the transformed cells than against the normal cells. Like the adult erythrocyte, the reactivity of the normal cells with antihematoside sera is
greatly increased by treatment with trypsin. It has been observed that
cells transformed with viruses synthesize less glycoprotein than do normal cells (Wu et al., 1969; Meezan et al., 1969). This may be related to
the increased accessibility of hematoside to antibody after transformation, in that the hematoside of normal cells may be covered by
glycoprotein.

The binding site for a glycoprotein from wheat germ which specifically
agglutinates many tumor cells and virally transformed cells appears to
be another case in which a carbohydrate structure is masked by a trypsin-
sensitive covering. For example, hamster or mouse cells in culture are
agglutinable by the glycoprotein after transformation with polyoma virus
while the untransformed parent cells are not (Burger, 1969). However,
treatment of the parent cells with trypsin renders them agglutinable and
inhibition of agglutination is prevented by the same sugar haptens that
inhibit agglutination of the transformed cells. Also, the amount of [3]H-
labeled agglutinin bound by untransformed cells is much less than that

bound by transformed cells but rises to the same level after trypsinization. These results indicate that normal cells contain the same number of binding sites in a "cryptic" form. Of all the sugars tested, only N-acetyl-glucosamine and its disaccharide derivative, GlcNAc-($\beta1 \rightarrow 4$)-GlcNAc (di-N-acetylchitobiose), are effective hapten inhibitors of agglutination suggesting that N-acetylchitobiose, which is about five times more effective as an inhibitor than the monosaccharide, occurs in animal glyco-proteins as part of the carbohydrate chain of thyroglobulin (Spiro, 1970).

Results obtained with the agglutinin conconavalin A, which is derived from jack bean, are similar to those obtained with the wheat germ ag-glutinin (Inbar and Sachs, 1969a,b). Conconavalin A agglutinates leu-kemic cells and cells in culture which have been transformed by viruses, by carcinogens or by x-irradiation. Normal cells are not agglutinated under the same conditions unless treated with trypsin. Quantitative studies on the binding of labeled agglutinin indicate that about 85% of the binding sites on the normal cell are in a cryptic form. The site prob-ably contains α-linked glucose since conconavalin A binds α-methylgluco-side and this sugar derivative is a potent inhibitor of agglutination.

The appearance of new determinants also occurs if an internal sugar which is normally covered by other sugars becomes the nonreducing terminal residue of the chain when the outer sugars are either removed or not put on. The first example of this type of unmasking was the dem-onstration that red cells exposed to certain bacteria were agglutinable by all human sera (Thomsen, 1927; Friedenreich, 1928). The new deter-minant (the "T" agglutinogen) arises by the action of sialidases which remove terminal sialic acid residues from carbohydrate chains and expose underlying structures, probably galactosyl residues (Uhlenbruck et al., 1969). Similarly, incomplete synthesis of carbohydrate chains has been proposed to account for changes in serologic specificity observed in some tumor cells and transformed cells as opposed to normal cells (Hakomori et al., 1968). For example, hematoside is the main glycolipid of normal fibroblasts while its precursor, cytolipin H, is a minor component. In transformed fibroblasts, the reverse is true; there is little hematoside and greatly increased cytolipin H (Hakomori and Murakami, 1968). This observation is of special interest because of the importance of cytolipin H for the serologic properties of many tumors as mentioned above. Also, suggestive of the incomplete synthesis of heterosaccharide chains by malignant cells is the finding that human adenocarcinomas no longer contain glycolipids with blood group A or B activities (Hakomori et al., 1967). Instead, they accumulate glycolipids that have O(H) specificity and that normally may be an intermediate in the synthesis of the more complex A- or B-active structures (see Fig. 3).

Possibly the absence of specific transferases may account for incomplete synthesis, but then the question becomes one of explaining why the enzymes are produced in some situations and not in others. The underlying reason for this variability of expression is unknown. Perhaps in some cases it is the loss of genetic material, while in other cases it may be part of the larger problem of the mechanisms responsible for differential gene expression during development.

IV. FUNCTION OF CELL SURFACE CARBOHYDRATES

The title of this section is somewhat misleading since the biological role of the carbohydrates of cell surfaces is largely unknown. However, as mentioned in the introduction, the very nature of their structural complexity suggests a participation in phenomena that require cell specificity, such as the sorting out of mixed dissociated cells, the selective accumulation of certain cells in areas of inflammation, or the discrimination of self and not self. Unique patterns of surface sugars could serve as "recognition" sites which would be recognized by complementary structures (presumably protein) on other cells or macromolecules. Carbohydrates would be ideally suited for such a role because of the enormous number of specific structures that can be formed from relatively few monomeric units. Four different monosaccharides in a heterosaccharide chain can be arranged in many more combinations than can twenty amino acids in a polypeptide of similar size. The monosaccharides can be linked to each other through any of several hydroxyl groups, the linkages can be either α or β and, in addition, extensive branching is possible.

One phenomenon that evidently requires cellular specificity and recognition is the process of morphogenesis (Curtis, 1967; Moscona, 1968). Suggestive evidence that complex carbohydrates are involved in this process, at least for tissue formation, comes from studies on mouse teratoma cells (Oppenheimer et al., 1969). When single teratoma cells are grown in a glucose-balanced salt solution, they fail to aggregate unless glutamine is added to the growth medium. Glutamine can be specifically replaced by glucosamine and mannosamine suggesting that glutamine is required because of its participation in the formation of glucosamine 6-phosphate from fructose 6-phosphate. Since glucosamine 6-phosphate is the precursor of both galactosamine and sialic acid (Fig. 2), possibly the crucial event in the conversion of nonadhesive to adhesive teratoma cells is the formation of certain complex saccharides.

Several studies on the nature of the receptors on lymphocytes that bind phytohemagglutinin as described in Section III indicate that surface carbohydrates may play a role in the immune response. The binding of phytohemagglutinin by these cells is the initial step in stimulating lymphocytes to undergo mitosis *in vitro;* while this cannot be considered as a "function" for their surface carbohydrates, it may possibly be a model that mimics the mechanism of antigenic stimulation of lymphocyte proliferation *in vivo.*

The curious distribution of glucose in animals has been used to support the notion that surface carbohydrates function as recognition sites (Gesner and Ginsburg, 1964). This hexose is the most abundant sugar found in nature and occurs in nearly every complex saccharide of plants and bacteria. In animals it comprises the reserve polysaccharide glycogen and substantial amounts are found free in body fluids. Yet with rare exceptions (chiefly collagenlike structural proteins) it is not found in mammalian glycoproteins. In glycolipids it occurs as the glycosyl residue closest to the lipid moiety and is never found in the distal parts of the heterosaccharide chains which extend into the aqueous environment. The exclusion of glucose would have a rational basis if the carbohydrate structures of cell surface were indeed reacting with complementary sites; the efficiency of a recognition surface based on glucosyl residues would be impaired by free glucose much as haptens interfere with antigen–antibody reactions. Evolutionary selection against this impairment would tend to exclude glucose as a component of these surfaces.

References

Adams, E. P., and Gray, G. M. (1968). *Chem. Phys. Lipids* **2**, 147.

Bahl, O. P. (1969). *J. Biol. Chem.* **244**, 575.

Burger, M. M. (1969). *Proc. Nat. Acad. Sci. U. S.* **62**, 994.

Brew, K., Vanaman, T. C., and Hill, R. L. (1968). *Proc. Nat. Acad. Sci. U. S.* **59**, 491

Caccam, J. F., Jackson, J. F., and Eylar, E. H. (1969). *Biochem. Biophys. Res. Commun.* **35**, 505.

Carlson, D. M. (1968). *J. Biol. Chem.* **243**, 616.

Ceppellini, R. (1954). "In Proceedings of the 5th International Congress of Blood Transfusion, Paris," p. 207.

Clamp, J. R., Dawson, G., and Spragg, B. P. (1968). *Biochem. J.* **106**, 16p.

Curtis, A. S. G. (1967). "The Cell Surface: Its Molecular Role in Morphogenesis." Academic Press, New York.

Dawson, G., and Clamp, J. R. (1968). *Biochem. J.* **107**, 341.

Fogel, M., and Sachs, L. (1964). *Exp. Cell Res.* **34**, 448.

Franks, D. (1968). *Biol. Rev.* **43**, 17.

Franks, D., and Dawson, A. (1966). *Exp. Cell Res.* **42**, 543.

Friedenreich, V. (1928). *Acta Pathol. Microbiol. Scand.* **5**, 59.

Gesner, B. M., and Ginsburg, V. (1964). *Proc. Nat. Acad. Sci. U. S.* **52**, 750.

Ginsburg, V. (1964). *Advan. Enzymol.* **26**, 35.

Gottschalk, A., ed. (1966). "Glycoproteins." Elsevier, New York.
Gottschalk, A. (1969). *Nature* (*London*) **222**, 452.
Hakomori, S. (1969). *Vox Sang.* **16**, 478.
Hakomori, S., and Murakami, W. T. (1968). *Proc. Nat. Acad. Sci. U. S.* **59**, 254.
Hakomori, S., Koscielak, J., Bloch, K., and Jeanloz, R. W. (1967). *J. Immunol.* **98**, 31.
Hakomori, S., Teather, C., and Andrews, H. (1968). *Biochem. Biophys. Res. Commun.* **33**, 563.
Inbar, M., and Sachs, L. (1969a). *Proc. Nat. Acad. Sci. U. S.* **63**, 1418.
Inbar, M., and Sachs, L. (1968b). *Nature* (*London*) **223**, 710.
Joffe, S., Rapport, M. M., and Graf, L. (1963). *Nature* (*London*) **197**, 60.
Kobata, A., and Ginsburg, V. (1969). *J. Biol. Chem.* **244**, 5496.
Kobata, A., and Ginsburg, V. (1970). *J. Biol. Chem.*, **245**, 1484.
Kobata, A., Grollman, E. F., and Ginsburg, V. (1968). *Biochem. Biophys. Res. Commun.* **32**, 272.
Kornfeld, S., and Kornfeld, R. (1969). *Proc. Nat. Acad. Sci. U. S.* **63**, 1439.
Kraemer, P. M. (1971). In "Biochemistry and Physiology of the Cell Periphery" (L. Manson, ed.), Vol. I, p. 67. Plenum Press, New York.
Liu, C., Patterson, B. W., Lapp, D., and Elbein, A. D. (1969). *J. Biol. Chem.* **244**, 3728.
Lindahl, U., and Rodén, L. (1966). *J. Biol. Chem.* **241**, 2113.
Lloyd, K. O., and Kabat, E. A. (1968). *Proc. Nat. Acad. Sci. U. S.* **61**, 1470.
Makita, A., Suzuki, C., and Yosizawa, Z. (1966). *J. Biochem.* **60**, 502.
Maley, F., McGarrahan, J. F., and DelGiacco, R. (1966). *Biochem. Biophys. Res. Commun.* **23**, 85.
Marchesi, V. T. (1969). In "Red Cell Membranes" (G. A. Jamieson and T. J. Greenwalt, eds.), p. 117. Lippincott, Philadelphia, Pennsylvania.
Marcus, D. M. (1969). *N. Engl. J. Med.* **280**, 994.
Meezan, E., Wu, H. C., Black, P. H., and Robbins, P. W. (1969). *Biochemistry* **8**, 2518.
Morawiecki, A. (1964). *Biochim. Biophys. Acta* **83**, 339.
Moscona, A. A. (1968). *Develop. Biol.* **18**, 250.
Neufeld, E. F., and Ginsburg, V. (1965). *Annu. Rev. Biochem.* **34**, 297.
Nordin, J. H., and Hansen, R. G. (1963). *J. Biol. Chem.* **238**, 489.
Oppenheimer, S. B., Edidin, M., Orr, C. W., and Roseman, S. (1969). *Proc. Nat. Acad. Sci. U. S.* **63**, 1395.
Osborn, M. J. (1969). *Annu. Rev. Biochem.* **38**, 501.
Pease, D. C. (1966). *J. Ultrastruct. Res.* **15**, 555.
Preiss, J. (1969). In "Current Topics in Cellular Regulation" (B. L. Horecker and E. R. Stadtman, eds.), Vol. 1, p. 125. Academic Press, New York.
Quintarelli, C., ed., (1968). "The Chemistry of Mucopolysaccharides." Little, Brown, Boston, Massachusetts.
Race, R. R., and Sanger, R. (1958). "Blood Groups in Man," 3rd ed. Blackwell, Oxford.
Rambourg, A., Neutra, M., and Leblond, C. P. (1966). *Anat. Rec.* **154**, 41.
Rapport, M. M., and Graf, L. (1969). *Progr. Allergy* **13**, 273.
Rapport, M. M., Graf, L., Skipski, V. P., and Alonzo, N. F. (1959). *Cancer* **12**, 438.
Rapport, M. M., Schneider, H., and Graf, L. (1962). *J. Biol. Chem.* **237**, 1056.
Revel, J-P., and Ito, S. (1967). In "The Specificity of Cell Surfaces" (B. D. Davis and L. Warren, eds.), p. 211. Prentice-Hall, Englewood Cliffs, New Jersey.

Roseman, S. (1968). In "Biochemistry of Glycoproteins and Related Substances" (E. Rossi and E. Stoll, eds.), p. 244. S. Karger, Basel.
Sarcione, E. J. (1964). J. Biol. Chem. 239, 1686.
Sherr, C. J., and Uhr, J. W. (1969). Proc. Nat. Acad. Sci. U. S. 64, 381.
Smyth, D. S., and Utsumi, S. (1967). Nature (London) 216, 335.
Spiro, R. G. (1962). J. Biol. Chem. 237, 646.
Spiro, R. G. (1967). J. Biol. Chem. 242, 1923.
Spiro, R. G. (1969). N. Engl. J. Med. 281, 991, 1043.
Spiro, R. G. (1970). Fed. Proc. Fed. Amer. Soc. Exp. Biol., 29, 600A.
Springer, G. F. (1963). Bacteriol. Rev. 27, 191.
Sweeley, C. C., and Dawson, G. (1969). In "Red Cell Membranes" (G. A. Jamieson and T. J. Greenwalt, eds.), p. 172. Lippincott, Philadelphia, Pennsylvania.
Szulman, A. E. (1965). J. Histochem. Cytochem. 13, 752.
Tal, C. (1965). Proc. Nat. Acad. Sci. U. S. 54, 1318.
Thomsen, O. (1927). Z. Immunitäts Forsch. 52, 85.
Thomson, D. M. P., Krupey, J., Freedman, S. O., and Gold, P. (1969). Proc. Nat. Acad. Sci. U. S. 64, 161.
Uhlenbruck, G., Pardoe, G. I., and Bird, G. W. G. (1969). Z. Immunitätsforsch. Allerg. Klin. Immunol. 138, 423.
Wardi, A. H., Allen, W. S., Turner, D. L., and Stary, Z. (1969). Biochim. Biophys. Acta 192, 151.
Watkins, W. M. (1966). In "Glycoproteins" (A. Gottschalk, ed.), p. 462. Elsevier, New York.
Watkins, W. M., Koscielak, J., and Morgan, W. T. J. (1962). In "Proceedings of the 9th Congress on Blood Transfusion," p. 230. Karger, Basel.
Winzler, R. J. (1969). In "Red Cell Membranes" (G. A. Jamieson and T. J. Greenwalt, eds.), p. 157. Lippincott, Philadelphia, Pennsylvania.
Wu, H. C., Meezan, E., Black, P. H., and Robbins, P. W. (1969). Biochemistry 8, 2509.

AUTHOR INDEX

Numbers in italics refer to the pages on which the complete references are listed.

Eberhard, S. J., 330, *339*
Eddy, A. A., 328, *334*
Edelhoch, H., 51, 52, *80*
Edelman, G. M., 54, 55, 79, 195, *220*
Edidin, M., 209, *219*, 456, *458*
Edsall, J. T., 155, 156, *218*
Edstrom, R. D., 256, *283*, 375, 376, *395*, 396
Edwards, D., 57, *78*
Egan, J. B., 307, 308, 310, *334, 335*
Eicholz, A., 329, *334*
Ekwall, P., 114, *144*
Elbein, A. D., 448, *458*
Ellingson, J. S., 274, *283*
Elliott, W. B., 408, *436*
Emsley, J. W., 24, 26, *78*
Endo, A., 162, 197, *221*, 255, 260, *283*, 376, *395*
Engelman, D. M., 8, 9, 163, 165, 172, 194, 208, *218, 221*
Englesberg, E., 238, *246*, 308, 311, 312, 316, *334, 335, 338*
Entine, G., 104, *142*
Epstein, C. J., 148, 162, *218*
Ernster, L., 408, 409, 410, 411, 412, 419, 426, 429, *434, 435, 437*
Erwin, V. G., 409, *437*
Esposito, G., 328, *334*
Estabrook, R. W., 404, 419, *438*
Evans, M. W., 150, 151, *218*
Eylar, E. H., 41, *78*, 347, 378, 379, 380, 382, *394, 395, 396*, 449, *457*

F

Faber, R. J., 87, *142*
Faelli, A., 328, *334*
Fahn, S., 323, 324, 333, *334, 335*
Falle, H. R., 95, 114, *142*
Farnham, S. B., 153, *219*
Farrant, J. L., 175, *218*
Fasman, G. D., 185, 186, *219*
Fast, P. G., 65, *77*
Faure, M., 7, 9, 17, 18, *79*, 157, 167, *219*, 278, *283*
Feeny, J., 24, 26, *78*
Fennessey, P., 347, 359, 370, *400*
Fenster, L. J., 274, *283*, 326, *335*
Fernald, G. W., 328, *334*
Fernández-Morán, H., 404, *434*
Fernbach, B. R., 191, *222*

Ferruti, P., 95, 104, 114, *142*
Fessenden-Raden, J. M., 419, 420, 424, 426, 428, 429, 432, *434, 437*
Fielding, P., 387, *395*
Finean, J. B., 17, *78*, 175, 179, 192, *218*, 264, *283*
Fisher, W. D., 233, *245*
Flanagan, M. T., 64, *78*
Fleischer, B., 75, *78*, 174, 175, 197, *218*, 255, 276, 278, 282, *283*, 402, *434*
Fleischer, S., 75, *78*, 174, 175, 197, *218*, 252, 255, 275, 276, 277, 278, 282, *283*, 402, 417, *434*
Flook, A. G., 28, 29, 30, 31, 32, *80*
Fluck, D. J., *78*
Förster, T., 55, *78*
Fogel, M., 453, *457*
Folch, J., 58, *78*
Folch-Pi, J., 180, *218*
Fontell, K., 114, *144*
Forrester, A. R., 101, *142*
Forstner, G., 329, *334*
Fortes, P. A. G., 405, *436*
Foster, D. W., 274, *283*
Fowler, L. R., 423, *435*
Fox, C. F., 197, *218*, 280, *283*, 294, 295, 296, 299, 300, 301, 302, 303, 333, *335, 336*, 340, 387, *395*
Fraenkel, D. G., 307, *340*
Fraenkel, G. K., *78*, 92, *142*
Frank, H. S., 150, 151, *218*, 228, *247*
Frank, L., 328, *335*
Franks, D., 453, *457*
Franks, F., 155, *219*
Franzen, J. S., 152, *219*
Frattali, V., 51, 52, *80*
Frazier, H. S., 321, *335*
Freed, J. H., *78*, 92, *142*
Freedman, R. B., 54, 58, 59, 67, *78*
Freedman, S. O., 453, *459*
Freeman, K., 146, 167, *219*, 420, *435*
Freer, J. H., 15, *80*
Freese, E., 242, *245*
Frerman, F. E., 347, 365, 370, 372, *395*
Frey-Wyssling, A., 176, *220*, 403, *434*
Friedenreich, V., 455, *457*
Friedman, D. L., 392, *395*
Frye, C. D., 209, *219*
Fuchs, E., 386, *395*
Fuisz, R. E., 328, *334, 339*

Schellman, C., 184, *221*
Schellman, J. A., 152, 184, 185, *221*
Schenkein, I., 311, *339*, 378, *399*
Scher, M., 347, 364, 366, 367, 370, 371, 372, 374, *399*
Schiefer, H.-G., 412, *437*
Schleif, R., 316, *339*
Schlesinger, M. J., 231, *247*
Schlessinger, D., 241, *246*
Schmitt, F. O., 15, *77*
Schnaitman, C., 272, *284*, 409, *437*
Schneider, H., 110, 115, 117, 119, 126, *142, 143, 144*, 452, *458*
Schneider, W. G., 24, *80*
Scholander, P. F., 315, *339*
Schoner, W., 322, *339*
Schultz, S. G., 328, *334, 335, 339*
Schwartz, A., 321, 322, 323, 324, 326, 337, *339*, 405, *436*
Schwartz, J. H., 237, *245*
Seelig, J., 94, 109, *144*
Seki, S., 30, *80*
Sekuzu, I., 278, *283, 284*
Sela, B. A., 392, 393, *399*
Sen, A. K., 320, 322, 324, 326, *333, 338, 339, 340*
Seubert, W., 322, *339*
Sezuki, I., 197, *219*
Shapiro, A. L., 306, *339*
Shapiro, B., 205, *221*, 389, *399*
Sharon, N., 351, 392, 393, 397, *399*
Shaw, D. R. D., 368, *396*
Shaw, T. I., 319, 323, *333*
Sheetz, M., 169, 191, 193, *219*
Sheffield, H. G., 406, *438*
Sheimin, R., *247*
Shen, L., 382, *399*
Sherman, F., 414, *437*
Sherr, C. J., 449, *459*
Shetlar, M. R., 378, *397*
Shimada, A., 354, *394*
Shimai, R., 354, *394*
Shinagawa, H., 316, *337*
Shipley, G. G., 17, 18, 28, 78, *80*
Shiraki, M., 186, *219*
Shockman, G. D., 15, *80*
Shorey, C. D., 240, *245*
Shortman, K., *247*
Showacre, J. L., 66, *78*
Siccardi, A., 389, *399*

Siegel, G. J., 324, *339*
Siekevitz, P., 215, *218*, 378, *398*
Siewert, G., 255, 270, *283*, 348, 351, 353, 354, 372, *399*
Silbert, D., 280, *284*, 303, *339*
Silbert, J. E., 381, *395, 399*
Silman, H. I., 146, 165, 167, *219, 220*, 420, *435*
Simkin, J. L., 378, *399*
Simoni, R. D., 306, 307, 312, *339*
Simpkins, H., 169, 186, 191, 193, *219*
Simpson, R. T., 230, *247*
Singer, I., 62, 63, *80*
Singer, L. A., 100, *144*
Singer, S. J., 152, 154, 157, 159, 168, 169, 170, 173, 174, 175, 186, 188, 189, 190, 191, 193, *218, 219, 220, 221*
Singer, T., 276, *283*, 421, *433, 437*
Sinha, R. K., 350, 354, *399*
Sinohara, H., 379, *399*
Sinsheimer, R. L., 386, 387, *395, 399*
Sirak, H. D., 407, *437*
Sistrom, W. R., 288, *339*
Sjöstrand, F. S., 403, *437*
Skipski, V. P., 452, *458*
Skou, J. C., 321, 322, 323, 324, 325, 326, *336, 339, 340*
Skoulios, A., 264, *283*
Sky-Peck, H. H., 379, *399*
Slater, E. C., 323, *340*, 421, *433, 437*
Slautterback, D. B., 252, 255, 275, 276, *283*, 417, *434*
Slichter, C. E., 22, *80*
Small, D. M., 30, *80*, 119, *142*
Smirnova, T. A., 240, *246*
Smith, A. L., 421, *434*
Smith, D., 238, *246*, 385, 391, *399*
Smith, E. E. B., 364, *399*
Smith, H., 3, 9, 236, *247*, 287, *340*
Smith, I. C. P., 85, 101, 110, 115, 117, 119, 122, 126, *142, 143, 144*
Smith, L., 433, *436*
Smith, M. F., 306, *339*
Smith, Z. G., 382, *396, 400*
Smoly, J. M., 410, *434*
Smyth, D. S., 444, *459*
Snipes, W., 125, *143*
Snyder, R. W., 387, *399*
Söll, D., 348, 352, 353, *394, 398*

478

AUTHOR INDEX

Thompson, T. E., 33, *81,* 107, *144,* 170, 213, *219,* 332, *338,* 403, 417, *437*
Thompson, W., 404, 412, 413, *436*
Thomsen, O., 455, *459*
Thomson, D. M. P., 453, *459*
Thomson, P. J., 328, *334*
Thomson, R. H., 101, *142*
Thorndike, J., 352, *399*
Thornley, M. J., 227, 228, *246*
Tien, H. T., 332, *337*
Tillack, T. W., 177, *221*
Timasheff, S. N., 192, *221*
Tipper, D. J., 238, *247,* 351, 353, 354, 395, *400*
Tisdale, H., 420, 421, *434*
Tobin, T., 326, *338, 340*
Toennies, G., 15, *80*
Tomita, K., 22, *79*
Tomkins, G. M., 288, *333*
Tonomura, Y., 104, *144*
Torriani, A., 225, 230, 231, *247*
Tosteson, D. C., 331, *340*
Tourtellotte, M. E., 75, *80,* 124, *144,* 194, *221,* 281, *284*
Toury, R., 409, *435*
Townend, R., 192, *221*
Tremblay, G. Y., 386, 387, *395, 400*
Trevithick, J. R., 243, *247*
Troy, F. A., 347, 365, 370, 372, *395*
Truman, D. E. S., 215, *221*
Tsong, T. Y., 194, 195, *220*
Tucker, A. N., 279, *284*
Tung, Y., 210, *218,* 282, *282*
Turner, D. L., 443, *459*
Tweet, A. G., 65, *78, 81*
Tyler, D. D., 404, 419, *438*
Tzagoloff, A., 254, *283, 284,* 430, *436*

U

Uchida, T., 362, 363, *398, 400*
Udenfriend, S., 66, *81*
Uesugi, S., 322, 325, 326, *340*
Uetake, O., 363, *400*
Uhlenbruck, G., 455, *459*
Uhr, J. W., 378, 379, *397, 399, 400,* 449, *459*
Umezawa, H., 238, *247*
Urbina, 76, *78*
Uretz, R. B., 68, *79*
Urry, D. W., 188, 189, 193, *221*

Ussing, H. H., *337*
Utahara, R., 238, *247*
Utsumi, K., 405, 407, *434, 436*
Utsumi, S., 444, *459*

V

Vagelos, P. R., 280, *284,* 303, *339*
Valentine, R. C., 205, *221*
Vallee, B. L., 230, *247*
Vallin, I., 408, *435*
Vanaman, T. C., 448, *457*
Van Deenen, L. L. M., 4, *9,* 34, 78, 121, *144,* 146, 192, 197, 198, *221*
Vandenheuvel, F. A., 15, *81,* 150, *221*
Vander kooi, J., 59, 63, 64, *81,* 195, 210, 212, *221*
van Groningen, H. E. M., 323, *340*
Vasington, F. D., 238, *247*
Veksli, Z., 28, 29, *81*
Velick, S. F., 255, *284*
Velluz, L., 183, *222*
Vidaver, G. A., 328, *340*
Villemez, C. L., 347, 374, *400*
Vinson, L. J., 74, *77*
Vinuela, E., 306, *339*
Vogell, W., 407, *436*
von Hippel, P. H., 157, *222*
von Hofsten, B., 235, *247*
von Ilberg, C., 322, *339*
Voynow, P., 384, *397*
Vreman, H. J., 323, *336*
Vureck, G., 58, 68, *78*

W

Waggoner, A. S., 79, 85, 100, 104, 105, 106, 107, 109, 115, 120, 123, 124, 125, 139, 141, *143, 144*
Wainio, W. W., 277, *283*
Wakil, S. J., 256, 273, 274, *283*
Waldner, H., 176, *220*
Walker, D. A., 128, *142*
Walker, L. M., 328, *338*
Wallach, D. F. H., 19, 20, 56, 67, *81,* 109, 110, *142,* 169, 188, 189, 190, 191, 193, *219, 222,* 255, *284,* 324, *340*
Walsh, P. M., 293, *335*
Wang, C. C., 332, *340*
Wang, H. H., 104, *142*

SUBJECT INDEX

A

Acridine orange, 212
ADP-glucose pyrophosphatase, 236, *see also* Periplasmic enzymes
Agglutinins, 452–455
  concanavalin A, 455
  reactions with surface antigen, 452
  tumor cells and, 454
  wheat germ, 455
Alamethicin
  interactions with phospholipids, 35
  NMR, 35
Amino acid transport, 327
Anesthetics, local, 61
Anilino-napthalene sulfonate (ANS), *see* Fluorescent probes
Asparaginase I, 237
Asparaginase II, *see* Periplasmic enzymes
Assembly of membranes, 264
  nitrate reductase system, 270
ATP, in lactose transport, 296
ATPase, *see also* Sodium-potassium ATPase
  bacterial, 267
  binding to membranes, 267
  effects of phospholipids, 274
  Mg^{++}-stimulated, 277
  microsomal, 274
Axonal membranes, 319, *see also* Sodium and potassium transport

B

Bacterial cell envelopes, 345, *see also* Lipopolysaccharides, Peptidoglycans, Teichoic acids
Bacterial membrane(s), *see also* Lipopolysaccharide
  channels, 228
  composition, 15
  EDTA and lysozyme treatment, 225
  murein-lipoprotein, 228
  osmotic shock, 226

outer membrane, 227
  plasma membrane, 227
  plasmolysis, 227, 229
Bacterial membrane vesicles, 310–314
  kinetics of substrate uptake, 313
  phospholipase effects, 311
  preparation of, 310
Basement membranes, *see* Glycoproteins
Bimolecular leaflets, 158, 264
  myelin, 159
  thermodynamic considerations, 158–162
  x-ray diffraction, 159
Binding proteins, 238, 314, *see also* Periplasmic enzymes
  amino acids, 238
  L-arabinose, 316
  arginine, 317
  galactose, 316
  histidine, 317
  leucine, 318
  leucine-isoleucine-valine, 317
  phosphate, 316
  sugars, 238
  sulfate, 315
Biogenesis of membranes, 213
Blood group substances, 449–452

C

Calorimetry, *see* Differential thermal analysis
Cardiolipin and isoprenoid alcohol phosphokinase, 270
Carrier lipids, *see* Glycosyl carrier lipids
Chaotropic agents, 255
Circular dichroism (CD), *see* Optical rotatory dispersion
Coenzyme Q, *see* Mitochondrial membranes
Collagen, 380, *see also* Glycoproteins
Concanavalin A, *see* Agglutinins
Cooperativity in membranes, 211
Counterflow, 291

481

Molecular Biology

An International Series of Monographs and Textbooks

Editors

BERNARD HORECKER

Department of Molecular Biology
Albert Einstein College of Medicine
Yeshiva University
Bronx, New York

NATHAN O. KAPLAN

Department of Chemistry
University of California
At San Diego
La Jolla, California

JULIUS MARMUR

Department of Biochemistry
Albert Einstein College of Medicine
Yeshiva University
Bronx, New York

HAROLD A. SCHERAGA

Department of Chemistry
Cornell University
Ithaca, New York

HAROLD A. SCHERAGA. Protein Structure. 1961.

STUART A. RICE AND MITSURU NAGASAWA. Polyelectrolyte Solutions: A Theoretical Introduction, *with a contribution by Herbert Morawetz.* 1961

SIDNEY UDENFRIEND. Fluorescence Assay in Biology and Medicine. Volume I—1962. Volume II—1969

J. HERBERT TAYLOR (Editor). Molecular Genetics. Part I—1963. Part II—1967

ARTHUR VEIS. The Macromolecular Chemistry of Gelatin. 1964

M. JOLY. A Physico-chemical Approach to the Denaturation of Proteins. 1965

SYDNEY J. LEACH (Editor). Physical Principles and Techniques of Protein Chemistry. Part A—1969. Part B—1970

KENDRIC C. SMITH AND PHILIP C. HANAWALT. Molecular Photobiology: Inactivation and Recovery. 1969

RONALD BENTLEY. Molecular Asymmetry in Biology. Volume I—1969. Volume II—1970

JACINTO STEINHARDT AND JACQUELINE A. REYNOLDS. Multiple Equilibria in Protein. 1969

DOUGLAS POLAND AND HAROLD A. SCHERAGA. Theory of Helix-Coil Transitions in Biopolymers. 1970

JOHN R. CANN. Interacting Macromolecules: The Theory and Practice of Their Electrophoresis, Ultracentrifugation, and Chromatography. 1970

WALTER W. WAINIO. The Mammalian Mitochondrial Respiratory Chain. 1970

LAWRENCE I. ROTHFIELD. (Editor). Structure and Function of Biological Membranes. 1971

DATE DUE

30 505 JOSTEN'S